introduction to
advanced physics

WQEIC

introduction to
advanced physics

DAVID BRODIE

JOHN MURRAY

Titles in this series:
Introduction to Advanced Biology 0 7195 7671 7
Introduction to Advanced Chemistry 0 7195 8587 2
Introduction to Advanced Physics 0 7195 8588 0

Further Advanced Chemistry 0 7195 8608 9
Further Advanced Physics 0 7195 8609 7

First published in 2000
by John Murray (Publishers) Ltd
50 Albemarle Street
London WIS 4BD

Layouts by Eric Drewery
Illustrations by Oxford Illustrators Ltd

Typeset in 10/12pt Gill Sans by Wearset, Boldon, Tyne and Wear
Printed and bound in Great Britain by Butler & Tanner, Frome and London

A CIP catalogue entry for this title is available from the British Library

ISBN 0 7195 8588 0

Contents

Introduction

It is a big step from GCSE level physics to post-16 work; so big that many students struggle in the early months, and a few never recover. This book is intended to provide help with this first step. Part A, *The Basics* (Chapters 1 to 22), takes students on from the familiar and into the new world of Advanced level study. It emphasises skills and themes as well as addressing specification content at an introductory level, providing a sound foundation on which to build. The longer chapters of Part B, *Becoming Advanced* (Chapters 23 to 37), then offer everything that is needed to take students to full AS level. In short, then, the book guides students through their first year post-16 and into their second, in two levels of work.

While every student, the most able as well as the least, will find it extremely helpful to make their first appraoch to each topic through the Part A chapters, no particular mode of working is assumed. Some teachers may find that they and their students prefer to work through all of Part A before proceeding at all to Part B. This has an advantage of 'spirality' – students will study a topic at an introductory level and then at a more advanced level a little later when their skills have moved on (see Figure 1). Others will prefer a topic-by-topic approach (Figure 2) – so that they deal, say, with elementary ideas of mechanics – *Variables of motion* (Chapter 10), *Newton's force* (Chapter 11*), Working with vectors* (Chapter 12), and so on, and then move straight on to higher levels through *Forces in equilibrium* (Chapter 30) and *Straight line motion* (Chapter 31).

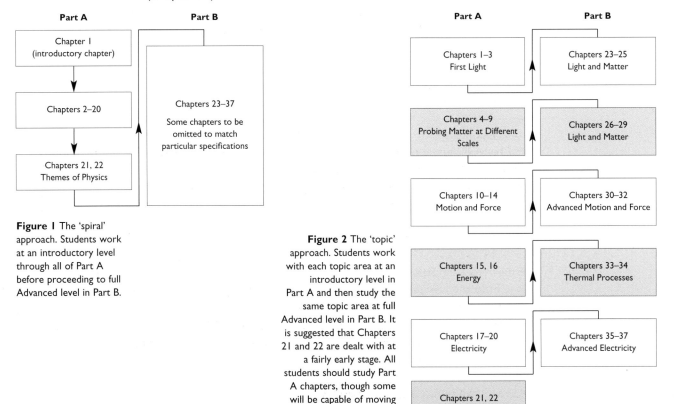

Figure 1 The 'spiral' approach. Students work at an introductory level through all of Part A before proceeding to full Advanced level in Part B.

Figure 2 The 'topic' approach. Students work with each topic area at an introductory level in Part A and then study the same topic area at full Advanced level in Part B. It is suggested that Chapters 21 and 22 are dealt with at a fairly early stage. All students should study Part A chapters, though some will be capable of moving through them more quickly than others.

The book contains everything that is needed for every AS specification. The different specifications vary considerably in their content, so any one centre will find that they do not need to study all chapters from Part B. The *Specification-matching matrix* following this Introduction shows which chapters are required for the different AS specifications. A level students will, in most cases, complete Part B chapters in the second year of study. The second book, *Further Advanced Physics*, will provide all the additional material that is needed for the second year of study, including a range of 'synoptic' chapters that will provide extra stimulation and support for this new component of the assessment programme.

The book sets out to provide clear explanation of concepts. More than this, it provides emphasis on themes and skills, including those relating to language and mathematics. Chapters 21 and 22 specifically address these fundamental general skill areas of physics, but there is support elsewhere, too. There are lists of *Key vocabulary* at the beginning of each chapter and students are encouraged to use these as the foundation on which to build their own frameworks of understanding. *The big questions* that open each chapter are not for students to answer, but serve to emphasise that physics is a process of inquiry. Each chapter also starts with a short *Background* section which provides a historical or social context for what follows.

The large numbers of different kinds of questions throughout the text are intended to provide plenty of opportunity to practise and develop ideas and skills, and to encourage mental engagement. These questions include:

- 'as-you-go' questions throughout each chapter, which provide basic practice or thought-provoking stimulation (or, where possible, a combination of these), and encourage independent use of the book as well as discussion;
- *Comprehension and application* exercises at the end of most chapters, which demand that students engage their minds with knowledge content of the chapter, apply their developing skills and have opportunity to consider physics in a range of broad contexts;
- *Extra skills tasks*, which reinforce key skills activities already provided through other questions and ensure that there is plenty of opportunity for developing the key skills of Communication, Information Technology and Application of Number.

There are no past examination questions with Part A chapters, since it is unreasonable to expect students to be able to cope with these in the first weeks and months of their course. There are plenty of them, however, at the end of each Part B chapter, and these deal with relevant content that is covered in both Part A and Part B chapters.

Answers to all numerical and other short-answer questions appear at the end of the book. In the case of examination questions, note that the answers given have not been provided by the examining boards.

I hope that I have, with the guidance, support and encouragement from the editors at John Murray Publishers, created a new kind of resource that helps to solve an age-old problem in physics education – emphasising foundation ideas and skills and then developing those to the appropriate level, thus making the transition from GCSE to post-16 study less daunting. If AS and A level study becomes, thereby, a more satisfying experience, in itself as well as in its importance for later study, then my year and a half of intensive effort will have been well rewarded.

In particular, I'd like to thank the following:

Katie Mackenzie Stuart, for the initial concept for this series, and for her continuing support;
Jane Roth and Julie Jones, for their meticulous and very hard work, and their awesome patience;
Marilyn Rawlings, for her painstaking search for excellent photographs;
Claire, for tolerance of unsocial working hours, and encouragement beyond the call of duty;
Tom, Eleanor and my parents, for their fond bemusement;
Eliza, for daring to suggest that physics is not everybody's favourite subject, but for being a sweetie anyway.

David Brodie

Specification-matching matrix

Board	AQA (NEAB)					AQA (AEB)				Edexcel						OCR				WJEC			
Module*	1	2	3	4	5	1	2	4	5	1	2	3C	4	5	6	A	B	CI	D	PH1	PH2	PH4	PH5
Chapter 1							✓																
Chapter 2			✓				✓						✓					✓		✓			
Chapter 3							✓						✓				✓				✓		
Chapter 4			✓							✓						✓							
Chapter 5			✓					✓		✓						✓							
Chapter 6							✓									✓			✓				✓
Chapter 7	✓						✓			✓									✓				✓
Chapter 8				✓			✓			✓									✓				✓
Chapter 9	✓			✓			✓			✓		✓							✓			✓	✓
Chapter 10		✓								✓						✓				✓			
Chapter 11		✓				✓				✓						✓				✓			
Chapter 12		✓				✓										✓				✓			
Chapter 13		✓				✓				✓						✓				✓			
Chapter 14		✓						✓		✓									✓			✓	
Chapter 15		✓				✓		✓		✓						✓			✓	✓		✓	
Chapter 16		✓				✓		✓		✓						✓			✓	✓		✓	
Chapter 17																	✓			✓			
Chapter 18			✓			✓					✓						✓			✓			
Chapter 19			✓			✓					✓						✓			✓			
Chapter 20			✓			✓					✓						✓			✓			
Chapter 21																							
Chapter 22																							
Chapter 23	✓																	✓		✓			
Chapter 24																							
Chapter 25				✓			✓						✓					✓	✓				
Chapter 26	✓						✓	✓					✓				✓			✓			✓
Chapter 27	✓						✓	✓					✓									✓	
Chapter 28	✓			✓			✓			✓		✓							✓	✓			✓
Chapter 29	✓						✓					✓											
Chapter 30		✓				✓				✓						✓				✓			
Chapter 31		✓				✓				✓						✓				✓			
Chapter 32		✓														✓							
Chapter 33		✓						✓		✓									✓			✓	
Chapter 34		✓						✓		✓									✓			✓	
Chapter 35			✓			✓				✓							✓				✓		
Chapter 36			✓			✓				✓						✓					✓		
Chapter 37			✓			✓	✓									✓							

* Modules in tinted columns make up the A2 part of the course

Acknowledgements

The author and publishers would like to thank the following for their contributions and advice:

John Gregson
Martin Hampshire
Malcolm Parry
Nicky Thomas
Neil Calder, CERN
Kitsou Dubois
Jane Croucher, King's College, London

Examination questions have been reproduced with kind permission from the following examining boards:

Assessment and Qualifications Alliance (AQA): The Associated Examining Board (AEB) and the Northern Examinations and Assessment Board (NEAB)
Edexcel Foundation (London Examinations)
International Baccalaureate Organisation
OCR

The answers in this book have not been provided by or approved by the examining boards. Their accuracy and the method of working are the sole responsibility of the author.

Thanks are due to the following for permission to reproduce copyright photographs and illustrations:

Cover The Wellcome Department of Cognitive Neurology, Functional Imaging Laboratory; **p.2** Fig. 1.1 C. N. R. I./Science Photo Library; **p.3** Fig. 1.3 J. Townson/Creation; **p.4** Fig. 1.4 A. Lambert; **p.7** Fig. 1.10 O. Andrews/Science Photo Library; **p.8** Fig. 1.15 Benelux Press/Ace Photo Agency; **p.9** Fig. 1.19 National Gallery, London; **p.10** Fig. 1.21 Ronald Grant Archive; **p.11** Fig. 2.1 A. Ronan/Image Select, Fig. 2.2 Science Photo Library; **p.12** Fig. 2.4 J. Birdsall; **p.13** Fig. 2.6a A. Lambert; **p.19** Fig. 3.1 Science Photo Library; **p.20** Fig. 3.3 both A. Lambert; **p.22** Fig. 3.7 P. Adams/Ace Photo Agency; **p.25** Fig. 3.10 F. Sauze/Science Photo Library; **p.30** Fig. 4.1 Image Bank; **p.31** Fig. 4.2 S. Allen/Image Bank; **p.35** Fig. 4.6 H. Reinhard/ B. & C. Alexander; **p.37** Fig. 5.1 Ancient Art & Architecture Collection, Fig. 5.2 R. Moreton/ Powerstock Zefa; **p.38** Fig. 5.4 J. Birdsall; **p.46** Figs.5.15, 5.16 & 5.17 J. Birdsall; **p.47** Fig. 6.1 C. Freeman/Royal Institution/Science Photo Library, Fig. 6.2 J. Townson/Creation; **p.48** Fig. 6.3 Rosenfeld Images/Science Photo Library; **p.51** Fig. 6.9 A. Lambert; **p.52** Fig. 6.11 P. Gould; **p.53** Figs.6.12a, 6.12b, 6.12c, 6.13a & 6.13b A. Lambert, Fig. 6.13c P. Gould; **p.55** Fig. 6.17 G. Tompkinson/Science Photo Library; **p.56** Fig. 7.1 N. Birch/Image Select; **p.57** Figs.7.2 & 7.3 A. Lambert; **p.58** Fig. 7.4 A. Lambert; **p.63** Fig. 8.1 Ronald Grant Archive, Fig. 8.2 Macmillan Cancer Relief; **p.64** Fig. 8.5 P. Gould; **p.68** Fig. 8.8 C. Powell, P. Fowler, D. Perkins/Science Photo Library; **p.69** Fig. 8.11 Particle Physics & Astronomy Research Council; **p.72** Fig. 9.1 J. Birdsall; **p.75** Fig. 9.5 Lawrence Berkeley Laboratory/Science Photo Library; **p.82** Fig. 10.1 J. Birdsall; **p.83** Fig. 10.2 Image Bank; **p.85** Fig. 10.6 A. Lambert; **p.88** Fig. 10.11b Mortimore/ Allsport UK; **p.89** Fig. 11.1 D. Santbech, *Problematum Astronomicorum*, Basel 1561; **p.92** Fig. 11.4 Daniels/Ardea London; **p. 93** Fig. 11.5 NASA/Science Photo Library, Fig. 11.6 Powerstock Zefa; **p.106** Fig. 13.1 A. Ronan/Image Select, Fig. 13.2 tl, tr & bl Science Photo Library, br A. Ronan/ Image Select, Fig. 13.3 Derby Museums & Art Gallery; **p.112** Fig. 13.12 A. Lambert; **p.114** Fig. 13.14

Arts Catalyst; **p.115** Fig. 14.1 Lady Blackett; **p.116** Fig. 14.2 Space Telescope Science Institute/ NASA/Science Photo Library, Fig. 14.3 A. Lambert; **p.124** Fig. 15.1 Stableford/Image Bank; **p.134** Fig. 15.16 Dr. J. Burgess/Science Photo Library; **p.141** Fig. 16.5 JPL Photo/NASA; **p.144** Fig. 17.1 Science Photo Library; **p.145** Fig. 17.2 Lawrence Berkeley Laboratory/Science Photo Library; **p.148** Fig. 17.12 *l* courtesy of the Royal Institution of Great Britain; **p.160** Fig. 19.2 A. Lambert; **p.171** Fig. 20.1 D. Davis/Tropix Photo Library; **p.174** Fig. 20.3**b** A. Lambert; **p.176** Fig. 20.7 Sutherland/Science Photo Library; **p.177** Fig. 20.10 A. Lambert; **p.186** Fig. 21.1 *l* Mary Evans Picture Library, *r* Science Photo Library; **p.188** Fig. 21.3 Steel Photography/Ace Photo Agency; **p.189** Fig. 21.5 Plailly/Science Photo Library; **p.201** Fig. 22.13 Colorsport; **p.204** Fig. 23.1 A. Lambert; **p.212** Fig. 23.16**a** A. Lambert, Fig. 23.16**b** Pasieka/Science Photo Library; **p.219** Fig. 23.30 A. Lambert; **p.223** Fig. 24.1 Lick Observatory/Science Photo Library; **p.227** Fig. 24.11 J. Townson/Creation; **p.230** Fig. 24.14 *both* A. Lambert; **p.234** Fig. 24.20 D. Brodie; **p.240** Fig. 25.1 Bousfield/Science & Society Photo Library; **p.244** Fig. 25.7 Aprahamian/ Sharples Stress Engineers/Science Photo Library; **p.250** Fig. 25.20 A. Lambert; **p.252** Fig. 25.23 A. Lambert; **p.253** Fig. 25.24 Last Resort; **p.255** Fig. 25.27 Department of Physics, University of Surrey, Fig. 25.28 Dr. J. Burgess/Science Photo Library; **p.261** Fig. 26.3 Science Photo Library; **p.271** Fig. 26.20 Dr. J. Burgess/Science Photo Library; **p.273** Fig. 27.1 Science Photo Library; **p.276** Fig. 27.6 *l* Department of Physics, Imperial College/Science Photo Library, Fig. 27.7 A. Ronan/Image Select, Fig. 27.8 *both* Department of Physics, Imperial College/Science Photo Library; **p.282** Fig. 27.18 Department of Physics, Imperial College/Science Photo Library, Fig. 27.19 Science Photo Library; **p.283** Fig. 27.22 B.Williams/Quadrant Picture Library; **p.288** Fig. 28.1 Edmaler/Science Photo Library; **p.292** Fig. 28.7 Science Photo Library, Fig. 28.8 Lawrence Berkeley Laboratory/Science Photo Library; **p.301** Fig. 28.17 Geoscience Features; **p.303** Fig. 28.18 British Geological Survey; **p.304** Fig. 28.20 Science Photo Library; **p.311** Fig. 29.3 C. Anderson/Science Photo Library; **p.312** Fig. 29.4 T. Beddow/Science Photo Library; **p.314** Fig. 29.6 CERN/Science Photo Library, Fig. 29.7 P. Loiez/CERN/Science Photo Library; **p.320** Fig. 29.14 Lawrence Berkeley Laboratory/Science Photo Library; **p.321** Fig. 29.15 Science Photo Library; **p.323** Fig. 29.17 CERN, Fig. 29.18 D. Parker & J. Baum/Science Photo Library; **p.326** Fig. 30.1 Leslie Garland Picture Library; **p.327** Fig. 30.2 D. Bolduc/Colorsport; **p.338** Fig. 31.1 Fox Photos/Hulton Getty; **p.352** Fig. 32.3 Z. V. M./Powerstock Zefa; **p.364** Fig. 33.1 Space Telescope Institute/Science Photo Library; **p.366** Fig. 33.4 Leslie Garland Picture Library; **p.376** Fig. 33.20 G. & M. Moss/Still Pictures; **p.381** Fig. 33.27 A. Ronan/Image Select; **p.400** Fig. 34.17**a** Powerstock Zefa, Fig. 34.17**b** MRP Photography; **p.406** Fig. 35.1 J. Birdsall; **p.421** Fig. 36.1 Science Photo Library; **p.427** Fig. 36.11 Unilab; **p.436** Fig. 37.1 J. Townson/Creation, Fig. 37.2 Montreal Neurological Institute, McGill University/C. N. R. I./Science Photo Library, Fig. 37.3 Image Select; **p.447** Fig. 37.24 Bridgeman Art Library, *The Betrayal of Images: Ceci n'est pas une pipe*, 1929, Los Angeles County Museum of Art, © D. A. C. S.; **p.448** Fig. 37.25 British Geological Society; **p.451** Fig. 37.29 Space Telescope Institute/NASA/Science Photo Library, Fig. 37.30 Science Photo Library; **p.452** Fig. 37.31 Schlumberger Oilfield Communications, Fig. 37.32 H. Turvey/Science Photo Library, Fig. 37.33 Deep Light Productions/Science Photo Library; **p.453** Fig. 37.34 Ouellette & Theroux/Publifoto Diffusion/Science Photo Library; **p.454** Fig. 37.36 D. Brodie.

l = left, *r* = right, *t* = top, *b* = bottom, *c* = centre

The publishers have made every effort to contact copyright holders. If any have been inadvertently overlooked the publishers will be pleased to make the necessary arrangements at the earliest opportunity.

PART A
THE BASICS

I
FIRST
LIGHT

1 Thinking about light

THE BIG QUESTIONS

- In what different ways is the travel of light affected by interactions with matter?
- How do these interactions influence what we see?
- What models can we use to help us to understand the journey of light to our eyes?

KEY VOCABULARY

absorption approximation diffuse reflection ideal intensity law law of reflection model normal parallel beam perspective radiation ray reflection replication retina scattering transmission

BACKGROUND

Light that has travelled for a hundred years from a distant star, or from a TV across the room, ends its journey in your eyes. Its energy is then taken from it and it ceases to exist as light. It is *absorbed*. It is from patterns in this light that your brain perceives the world. The ideas of physics are built from such perception.

Now, through our physics, through working with ideas of the nature of matter and light, we have learnt to build pictures of our active brains; fMRI scans – functional magnetic resonance images – show which parts of our brains are most active as we perform different tasks (Figure 1.1). They can show, for example, the regions that are working hardest when our vision is stimulated.

Figure 1.1
An fMRI scan provides a record of the activity of a human brain. In this case the photo shows a horizontal section through a normal brain.

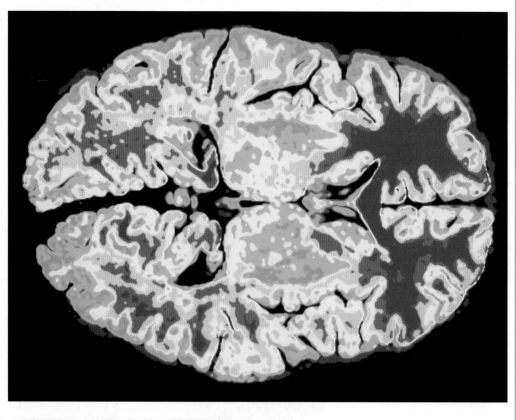

fMRI images are made possible by providing bursts of energy from the scanning equipment to atoms of the brain, and then detecting the distinctive responses of different kinds of atoms. Although the radio waves that emerge from the head are invisible to us, we can use detectors, together with computers, to construct visible pictures.

The retina

Light passes through the pupils of your eyes, and the lens system projects images of the outside world on to your **retinas**, the inner surfaces of your eyeballs (Figure 1.2).

A retina contains a layer of light-sensitive cells, in which the arrival of light causes chemical changes and so creates electrochemical impulses that travel along nerves to the brain. The chemical change quickly reverses so that cells are ready to transmit new impulses, or 'fire' again. (A bright flash of light leaves you temporarily blind – all the cells have fired and until the reactions can reverse they are not capable of firing again.)

Figure 1.2
Inside your eye.

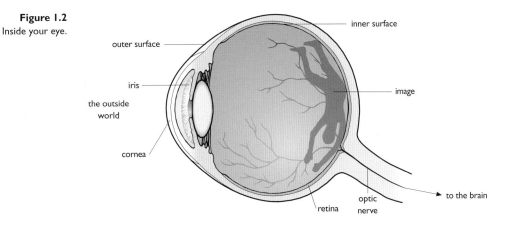

The journey of light to the retina

The changes that light stimulates in an area of retina, and therefore what you see, depends upon the nature of the light and its patterns in space and time. These patterns in turn depend on the nature of the source of the light – the Sun, a flame, a lamp, or a TV screen. They depend on whether or not the source is moving and whether or not it is continuous or flashing and flickering. They are also very much affected by the interactions between light and matter that took place during the journey of the light from its source to your eyes (Figure 1.3).

Light might set off, say from a filament in a lamp, and before it reaches your eyes it may have interacted with the glass of the lamp, with air, with a wooden table, a book, a wall, and your own skin. Many of these interactions will be **reflections**, in which light strikes a surface and does not pass into it, but travels back away from it. Not all of the light that falls on a surface is reflected; some is either absorbed or transmitted by the material. On **absorption** the light ceases to exist, and its energy is transferred to the absorbing material.

Figure 1.3
Light that reaches our eyes tells a story of where it came from and where it has been. The light reflected by a glass of cola to make it visible may have originated in the Sun, but it has been influenced by its interactions with the dark liquid.

Why does a dark liquid have light froth? Ideas about reflection and absorption provide an answer. The liquid itself is a fairly good absorber of light, so the intensity of the light decreases quite rapidly as it travels through the liquid. Before absorption can take place there is some small degree of reflection at the surface. Bubbles have a lot of surface area, and not much thickness of liquid in which absorption can take place. Reflection is then the dominant process, so the froth reflects a lot of the light that is incident upon it.

Figure 1.4
A retina is an absorbing surface, but it is also a reflecting surface. Some of the light from a bright flash returns to the camera to create 'red eye' in a photograph.

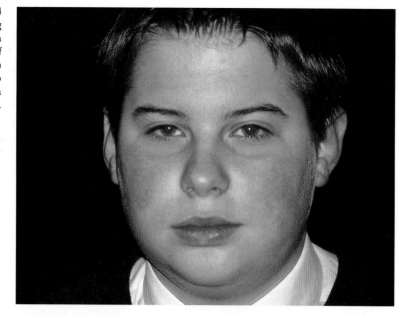

If the light of a lamp shines on smoke, there are reflections in all directions from the particles of smoke. Such reflection in many directions is called **scattering** of the light. Mirrors and smooth water surfaces do not scatter light but reflect it in a predictable way. Most other surfaces send light in directions that are, in practice, unpredictable; they scatter the light.

Many surfaces are selective in their reflection and absorption (Figure 1.4), so the light that leaves the surface appears different to our eyes, compared with the light that arrived from the lamp. We see colours, and talk about a red bus or a blue flower. The surfaces absorb and reflect different types of light (see also Chapter 3).

The air and glass absorb only a little of the light from the filament of the lamp. There may be some reflection at boundaries between air and glass, but otherwise the air and glass largely allow the light to travel through them; they transmit the light. Our own bodies are poor transmitters of light. It is as a result of this that we are accompanied by our shadows wherever we go in the light of the Sun (Figure 1.5).

Figure 1.5
Sunlight and its interactions with matter: **transmission**, absorption and reflection.

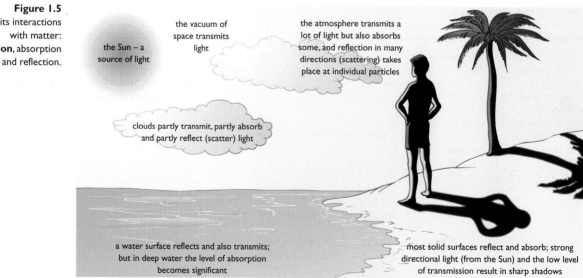

the vacuum of space transmits light

the Sun – a source of light

the atmosphere transmits a lot of light but also absorbs some, and reflection in many directions (scattering) takes place at individual particles

clouds partly transmit, partly absorb and partly reflect (scatter) light

a water surface reflects and also transmits; but in deep water the level of absorption becomes significant

most solid surfaces reflect and absorb; strong directional light (from the Sun) and the low level of transmission result in sharp shadows

Light interacts with matter and its resulting **intensity** is affected by a combination of reflection, absorption and transmission. Intensity of light is an indication of its brightness; it is the power (or energy transfer per second) of the light per unit of area.

Graphic representation of light intensity

A picture can convey information very quickly. We can even show quantities in pictures, such as in the form of a graph. For a beam of light that is not spreading – called a **parallel beam** of light – we can show graphically how its intensity varies as it interacts with matter (Figure 1.6).

Figure 1.6
Graph of the reflection, absorption and transmission of an incident beam of light.

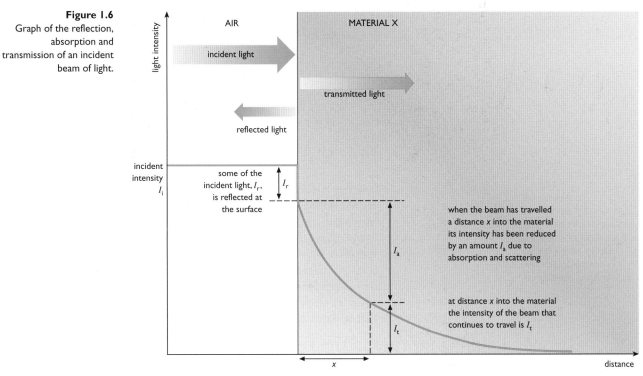

For any point inside material X at distance x from the surface, the transmitted light intensity is such that:

total incident light intensity	=	reflected light intensity	+	absorbed/scattered light intensity	+	transmitted light intensity

$$I_i = I_r + I_a + I_t$$

1 What is the relevance of absorption to
 a the appearance of a dark liquid
 b the action of a retina
 c the action of the film in a camera?
2 How do we know that the inner surfaces of the eye reflect as well as absorb light?
3 Make two sketches of Figure 1.3, one showing the effect of ordinary cola drink on the intensity of a beam of light, and another, drawn to the same scale, showing what happens when the cola is watered down.
4 **DISCUSS**
 Look at the fMRI scan (Figure 1.1), the photograph of the face (Figure 1.4), the illustration of the interactions of sunlight (Figure 1.5), and the graph showing the variation in intensity of a beam of light (Figure 1.6). Discuss which of these images is the most realistic representation of reality, and which is the least realistic. Which carries most information, and which carries least?

Physicists make much use of their eyes, and of visual representations like the graph. It shows what happens to the total intensity of a beam of light that arrives at a surface of a material. Note that reflection takes place strongly at the surface. Transmission, scattering and absorption take place within the body of the material.

Rays and ray diagrams – models of reality

Physics gives us the knowledge and skills to predict the behaviour of the world around us. To do this we represent the world in different ways – through language, through mathematics, and through images. Physics makes **models**. A model, whether in words or numbers or pictures, is a representation of some part of the real world. It is a useful aid for our thinking and for making predictions. We can use very different models to help us to think in different ways about just one aspect of the physical world. There are different ways, for example, of representing the spread of light from a source (Figure 1.7). None of these representations is the real thing – they are only models.

Figure 1.7
Alternative models of the same physical reality: radiation of light from a source. A ray is a representation of reality that has the benefit of simplicity.

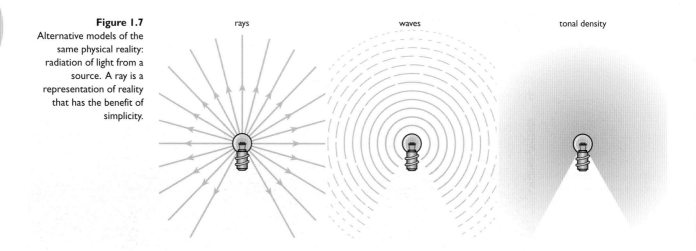

rays   waves   tonal density

Figure 1.8
Light strikes a mirror from all directions. To be able to think about reflection we reduce all of this light to that from just one direction, in a ray diagram.

ray   reflecting surface

A **ray** is a narrow line, drawn to show the pathway of light. An **ideal** ray has no thickness at all. An ideal is something that can exist in our minds but not in reality. However, a diagram showing rays of light is a useful model of reality. It does not attempt to show all of the light. It simplifies, showing just enough information to help us to think and predict – to make analysis possible. A ray can provide a valuable representation of reflection of light, for example (Figure 1.8). It is much easier to think about a single ray than to try to consider all of the light that strikes a surface.

Looking for rays in the real world

It is difficult to create a very thin 'ray' of real light, but simply by putting a slotted card in front of a lamp we can make a reasonably narrow beam (Figure 1.9a). Such a beam is approximately like an ideal ray. It is a useful practical **approximation** to the ray. We can then see what happens to the beam when it interacts with a reflecting surface, such as a mirror.

Figure 1.9
We can't, in practice, create ideal rays from a source of light, but we can use the ray concept to make worthwhile investigations.

An alternative approach is to line up a series of pins stuck vertically into a wooden board. Then when we look along the line we are looking along an imaginary ray (Figure 1.9b).

a   b

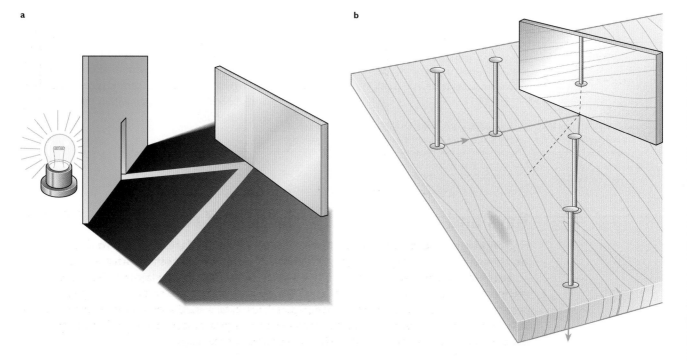

Rays and reflection

Using a thin beam (Figure 1.10) or a line of pins as above, we can examine real world behaviour of reflection by a mirror. We can record our observations in 'ray diagrams'. What we find is a simple pattern of behaviour that can be seen time and time again by different observers – the findings of this experiment are easily **replicated**.

Figure 1.10
A laser beam is a fine beam that spreads very little. The beam does have some width, but it is a very close approximation to an ideal thin ray.

When we draw a **normal** line, a line that is perpendicular to the reflecting surface, the angle between the normal and the incoming or incident ray is equal to the angle between the normal and the reflected ray. Such a repeatedly observed pattern of behaviour is called a **law**, and this is the **law of reflection**. Put briefly, the law of reflection states that the angle of reflection is equal to the angle of incidence (Figure 1.11).

Figure 1.11
The law of reflection: the ray model of light reveals a pattern of behaviour that seems to be universal – a pattern that we seem to replicate whenever we test it.

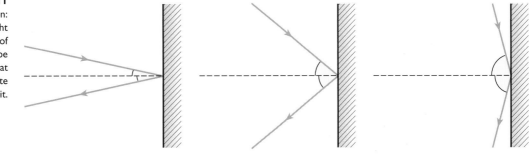

Mirror images – the ray model

By drawing a number of rays spreading from a point we can develop a model of the formation of an image by a mirror.

Rays of light spread outwards, or radiate, from their source. (**Radiation** is a word used to describe this process of spreading, but it is also sometimes used to describe what is spreading.)

Suppose that some spreading rays strike a reflecting surface, and the law of reflection is obeyed. The reflected rays continue to radiate, but the point from which they now *appear* to be spreading is behind the reflecting surface (Figure 1.12). In fact, the diagram suggests that it is as far behind the reflecting surface as the actual source (the original object) is in front. The ray model predicts the apparent source of the light – the position of the image.

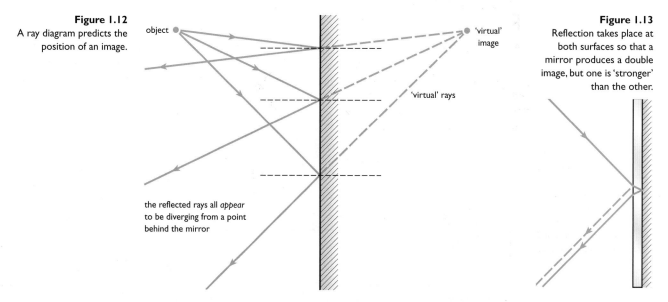

Figure 1.12
A ray diagram predicts the position of an image.

object

'virtual' image

'virtual' rays

the reflected rays all *appear* to be diverging from a point behind the mirror

Figure 1.13
Reflection takes place at both surfaces so that a mirror produces a double image, but one is 'stronger' than the other.

A glass mirror has two reflecting surfaces. The principal reflection, which creates the bright image we see, actually takes place at the back surface of the mirror and not at the front (Figure 1.13). For simplicity we usually treat a mirror as a single reflecting surface.

5 a Copy Figure 1.14 and add rays and normals to find out whether the person can see the top of her head and her toes.

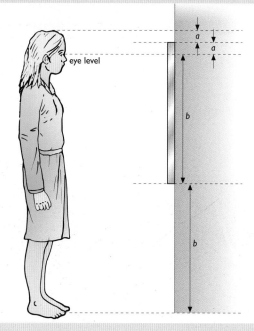

eye level

a

a

b

b

b

Figure 1.14

b What happens to the person's view of herself as she moves closer to the mirror? (You might need to draw another sketch to help you with this. You will then be applying the ray model to make a prediction about the world. Try to make a practical test, an experiment, of your prediction.)

6 Writing explanations is a difficult but very important skill. Write an explanation of why a dance floor mirror globe like those in Figure 1.15 sparkles as it turns. Include mention of the law of reflection.

Figure 1.15

7 What prediction does the ray model provide in the following situation? A ray strikes a mirror with an angle of incidence of 10° and so the angle of reflection is 10°. The mirror is then turned so that the angle of incidence is 20°. Through what angle does the reflected ray turn?

8 Two parallel plane mirrors produce multiple images of object X (Figure 1.16).
a Copy the diagram, allowing plenty of space 'behind' the mirrors, and sketch the positions of the first images produced by each mirror.
b Other images are all 'images of images' and not direct images of X. Sketch the positions of the second images produced by each mirror.

X

Figure 1.16

Rays and perspective

Around AD150, an Egyptian called Ptolemy used the concept of the ray to analyse reflection. A long time later, around AD1000, an Arab scholar called Alhazen (or Al-Haytham) took Ptolemy's ideas further, and made a thorough geometrical analysis. In Italy, another 400 years later, Alhazen's ideas made it possible for artists to paint with **perspective**. This involved showing near objects larger than far ones. Their paintings, for the first time, could make the same geometric patterns on our retinas as did the scenes they were showing.

Figure 1.17
A ray analysis shows that near and far objects make different angles at our eyes. The larger the angle they make, the larger the area of retina the image covers and the bigger the object seems.

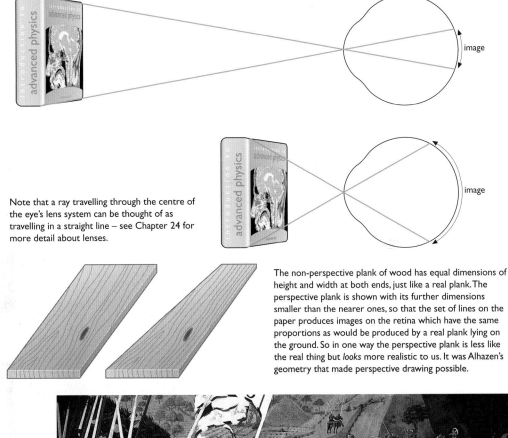

Note that a ray travelling through the centre of the eye's lens system can be thought of as travelling in a straight line – see Chapter 24 for more detail about lenses.

Figure 1.18
Which of these is more like the real thing?

The non-perspective plank of wood has equal dimensions of height and width at both ends, just like a real plank. The perspective plank is shown with its further dimensions smaller than the nearer ones, so that the set of lines on the paper produces images on the retina which have the same proportions as would be produced by a real plank lying on the ground. So in one way the perspective plank is less like the real thing but *looks* more realistic to us. It was Alhazen's geometry that made perspective drawing possible.

Figure 1.19
The Battle of San Romano, by Paolo Uccello c. 1440.

9 What happens to the size of a bus as it gets closer to you? What happens to the size of the images of the bus on your retinas? Draw rays to illustrate your answer.

10 DISCUSS
a While fixing your make-up at the bus stop you see, reflected in your plane mirror, the approaching bus. How does the speed of the 'virtual bus' in your mirror compare with the speed of the real bus?
b Sitting on the moving bus, and still fixing that make-up, you watch the reflection of a car that is overtaking the bus. How does your view of the speed of the image of the car compare with the actual speed of the car?

Paolo Uccello was fascinated by the power of perspective painting. Such paintings produce the same geometrical patterns on our retinas as the actual scenes do, and so they provide a convincing representation of depth and distance. But Uccello was not aware that we also use other clues to judge distance. Absorption and scattering of light by the air reduce the brightness of distant objects. The figures in the background are painted more brightly than they would actually appear.

● **Comprehension and application**

The silver screen

The light that you see reflected from a cinema screen originates at a high power lamp in the projector. It travels through the air, through the film, through the projector lens system and reaches the screen where the image is formed. The screen is silvery white, for maximum reflection of light. But the screen does not act as a mirror. Each small area on the screen scatters the light that arrives, with slightly different incident pathways, in very different directions. For each individual ideal ray the law of reflection is obeyed, but the screen is a somewhat rough surface and its normals lie in many directions. The resulting scattering is also called **diffuse reflection** (Figure 1.20).

Figure 1.20
Diffuse reflection.

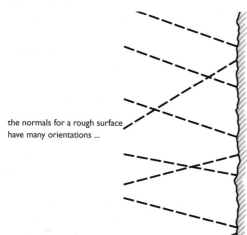

the normals for a rough surface have many orientations ...

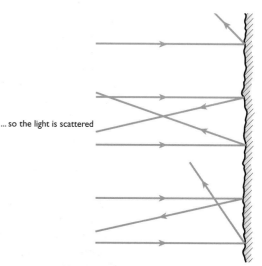

... so the light is scattered

11 Describe the effect of the film on the light, in terms of absorption and transmission.

12 **a** For health reasons, smoking is banned in cinemas. How would the travel of light be affected in a cinema that was heavily polluted with cigarette smoke?
   b If a beam is not spreading out (or converging) as it travels then its rays are parallel to each other. Sketch graphs showing the intensity of transmitted light for the journey of such a beam in clean and smoky cinemas.

13 Why is it important that the screen should be clean?

14 **a** Use the ray model to describe the important differences in behaviour between the cinema screen and a mirror.
   b What would you expect to happen if the cinema screen were replaced by a large mirror?

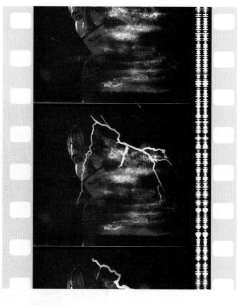

Figure 1.21
A frame of a film is composed of a layer of coloured materials on plastic backing. The coloured areas act as filters, absorbing and transmitting light. To produce a clear image, the projector lens system must make sure that light from a very small area of the film travels to only a small area of the screen.

● **Extra skills task**

Communication and Information Technology

Prepare a poster presentation suitable for your fellow students, explaining how light can be reflected from different surfaces, and the different effects that are seen. Include ray diagrams showing reflection from curved and flat surfaces.

Your display should include labelled, computer generated diagrams as well as word-processed text. Sources of information might include your textbooks or CD ROMs. Your poster should be clear, interesting and informative.

Wave models of light and sound

THE BIG QUESTIONS

● What have been the key observations that have caused ideas about light to change?
● What variables are associated with the travel of light and sound?
● How can we apply wave ideas to understand sound?

KEY VOCABULARY

amplitude   antinode   base quantity   base unit   cancellation of waves
compression   corpuscular theory   derived unit   diffraction
dimensions (of a quantity)   displacement   ether (an important word from the
history of physics)   frequency   fundamental   hertz   hypothesis   interference
interference pattern   inverse relationship   longitudinal travel   medium   node
overtone   rarefaction   refraction   reinforcement of waves   SI system
stationary wave   superposition   theory   transverse travel   wave theory
wavelength

BACKGROUND

What *is* light? At different times and in different places there have been different answers.
Aristotle, around 350BC, for example, suggested that we saw because 'light' spreads from our
eyes to what we see. Ptolemy, in about AD150, argued that the action took place the other way
round; that light travelled from the observed to the observer.

Almost exactly 2000 years after Aristotle, in 1637, René Descartes was still only speculating. He
suggested that light might be a 'pressure' which *either* existed in the space around visible objects
and took no time to travel, *or* was carried away from sources of light by travelling particles.

Later in the same century, Isaac Newton (Figure 2.1) favoured this particle or **'corpuscular
theory'** of light, whilst a Dutchman called Christiaan Huygens (Figure 2.2) suggested that light
travelled as impulses, or travelling disturbances. Newton's and Huygens' ideas were competing
theories. Perhaps because of Newton's reputation for his work in various areas of science, for
more than a hundred years the corpuscular theory was accepted as the better way of thinking
about the nature of light. But that was certainly not the end of the story of light and our
attempts to understand it.

Figure 2.1 (left)
Isaac Newton favoured
the particle or
corpuscular theory of
the travel of light.

Figure 2.2 (right)
Christiaan Huygens
developed a view of light
as travelling disturbances
which could be thought
of as being similar to
waves on the surface of
water.

Young, Fresnel and the new wave theory

An explanation is a description of processes that underlie what we see. A **theory** is a source of explanations for the observations we make of our world. A good theory can provide explanations for all related observations. Also, it is not contradicted by any observation.

The corpuscular theory of light provided explanations for reflection and refraction. However, the theory of the nature of light went through a major change after 1800. Thomas Young tried out the alternative idea, that light travels in a way that is similar to the travel of disturbances, or waves, on the surface of water.

Water waves are mechanical – they involve material that we can see and feel and that is subject to forces. Water waves show certain types of behaviour:

- reflection (obeying a law of equal angles)
- **refraction** – the speed of the waves differs in different depths of water, and the change in speed can result in a change in the direction of travel
- **diffraction** – waves spread out after passing an obstacle or passing through an opening
- **superposition**, leading to **interference** – waves from two (or more) sources add together creating patterns of reinforcement and cancellation
- absorption – waves lose energy as they travel, although water waves lose energy slowly.

Waves also have measurable properties such as speed and **amplitude** (and frequency and wavelength – see page 14). The amplitude of a water wave is the maximum displacement of the water surface from its resting position (Figure 2.3).

Figure 2.3
The amplitude of a water wave. The wave has a fixed amplitude, but at a point the wave has a varying **displacement**.

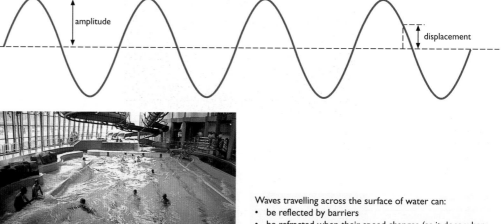

Figure 2.4
Water waves in action. Processes of reflection and superposition can create complex wave patterns.

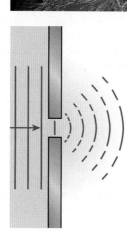

Waves travelling across the surface of water can:
- be reflected by barriers
- be refracted when their speed changes (as it does when the depth of water changes)
- be diffracted, when they pass rigid obstacles or go through gaps
- experience superposition, creating patterns of reinforcement and cancellation.

Figure 2.5
Diffraction of waves takes place at gaps in barriers and around obstacles, and this can lead to interference patterns.

Water waves spread out from gaps in a barrier, such as the gaps in a harbour sea wall. This spreading is called diffraction (Figure 2.5). Two sets of diffracted waves can cross over and add together, to produce an **interference pattern**. If the interference is produced by two unchanging sources of waves then a regular and fixed pattern of alternating **reinforcement** and **cancellation** is produced. In some places, where the crest of one wave meets a crest of the other, the waves add together to produce increased intensity of disturbance, or reinforcement. Where the crest of one wave meets a trough of the other, if they have the same amplitude, they cancel each other so that there is no intensity of disturbance at all. Alternating bands of reinforcement and cancellation make up the interference pattern. (You can find out more detail about this in Chapter 25.)

Young made the **hypothesis** that *if* light travels in a wave-like way then it should also experience diffraction and it should be possible to create interference patterns. He set up an experiment to test the hypothesis, using two closely spaced holes. Light travelled through the holes, and was seen to produce interference patterns of bright and dark spots.

His observations convinced Young that light travelled in a wave-like way. But people are slow to give up their strongly-held ideas and many still preferred the corpuscular theory. The debate continued for many years.

Twenty years later Augustin Fresnel developed Young's ideas into a mathematical description of wave travel, and came up with the surprising prediction that diffraction and interference effects should result in a bright spot in the centre of the shadow of a very small round object. Many thought that this was silly, until it was found to be exactly what happens (Figure 2.6).

Figure 2.6
The pattern of interference around a small obstacle matches Young's hypothesis and the predictions made by Fresnel's wave calculations. It provides strong support for the idea that light travels in a wave-like way.

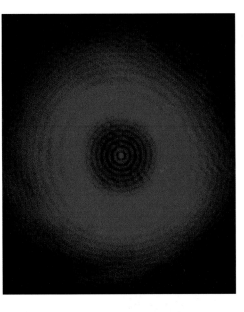
a

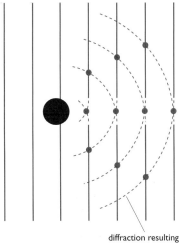
b

diffraction resulting in an interference pattern

Young used wave ideas to make the prediction that light might show diffraction and interference effects, and Fresnel developed the ideas to make the prediction that a small obstacle would have a bright spot at the centre of its shadow. Both predictions were confirmed by experiment, providing support for wave explanations of the travel of light. So the general accepted theory of light moved from the corpuscular theory to the **wave theory**. (It moved on again a hundred years later – see Chapter 26.)

A medium for the waves

Young's and Fresnel's observations that light shows all of the types of behaviour that water waves show, the notion that light travels as a wave, was not just tempting but irresistible. But a problem still existed. With water waves it was clear that it was the water itself that was vibrating as a wave travelled across its surface. The water surface is the **medium** for the waves. Nobody knew what might be vibrating to allow light waves to travel. Some suggested that space must be filled with some kind of medium for the transmission of light. They called this medium the **ether** of space. Scientist James Clerk Maxwell (see page 19) wrote in the *Encyclopaedia Britannica* in the mid-19th century:

> *'Ether, a material substance of a more subtle kind than visible bodies, supposed to exist in those parts of space which are apparently empty.'*

Note that Maxwell was careful with his words. He knew that the ether had never been observed directly, and he didn't say that the ether *did* exist but that people *supposed* that it existed. About 30 years after Maxwell died it was his work that led Einstein to show that the idea of the ether should be abandoned.

1 A swimmer rides down a flume into a pool and creates circular waves. Under what circumstances would you expect to observe
   a reflection
   b refraction
   c diffraction?
2 If there are two flumes side by side leading into a pool, and two swimmers hit the water at the same time, what patterns might you see due to superposition of the two sets of waves?
3 What was Young's hypothesis?
4 What was Fresnel's hypothesis?
5 The corpuscular theory of light provides a picture of how light can be reflected. Why does observation of diffraction of light force us to abandon the corpuscular theory?

Speed, frequency and wavelength

Waves on water have a measurable speed. They also have a **wavelength**, the length of one complete cycle, and a **frequency**, which is the number of complete waves generated by a source per second, or the number of complete waves passing a fixed point per second.

Wavelength is a distance measured in metres. Frequency is a measure of the number of events in each second, measured in cycles per second which can be written cps or s^{-1} (which means just 'per second'), but because frequency is a commonly used quantity, the s^{-1} is called the **hertz**, Hz.

Seconds, metres and hertz are all members of the same international system of units, called the **SI system**. Each unit in this system has a very strict definition. The second and the metre are **base units**, used for measurements of fundamental quantities – distance and time. The hertz is a **derived unit**, one whose definition is built up from base units.

A relationship between wavelength and frequency

Picture two trains, one with long coaches and the other with short trucks. If the trains have the same speed then more of the short trucks travel past a fixed point in each constant period of time. That is, the frequency with which the trucks go by is higher than that for the coaches. There is an **inverse relationship** between the length of the coach or truck and the frequency – as the length increases, the frequency decreases.

Now picture two waves of different wavelength and the same speed. More of the shorter waves will pass a fixed point in each second. That is, a shorter wavelength corresponds to a higher frequency. There is an inverse relationship between wavelength and frequency (Figure 2.7).

Figure 2.7
The inverse relationship between wavelength and frequency.

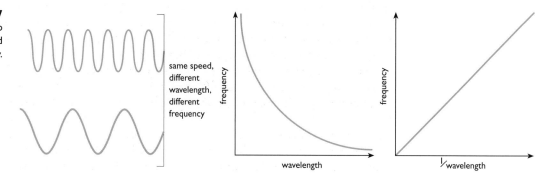

same speed, different wavelength, different frequency

For all types of wave,

speed = frequency $\times$ wavelength
$$v = f\lambda$$

Each quantity has a unit. The SI unit of speed is the metre per second, $m\ s^{-1}$. The SI unit of frequency is the hertz, Hz, and the SI unit of wavelength is the metre.

Dimensions of quantities

We can write each of the above quantities in terms of **dimensions**. The dimensions of a quantity show it in terms of **base quantities**, such as mass, length and time. (The other base quantities of the SI system are electric current, thermodynamic temperature, amount of substance and luminous intensity.)

Using [M] for mass, [L] for length and [T] for time, the dimensions of the quantities above are given in Table 2.1.

Quantity	SI unit	Dimensions
speed	$m\ s^{-1}$	$[L][T]^{-1}$
frequency	Hz	$[T]^{-1}$
wavelength	m	$[L]$

Table 2.1

Note that in the relationship $v = f\lambda$, the dimensions of the left-hand side are:

$$[L][T]^{-1}$$

and the dimensions of the right-hand side are:

$$[L] \times [T]^{-1} \quad \text{or simply} \quad [L][T]^{-1}$$

So the two sides of the formula have the same dimensions. This is true for all formulae.

6 The engine of a boat vibrates at 60 Hz so that ripples with a wavelength of 0.005 m spread outwards across the water. What is their speed?

7 A long rope is flicked at a steady frequency and waves travel along it at 3.2 m s^{-1}, with a wavelength of 1.6 m. What is the frequency?

8 State the units and dimensions of **a** area, **b** volume.

9 What is the approximate frequency of visible light if its speed is 3×10^8 m s^{-1} and its wavelength is in the region of 5×10^{-7} m?

The relationship for light waves

If light travels in a wave-like way then we should expect a particular sample of light to have a frequency and a wavelength as well as speed, and the relationship between these three variables should be the same as for other waves. This is found to be the case.

The speed of light is often symbolised as c, and its frequency as ν. So for light,

$$c = \nu\lambda$$

ν, like λ, is a Greek letter, pronounced 'new'.

Water waves, light waves and sound waves compared

We have compared light waves to waves on water. An obvious difference is that water waves spread across a two-dimensional surface, whereas light spreads from a source into three-dimensional space. But there are strong similarities – both light waves and water waves are reflected, refracted and diffracted, and both show superposition effects such as interference. Sound waves also show these types of behaviour.

Sound travels about a million times more slowly than light through air. That is, the speed of sound in air is a factor of 10^6 slower than that of light. We also know that sound needs a physical medium through which to travel, whereas light can travel through a vacuum. This leads us to the idea that sound waves involve mechanical vibrations of material. Sound waves are created by vibration of a physical source – resulting in **compressions** and **rarefactions** of the surrounding air or other medium. The vibration pushes particles of the medium closer together to create compressions, and then further apart to create rarefactions. Due to interactions between the particles of the medium, these alternating compressions and rarefactions spread. Gases are poor carriers or conductors of sound, because of the weak particle-to-particle interactions, compared with liquids and solids. Whales and dolphins have hearing that is more sophisticated than ours – they live in an environment in which sound travels quickly and over long distances.

Figure 2.8
Sound waves, water waves and their travel.

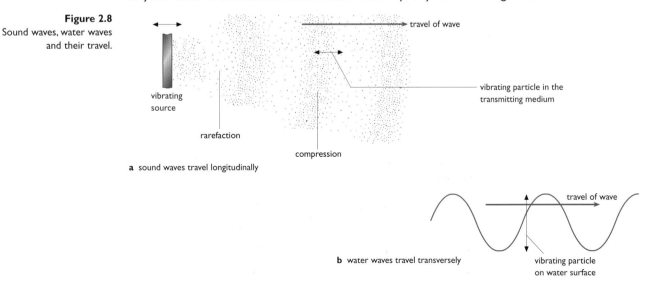

travel of wave

vibrating source

rarefaction

compression

vibrating particle in the transmitting medium

a sound waves travel longitudinally

travel of wave

b water waves travel transversely

vibrating particle on water surface

The mode of travel of the compressions and rarefactions that make up a sound wave is different to the way that water waves travel across a surface (Figure 2.8). For the sound waves, the vibration of the particles of the medium is parallel to the direction in which the wave is travelling. This is described as **longitudinal travel**. The particles of the water vibrate in a direction that is perpendicular to the direction in which the waves are moving – the particles on the surface move up and down (approximately) as the wave travels horizontally along. This is **transverse travel**.

The generation of sound waves by stationary waves in strings

Most musical instruments generate sound by vibration of strings, or air in pipes. A string is displaced at some point along its length by being plucked or hit. This generates a wave that travels rapidly along the string itself. It is a transverse wave – the vibration of the particles of the string is perpendicular to the direction of travel, which is along the string. The ends of the string are rigidly fixed, and energy can pass only very slowly from the string to these fixed points. Instead the energy remains in the string, and the wave is reflected with its frequency unchanged. A consequence of the reflections from the two ends of the string is that waves on a string add together – superposition takes place. It is this that leads to a **stationary wave** or standing wave. How waves of the same frequency create a standing wave as a result of reflection is best shown by graphic representation (Figure 2.9).

Figure 2.9
Formation of a stationary wave: **a** shows how a single pulse combines with its own reflection, and **b** shows how a continuous train of waves combines with its own reflection.

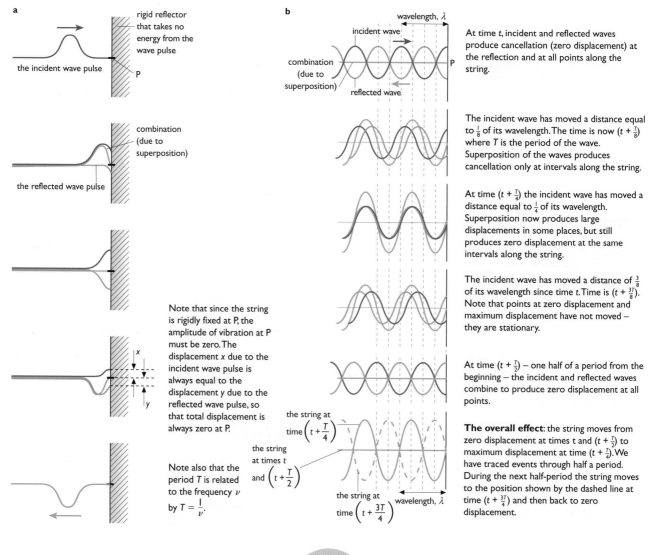

a

rigid reflector that takes no energy from the wave pulse

the incident wave pulse

P

combination (due to superposition)

the reflected wave pulse

Note that since the string is rigidly fixed at P, the amplitude of vibration at P must be zero. The displacement x due to the incident wave pulse is always equal to the displacement y due to the reflected wave pulse, so that total displacement is always zero at P.

Note also that the period T is related to the frequency ν by $T = \frac{1}{\nu}$.

b

wavelength, λ

incident wave

combination (due to superposition)

P

reflected wave

the string at times t and $\left(t + \frac{T}{2}\right)$

the string at time $\left(t + \frac{T}{4}\right)$

the string at time $\left(t + \frac{3T}{4}\right)$

wavelength, λ

At time t, incident and reflected waves produce cancellation (zero displacement) at the reflection and at all points along the string.

The incident wave has moved a distance equal to $\frac{1}{8}$ of its wavelength. The time is now $\left(t + \frac{T}{8}\right)$ where T is the period of the wave. Superposition of the waves produces cancellation only at intervals along the string.

At time $\left(t + \frac{T}{4}\right)$ the incident wave has moved a distance equal to $\frac{1}{4}$ of its wavelength. Superposition now produces large displacements in some places, but still produces zero displacement at the same intervals along the string.

The incident wave has moved a distance of $\frac{3}{8}$ of its wavelength since time t. Time is $\left(t + \frac{3T}{8}\right)$. Note that points at zero displacement and maximum displacement have not moved – they are stationary.

At time $\left(t + \frac{T}{2}\right)$ – one half of a period from the beginning – the incident and reflected waves combine to produce zero displacement at all points.

The overall effect: the string moves from zero displacement at times t and $\left(t + \frac{T}{2}\right)$ to maximum displacement at time $\left(t + \frac{T}{4}\right)$. We have traced events through half a period. During the next half-period the string moves to the position shown by the dashed line at time $\left(t + \frac{3T}{4}\right)$ and then back to zero displacement.

Note that all points at which superposition results in 'cancellation' or zero amplitude of vibration are called **nodes**. Points at which amplitude is maximum are called **antinodes** (Figure 2.10).

Figure 2.10
Nodes and antinodes.

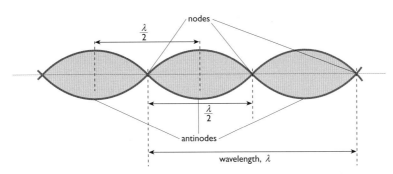

Note that distance between adjacent nodes is $\frac{\lambda}{2}$. Also, distance between any adjacent pair of antinodes is $\frac{\lambda}{2}$.

Fundamental frequency and overtones

A vibrating string must have nodes at its two fixed ends. But the number of nodes and antinodes along the length of the string can vary. The simplest mode of vibration has just one antinode between the two ends. This is called the **fundamental** mode of vibration. But it is also possible for a string to vibrate with a node at its centre and with two antinodes – this is the first **overtone**. The second overtone has two nodes as well as the two end nodes, and has three antinodes, as shown in Figure 2.11.

Figure 2.11
Fundamental, first, second and third overtones.

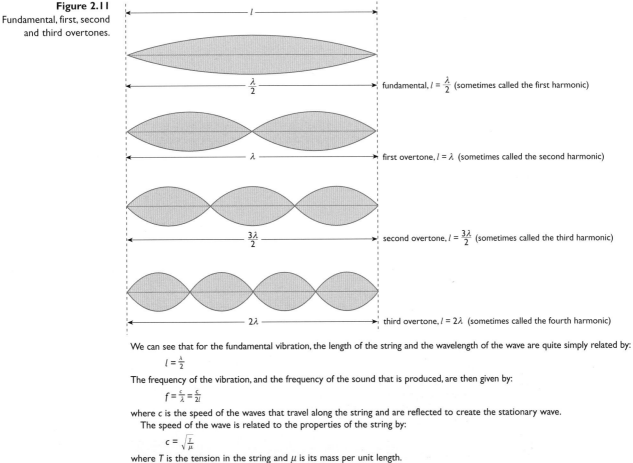

fundamental, $l = \frac{\lambda}{2}$ (sometimes called the first harmonic)

first overtone, $l = \lambda$ (sometimes called the second harmonic)

second overtone, $l = \frac{3\lambda}{2}$ (sometimes called the third harmonic)

third overtone, $l = 2\lambda$ (sometimes called the fourth harmonic)

We can see that for the fundamental vibration, the length of the string and the wavelength of the wave are quite simply related by:

$$l = \frac{\lambda}{2}$$

The frequency of the vibration, and the frequency of the sound that is produced, are then given by:

$$f = \frac{c}{\lambda} = \frac{c}{2l}$$

where c is the speed of the waves that travel along the string and are reflected to create the stationary wave.

The speed of the wave is related to the properties of the string by:

$$c = \sqrt{\frac{T}{\mu}}$$

where T is the tension in the string and μ is its mass per unit length.

So the frequency of the fundamental is given by:

$$f = \frac{\sqrt{\frac{T}{\mu}}}{2l} = \frac{1}{2l}\sqrt{\frac{T}{\mu}}$$

● More stationary waves

Figure 2.12
Stationary waves in pipes.

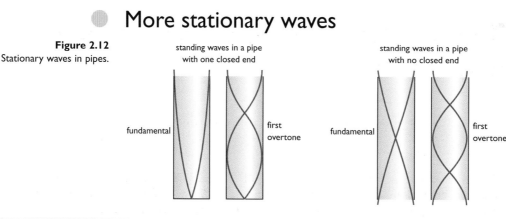

Note that the stationary waves in the air in the pipe are longitudinal waves, but they are represented here by comparing them to the transverse waves that exist on strings.

10 a Describe the differences between the travel of a wave along a string and along a column of air in a pipe.
b Explain the roles of reflection and superposition in the formation of a standing wave on a string. Use sketch drawings to help you to give *full* explanations.
11 a List the relationships between length of string and wavelength for the fundamental and first, second and third overtones.
b Given that $v = f\lambda$ list the corresponding relationships between frequency and length of string.
c The speed of waves along the string is the same for all of these modes of vibration. Give the ratio:

overtone frequency/fundamental frequency

for first, second and third overtones.
12 From an experimental investigation of the relationship between fundamental frequency and length of string, what quantities should be plotted in order to produce a straight line graph?

Stationary waves can exist in pipes, but here there is always a node at any closed end of a pipe and an antinode at any open end.

It is not only in sound that standing waves can be set up. A microwave transmitter facing a reflecting surface will produce patterns of nodes and antinodes. The microwaves have wavelengths of a few centimetres, so they are suitable for studying wave behaviour. Microwaves belong to the spectrum of light waves, as discussed in the next chapter.

● **Extra skills task**   Communication

Imagine that, after a little adjustment of your time co-ordinates, you walk into a room in which Isaac Newton and Christiaan Huygens are discussing whether corpuscle (particle) ideas or wave ideas provide the most complete explanation of the behaviour of light. Doing your best to ignore the fleas in their wigs, you cautiously place a copy of Figure 2.6 on the table in front of them. What effect would you expect this to have on:

a their conversation
b the course of the history of science (for the moment, suppose that 'science' is the seeking of an understanding, through experiments, of the underlying nature behind the observations we make)
c the course of the history of technology (for the moment, suppose that 'technology' is the using of brains and hands to provide material benefit. Radio broadcasting and optic fibres are communications technologies)
d the course of history (the affluence of individuals, social movements, international affairs, wars, politics and so on)?

Discuss your ideas with one or two other people and then take part in a group discussion of the listed points. You should also consider general issues, such as the relationship between scientific knowledge and technology, and the impact of physics on how we think and how we live. The discussion will need to be chaired, and this can be done by a student or a teacher. Everybody should have the opportunity to contribute ideas.

3 The electromagnetic spectrum and colour

THE BIG QUESTIONS	● How much of light is visible to us, and how much is invisible?
	● How are our sensations of colour related to the observable and measurable behaviour of light?
KEY VOCABULARY	addition of colour amplitude modulation, AM analogue bandwidth bit digital dispersion electromagnetic spectrum filter frequency modulation, FM infrared pigment primary colours radio waves subtraction of colours ultraviolet wavefront X-rays

BACKGROUND
Perhaps no scientist has made a bigger contribution to our understanding of light than the Scottish physicist James Clerk Maxwell (Figure 3.1). He developed ideas which predicted that the light we see is just part of a very wide **electromagnetic spectrum**, and that the speed of light through space is constant.

Figure 3.1
James Clerk Maxwell, 1831–1879.

● The visible spectrum and its wavelengths

An interference pattern produced by two identical sources of waves has alternating bands, or fringes, of reinforcement and cancellation. A simple wave diagram shows that waves of different wavelengths produce patterns of interference with different widths or fringes (Figures 3.2 and 3.3). The smaller the wavelength, the narrower the spacing of interference fringes.

Figure 3.2
Waves of different wavelengths produce different interference patterns.

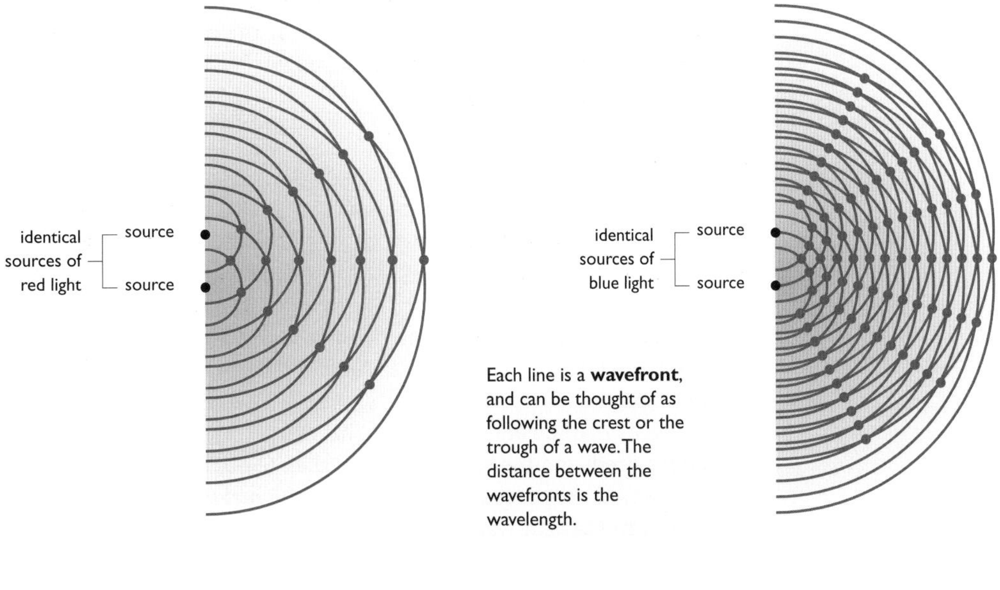

identical sources of red light ⎡ source ⎣ source

identical sources of blue light ⎡ source ⎣ source

Each line is a **wavefront**, and can be thought of as following the crest or the trough of a wave. The distance between the wavefronts is the wavelength.

The pattern produced by violet light is different from the pattern produced by red light, showing that violet light has a short wavelength compared with that of red light. Wavelength decreases from red through the other colours of the spectrum – orange, yellow, green, blue, indigo, violet.

Figure 3.3
Interference fringes produced by the effect of two closely spaced slits on light of different colours. The smaller the wavelength of the light, the narrower the spacing of the fringes.

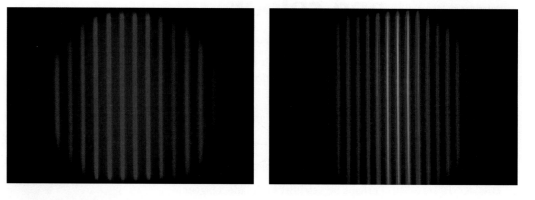

Dispersion of white light by a prism or a raindrop (Figure 3.4) shows us that white light can be separated into bands which we see as having different colours. The continuous spectrum that you see in a rainbow or when a prism disperses light is a result of continuous variation of wavelength. The degree of refraction that takes place at the surfaces of water drops and prisms is dependent on wavelength.

1 **a** What is the approximate range of wavelengths of visible light?
   b Use $c = \nu\lambda$ to calculate the approximate highest and lowest frequencies of visible light. ($c = 3 \times 10^8$ m s^{-1})
2 Using the same interference equipment, which will produce the narrowest fringes, yellow light or blue light?
3 Which experiences strongest refraction when passing between air and water, red light or violet light?

Figure 3.4
A prism or a raindrop disperses white light into the colours of the spectrum. White light is a mixture of wavelengths that vary *continuously* from a low of about 4×10^{-7} m to a high of about 7×10^{-7} m.

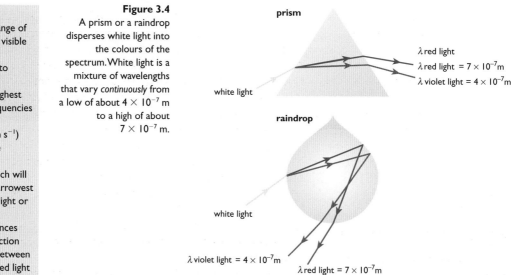

prism

λ red light
λ red light $= 7 \times 10^{-7}$m
λ violet light $= 4 \times 10^{-7}$m
white light

raindrop

white light

λ violet light $= 4 \times 10^{-7}$m
λ red light $= 7 \times 10^{-7}$m

Invisible radiations

Two 18th century astronomers, William and Caroline Herschel, noticed that they could feel warmth beyond the red end of the spectrum. Then when photographic technology emerged about half a century later, photosensitive chemicals were found to be affected by radiation that was invisible, beyond the violet end of the spectrum. These two 'new' radiations are now called **infrared** and **ultraviolet**.

However, the biggest change in thinking about light waves came after 1872, when James Clerk Maxwell developed a mathematical description of electric and magnetic fields and how they travel through space.

Maxwell's mathematics suggested that electricity and magnetism were deeply related. His equations predicted the speed at which their 'electromagnetic' effects could spread into space, and this was in very close agreement with the measured value of the speed of light. It seemed that the travel of light was due to changing electric and magnetic fields (see Figure 3.5).

Figure 3.5
Electric and magnetic fields are regions of space surrounding charged particles or magnets, in which other bodies may experience a force. The fields are represented by field lines which show the directions of the force that can act at different points.

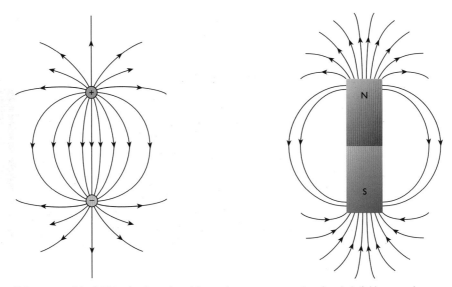

If the source of the field, i.e. the charged particles or the magnet, moves, then the whole field pattern does not move rigidly along with it; it takes time for the effect of the movement to spread through space. Maxwell calculated a prediction of the speed with which electric and magnetic field effects can move. The value he obtained was in close agreement with the experimentally determined speed of light.

Maxwell's theory also predicted that the oscillating electric and magnetic fields could have a huge range of wavelengths and frequencies, much bigger than the range of wavelengths and frequencies of visible light. This was in agreement with the observations of infrared and ultraviolet radiations that had already been made. But Maxwell's ideas won even greater support after 1888, when Heinrich Hertz succeeded in sending electromagnetic signals from one side of a laboratory to another, and showed that these signals could be reflected, refracted and diffracted. Hertz had discovered **radio waves** – long wavelength radiation as predicted by Maxwell.

In 1895 **X-rays** were discovered, and their ability to travel through human flesh caused a public sensation. X-rays could be generated by firing a beam of 'cathode rays' at a metal target in an evacuated glass tube. By 1898, X-rays were used as fairground attractions and by army surgeons to find bullets in wounded soldiers. Ignorance of the hazardous nature of X-rays resulted in people, such as the medical users and fairground operators, developing cancers. However, X-rays were recognised as electromagnetic radiation, with the same speed as radio waves, infrared, visible light and ultraviolet radiation, but with higher frequency.

Gamma radiation is in itself the same as high-frequency X-radiation, the only difference being that gamma radiation emerges from radioactive material.

---

4 Using the data from Figure 3.6, calculate the ratio of
a a typical X-ray frequency to the frequency of visible light
b the frequency of visible light to that of typical 'long wave' radio waves
c a typical X-ray frequency to that of typical 'long wave' radio waves
d a typical X-ray wavelength to that of typical 'long wave' radio waves?

Figure 3.6
The whole electromagnetic spectrum.

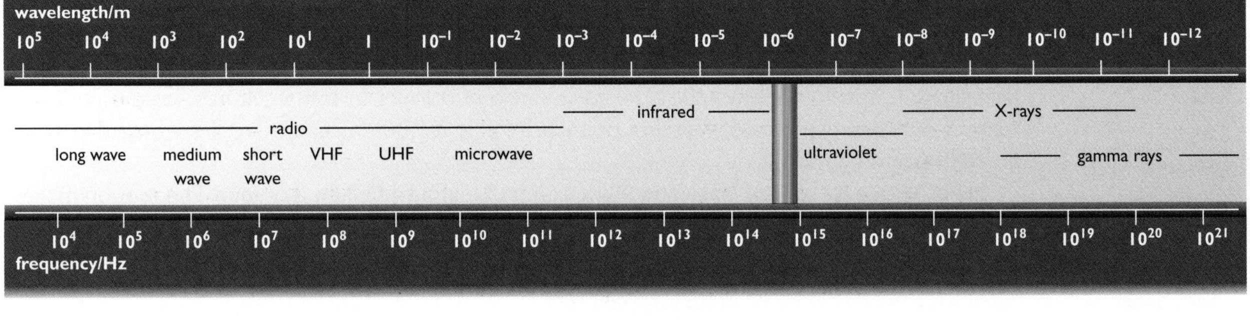

wavelength/m																	
10^5	10^4	10^3	10^2	10^1	1	10^{-1}	10^{-2}	10^{-3}	10^{-4}	10^{-5}	10^{-6}	10^{-7}	10^{-8}	10^{-9}	10^{-10}	10^{-11}	10^{-12}

infrared — X-rays

radio — ultraviolet — gamma rays

long wave · medium wave · short wave · VHF · UHF · microwave

10^4	10^5	10^6	10^7	10^8	10^9	10^{10}	10^{11}	10^{12}	10^{13}	10^{14}	10^{15}	10^{16}	10^{17}	10^{18}	10^{19}	10^{20}	10^{21}

frequency/Hz

Wavelength and the scattering of light

Light from the Sun is reflected by individual particles in the atmosphere, and since this reflection takes place in all directions the light is scattered. Light at the blue end of the spectrum experiences rather more of this particle-by-particle scattering. A larger proportion of the blue light than red light reaches your eyes after being scattered from all directions of the sky, so the sky appears blue (Figure 3.7).

In the evening the Sun is low and the light must shine through a large distance of atmosphere before reaching your eyes. Much of the blue light is scattered out of the direct beam of sunlight (Figure 3.8). The light that remains in the beam no longer has an even mix of wavelengths; it no longer appears anything like white. The least scattered light, red light, is now the strongest component of the beam, followed by orange and yellow.

Figure 3.7
Why is the sky blue but a sunset yellow, orange or red?

Figure 3.8
Atmospheric scattering: red light is scattered less than light of other colours. Blue, indigo and violet experience most scattering.

Colour vision

Seeing in colour is a set of experiences or sensations that take place in your mind. When you were small you learnt to give names to these experiences. We all learnt from different people, but even so we all give the same names to experiences that correspond to the same outward realities. We have no way of knowing what another person experiences when they see, say, a blue surface, but most people will agree on the colour's name. (It is interesting that people from different cultures do not always list the same range of colours. In some cultures, for example, there is no word that exactly matches our word for green but instead there are words that have no equivalent in English.)

A disc with a pattern of black and white can look coloured when it is spun. The motion makes your brain construct colour experience from a black and white surface. So perception of colour is complex. However, the principal method that your brain uses to create colour sensations is based on signals it receives from different types of cell in the retina.

Figure 3.9
Relative spectral sensitivities of **a** the rod cells, and **b** the three types of cone cell.

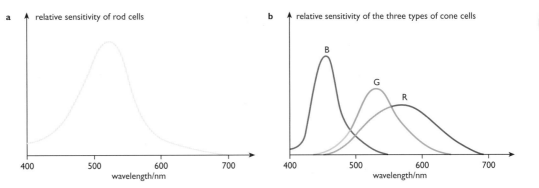

a relative sensitivity of rod cells

b relative sensitivity of the three types of cone cells

5 Why does colour vision deteriorate so much in low light?

6 Red–green colour blindness is quite common. People who are affected have great difficulty in telling the difference between red light and green light. Suggest, in terms of the person's visual system, a possible cause.

Photosensitive cells in the retina are of two different shapes – rods and cones. The rods are all the same; they are more sensitive than the cones to low levels of light so that they work well at night but they do not respond differently to different wavelengths (Figure 3.9a). They do not provide the brain with information to generate the wavelength-dependent sensations that we call colour.

Cones are not as sensitive as rods; they need brighter light to operate. But there are three types of cone, and each type responds to a different range of wavelengths of light (Figure 3.9b) and sends a signal to the brain so that the brain responds differently to different wavelengths. In this way the brain produces the conscious sensations of colour.

Colour mixing by addition

Figure 3.10
A TV screen has just three kinds of dot – red, green and blue, to correspond with the three kinds of wavelength-sensitive cone cells in the human eye. All other colours on a TV screen, including white, are made from combinations of red, green and blue light.

A TV screen is not a reflecting surface but a source of light. It produces light of three colours – red, green and blue. The relative brightness of the three types of coloured dots varies. Your eyes and brain add these three together to produce a full range of colours (see Figure 3.10). The colour that you see is produced by adding together light of all three colours, so this is called colour mixing by **addition**.

We can also add coloured light from three different sources. Light from a red lamp, a blue lamp and a green lamp can add together on a white screen (Figure 3.11). Red, blue and green are **primary colours**, corresponding to the three types of cone in your eyes. Where pairs of lights overlap they produce secondary colours – yellow, cyan and magenta. Where all three coloured lights overlap they add together to appear as white light.

Figure 3.11
Different combinations of the three primary wavelengths can produce the full range of visual sensations that we call colour.

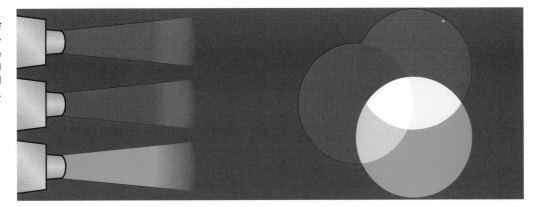

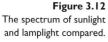

Sunlight and lamplight

Light emitted by the Sun is made of a much wider range of wavelengths than the visible spectrum, but most non-visible radiation is absorbed by the atmosphere. Sunlight that falls on the Earth's surface is, approximately, an even mix of the full visible wavelength range – it is white light.

Figure 3.12
The spectrum of sunlight and lamplight compared.

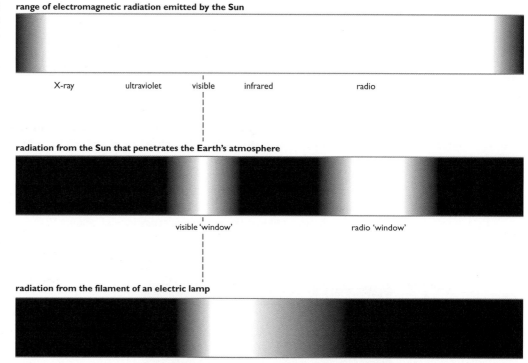

The filament in an electric lamp is also a source of light, but it is not nearly as hot as the Sun. Its light does not reach so far towards the high-frequency, short-wavelength end of the visible electromagnetic spectrum. Compared with sunlight, lamplight contains a bigger proportion of red, orange and yellow light than blue, indigo or violet light (Figure 3.12). An object viewed in lamplight looks more orange than it does when seen in sunlight. When colour matching clothes or paints it is important to have a good source of white light.

Subtraction of colour – filters

When a **filter** is illuminated by white light it absorbs some wavelengths and transmits others. The absorbed wavelengths are **subtracted** from the beam of light.

A red filter transmits red light and absorbs other colours (Figure 3.13). If the transmitted light then shines onto a white screen, we see the red light reflected. A cinema screen is always white. A frame of film is a pattern of filters, producing patterns of reflected colours when we look at the white screen.

Figure 3.13
The action of a red filter on white light.

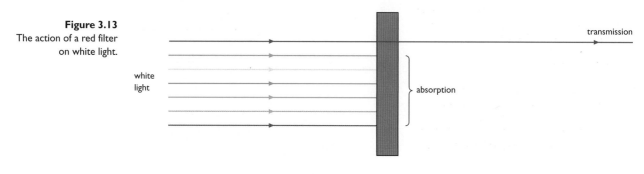

Subtraction of colour – pigments

Pigments, such as paints or dyes, absorb some wavelengths of light and reflect others. The absorbed wavelengths are subtracted from the light that falls on the pigmented surface. We see only the reflected light.

A red pigment reflects red light and absorbs other colours (Figure 3.14). If white light falls on the surface we see only red. If blue (or green) light falls on the surface we see no reflected light (if absorption is complete). Chlorophyll in plants acts as a green pigment (Figure 3.15).

Figure 3.14
The action of red pigment.

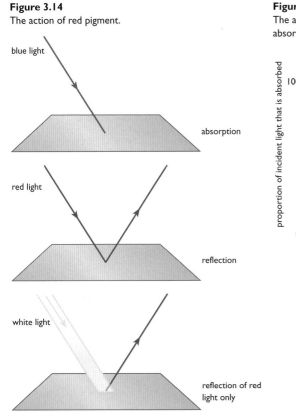

Figure 3.15
The absorption of light by chlorophyll, the green pigment in plants. The light that is not absorbed by a leaf is reflected.

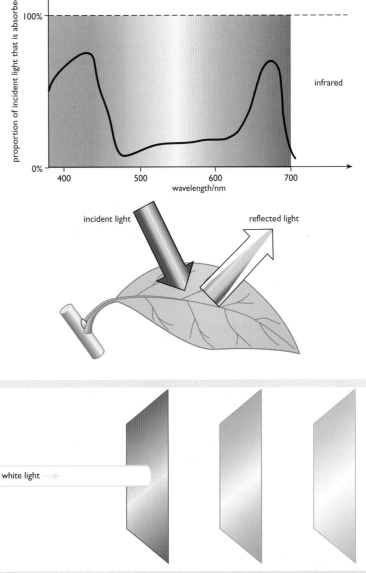

7 How will a blue surface appear
  a at night when it is too dark for your cones to work
  b when illuminated by blue light
  c when illuminated by red light
  d when illuminated by a filament lamp (compared with its appearance in sunlight)?
8 Imagine that you need to check two shades of green paint to see how similar they are. How could you do this?
9 Three filters, a red, a green and a blue, are set up in a row and a beam of white light is shone at them, reaching the red first (Figure 3.16).
  What light reaches
  a the green filter
  b the blue filter?
10 What evidence is there that the chemicals used in photography have a sensitivity to light different from that of the human eye?
11 Sketch a graph of the proportion of incident light that is *reflected* by chlorophyll, versus wavelength.

Figure 3.16

● **Comprehension and application**

Radio waves – introduction to analogue and digital broadcasting

Figure 3.17
Radio frequencies.

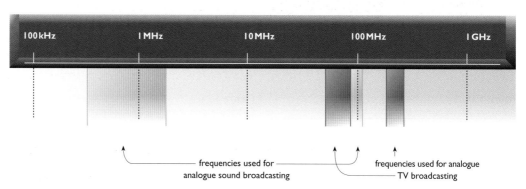

kilo ≡ 1000
mega ≡ 1 000 000
giga ≡ 1 000 000 000

1 kHz = 1000 Hz
1 MHz = 1 000 000 Hz
1 GHz = 1 000 000 000 Hz

Radio waves carry information. The information is spread, or broadcast, at the speed of light, usually in all directions from the broadcasting station. Each broadcasting station must have its own frequency range. If different signals were broadcast using the same or overlapping frequency ranges then the receiver would be seriously affected by the interference caused by superposition of the waves.

The information can be in **digital** form – in patterns of 'on' and 'off'. Or it can be 'encoded' in patterns in the waves; these **analogue** patterns are called modulations. The waves can be modulated in one of two ways – either by patterns in the amplitude of the fixed wavelength radio wave, **amplitude modulation**, **AM**, or by patterns in the frequency of the fixed amplitude wave, **frequency modulation**, **FM** (Figure 3.18).

Figure 3.18
The modulation of a radio wave, or the pattern of the digital signal, conveys the information.

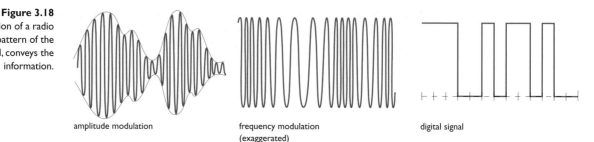

amplitude modulation   frequency modulation (exaggerated)   digital signal

AM and FM radio are both analogue methods, encoding information in continuously varying patterns in the waves. Mobile phones and digital broadcasting use digital transmissions, sending information as a pattern of on–off bursts or **bits**. The patterns of bits carry information in binary code (Figure 3.19).

Figure 3.19
Bursts of waves provide bits which carry the binary code of a digital signal.

Binary code can be converted into radio waves, for example the code 0101 becomes:

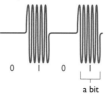

Each bit lasts for many periods or cycles of the radio wave. For the sake of simplicity, only a few such periods are shown here. For example, each bit might last for about 15 μs, while the radio wave has a frequency of 1 MHz.

A frequency-modulated, or FM, signal needs a range of frequencies within which frequency can vary. The wider this range or band of frequencies is, the more complex the modulation can be. The capacity of a system for carrying complex information is measured in terms of **bandwidth**. To transmit good sound quality a bandwidth of about 18 kHz is the minimum requirement. Transmission of television, consisting of both sound and complex, rapidly changing pictures, needs a much bigger bandwidth, usually 8 MHz (Figure 3.20).

Figure 3.20
A section of the ultra-high-frequency radio spectrum showing the 8 MHz bandwidth, from 604 to 612 MHz, of a typical analogue TV broadcasting station, or 'channel'.

The need for a significant bandwidth means that analogue TV transmission can only use ultra-high-frequency waves; medium-frequency waves of 1 MHz, for example, would be no good if the frequency had to vary by 8 MHz to carry its signal.

Digital broadcasting makes it possible to send complex information without using a wide band of frequencies. This makes it possible to have far more stations without interference taking place, so providing increased choice of stations for viewers and listeners.

12 Frequencies used for broadcasting range from about 100 kHz to 1 GHz. What is the corresponding range of wavelengths?

13 If a station broadcasts 108 MHz and uses a bandwidth of 18 kHz, by what proportion does the frequency of the signal vary?

14 Use a diagram to show why it would not be a good idea to try to use 1 MHz waves to transmit analogue TV signals.

15 Why do TV stations need a wider bandwidth than radio stations?

16 Why don't we use X-rays for broadcasting? If we could, what advantages might an X-ray broadcasting system have?

17 How can a digital station transmit information rapidly without using a wide range of frequencies?

● **Extra skills task**   ## Communication and Information Technology

Use the Internet to research information on digital broadcasting. Prepare a 5–10 minute verbal presentation, using visual aids such as OHPs and slides, suitable for your fellow students. Explain the differences between analogue and digital broadcasts. To cover this fully, several questions will need to be answered. For example, why are old analogue televisions incompatible with digital broadcasting? Will reception be improved country-wide, and if so, how is this achieved? How have recent technological advances made digital broadcasts possible?

II
PROBING MATTER AT DIFFERENT SCALES

4 Density – a property of a material

THE BIG QUESTION
● How can we describe and compare materials in a way that is most useful to engineers and designers?

KEY VOCABULARY
anomalous expansion of water   brittleness   deformation   density
elastic solid   hardness   physical property   plastic solid   qualitative
quantitative   ratio   significant figures   toughness

BACKGROUND

Figure 4.1
Participants in a material-rich, number-driven society.

You, like these people in a music shop, are a participant in a material-rich society. It is a world you actively choose every time you buy CDs or clothes, travel, open a pack of food, use a computer or television, or go out for entertainment.

When an innovative bicycle or a computer keyboard, for example, is at the design stage, it is essential for the designer to know how the different materials it could be made from will behave. We can describe materials in words, in terms of their qualities, or we can compare them using numbers. The **qualitative** 'feel' of a material is useful, but mathematical or **quantitative** relationships provide the precise predictions that are needed for making and using a product.

Scientists and engineers use mathematical relationships to define **physical properties** of materials. Physical properties have constant values for named materials, provided that conditions such as temperature do not change.

● Solid descriptions

Most of the objects we buy and use are solid. Words like 'brittleness' and 'hardness' provide useful descriptions of the behaviour of solids:

* **brittleness** – a low ability to withstand **deformation** (change of shape due to an applied force) and a tendency to snap or fracture suddenly
* **toughness** – the ability to withstand temporary deformation without fracturing (the opposite of brittleness)
* **hardness** – resistance to scratching or abrasion. A harder material will scratch a softer one, but not the other way round; this makes it possible to place solids in a list or hierarchy of hardness, from hardest to softest
* **elastic** and **plastic** – these words have opposite meanings. An elastic solid returns to its original shape after experiencing a deforming force which is then removed; a plastic solid keeps its new shape after the force has stopped acting.

Glass is brittle, for example, whereas rubber is tough. Glass is harder than rubber. Glass and rubber are both elastic at room temperature, though the elasticity of glass is hard to test because of its brittleness. Both materials become more plastic when heated.

It is possible to make measurements that indicate toughness, brittleness and hardness, but these concepts are also useful as qualitative (non-numerical) descriptions.

Note that the word 'plastic' in everyday use means a synthetic material that at some stage during the manufacturing processes can be shaped by forces. They can, for example, be stretched into fibres or forced into moulds. In the process of taking the new shape it can be described as plastic. But once shaped, these materials are usually not very plastic – they do not change shape easily. In fact, most are quite elastic solids, the opposite of plastic. However, the name 'plastic' has stuck.

Adding quantity to material descriptions

It is very often useful to quantify – to express behaviour in numbers and units. For example, we can measure the mass of a body or the resistance of a wire. A particular piece of copper wire has particular mass and resistance. Another sample of copper will be likely to have a different mass and resistance.

We can also measure quantities that do not apply to any one particular sample but to *all* samples of the material. Density is an example of such a quantity. Every piece of pure copper, whatever its size or shape, has the same density at any specified temperature. Mass and resistance are properties of the piece of wire, but density is a physical property of the material.

Physical properties of materials describe all kinds of behaviour, such as mechanical, optical, electrical and thermal behaviour. Mechanical properties include density and Young's modulus; other properties include refractive index, electrical resistivity, thermal conductivity, thermal expansivity and specific heat capacity. The definitions of these quantities can be found in the appropriate chapters later in the book. Here we are simply making the point that physical properties provide useful means of comparison of materials. Each material has its own collection of values that create a unique picture or profile of itself (Figures 4.2 and 4.3).

Figure 4.2
Aluminium is abundant in the crust of the Earth, but to extract the metal from the ore is an expensive process that requires large resources of energy. The process is commercially worthwhile because aluminium offers useful physical and chemical properties. It conducts electricity well, it resists corrosion, it forms strong alloys with other metals, and it has low density.

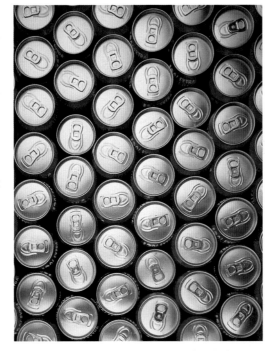

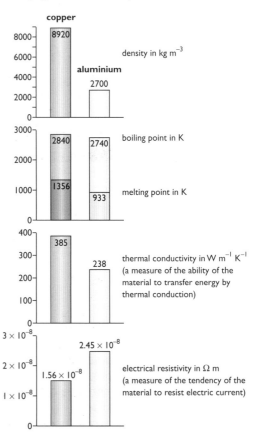

Figure 4.3
Each material has its own characteristic set of physical properties.

Density

A large sample of aluminium has large mass and large volume. A small piece of the aluminium cut away from the large block will have smaller mass and smaller volume. Both of the samples will have the same **ratio** of mass to volume provided their temperatures stay the same. That is, division of mass by volume gives the same answer for both samples. Whatever their size, they have the same **density** when they have the same temperature.

The density of a material is defined as the ratio of mass to volume for any sample of the material. Note that since density changes with temperature, the value that you find in a table of densities of materials will apply at a certain fixed temperature.

We can write the definition of density in a more abbreviated, mathematical notation:

$$\text{density of a material at a fixed temperature} = \frac{\text{mass of any sample of the material}}{\text{volume of the sample}}$$

Abbreviating further, we can use an obvious m for mass and V for volume. (The V is usually written in capitals to avoid possible confusion with velocity.) The letter d is not used for density because of its use in calculus. It is standard to use the Greek letter rho, ρ. Thus the equation can be written:

$$\rho = \frac{m}{V}$$

Note that since the SI unit of mass is the kilogram, kg, and the unit of volume is the cubic metre, m^3, the unit of density is the kilogram per cubic metre, $kg\ m^{-3}$. The units and dimensions are shown in Table 4.1.

Table 4.1	Quantity	SI unit	Dimension
	mass	kg	$[M]$
	volume	m^3	$[L]^3$
	density	$kg\ m^{-3}$	$[M]\,[L]^{-3}$

Table 4.2 gives the densities of some materials, all at 20 °C.

	Name of substance	Density/$kg\ m^{-3}$
Table 4.2	air	1.29 at atmospheric pressure
Some densities at 20 °C.	carbon dioxide	1.98 at atmospheric pressure
	copper	8920
	ethanol	790
	glycerol	1260
	gold	19 300
	hydrogen	0.0899 at atmospheric pressure
	mercury	13 600
	oak	720 at 18% moisture
	oxygen	1.43 at atmospheric pressure
	water	998

The table holds more information than is obvious at first sight. Metals tend to have high density, for example. Both solids and liquids have wide ranges of densities. Gases have low density. Also, for gases, it is necessary to specify the pressure as well as the temperature, because gas density is very dependent on pressure. Solids and liquids, however, cannot be easily compressed into smaller volumes – pressure has so little effect on their density that we needn't consider it for almost all practical purposes.

1 Sort the following into properties of materials and properties of particular samples:
  - density
  - brittleness
  - mass
  - resistance
  - volume
  - resistivity.
2 Why isn't it necessary to specify the pressure for solids and liquids when quoting density values?
3 Which would you expect to be denser, seasoned (slowly dried) oak or unseasoned oak? Explain your answer.
4 Which of these numbers does not show three significant figures?
  501   501.5   1.43   0.0436
5 In the formula $\rho = m/V$, what are the dimensions each side of the $=$ sign?

Table 4.2 shows density to three **significant figures**. Significant figures are the digits in a number, all of which are justified by the original measurement. The density of water, for example, is given as 998 kg m^{-3} and in doing so the table is saying that it is known that the density is not 997 or 999 kg m^{-3}. For gold it is saying that the density is not 19 200 or 19 400 kg m^{-3}, but that the value lies between 19 250 and 19 349 kg m^{-3}. The table suggests that the precision of measurement is not precise enough to say whether the density of gold is, say, 19 290 or 19 310 kg m^{-3}, which are measurements to four significant figures.

Some exceptional densities

A collection of materials in a particular environment can be said to have an average density. Table 4.3 compares some widely differing average densities.

Table 4.3 Approximate values of some average densities.

	Average density/kg m^{-3}
the Universe	10^{-27}
a person	10^{3}
a neutron star	10^{17}

6 By what factor is a neutron star denser than the Universe in general?
7 **a** By what factor, approximately, is copper denser than water? (See Table 4.2).
  b Estimate by what factor a copper washer is denser than the Universe.

Notice that you are 10^{30} times more dense than the Universe as a whole. But a neutron star is roughly 10^{14} times denser than you. That means if you could squash yourself to the same density as a neutron star, instead of your present volume of rather less than 0.1 m^{3}, you'd have a volume of less than 10^{-15} m^{3}. That's less than 10^{-6} mm^{3} – a speck about one hundredth of a millimetre across.

● Unit conversions

Figure 4.4 How many cm^3 in 1 m^3?

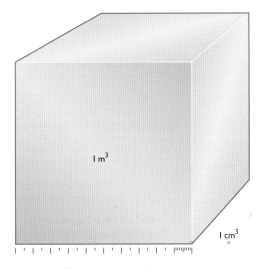

1 m^3

1 cm^3

There are one hundred centimetres in every metre but a *million* cubic centimetres in every cubic metre (Figure 4.4). This may be surprising, since we think of a hundred as a very easily managed number but a million as a number that cannot be pictured all at once. Nevertheless, the calculation shows:

$$100 \times 100 \times 100 = 1\ 000\ 000$$

or

$$10^2 \times 10^2 \times 10^2 = 10^6$$

So

$$10^6 \text{ cm}^3 = 1 \text{ m}^3$$

This means that

$$\text{volume in cm}^3 = \text{volume in m}^3 \times 10^6$$

For many everyday objects, the cubic metre is too large to be a convenient unit, and we often measure volume in cubic centimetres. We then normally measure the mass of the object concerned in grams.

There are one thousand grams in every kilogram.

$$1000 \text{ g} = 10^3 \text{ g} = 1 \text{ kg}$$

So

$$\text{mass in g} = \text{mass in kg} \times 10^3$$

When measurements are made in cubic centimetres and grams then the unit of density will be the gram per cubic centimetre, g cm^{-3}:

$$\text{density in g cm}^{-3} = \frac{\text{mass in g}}{\text{volume in cm}^3} = \frac{\text{mass in kg} \times 10^3}{\text{volume in m}^3 \times 10^6} = \frac{\text{mass in kg}}{\text{volume in m}^3} \times 10^{-3}$$

So to convert density in kg m^{-3} to density in g cm^{-3} we divide by a thousand. The density of water at 20 °C is 998 kg m^{-3} or 0.998 g cm^{-3}.

The fact that the density of water is so close to 1 g cm^{-3} is not an accident. When the metric system was first devised in France about 200 years ago the gram was the basic unit of mass. It was defined as the mass of 1 cm^3 of water at 4 °C, which made the density of water 1.00 g cm^{-3}. (This value is given to three significant figures – the zeroes are significant because they tell us about the precision of the value. It is not 0.99 or 1.01, but 1.00 g cm^{-3}.) Since then the formal definitions of units of mass and length have changed and also the density of water is most often given at 20 °C rather than 4 °C. The density of water is 0.998 g cm^{-3} at 20 °C.

Litres and millilitres

The historical definitions of a cubic centimetre and a millilitre are different, but their sizes are the same. In physics, we measure volumes in cubic metres or cubic centimetres. However, litres (1 litre is a thousandth of a cubic metre) and millilitres are used in everyday life.

8 There are 10^3 mm in 1 m. How many cubic millimetres are there in a cubic metre? Express your answer in powers of ten.

9 What is the length of the side of cube that has a volume of
  a 10^6 m^3
  b 10^{-6} m^3
  c 10^6 mm^3
  d 10^{-6} mm^3?

10 What is the density of aluminium in g cm^{-3}? (Refer to Figure 4.3, page 31.)

11 What is the average density of the Universe in g cm^{-3}? (Refer to Table 4.3, page 33.)

12 Assuming you have an average density as given in Table 4.3, estimate the mass of
  a 1 cm^3 of your flesh
  b an ear
  c your head.

● Density and temperature

Nearly all substances expand as their temperature rises. That is, for a sample of nearly any material, its volume increases but its mass stays the same. The ratio of mass to volume, the density, therefore must decrease – thermal expansion reduces density. For reliable density comparisons between materials it is therefore necessary to quote values that apply at the same temperature.

Water's very unusual pattern of variation of density with temperature (Figure 4.5) – called the **anomalous expansion of water** – means that liquid water is denser than ice.

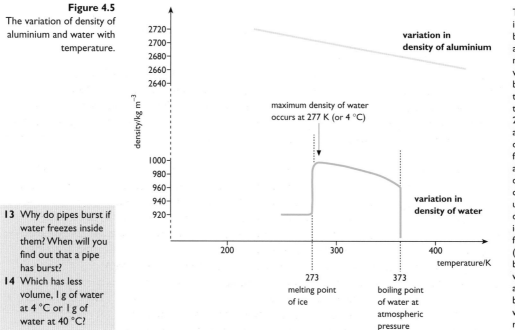

Figure 4.5
The variation of density of aluminium and water with temperature.

The graphs provide a visual impression of the difference in behaviour of water and aluminium. Aluminium has a much higher density than water, so the graph uses a broken density axis to cover the range of densities from less than 1000 kg m^{-3} to more than 2700 kg m^{-3}. The line for aluminium is almost, though not quite, a straight line, showing a fairly steady decrease in density as temperature rises. Water does not just behave in a more complicated way but in a very unusual way. The liquid is denser than the solid, so that ice floats on water and water freezes from the top down (other liquids freeze from the bottom up). Stranger still, water has a maximum density at 4 °C. Water at 4 °C sinks below both warmer and colder water, and in deep water it remains well insulated.

13 Why do pipes burst if water freezes inside them? When will you find out that a pipe has burst?

14 Which has less volume, 1 g of water at 4 °C or 1 g of water at 40 °C?

Figure 4.6
The sea freezes at the surface but life can go on in the liquid below, because of the anomalous behaviour of water.

● **Comprehension and application**

Fuel transport and safety

Oil contains a mixture of materials, which have structures based on chains of carbon atoms. Chains have different lengths, and can have branches. In the oil industry, these different components of the oil are called fractions. They are all potential fuels, and some fractions are raw materials for making a wide range of products, from plastics to pharmaceuticals. A principal task of an oil refinery is to separate the oil into its fractions, which it does by taking advantage of the different boiling points.

Fractions with simple structures of just a few carbon atoms are gases at 298 K (which is 25 °C) and at atmospheric pressure. These fractions have lower density than the fractions with longer and more complex carbon chains. In natural crude oil reserves under the ground these low density fractions may collect in the rock above the oil, or they may have seeped away. Where the oil is under high pressure some of these less dense fractions may remain as liquid even though the temperature is above their boiling point.

A liquid occupies much less volume than the same mass of the same material in the gaseous state, so it makes sense to transport material in liquid rather than gaseous form. Propane and butane are two gaseous oil fractions, also called petroleum gas, that can be liquefied by pressure at everyday temperatures. The liquids can be transported in pressurised canisters, providing a convenient fuel. Butane, for example, is used for camping stoves and lanterns.

Table 4.4
Properties of some oil fractions.

Name of fraction	Boiling point/K	Density of liquid/ kg m^{-3}	Density of gas at 298 K and atmospheric pressure/ kg m^{-3}
methane	109.1	466	0.72
ethane	184.5	572	1.36
propane	231.0	585	1.99
butane	272.6	601	2.62
pentane	309.2	626	–
hexane	342.1	660	–
heptane	371.5	684	–
octane	398.8	703	–
air	–	–	1.26

15 Explain why there are blank spaces in Table 4.4.

16 a Calculate the volumes of road tankers required to transport 10 tonnes (10 000 kg) of:
   i propane (as a liquid)
   ii octane.
   b Sketch graphs of mass (*y*-axis) against volume (*x*-axis) for propane (liquid) and octane. What is the significance of the gradients of these graphs?
   c Calculate the volume occupied by 10 tonnes of methane gas.

17 Methane is used as domestic fuel; it is often called 'natural gas' or just 'gas'. Why must it be transported by pipeline and not by road tanker?

18 Imagine leaks of methane and butane into the open air on windless days. Describe the different ways in which you would expect the two gases to spread through the air.

19 Many mining tragedies have been caused by leaks of methane into coal workings from the seams of coal. Methane could explode or it could simply displace air and cause suffocation. Miners used to hang up cages holding canaries. If the canary showed signs of suffocation (for example, if the canary stopped singing) then that was a warning. Where would have been the best place to hang such a cage?

● **Extra skills task**   Application of Number

1 Taking the density of water as 998 kg m^{-3}, express the densities of each of the liquids in Table 4.4 as a percentage of that of water. Give answers correct to three significant figures.

2 a What is the density of methane gas (at atmospheric pressure and 298 K) as a percentage of that of methane liquid?
   b What is the volume of 1 tonne of methane liquid as a percentage of 1 tonne of the gas (at atmospheric pressure and 298 K)?

5 Matter and force

THE BIG QUESTIONS
- How can we analyse the behaviour of different samples of a material, such as different copper wires, when they are subject to a force that deforms them?
- How can we analyse the behaviour of different materials, such as copper and steel, when they are subject to a force that deforms them?

KEY VOCABULARY

breaking stress   compression   constant of proportionality   deformation dimensionless quantity   ductile   extension   gradient   Hooke's Law input (or independent) variable   inverse proportionality   limit of proportionality output (or dependent) variable   proportionality   spring constant   strain stress   tension   Vernier scale   Young's modulus

BACKGROUND

This Roman aqueduct (Figure 5.1), which crosses the River Gard in France, is an impressive structure because the river is wide here. What's more impressive is that it was built about 2000 years ago. To build such a bridge, people must have lived in a stable society as we do. They used stones for their bridge, and we have to assume that they had learnt how to build structures by going out and doing it. That's fine, but the more you can plan ahead, the more you can achieve.

Figure 5.1
Remains of a Roman aqueduct in France.

Nowadays, before a bridge is built there are teams of engineers working with the help of computer modelling to consider just how the various materials of a planned bridge and its environment will interact. We can use a wider variety of materials than the engineers of two millennia ago. We can build bigger bridges to cross wider rivers, and in places where wind and tide are stronger. With detailed engineering we know that we can cross our bridges in safety.

Several thousand people now cross the 'Second Severn Crossing' (Figure 5.2) between England and Wales every single day. Will the remains of our bridge be found by archaeologists 2000 years from now? That's something we can never know.

Figure 5.2
The Second Severn Crossing under construction.

The building of this bridge involved meticulous engineering. The effects of forces, such as those exerted by the wind and the tides, and those of tension in the cables, had to be carefully predicted using knowledge of force and materials.

Responses to external forces by solids, liquids and gases

Forces of **compression** tend to squash materials while forces of **tension** tend to stretch them (Figure 5.3). The resulting change in volume or shape of the material is called **deformation**. Compression and tension are deforming forces. Of course, different materials behave differently, and the most obvious way to classify materials is into three states of matter – solid, liquid and gas – which respond in different characteristic ways to deforming forces (see Figure 5.4).

Solids and liquids both experience very small changes in volume when subject to deforming forces. Liquids, however, offer little resistance to these forces when it comes to changing shape. They change shape, without changing volume, very easily. That is the chief way in which we tell a liquid from a solid. Gases can experience large changes in volume when deforming forces are applied. Compared with solids and liquids, gases are easily compressed and expanded.

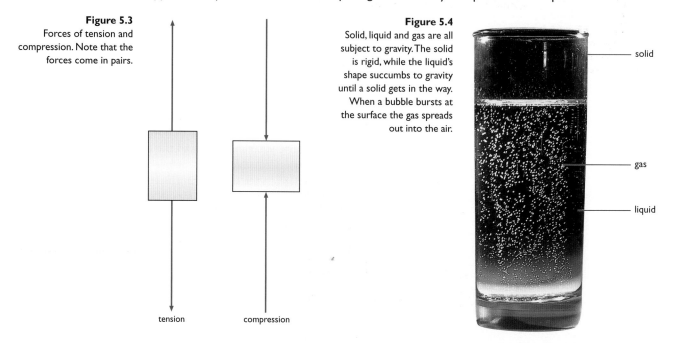

Figure 5.3
Forces of tension and compression. Note that the forces come in pairs.

tension   compression

Figure 5.4
Solid, liquid and gas are all subject to gravity. The solid is rigid, while the liquid's shape succumbs to gravity until a solid gets in the way. When a bubble bursts at the surface the gas spreads out into the air.

solid

gas

liquid

Tension and extension

Forces of tension produce most effect when they are pulling on wires or threads; these are objects with small cross-sectional areas compared with their lengths. Under tension, a force acts from each end of a wire or thread. These forces are equal. (They must be equal so that they balance; unbalanced forces would produce acceleration.) The size of either of these forces is the tension acting (Figure 5.5).

The tension produces an **extension**, or an increase in length. Here we can use standard mathematical notation to represent change, by using the Greek letter Δ (delta). If the original length of a wire is l, then any change in its length, or extension, is Δl.

1 For a rope mooring a ship, predict the effect on Δl of using:
a a thicker rope (with larger cross-sectional area)
b a longer rope.

Figure 5.5
The forces are equal. The tension acting on the rope is equal to either one of the forces.

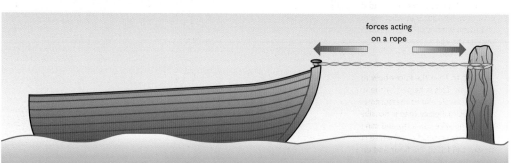

forces acting on a rope

Tension, extension and proportionality

We are ready to search for a quantitative relationship between a deforming force (tension) and the response of a material (extension). This is most easily tried in practice for a material that produces a large extension for just a small tension, such as rubber. The force of gravity (balanced by the action of a firm support) provides the tension. This force is applied and measured by using standard masses (and remembering that each mass of 100 g applies a force of about 1 newton). The extension is easily measured with a rule clamped alongside the rubber, which can be a simple rubber band (Figure 5.6a).

Measurements on rubber produce an extension–tension curve, not a straight line (Figure 5.6b). Also, it is not the most useful of graphs – the behaviour of *metals* under tension is of more interest because of their engineering importance. However, to plot extension–tension graphs for metals takes rather more care (Figure 5.7).

Figure 5.6
The experimental set-up and the resulting extension–tension graph for rubber.

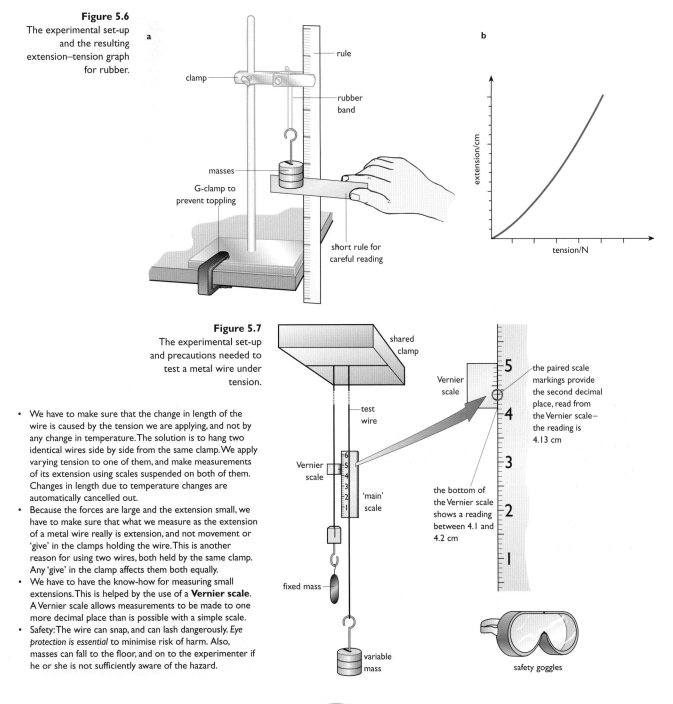

Figure 5.7
The experimental set-up and precautions needed to test a metal wire under tension.

- We have to make sure that the change in length of the wire is caused by the tension we are applying, and not by any change in temperature. The solution is to hang two identical wires side by side from the same clamp. We apply varying tension to one of them, and make measurements of its extension using scales suspended on both of them. Changes in length due to temperature changes are automatically cancelled out.
- Because the forces are large and the extension small, we have to make sure that what we measure as the extension of a metal wire really is extension, and not movement or 'give' in the clamps holding the wire. This is another reason for using two wires, both held by the same clamp. Any 'give' in the clamp affects them both equally.
- We have to have the know-how for measuring small extensions. This is helped by the use of a **Vernier scale**. A Vernier scale allows measurements to be made to one more decimal place than is possible with a simple scale.
- Safety: The wire can snap, and can lash dangerously. *Eye protection is essential* to minimise risk of harm. Also, masses can fall to the floor, and on to the experimenter if he or she is not sufficiently aware of the hazard.

For a metal, once a reliable set of measurements of tension and extension has been achieved they reveal a simple relationship. When plotted as a graph they form a straight line (although the pattern may begin to curve when tension and extension become large – discussed later in this chapter). Straight lines, especially when they pass through the origin of the graph, are welcome results to investigations of relationships. A straight line is easily identified as such and it allows us to see, at a glance, the nature of the relationship. A straight line that passes through the origin (the point described by the co-ordinates (0, 0)) is particularly useful. It shows that the **output (or dependent) variable** (the extension in this case) is proportional to the **input (or independent) variable** (the tension). Whatever change happens to the tension, the extension changes by the same proportion (provided the line remains straight). A straight line through the origin shows **proportionality**: extension is proportional to tension.

Gradient of a straight line graph

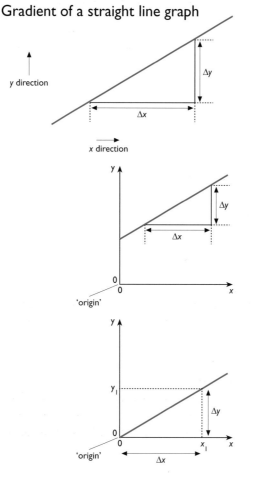

Figure 5.8
Calculating the gradient of a straight line graph.

The gradient of a straight line on a graph is determined from a triangle. In general, the larger the triangle the better, in order to obtain good measurements of the height and the base (with minimum % error).

height $= \Delta y$ (a change in y)
base $= \Delta x$ (a change in x)

$$\text{gradient} = \frac{\text{height}}{\text{base}} = \frac{\Delta y}{\Delta x}$$

Here the straight line does not pass through the origin so y is not proportional to x. That is, a change in x does not produce a change in y by the same proportion. However, we can still say that:

$$\text{gradient} = \frac{\text{height}}{\text{base}} = \frac{\Delta y}{\Delta x}$$

A straight line through the origin (0, 0) is a special case: y is proportional to x. That is, a change in x results in a change in y by an identical proportion. The triangle can be drawn with one of its points at the origin. Then

$$\text{gradient} = \frac{\text{height}}{\text{base}} = \frac{\Delta y}{\Delta x} = \frac{y_1 - 0}{x_1 - 0} = \frac{y_1}{x_1}$$

For a straight line through the origin, the gradient is equal to the ratio of any corresponding pair of values of x and y.

The graphs in Figure 5.8 all have the same **gradient**. It is calculated by dividing any change in y by the corresponding change in x:

$$\text{gradient} = \frac{\text{change in } y}{\text{change in } x} = \frac{\Delta y}{\Delta x}$$

Δy and Δx can be found by drawing a triangle: Δy is the height of the triangle and Δx is the base.

Note that we can say more when the graph passes through the origin. We can draw a triangle with one point at the origin. Then Δy and Δx are then corresponding values of y and x:

$$\Delta y = y \quad \text{and} \quad \Delta x = x$$

So, for a straight line passing through the origin,

$$\text{gradient} = \frac{\text{change in } y}{\text{change in } x} = \frac{\Delta y}{\Delta x} = \frac{y}{x}$$

We can rearrange this to:

$$y = \text{gradient} \times x$$

Remember that this only applies when the graph is a straight line passing through the origin. That is, it applies when y is proportional to x. So we can say that,

for proportionality, $y = \text{gradient} \times x$

The gradient of a straight line never changes. It is constant. The gradient of a straight line graph that passes through the origin is also known as a **constant of proportionality**.

Tension, extension and gradient

The gradient of a graph is a quantity that can carry useful information. A graph of extension plotted against tension for a metal wire is a straight line through the origin (provided that the wire is not stretched too far). So we can say not just that gradient $= \Delta y/\Delta x$ but that the gradient $= y/x$. Now y is the extension, which we will call Δl, and x is the applied tension or force, F. So,

$$\text{gradient} = \frac{y}{x} = \frac{\text{extension}}{\text{tension}} = \frac{\Delta l}{F}$$

or

$$\Delta l = \text{gradient} \times F$$

If we call the gradient c, for constant, then

$$\Delta l = cF$$

and so,

$$c = \frac{\Delta l}{F}$$

Since the SI unit of extension, Δl, is the metre and the SI unit of tension is the newton, the SI unit of the gradient, c, is the metre per newton, which can be written as m N^{-1}.

Different wires provide graphs with different gradients. So the gradients of the graphs provide a way of comparing wires. Note, however, that the gradient is *not* a property of a material, because it applies to a particular wire.

Limit of proportionality

When a metal wire is subject to a large tension, the gradient of the extension–tension graph ceases to be constant. The graph starts to curve. Extension is no longer proportional to tension. The point on the graph at which this curvature starts is called the **limit of proportionality** (Figure 5.9).

> **2** Sketch a graph with a constant gradient showing proportionality.
>
> **3** Some straight line graphs do not show proportionality. Sketch such a graph.
>
> **4 a** Draw a straight line through the origin for which $\Delta y/\Delta x = 0.75$.
> **b** What is the value of y for which $x = 3.2$?

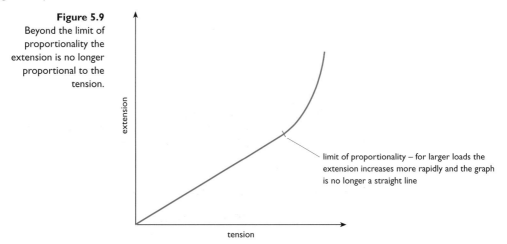

Figure 5.9
Beyond the limit of proportionality the extension is no longer proportional to the tension.

limit of proportionality – for larger loads the extension increases more rapidly and the graph is no longer a straight line

extension

tension

Changing the variables around

In hanging masses from a wire and then measuring the resulting extension we are making the tension the input variable. The extension is the output variable. The convention is to plot the input variable on the *x*-axis and the output variable on the *y*-axis.

However, it would be possible to do the experiment 'the other way round'. That is, to decide in advance what extensions we would measure, and then to find out what tensions were required to produce the extensions we've already chosen. Now extension is the input variable and tension is the output variable.

Again, we could plot a graph to 'see' the relationship. We would get a straight line that passed through the origin. Proportionality seems to apply whichever way round we do the experiment.

Now we can say:

tension is proportional to extension

$$F = \text{gradient} \times \Delta l$$

If we call this gradient *k*, then

$$F = k\,\Delta l$$

and so,

$$k = \frac{F}{\Delta l}$$

But this new gradient, *k*, will not be the same as the gradient, *c*, of the graph obtained before. Taking the equation on the previous page,

$$c = \frac{\Delta l}{F}$$

we see that one gradient is the inverse of the other, that is,

$$k = \frac{1}{c}$$

Gradients of graphs of proportional relationships with the axes switched always have this inverse relationship (Figure 5.10). Note that the SI unit of the gradient *c* is the metre per newton, $m\,N^{-1}$, while the SI unit of the gradient *k* is the newton per metre, $N\,m^{-1}$.

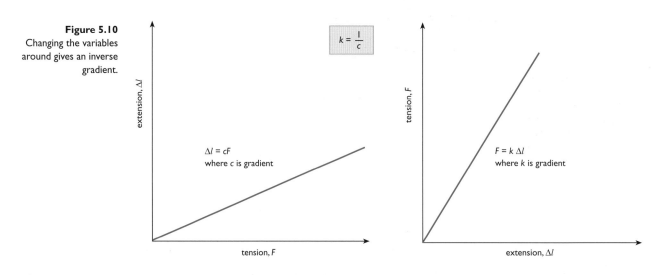

Figure 5.10
Changing the variables around gives an inverse gradient.

The spring constant

5 Rearrange the spring constant equation to make Δl the subject.
6 A spring stretches 0.02 m for each newton of tension applied. What is its spring constant?
7 What is the spring constant of a wire that stretches 2×10^{-4} m for each newton of force?
8 The unit of extension is the metre. What is its dimension in terms of mass [M], length [L], time [T] and/or combinations of these? (You can read more about dimensions on page 14 and in Chapter 22.)
9 Suppose that $\Delta l = 0.4$ mm and $F = 24$ N, and the limit of proportionality has not been exceeded. What is the value of the spring constant **a** in N mm^{-1}, **b** in N m^{-1}?

Whichever way round we take the measurements of tension and extension and plot the graph, we see a gradient that is constant. This is true whether we are working with wires or springs. Using the equation

$$F = k \times \Delta l$$

the gradient k is sometimes called the **spring constant**. Every spring and every wire has its own spring constant. Note that when we rearrange the equation to make k the subject:

$$k = \frac{F}{\Delta l}$$

The spring constant has units N m^{-1}.

Extension and other variables

We already know that, provided the limit of proportionality is not exceeded, the extension of a wire is proportional to the tension applied to it. But the amount of extension of a wire does not depend only on the tension acting on it. There are three other variables involved.

- *The longer the wire*, the more it extends for a given force, and in fact:

    extension is proportional to original (unextended) length (for a given force and cross-sectional area)

    Δl is proportional to l

- *The thicker the wire* (that is, the bigger its cross-sectional area), the less it extends for a given force. Measurements reveal the relationship:

    extension is **inversely proportional** to cross-sectional area (for a given force and length)

If you were to perform an experiment in which all other variables were fixed, you would find that as the cross-sectional area A increased, the extension Δl would decrease. A graph of Δl against A would be a curve. But if you plotted extension against the *inverse* of the cross-sectional area, Δl against $1/A$, then the graph would be a straight line passing through the origin (Figure 5.11).

Δl is inversely proportional to A

and, from the straight line graph,

Δl is proportional to $\dfrac{1}{A}$

Figure 5.11
Inverse proportionality produces a curve when the two quantities are plotted. But by taking the inverse of one of the quantities we get a straight line through the origin.

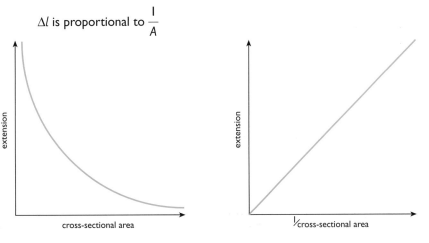

10 Explain why the spring constant is not a physical property of a material.

- *The material of the wire* has a very significant influence. We can quantify this influence by defining a measurable physical property of the material. This physical constant is called **Young's modulus**, E, and we will define it shortly.

Stresses, strains and a physical constant

The ratio of force to cross-sectional area is called **stress**:

$$\text{stress} = \frac{F}{A}$$

The unit of stress is the newton per square metre, N m^{-2}.

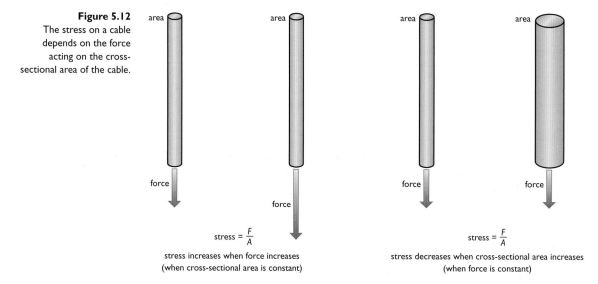

Figure 5.12
The stress on a cable depends on the force acting on the cross-sectional area of the cable.

stress $= \dfrac{F}{A}$

stress increases when force increases
(when cross-sectional area is constant)

stress $= \dfrac{F}{A}$

stress decreases when cross-sectional area increases
(when force is constant)

The ratio of extension to original length is called the resulting **strain**:

$$\text{strain} = \frac{\Delta l}{l}$$

Strain is a ratio of two lengths, both measured in metres. It therefore has no units itself. Also, its dimensions are $[L][L]^{-1}$. In other words, it has no dimensions at all. It is a **dimensionless quantity**.

Figure 5.13
The strain on a wire depends on the original length of the wire.

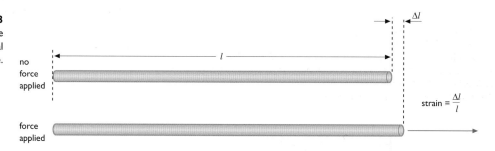

no force applied

force applied

strain $= \dfrac{\Delta l}{l}$

Note that

$$\frac{\text{stress}}{\text{strain}} = \frac{F/A}{\Delta l/l} = \frac{F \times l}{A \times \Delta l}$$

Measurements show that this ratio, stress/strain, is the same for all wires of the same metal, provided that they are not overstretched (past the limit of proportionality). That is, for all wires of a particular metal, a graph of strain against stress is a straight line until the limit of proportionality is reached. The ratio is a physical constant of the metal in question.

This physical constant, related to the gradient of the strain–stress graph (Figure 5.14), is the Young's modulus of the metal:

$$\text{Young's modulus} = \frac{\text{stress}}{\text{strain}}$$

Figure 5.14
The graph of strain against stress is a straight line through the origin (within the limit of proportionality); Young's modulus for the material is equal to the inverse of the gradient.

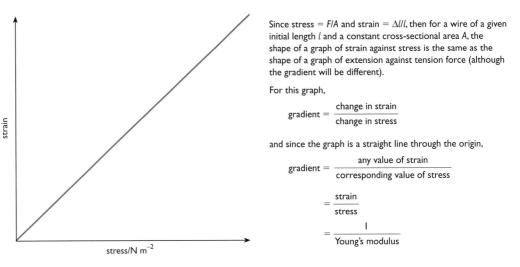

Since stress = F/A and strain = $\Delta l/l$, then for a wire of a given initial length l and a constant cross-sectional area A, the shape of a graph of strain against stress is the same as the shape of a graph of extension against tension force (although the gradient will be different).

For this graph,

$$\text{gradient} = \frac{\text{change in strain}}{\text{change in stress}}$$

and since the graph is a straight line through the origin,

$$\text{gradient} = \frac{\text{any value of strain}}{\text{corresponding value of stress}}$$

$$= \frac{\text{strain}}{\text{stress}}$$

$$= \frac{1}{\text{Young's modulus}}$$

The unit of Young's modulus is the same as that of stress, since strain is a dimensionless quantity with no units. That is, the unit of Young's modulus is the $N\,m^{-2}$. Some typical values of Young's modulus are given in Table 5.1.

Table 5.1
Values of Young's modulus for three metals.

Metal	Young's modulus/$N\,m^{-2}$
aluminium	7.0×10^{10}
copper	1.3×10^{10}
steel	2.1×10^{11}

11 **a** If you were to plot extension Δl against tension F for two different wires of the same material, would your graphs have the same gradients?
  b If you were to plot strain against stress for two different wires of the same material, would your graphs have the same gradients?
12 A wire 1.5 m long extends by 1.5 mm when a tension of 30 N acts. The Young's modulus of the metal is $2 \times 10^{11}\,N\,m^{-2}$. What is the cross-sectional area of the wire in
  a square metres
  b square millimetres?
13 Sketch graphs, on the same axes, of *stress* (*y*-axis) against *strain* (*x*-axis) for aluminium, copper and steel. (Refer to Table 5.1.)
14 **a** What is the ratio of the Young's moduli of copper and aluminium?
  b Which will stretch more, a copper wire or an aluminium wire of the same size experiencing the same tension?
  c What will be the ratio of the extensions?

Breaking stress (also sometimes called ultimate tensile strength) is the stress required to break a material under tension. Different samples of a material break at different force, but stress is the force acting normally per unit of cross-sectional area. Breaking stress is a constant for a pure material at constant temperature, but is strongly dependent on temperature and on the presence of impurities.

Hooke's Law

The observation that we can make from the straight line graph of strain against stress, that:

strain is proportional to stress (provided that the limit of proportionality is not exceeded)

is called **Hooke's Law**.

Disobeying Hooke's Law

Copper is a **ductile** material. It obeys Hooke's Law only when the applied stress is small and, as stress is increased, it quickly loses this proportionality and soon after it stops being elastic. The shape of the copper is then permanently changed and it behaves in a plastic way. The ease with which copper can be pulled, bent or twisted into different shapes makes it suitable as a material for jewellery, in ways that are not possible with less ductile materials, such as steel (Figures 5.15 and 5.16).

Figure 5.15
Copper jewellery. Copper becomes plastic at quite low stress.

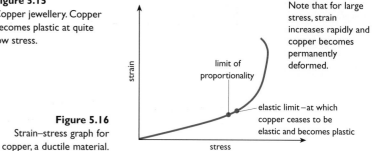

Note that for large stress, strain increases rapidly and copper becomes permanently deformed.

limit of proportionality

elastic limit — at which copper ceases to be elastic and becomes plastic

Figure 5.16
Strain–stress graph for copper, a ductile material.

Ceramics are materials made by changing the structure of rocks, sands and clays by heating. They are not ductile and do not become plastic under large stress. They are, however, brittle. Some ceramics can have a high breaking stress, but nevertheless, their response to large stress is to break rather than to yield to the stress by changing shape (Figures 5.17 and 5.18).

Figure 5.17
When subject to stress, ceramic material breaks rather than changing shape.

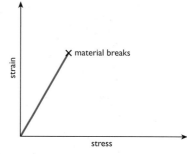

material breaks

Figure 5.18
Strain–stress graph for ceramic, a brittle material.

Rubber is a natural polymeric substance, in contrast to nylon and polythene which are synthetic polymers. Both natural and synthetic materials are made of molecules which are long chains, and these long chains can change shape and sometimes slide past each other when the material is subject to a deforming stress. They are essentially elastic, but their behaviour is more complex than metals for example, which have simpler structures, and polymers show little obedience to Hooke's Law (Figures 5.19 and 5.20).

Figure 5.19
A wrap-around layer of molecules, each of which can change shape, allows a wet suit to be made of a continuous sheet and still be flexible.

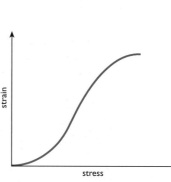

Figure 5.20
Strain–stress graph for rubber.

15 What is the difference in the effect of an applied force on a ductile and a brittle material?

16 Explain why it isn't possible to plot strain–stress graphs for copper and glass on the same axes.

17 The limit of proportionality of copper occurs at a stress of $1.8 \times 10^{8}\,\mathrm{N\,m^{-2}}$ and the Young's modulus of copper is $1.3 \times 10^{11}\,\mathrm{N\,m^{-2}}$. What will be the extension of any 2 m long copper cable when it reaches its limit of proportionality?

● **Extra skills task** Information Technology and Application of Number
Carry out an experiment to test Hooke's Law for copper springs, varying the diameter and length of the wire used. As in all practical work, first consider all possible safety hazards and take action to minimise risk of harm. Use a spreadsheet program to calculate the spring constant for each spring, for low values of stress. Process the data further by using the spreadsheet program to produce graphs that display the trends seen in your experiment, both for the individual springs and for the set of springs. Comment on the springs' obedience of Hooke's Law, and on the relationships between spring constant and the nature of the springs.

6 Particle theory of matter

THE BIG QUESTIONS
- What evidence is there that matter has a particle nature and is not continuous?
- What aspects of the behaviour of materials can be explained and predicted by particle theory?

KEY VOCABULARY
amorphous solid   atmospheric pressure   bonds   Brownian motion
buckyballs or buckminsterfullerene   diffusion   falsification   monomer   polymer
pressure   thermoplastic polymer   thermosetting polymer   X-ray diffraction

BACKGROUND

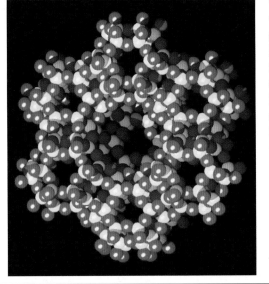

Figure 6.1
This is a computer model of zeolite – a large molecule used by nanotechnologists. It is, of course, just a model. The real thing is fundamentally invisible in this detail.

We cannot see particles when we look at a wooden table, butter, water or air. We see and feel apparent continuity. Yet when we look at the behaviour of a material we are forced to think harder about its nature. We are forced into imagining that it is made of invisible particles. Particle theory is a rich source of explanations and predictions of the behaviour of matter.

Chemists, metallurgists and other materials technologists can apply ideas of the particle nature of matter not just to explain the behaviour of matter but to manipulate materials. Companies invest in new technologies, such as nanotechnology, so that they can make new products, to make money and survive. In the fast-growing study of nanotechnology (technology at a scale of 10^{-9} metres), it's possible to make molecules interact physically and chemically to such an extent that some kinds of new molecules can be built to design.

A theory that has yet to be falsified

Small droplets in a spray or mist evaporate. The material becomes invisible, but we can still detect it with our noses as it spreads through the air (Figure 6.2). Particle theory claims that this mixing or **diffusion** of materials, such as air and deodorant, is possible because the particles of the two materials are in constant motion and have spaces between them.

Figure 6.2
After spraying, the material diffuses into the air.

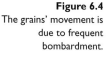

Figure 6.3
Computer microchips are made by allowing impurities to diffuse into small blocks of silicon inside reactors like these.

Figure 6.4
The grains' movement is due to frequent bombardment.

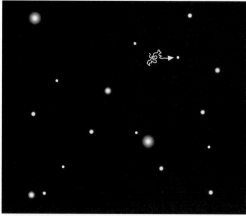

When we look at matter in the human-sized (or macroscopic) world it appears continuous and not particulate. It is hard to imagine how materials with a continuous nature could mix together, while it is easy to imagine how two collections of particles can blend. Diffusion of one material into another (Figure 6.3) strongly supports the theory that matter, on the sub-visible scale, is broken into particles and is not a continuum.

Brownian motion is the motion of small grains visible under the microscope (Figure 6.4), like grains of smoke or fine pollen, with frequent random changes in direction. It was first recorded early in the 1800s by Robert Brown. Albert Einstein, nearly a century later, did the maths to show that Brownian motion observations agree with the theory that the changes in direction are caused by collisions with invisibly small but energetic particles of the material surrounding the grains. The matching of Robert Brown's observation with particle theory is a persuasive reason for accepting that it is a strong theory.

A theory is a source of explanations and predictions. A good theory is a rich source, and is consistent with all of the observations. Of course, it is possible that some new observations will prove a theory to be inconsistent with how the world behaves. The theory would then be **falsified**. But so much examination of matter has been made, and not one observation seems to falsify particle theory. It is accepted as a strong theory.

I DISCUSS
Can you imagine an observation that would falsify particle theory? (Be as imaginative as you like! And if you can manage to make such an observation then perhaps you should skip the rest of this course and move straight to a Nobel Prize – your name will be remembered for generations to come.)

2 DISCUSS
Do either Brownian motion or diffusion provide certain proof that matter has a particle nature?

Particle theory explanations of the states of matter

We experience water in all three states. The solid is hard to compress and rigid in shape. The liquid is hard to compress but has no rigidity of shape. The gas is easy to compress and also has no rigidity of shape. Particle theory can provide explanations of these differences if we add to it the idea that the particles interact and we consider this interaction in terms of forces and energy.

We can think of the three states of matter as three different energy conditions of the particles. We know that to turn a solid to a liquid, even without raising its temperature such as when ice at 0 °C becomes water at 0 °C, we must supply energy to it. And energy is needed again to turn it into a gas, also without changing its temperature, as when water at 100 °C becomes steam at 100 °C (Figure 6.5). In solids, particles each have relatively little average energy, rather more energy when the material is liquid, and more again when it is a gas. These different energy conditions then relate to the forces that can act.

Figure 6.5
The three states of matter are three different energy conditions.

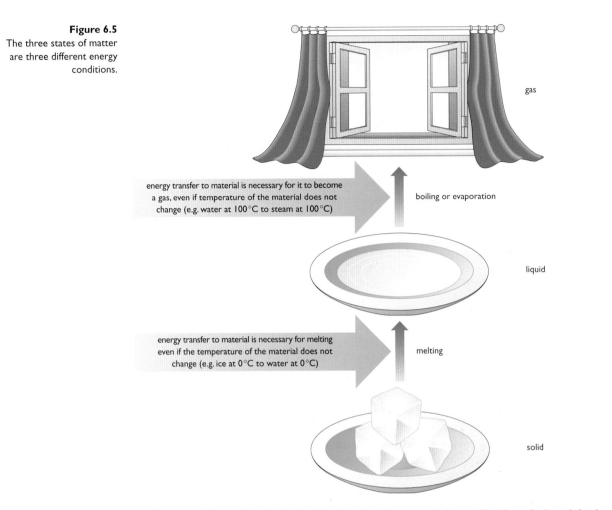

energy transfer to material is necessary for it to become a gas, even if temperature of the material does not change (e.g. water at 100 °C to steam at 100 °C)

boiling or evaporation

gas

liquid

energy transfer to material is necessary for melting even if the temperature of the material does not change (e.g. ice at 0 °C to water at 0 °C)

melting

solid

Attractive forces that hold particles in place in a material are called **bonds**. In solids, the forces are attractive until the particles get very close together. Then the force switches to repulsive mode (Figure 6.6). We can suppose that this is true because if particles go on and on attracting each other as they get closer then they will merge into each other and the density of the material will rise and rise towards infinity. Whenever you 'touch' an object, some particles in your skin get so close to some particles of the object surface that repulsive forces act.

Figure 6.6
Without the attractive forces between particles there would be no solids (or liquids). Without the repulsive forces particles would move closer and closer together and merge into a tiny point of very high density.

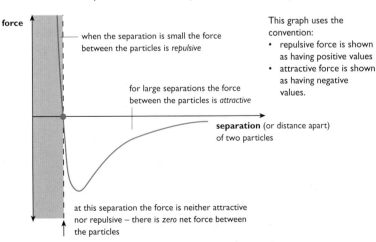

force

when the separation is small the force between the particles is *repulsive*

for large separations the force between the particles is *attractive*

This graph uses the convention:
• repulsive force is shown as having positive values
• attractive force is shown as having negative values.

separation (or distance apart) of two particles

at this separation the force is neither attractive nor repulsive – there is *zero* net force between the particles

The particles in a solid with relatively low energy cannot break free of their medium-range mutual attractions but are also held apart by their close-range mutual repulsions. Each one is held in a fixed general location and all it can do is vibrate.

3 Some people hang on to the idea that particles in a liquid are significantly further apart than particles in a solid. How could you use a container of ice cubes to demonstrate that this is not true?

4 Why does steam at 100 °C, from a kettle spout, deliver more energy to your hand (if you are foolish enough to place your hand over the spout) than the same mass of water at 100 °C?

5 Why do we believe that forces between particles are repulsive at short range but attractive at longer range?

6 When you press on a surface such as a table or a pen, what prevents your finger from merging with it?

7 Imagine that you could create a chamber in which there were no attractive forces between particles, only repulsive forces. At first there is a vacuum in your chamber, and then you introduce a small solid object. How will it behave in the chamber?

What would happen if you'd built your chamber so that only attractive forces existed inside, and there were no repulsive forces?

Liquid is a more subtle state, and is relatively uncommon; very few of the chemical elements are liquids at room temperature. The particles in a liquid have enough energy to break free from one-to-one forces with their neighbours, but then soon form new attractions with new particles. As in solids, repulsive forces play their part when an external force tries to push particles close together, and so liquids change only very little in volume when they are subject to pressure.

Particle-to-particle distance is much the same in liquids as in solids – a point that is consistent with the observation that only small changes of volume take place on melting and freezing. However, the continuous breaking and making of the bonds between liquid particles makes rigidity impossible.

There are only extremely weak forces between the particles of a gas, and each particle is (almost) free to go its own way. Gases spread out to fill all the available space. They are also easily compressed. The particles are separated by comparatively large distances, and are much too far apart for repulsive forces to come into play except during direct collisions. There is only a void between them. There is nothing to stop them getting closer together except their own energetic motion.

The pressure exerted by a gas

Though a gas has no rigidity, it can still exert **pressure** on surfaces. Gas pressure can inflate a balloon or a tyre, or crush an evacuated container.

Again, there is a particle explanation for this. When a fast moving particle of a gas collides with a surface both the particle and the surface experience force. The small particle bounces back away from the surface. The surface may show little effect when hit by just one particle, but a surface which is experiencing bombardment by many particles will be subject to a significant force. The total force acting per unit area of the surface is pressure:

$$\text{pressure} = \frac{\text{force}}{\text{area}} = \frac{F}{A}$$

Note that since the unit of force is the newton, N, and the unit of area is the square metre, m², the unit of pressure is the newton per square metre, N m⁻². In the SI system this is given its own name, the pascal, which is abbreviated to Pa.

The walls of balloons, tyres and other containers have two sides, with gas on both sides. For a container such as an open bottle, the gas on both sides is air and the pressures it exerts on the inner and outer surfaces are the same. For an inflated balloon or a tyre, the pressure on the inner surface of the wall is bigger than the pressure on the outer surface (Figure 6.7). The inequality of the forces on unit area compensates for the tension in the wall itself.

Figure 6.7
Inside a tyre the pressure must be greater than that outside.

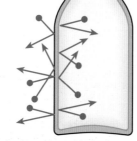

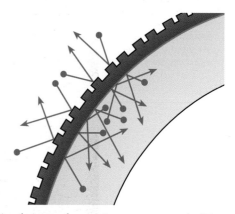

for an open bottle, equal areas of the inner and outer surfaces experience the same rate of collisions (at the same temperature)

for a tyre, the inner surface experiences a greater rate of collisions (at the same temperature)

The statement that gases are easily compressed is certainly true when comparing them with solids and liquids. But when pumping up a bicycle tyre it can seem not quite so true. A force must be exerted. The reason for this is that the piston or plunger on the pump has two sides and is experiencing unequal pressure from each side. On the outside the pressure is exerted by the air of the atmosphere – **atmospheric pressure**. On the inside the pressure is exerted by the air in the tyre that is already compressed and at a pressure higher than that of the atmosphere (Figure 6.8). At the same temperature, the particles inside the tyre have the same range of speeds as the particles outside but are closer together. The inner face of the plunger is experiencing more collisions, and so more force, than is the outer face. The areas of the inner and outer sides of the plunger are the same, so the inner side is experiencing more pressure. Pumping up the tyre requires force to compensate for this.

8 If it's not repulsive force that makes you have to work to compress a gas, and push its particles closer together, then what is it?

9 Atmospheric pressure is approximately 1×10^5 Pa.
   a Estimate the total force acting on your skin.
   b Why are you not crushed?
   c What would happen if atmospheric pressure were removed from your skin (for example, if you stepped outside a spacecraft without a space suit)?

Figure 6.8
The pressure on the inner side of a bicycle pump plunger is greater than atmospheric pressure.

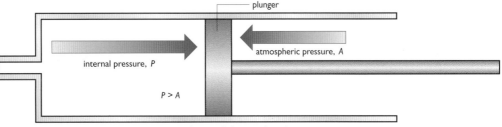

internal pressure, P

atmospheric pressure, A

$P > A$

a simple pump (valves not shown)

Probing matter with X-rays

Visible light waves are diffracted when they radiate past very small grains or through very small gaps that have a similar size to the wavelength of the light, about 5×10^{-7} m (Figure 6.9). Light waves diffracted by many such small grains or gaps add together by superposition and produce interference effects (see Chapters 2 and 3). If the grains or gaps are randomly arranged then superposition destroys any clear pattern. If the grains or gaps are part of a regular array then superposition produces reinforcement of the diffracted waves in certain directions resulting in a clear pattern. But visible light waves do not produce clear patterns when diffracted by atoms, and the gaps between them, in a crystal. The atoms and the gaps are too small compared with the wavelength of the light.

Figure 6.9 Look at the single-colour (or single-wavelength) light from a sodium street lamp through the mesh of a nylon umbrella and you will observe diffraction effects. The regular pattern in the fabric produces regular patterns in the light.

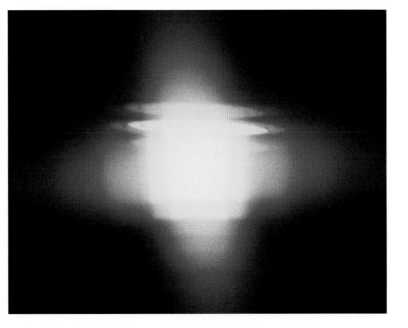

X-rays have a much smaller wavelength than visible light – less than about 10^{-8} m (see Figure 3.6) – but they are also diffracted by grains and gaps that have a size comparable to their wavelength. They produce clear diffraction patterns, which can be photographed, when a narrow beam of a single wavelength passes through a small crystal (see Figures 6.10 and 6.11). This **X-ray diffraction** differs from diffraction of visible light only in the wavelength of the waves, and therefore in the size of the obstacles and gaps that produce clear diffraction patterns.

Crystals contain many particles and many gaps. Waves diffracted at each of these gaps add together, as described by the principle of superposition, to produce a pattern. The nature of this diffraction pattern depends on a number of factors, such as the orientation of the particles and any variations in the distances between them. Although complex, the diffraction pattern can be analysed to work out the arrangement of the particles in the crystal – the crystal structure.

10 Why don't crystals produce diffraction patterns with visible light?

11 What is the feature of X-rays that makes them ideal for studying crystals?

Figure 6.10
The practical arrangement for X-ray diffraction by a crystal.

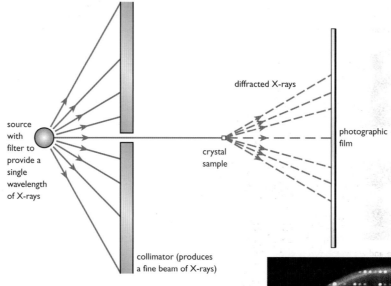

source with filter to provide a single wavelength of X-rays

collimator (produces a fine beam of X-rays)

crystal sample

diffracted X-rays

photographic film

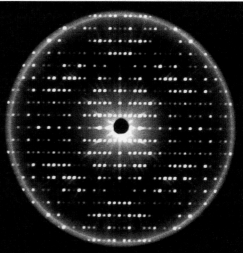

Figure 6.11 The spacings of atoms in a crystal are small enough and regular enough to produce diffraction of X-rays. The X-rays have a single wavelength, and the pattern that is produced depends on the patterns amongst the atoms.

Some crystal structures

Diamond crystals are hard and strong. However, the fact that they are hard and strong relative to other crystals is not merely because of the strength of the bonds holding the carbon atoms together. It is also partly because of the tetrahedral structure of the atoms, and because a single diamond crystal is just that – a continuous array of atoms without boundaries or flaws where cracks could grow if the crystal were subject to forces of tension or compression.

If it were possible to build a car out of a single crystal of metal it would be very robust indeed. Unfortunately, metal crystals are usually very small, and they are often flawed. Figures 6.12 and 6.13 show models of a variety of crystals.

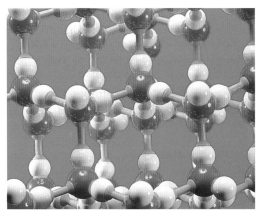

a An ice crystal – two kinds of atom in a repeating array.

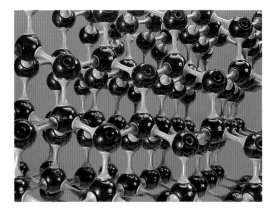

b Diamond crystals can be large, and within the crystal each atom is linked to four neighbours, the same pattern repeating over and over again throughout the crystal.

c Molten igneous rock creeps through cracks in the rock deep underground and slowly solidifies. The slowness of the process allows each atom to settle where the forces of attraction with its neighbours are in balance. The result is a crystal array of quartz.

Figure 6.12 (left)
Models of crystals: **a** ice,
b diamond, **c** quartz.

Figure 6.13 (right)
Looking at metal crystals.

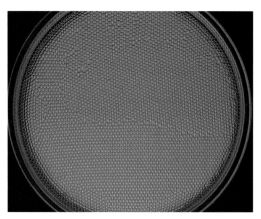

a By bubbling gas steadily through a glass jet into soapy water, a layer of bubbles all of the same size can be made in a dish. The bubbles and the way they arrange themselves make a good (if only two-dimensional) model of a crystal structure. Crystal boundaries and flaws can be seen.

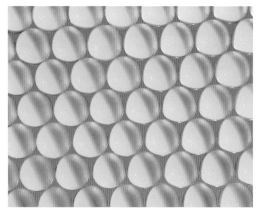

b Metal atoms bond together into crystals, but the crystals are rarely large.

c Metal crystals are sometimes visible. In the very thin layer of zinc on a galvanised surface each zinc crystal reflects the light differently, and can be seen.

Non-crystalline solids

12 What determines whether silicon and oxygen atoms will form into a crystal array or into an amorphous structure?

Glass and quartz are chemically the same – silicon and oxygen bonded together. The difference lies in their structure. Quartz is a common material, an igneous rock. The quartz that we find in white bands in rocks has usually solidified very slowly deep underground. Almost one by one the particles fall into the places in which the forces on them are in balance. They fall into a neat array (Figure 6.12c). But if quartz is melted and then cooled quickly, the particles lose energy before they have time to line up in orderly crystal arrays. The result is glass, which is called an **amorphous** (meaning 'shapeless') solid (Figure 6.14). Instead of being composed of a uniform crystal, glass has regions of greater and lesser strength. When subject to shear forces, cracks soon spread through regions of weakness, and the material snaps – glass is very brittle.

Figure 6.14
From the outside a piece of glass looks like a continuous crystal, but in fact it has an irregular, non-crystalline structure that makes it brittle. This diagram shows the structure of glass simplified to two dimensions, with lines representing the interatomic bonds.

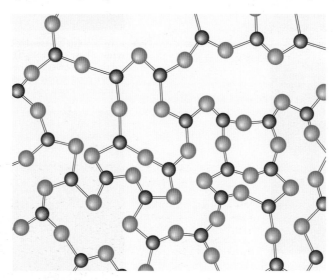

Polymers

Figure 6.15
Tar and polythene are chains of carbon atoms each bonded to two hydrogen atoms, as shown here. The difference in behaviour is due to the length of the chains. Tar has chains of a few tens of carbon atoms. The much longer chains in polythene are tangled together.

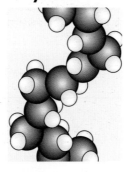

Polymers have long chain molecules; each 'link' of the chain is called a **monomer**. Tar is a material with relatively short chains (a few tens of carbon atoms with hydrogen atoms attached) that can slide past each other (see Figure 6.15). Longer chains, as in polythene, may be sufficiently tangled together to make the material solid even though the forces between the chains are small. Polythene is a **thermoplastic polymer**, which becomes soft and able to be moulded (i.e. 'plastic') when it is reheated. In **thermosetting polymers** like polyurethane there are stronger forces between the chains so that the material is stiff, and remains stiff when it is reheated.

● ●

● **Comprehension and application**

Buckyballs

Carbon is a material in which atoms, all of them chemically identical, can form different arrangements. In diamond they form a strong tetrahedral structure – vast numbers of atoms in neat arrays. Despite the strong and hard nature of the structure, though, the material is still carbon. When heated enough, in air, it burns.

Graphite also has a crystal structure, but the carbon atoms are gathered in smaller and flatter crystals that can easily slide over each other (Figure 6.16a), making it easy to leave a trail of graphite across the surface of paper just by dragging a stick of graphite (a pencil) across it.

Every time you burn a candle you create an entire range of carbon structures. There may be some individual carbon atoms, some graphite crystals, and perhaps some very tiny tetrahedral diamond structures. There will also be '**buckyballs**'. These were only discovered in the 1980s. Sir Harry Kroto (Figure 6.17) and other scientists vaporised carbon with lasers; they reflected radio waves from it and they did chemical tests – and they found that the sooty material contained clusters of carbon atoms with 60 or more atoms. The scientists wondered how these arrays of atoms might be orientated, and realised that the new carbon clusters were hollow round cages of carbon atoms. The structure reminded them of the sections of a football, and of the big domes built by architect Richard Buckminster Fuller. The giant carbon structure was named **buckminsterfullerene** (Figure 6.16b), and now the clusters are usually referred to as buckyballs. Discovery of a new type of material raises a question – is it useful? Buckyball material makes a good lubricant, because the balls roll. It also has uses as a chemical catalyst, a superconductor and in the manufacture of drugs.

Figure 6.16
The structures of
a graphite and
b buckminsterfullerene,
C_{60}.

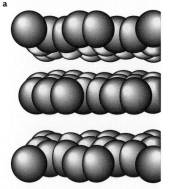

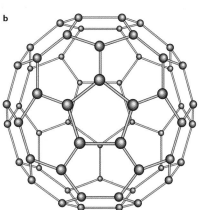

13 Explain the differences in the physical properties of
  a glass and quartz
  b graphite and diamond.
14 In the scientists' models of possible structures for the unknown form of carbon, what do you think were represented by
  a the jelly babies (see caption on the right)
  b the toothpicks?
15 Why is the work of Sir Harry Kroto valued so much that he has been made a knight?
16 **DISCUSS**
  Do you accept particle theory as:
  a absolute truth
  b the best theory we've got and 'true' as far as we know and for all practical purposes
  c useful – whether it's 'true' or not doesn't matter
  d lacking enough supporting evidence to be accepted as a source of reliable explanation
  e absolutely untrue?
  Justify your position.

Figure 6.17
Sir Harry Kroto.

In the 1980s a scientist called Harry Kroto paid his own air fare to Texas, to use a laser for vaporising carbon. This produced a form of carbon with physical properties different from those of graphite or diamond, but more like soot. With colleagues he tried to work out the structure. Their sophisticated techniques included the use of jelly babies and toothpicks to build models of possible structures. Now these buckyballs have many uses, and Harry Kroto is Sir Harry Kroto.

● **Extra skills task**

Information Technology and Communication

Carbon exists in several forms including diamond, graphite and buckyballs. Use a computer (e.g. the Internet, CD ROMs), scientific articles from journals or books and other sources to research one of these forms of carbon.

You should carry out efficient searches on the computer using appropriate tools, such as Internet search engines and database query techniques. You may also be able to use CD ROM indexes provided by many journals to find suitable articles in scientific journals.

Use the information to prepare a research report and present this to your group. Specialise in one or two aspects of your chosen material, such as details of structure and other physical properties, related commercial usage, historical details of discovery and development, and so on. You should relate some of the physical properties and consequent uses directly to the structure of the material.

Images found in your research should be put to good use. Don't forget to provide a bibliography to give credit to the originators of the material you have gathered.

7 First explorations of atomic structure

THE BIG QUESTIONS
- What evidence can we see in the macroscopic (human-sized) world that atoms are not fundamental particles but have a structure of even smaller particles?
- What are these main components of atoms?

KEY VOCABULARY alpha radiation   anode   cathode   cathode rays   charge   electric field   electrolysis   electron   excited (gas)   fluoresce   magnetic field   neutral   nuclear or planetary model   nucleus   photoelectric effect   'plum pudding' model   thermionic emission   vacuum tube

BACKGROUND Our ideas about the world at the level of atoms were created by bringing together observations and ideas from different areas of physics and chemistry of the 19th and early 20th centuries. They were made possible, most of all, by the new vacuum technology of the mid-19th century, which allowed low pressure gases and the light that they emit to be studied.

Vacuum technology led to the discovery of the electron. This technology was then combined with the new knowledge of the electron to create electronic valves that were key components of radios and of early computers (Figure 7.1). A modern TV tube still blasts electrons at a screen from the opposite end of a glass vacuum tube.

Figure 7.1
This is a 'triode valve', made in 1919: a glass tube with a vacuum inside to allow electrons to travel. Nowadays, we use transistors for the same purposes.

The 20th century was the 'electronic century'. From the first valves for radios, through TV, to electron microscopes, telecommunications and computers – knowledge of the electron, and application of the knowledge, reshaped the lives of everyone.

Excited gases

A flame is hot gas. It is also a source of light and its colour can vary, depending on the atoms that are present in the gas. Chemists use this as a way of identifying elements in an unknown material, in what they call 'flame tests' (Figure 7.2). Sodium produces a strong yellow colour, the presence of copper in a flame produces a vivid green colour, and a flame containing potassium will be a lilac colour, assuming that this is not hidden by colours produced by other elements.

Figure 7.2
The colour of a flame depends on the elements that are in the hot gas. These flames contain **a** sodium and **b** potassium. Note that the colour of the flame containing atoms of sodium is the same as the colour of many street lights.

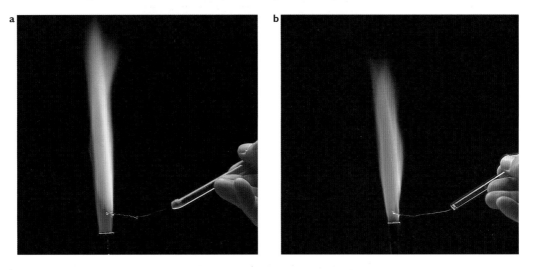

The same materials can exist as gases at low pressure (very much less than atmospheric pressure) inside glass tubes. These are called **vacuum tubes**, because of the very low pressure. The gases can be given energy, or **excited**, not thermally but electrically by applying a high voltage to them. Then the gases emit the same colours as are seen in flames. This principle is used in sodium street lights (Figure 7.3).

Figure 7.3
This street light is a tube containing sodium gas at very low pressure. The gas in the tube is excited – given extra energy – by an applied voltage, so that it emits light.

1 If street lights contained copper instead of sodium, what colour would they be?
2 How are sodium atoms given energy
   a in a flame
   b in a street light?

Each element, it seems, has its own unique relationship with light. Many observations and experiments were made on this phenomenon in the mid-19th century. But at that time nothing was known about atomic structure, and in fact even the existence of atoms was very much a matter of debate amongst the international science community. It was some time before a consistent theory could supply an explanation for what goes on in individual atoms when they emit and absorb electromagnetic radiation. This chapter provides the beginning of the story of the development of that theory.

Cathode rays

Figure 7.4
Cathode rays, first
noticed in the evacuated
glass tubes of 19th century
experimenters, start the
story of our knowledge of
matter on the atomic and
sub-atomic scale. They also
provided the start of radio,
television and computer
technology.

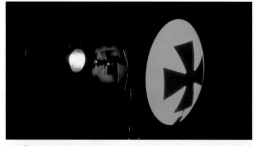

In the investigations of gases in vacuum tubes, another new phenomenon was discovered. An invisible radiation was found to spread from the negatively charged metal plate, the **cathode**, in the glass tubes. These **cathode rays** could make the glass walls of the tube glow, and could make some materials glow brightly, or **fluoresce** (Figure 7.4). The surfaces that the cathode rays hit also accumulated electric charge.

Controversy developed over the nature of these cathode rays. Some believed that the cathode rays were some new form of electromagnetic radiation, or something similar but so far undiscovered. Others thought that the rays were streams of particles. The cathode rays were studied under many different conditions, including under the influence of **electric fields** and **magnetic fields**.

Electric fields are regions in which charged particles experience electric force. Every charged object, from a single particle to a thundercloud, generates an electric field in the space around it. Magnetic fields are regions in which charged particles that are moving (including the electrons flowing in a wire in a circuit) experience magnetic force. Every magnet and every wire in which a current is flowing generates a magnetic field in the space around it.

Some support for the notion that cathode rays were streams of particles came from the observation that the pathways of the rays could be bent by a magnetic field, as would be expected for moving particles carrying electric **charge**. (Electric charge is the ability of a body to exert and experience electric force and, when moving, magnetic force.) At a point in an electric field or a magnetic field, some bodies experience force in one direction and some bodies experience force in the opposite direction. This leads us to the idea that bodies can have two different types of charge. The names chosen for these are positive and negative charge.

A particle hypothesis

An Irish physicist called G. Johnstone Stoney had developed a hypothesis concerning **electrolysis**. (Electrolysis is the flow of current through solutions and the resulting deposition of some of the materials from them.) Stoney's hypothesis was based on the idea of the existence, within matter, of negatively charged particles. He went so far as to give a name to his charge-carrying particles – **electrons**. Electrolysis and cathode rays were previously unlinked areas of physics. At that time there was not much evidence to support the idea that cathode rays and Stoney's electrons were the same thing. However, the particle idea was put to the test.

Settling the cathode ray controversy

Some scientists continued to suggest, especially after the discovery of X-rays in 1895, that cathode rays were similar to electromagnetic radiation. Others favoured the hypothesis that they were streams of particles. Joseph John Thomson, a scientist working in Cambridge, wrote the following introduction to his account of his experiments on cathode rays in 1897:

'The experiments discussed in this paper were undertaken in the hope of gaining some information as to the nature of cathode rays. According to the almost unanimous opinion of German physicists they are due to some process in the ether. Another view is that they are in fact wholly material, and that they mark the paths of particles of matter charged with negative electricity. The following experiments were carried out to test some of the consequences of the electrified particle theory.'

(from the *London, Edinburgh and Dublin Philosophical Magazine and Journal of Science*, October 1897)

The hypothesis that cathode rays were streams of particles predicted that they would obey the laws of mechanics. They would be subject to forces and matching acceleration in the same way as other bodies. Thomson fired cathode ray beams through electric fields, magnetic fields, and combinations of these, and he measured the ways in which the beam was deflected. (See page 260 for more detail of Thomson's experiment.) The deflection of the beam could certainly be accounted for by supposing that the beam was a stream of particles.

3 Why do we believe that there are two types of electric charge?

4 What are the effects of cathode rays on a material that they hit?

5 Imagine that you are a 19th century scientist. You fire some cathode rays at a surface and notice that the surface has negative charge. How many possible causes of this would you be able to think of?

Thomson was actually able to work out what the ratio of the charge to the mass of these particles should be. What's more, using Stoney's estimate for the size of the charge, he could then actually obtain a value for the mass. The calculated mass was very small indeed – much, much smaller than the mass of an atom. But the fact that Thomson was able to work out a mass persuaded most physicists that cathode rays could usefully be thought of as streams of extremely small particles, and Stoney's name for these – electrons – has stayed with them ever since.

6 a Before Thomson's experiment, what were the two competing ideas about the nature of cathode rays?
b What did Thomson's experiment show that supported one idea rather than the other?

The electron theory as a source of further explanation

Within a few years, Thomson and others had shown that various phenomena could be explained in terms of electrons, including:

- **Thermionic emission** – the flow of negative charge into the space surrounding a heated negatively charged metal surface. The negative charge could now be explained as being carried by electrons. In fact, thermionic emission was the source of the electrons in cathode ray tubes (see below).
- The **photoelectric effect** – also a flow of negative charge into the space around a metal surface. The difference from thermionic emission is that photoelectric emission is caused by electromagnetic radiation supplying energy to the metal. (The photoelectric effect has special significance in the development of quantum ideas – see Chapter 26.)

Vacuum tubes and thermionic emission of electrons

Cathode rays in a vacuum tube involve mass movement of electrons which have escaped into the vacuum from the metal surfaces of the electrodes (Figure 7.5). Such escape is called thermionic emission, because the effect increases with temperature. Thermionic emission can be thought of as the liberation of electrons from within the metal, rather as evaporation is the liberation of molecules from a liquid.

7 a Outline the similarities and differences of thermionic emission and the photoelectric effect.
b Which of the effects is essential to the working of a TV tube? In what way is it essential?

Due to their negative charge, any electrons escaping from the negative electrode, the cathode, will accelerate towards the positive electrode, the **anode**. In the cathode ray oscilloscope, the TV tube and the electron microscope, the cathode is heated by a heating filament and so large numbers of electrons escape. A constant flow of electrons from cathode to anode can take place.

Figure 7.5
A vacuum tube is an evacuated glass tube with positive and negative electrodes – the anode and cathode, respectively. These are connected to a power supply by way of wires passing through the glass. The vacuum isn't necessary for the thermionic emission, but is needed for the emitted electrons to be able to move freely without collisions with molecules of gas.

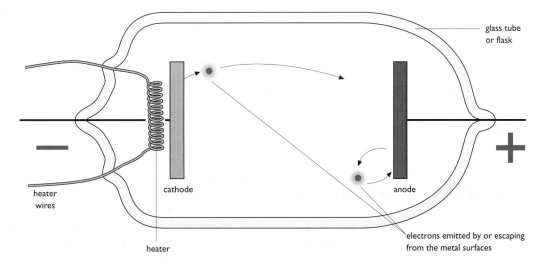

glass tube or flask

heater wires

cathode

anode

heater

electrons emitted by or escaping from the metal surfaces

A new model of atoms

By the end of the 19th century the existence of atoms was still not accepted by everybody, but atoms provided such a consistent source of explanations of a whole range of phenomena from chemistry and physics that most scientists were happy to treat the idea as a good working assumption.

It followed that if matter consisted of atoms then the electrons emitted in thermionic emission and the photoelectric effect must emerge from these atoms. On the available evidence, scientists were able to build themselves a simple model of what the structure of atoms might be like. Atoms are, in general, electrically **neutral**, having neither net positive nor net negative charge. If they contain some charged particles then these must be balanced by the presence of particles of the opposite charge. Since electrons carry negative charge, the rest of the atom – the part left behind in thermionic emission and the photoelectric effect – must be positive. And if electrons are very small and have very tiny mass – much, much smaller than the mass of an atom – the positive part of the atom must make up most of its mass. So the scientists envisaged the electrons embedded in the positively charged component. Assuming that the density of this positively charged component is the same as the density of the atom as a whole, and of the whole material, they visualised atoms as illustrated in Figure 7.6.

This visual model, sometimes called the **'plum pudding' model** (the electrons were the plums and the positive component of the atoms was the spongy material around the plums), lasted about a dozen years before it was shown to be of little use.

Figure 7.6
'Plum pudding' atoms – electrons embedded in a positively charged 'sponge'. The density of each atom is similar to the density of the material.

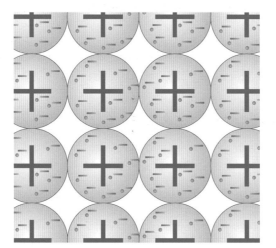

Alpha scattering and Rutherford's model

In 1909, Hans Geiger and Ernest Marsden, working with Ernest Rutherford at Manchester University, fired **alpha radiation** at gold foil. Alpha radiation was known to be very energetic radiation emitted by radioactive material. It was also known that the radiation travelled in individual units or packages, which we call alpha particles (see Chapter 8). Geiger and Marsden used a screen that scintillated – gave a little spark of light – when an alpha particle hit it. The screen was attached to a microscope that they could move in an arc around the gold foil. The faint scintillations were hard to see, and Geiger and Marsden had to sit in the dark before looking for them, so that their eyes could adjust to the low light intensity.

The alpha particles were known to be energetic, and Rutherford and his team expected the gold atoms to put up little resistance to them. They expected the alpha particles to shoot straight through the gold foil. And, that *is* what happened for nearly all of the alpha particles; but a few of them were deflected, and some, fewer still, were deflected through more than 90° (Figure 7.7). Geiger and Marsden counted the scintillations at the different angles. Even though significant deflection was comparatively rare, Rutherford knew that it couldn't be explained by the existing 'plum pudding' model of the atom.

8 The mass of the electrons seems to be a tiny proportion of the mass of the whole atom. How does the charge carried by the electrons in an atom compare with the charge carried by the rest of the atom?

9 In what ways were Geiger and Marsden's observations inconsistent with the 'plum pudding' atom?

Figure 7.7
Geiger and Marsden's
apparatus.

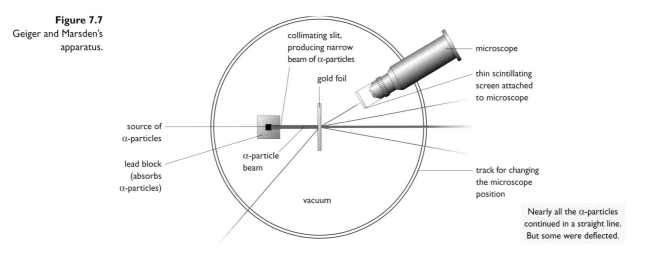

collimating slit,
producing narrow
beam of α-particles

microscope

thin scintillating
screen attached
to microscope

gold foil

source of
α-particles

lead block
(absorbs
α-particles)

α-particle
beam

track for changing
the microscope
position

vacuum

Nearly all the α-particles
continued in a straight line.
But some were deflected.

Rutherford's nuclear model

Rutherford and his team looked at the distribution of the scintillations around the circle surrounding the gold foil, and made some careful calculations. By 1911 they showed that the scattering of some alpha particles through large angles suggested that the gold contained positive charge (to repel the positive alpha particles) that must be compressed into a small volume (to be dense enough to exert a significant force). They realised that the positive charge of an atom could not be spread throughout the whole atom, but must be compressed into a comparatively very small volume, and that the rest of the atom was mostly empty space. What they were saying was that the positive charge, and most of the atomic mass, was all contained in a **nucleus**.

So where did that leave the electrons? Rutherford wasn't sure. He suggested that they were in orbit around the nucleus, taking up the rest of the space occupied by the atom (Figure 7.8). He was sure that the previous visual model of the atom, the 'plum pudding' model, was of little value and he replaced it with the new **nuclear or planetary model**.

Figure 7.8
Simple nuclear or
'planetary' model atoms.
Rutherford knew that the
nucleus was small
compared with the size of
the whole atom.

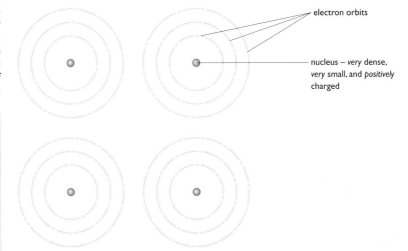

electron orbits

nucleus – *very* dense,
very small, and *positively*
charged

10 What was Rutherford's solution to the inconsistency between the 'plum pudding' model of the atom and the observations made by Geiger and Marsden?

11 Rutherford knew that his new model of the atom was not the whole story. How did he know this?

12 **DISCUSS**
If you had been Rutherford, would you have told the world about your ideas, knowing that they only provided an incomplete picture?

Rutherford and other physicists knew that this new model couldn't provide a complete theory of the atom, however, because it seemed to disobey other laws. They knew that accelerating charged particles emit radiation. That was the principle of creation of radio waves – a phenomenon which at that time was soon to provide a new means of communication that would change the world. Electrons in orbit, like anything in circular motion, are accelerating. In atoms the orbiting electrons would be expected to emit radiation steadily. And in doing so they should lose energy and spiral into the nucleus. Obviously this didn't happen, because if it did then no atom would last very long at all. The new idea of the atom didn't match all the physicists' expectations, but it was the best idea yet available.

● **Extra skills task** ## Application of Number and Information Technology

Geiger and Marsden's data

Rutherford recognised that the pattern of scattering of alpha particles by gold foil suggested that the positive charge of the atom was all concentrated in a very small volume, surrounded by a much larger volume occupied by negative charge. He calculated that if this idea of a nuclear atom were correct then the pattern of alpha scattering would obey the following relationship:

$$N = K \frac{1}{\sin^4(\phi/2)}$$

where N is the number of alpha particles producing scintillations at angle ϕ in a chosen period of time, and K is a constant. It was this relationship that Geiger and Marsden put to the test. Some of their results are shown in Table 7.1.

Table 7.1
Some of Geiger and Marsden's data for gold foil.

Angle ϕ	$1/\sin^4(\phi/2)$	N
45°	46.6	1435
75°	7.25	211
120°	1.79	51.9
150°	1.15	33.1

A graph that follows the general pattern $y = mx$ is a straight line passing through the origin, with gradient m. Plot a graph of N (y-axis) against $1/\sin^4(\phi/2)$ (x-axis).

Does this support or falsify Rutherford's relationship? Is this a graph of the form $y = mx$? What is the value of K?

A fuller experiment by Geiger and Marsden was designed to investigate the way the particles were scattered according to the atomic mass of the foil they were fired into. Some of their results are shown in Table 7.2.

Table 7.2
Some of Geiger and Marsden's data for other metals.

Substance	Atomic mass A /atomic mass units	Number of scintillations N at a particular angle in a chosen period of time
gold	197	581
tin	119	270
silver	107.9	198
copper	63.6	115
aluminium	27.1	34.6

They proposed that there was a relationship between these quantities: $N^x A^y = c$, where x and y, as well as c, may have any value, including non-integer (non-whole number) and negative values. Note that $N^x = c/A^y$.

Try various values of x and y until you obtain a straight line graph, linking N and A, which has a gradient equal to the constant. Give the values of x and y. Use a computer as appropriate.

Ionising radiation

THE BIG QUESTIONS
- How can we detect radioactive emissions?
- How can we tell that there are different kinds of radioactive emissions?

KEY VOCABULARY
absorption   alpha radiation   avalanche effect   beta radiation   cloud chamber
discrete   electrode   G–M tube   gamma radiation   intensity (of radiation)
inverse square law   ion   ion pair   ionisation   ionisation chamber
medical tracer   penetration   photographic emulsion   radiation sources
saturated vapour   spark chamber   spark detector

BACKGROUND
Smoke detectors and **medical tracers** (materials injected into our bodies to produce images of the interior) use the ability of ionising radiation to penetrate through matter. However, as its name suggests, ionising radiation does not merely penetrate but also has fundamental effects on matter. Food irradiation is an example of its use to kill bacteria, while placing a source of radiation near to or inside a tumour can kill the cancerous cells.

The radiation itself doesn't care about humans one way or the other; it can do harm (Figure 8.1) and it can do good (Figure 8.2). It is entirely up to us which way we choose to use it.

Figure 8.1 (left) Actor John Wayne died of lung cancer. It may have been caused by cigarettes, or it may have been caused by exposure to ionising radiation during filming of a 'western' on an old nuclear bomb test site.

Figure 8.2 (right) Ionising radiation has been used as part of this mother's cancer treatment.

● Ionising effects of radioactive emissions

Most of the objects around us are very slightly radioactive, but some are more strongly radioactive. These objects contain certain materials that are **radiation sources**. The high-energy radiation that emerges from them does not discriminate between matter that is in living material and matter that is not. The energy that it carries transfers to the atoms of the material through

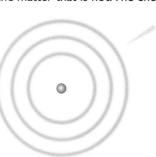

Figure 8.3
Ionisation: ionising radiation supplies energy that allows an electron to escape from an atom. The atom then has an imbalance of positive and negative charge – it has become an ion.

which it passes, and it ionises them. That is, some of the atoms of the material gain energy from the radiation, and the energy allows electrons to escape from them. With electrons missing, these atoms have more positive than negative charge. They have become positive **ions** (Figure 8.3).

Ionisation can change the chemical behaviour of atoms. Within the complex molecules of our bodies such changes can be harmful. Ionisation can cause mutations, through changes in DNA, and it can cause changes that may lead, usually after many years, to cancer.

Ionisation and detection – the cloud chamber

It is possible to fill a space with **saturated vapour** of water or alcohol. A saturated vapour will begin to condense when it is cooled. However, the condensation will only go as far as the formation of very small invisible droplets, which will be stable unless cooling continues or further condensation into larger drops is encouraged by the presence of dust or ions. The travel of a heavily ionising alpha particle through such a cooled vapour will result in a trail of drops that are big enough to form a visible track. This is the principle of a **cloud chamber**.

Figure 8.4 shows two types of cloud chamber, designed to view the tracks of ionising particles.

Figure 8.4
Sections through
a a diffusion cloud chamber, and
b an expansion cloud chamber.

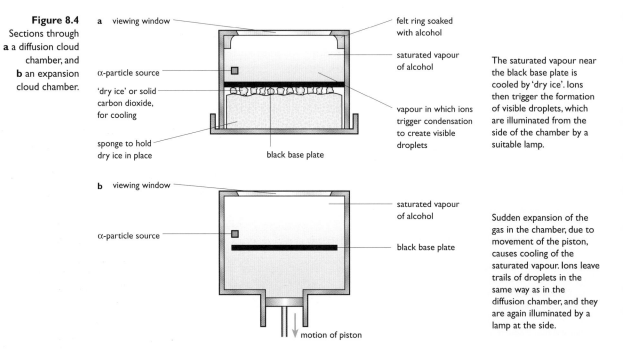

a viewing window — felt ring soaked with alcohol — saturated vapour of alcohol — α-particle source — 'dry ice' or solid carbon dioxide, for cooling — vapour in which ions trigger condensation to create visible droplets — sponge to hold dry ice in place — black base plate

The saturated vapour near the black base plate is cooled by 'dry ice'. Ions then trigger the formation of visible droplets, which are illuminated from the side of the chamber by a suitable lamp.

b viewing window — saturated vapour of alcohol — α-particle source — black base plate — motion of piston

Sudden expansion of the gas in the chamber, due to movement of the piston, causes cooling of the saturated vapour. Ions leave trails of droplets in the same way as in the diffusion chamber, and they are again illuminated by a lamp at the side.

What we see in cloud chambers (Figure 8.5) tells us that ionising radiations travel in **discrete** packages. That is, the radiation is not continuous like ripples spreading on a pond, but behaves as if it is composed of individual particles. We will return to this point later.

Figure 8.5
Cloud chamber tracks show that ionising radiation is composed of individual particles.

Three levels of ionising ability

Cloud chambers provide one way of comparing the ionising abilities of different radiations. The radiation emitted by different radioactive materials can be compared and classified into three groups according to their ionising ability (Table 8.1).

Table 8.1
Classifying by ionising ability.

Ionising ability	Name of radiation
very strongly ionising – creates dense tracks of ions in its paths	alpha radiation
quite weakly ionising – leaves only fairly weak tracks of ions	beta radiation
weakly ionising – leaves only scattered ionisation events	gamma radiation

Identifying the radiations

Having discovered that there are three main types of ionising radiation, we can take different radiation sources and make more observations. We discover that the radiations respond differently to electric and magnetic fields.

We find that **gamma radiation** is not affected by electric and magnetic fields, but the pathways of **alpha** and **beta radiations** are deflected (Table 8.2). This leads us to the idea that they are streams of charged particles.

Table 8.2
Effects of electric and magnetic fields on the three types of radiation.

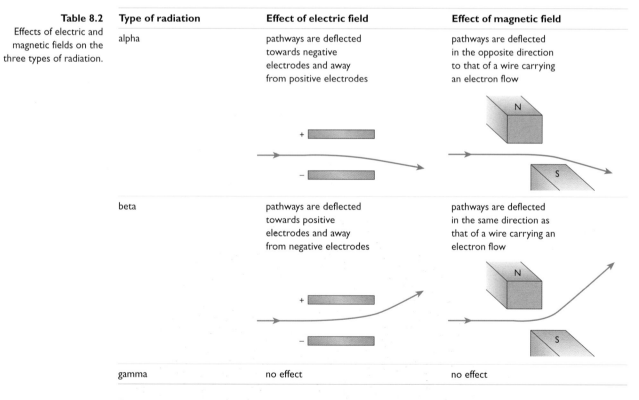

Type of radiation	Effect of electric field	Effect of magnetic field
alpha	pathways are deflected towards negative electrodes and away from positive electrodes	pathways are deflected in the opposite direction to that of a wire carrying an electron flow
beta	pathways are deflected towards positive electrodes and away from negative electrodes	pathways are deflected in the same direction as that of a wire carrying an electron flow
gamma	no effect	no effect

From the deflections we can see that alpha and beta radiations carry electric charge (of opposite sign) while gamma radiation does not. Alpha and beta radiations behave as beams of charged particles, charged positively and negatively, respectively.

The more mass a particle has at a particular speed, the less its path is deflected by electric or magnetic fields of given strength. We can use the deflections to compare the masses of the particles. Alpha particles are much more massive than beta particles. Gamma radiation carries neither mass nor charge. Table 8.3 summarises our knowledge of the radiations so far.

Table 8.3	alpha radiation	streams of particles with relatively large mass and with positive charge
The nature of the three types of radiation.	beta radiation	streams of particles with relatively small mass and with negative charge
	gamma radiation	no mass and no charge (photons*)

*Photons can be thought of as carriers of electromagnetic radiation energy – gamma rays are a kind of electromagnetic radiation, like X-rays. Read more about photons in Chapter 26.

Three levels of penetrating ability

1 What behaviour of gamma radiation tells us that it has no charge?
2 Gamma radiation is the most penetrative. What does this tell you about the rate at which gamma radiation transfers energy to the medium through which it travels, in comparison with alpha and beta radiations?
3 Present all the characteristics of alpha, beta and gamma radiation in a single summarising table.
4 Alpha radiation creates strong separate straight tracks in a cloud chamber, rather than causing a general clouding. What does this tell us about the nature of alpha radiation?

When alpha and beta particles run out of energy they stop moving, and when gamma rays run out of energy they cease to exist. These are the processes that we call **absorption**. All three types of radiation interact with particles of the medium through which they travel, and many of these interactions are ionisation events. The more the radiations ionise as they travel, the more rapidly they transfer their energy to the material through which they are travelling. Alpha particles, the most strongly ionising, transfer their energy most quickly. Alpha radiation **penetrates** only small distances into material. A beam of gamma radiation transfers its energy most slowly, and penetrates the furthest. So, there is an inverse correlation between ionising and penetrating abilities – one increases as the other decreases.

Intensity of radiation and distance from the source

Intensity of radiation is the rate at which it transfers energy through each unit of area normal to its direction of travel. It is defined by the formula:

$$\text{intensity of radiation} = \frac{\text{rate of transfer of energy}}{\text{normal area through which the energy transfers}}$$

The SI unit of the rate of transfer of energy is the watt, W (which is equivalent to 1 joule per second), and the unit of area is the square metre. So intensity is measured in watts per square metre, $W\,m^{-2}$.

The intensity of any radiation that travels into the space around its source decreases as the distance increases. In relatively simple cases this is because the energy becomes spread across an ever-increasing area. This applies to light spreading from a star, and to gamma radiation spreading from a source. In both cases the radiation is not absorbed significantly as it travels.

The behaviour of the intensity of alpha and beta radiations as they travel from their sources is more complex (Figure 8.6). Not only does their intensity decrease as their energy spreads, but unless they are travelling through a vacuum they experience very significant absorption by the medium through which they travel. For alpha radiation the intensity in air soon falls sharply to zero, whilst the intensity of beta radiation decreases more gradually. For gamma radiation (as for light from a star) we can consider a three-dimensional sector of space around the source, with a square cross section. We can think of energy transferring through this cross-sectional area, and the direction of the transfer is normal (at 90°) to the area.

Suppose that at a distance x from the source the area of the square cross section is A, the sides of the square having length y.

No absorption is taking place, so the energy travels through any cross-sectional area along our three-dimensional sector at the same rate, no matter how far or how near the area is to the source. So, in the relationship,

$$\text{intensity} = \frac{\text{rate of transfer of energy}}{\text{area}}$$

the rate of transfer of energy can be considered to be constant.

$$\text{intensity} = \frac{\text{constant}}{\text{area}} = \frac{c}{A}$$

Figure 8.6
Variation of intensity of radiation in air for sources of **a** alpha radiation, **b** beta radiation and **c** gamma radiation.

Alpha particles are stopped and alpha radiation is absorbed by a few centimetres of air. The intensity of the radiation falls sharply to zero.

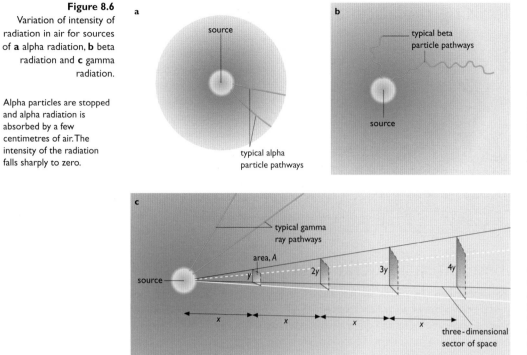

Air also absorbs beta particles, but most beta particles travel further than alpha particles. The intensity of the radiation fades more gradually than for alpha radiation.

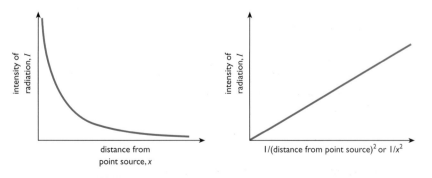

Gamma radiation experiences little absorption by the air, but intensity falls with distance from the source, as the energy radiation 'spreads' over an ever-increasing area.

We can say that intensity is inversely proportional to the area. But note that the area, A, is related to the length of the side of the square, y.

$$A = y^2$$

Note also that when x increases and becomes $2x$, the side of the square increases from y to $2y$. In general, the length of the side, y, is proportional to the distance from the source. So,

$$y = kx \quad \text{where } k \text{ is the constant of proportionality}$$
$$A = [kx]^2 = k^2x^2$$

which means that

$$\text{intensity} = c/k^2x^2$$

since c and k are both constant, the quantity c/k^2 is also constant. We could call this C. Then,

$$\text{intensity} = C/x^2$$

The intensity is proportional to the inverse of the square of the distance from the source. This finding, obeyed by gamma radiation in air, and light in space, is called the **inverse square law** (Figure 8.7).

5 Explain why alpha and beta radiations do not obey the inverse square law.

6 Lead is a much better absorber of gamma radiation than air. Would gamma radiation travelling from a source embedded in a large block of lead obey the inverse square law?

Figure 8.7
Alternative graphical relationships of the same reality – inverse square law behaviour.

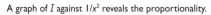

A graph of I against x shows that intensity falls quite rapidly with x, due to spreading of the radiation through space.

A graph of I against $1/x^2$ reveals the proportionality.

Ionising radiation and photographic emulsion

Figure 8.8
Photographic emulsion can reveal trails of ionisation.

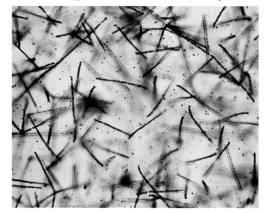

A simple form of **photographic emulsion** is made of grains of silver bromide embedded in gelatin (Figure 8.8). Areas of the emulsion that have been exposed to light become black. Ionising radiation has the same effect as light. The difference is that a single particle of alpha or beta radiation or a gamma ray photon has enough energy to leave a trail of chemical change behind it, whereas the emulsion has to be exposed to ordinary light for some time before enough change takes place to create a meaningful image.

The ionisation chamber

An **ionisation chamber** is an enclosed space containing air, with two **electrodes**, which are metal bodies between which a potential difference or voltage (see Chapter 18) is applied. A simple way to apply the potential difference is to make the body of the chamber out of metal so that it can act as one electrode itself and to use a rod in the centre of the chamber as the other electrode (Figure 8.9a).

Each ionisation event within the chamber creates a free electron and a positive ion; these are called an **ion pair**. They experience electric force due to the applied potential difference. The electron accelerates towards the positive electrode of the chamber, and the ion accelerates towards the negative electrode. When the particles reach the electrodes they induce movement of electrons within the metal. (For example, an electron from within the negative electrode will join with an arriving positive ion and neutralise it.) If enough ionisation is taking place in the chamber then this flow of charged particles can create a continuous current in an external circuit that is just big enough to measure – typically about 10^{-9} A. An ionisation chamber does not count individual ionisation events, but gives a measure of the *level of ionisation* that is taking place within it.

Figure 8.9
An ionisation chamber. Radiation causes ionisation in the air in the chamber, creating ion pairs which are accelerated by the voltage. A current can be detected by a sensitive meter in an external circuit.

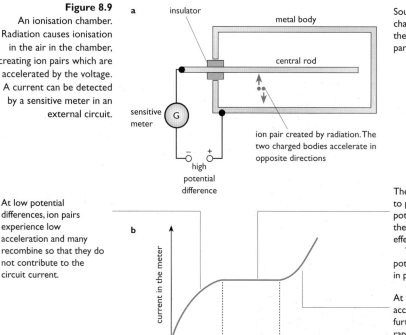

Sources of radiation are normally introduced into the chamber where the radiation causes ionisation. The size of the current depends on the rate of creation of free charged particles in the chamber.

At low potential differences, ion pairs experience low acceleration and many recombine so that they do not contribute to the circuit current.

The applied potential difference in this range is large enough to prevent recombination. Further small increases in potential difference within this range do not then increase the number of free charged particles and so have very little effect on current.

The ionisation chamber is operated with the applied potential difference within this 'plateau' range, where a change in potential difference produces very little change in current.

At large applied potential differences the ion pairs are accelerated so much that they ionise other atoms. These further ion pairs are also accelerated and can ionise. The rapid increase in the number of free charged particles is an 'avalanche effect'.

The size of the applied potential difference is important. If it is too small then the electron and ion do not experience much acceleration, and as they move slowly towards the electrodes they are likely to meet particles of opposite charge and recombine (become neutralised). At suitably higher potential differences the electrons and ions are moving too fast for this to happen, so that almost every ion pair that is created in the chamber contributes to the current in the circuit. However, if the particles accelerate too much then they can have enough energy to cause ionisation themselves when they collide with air molecules. This produces more and more ion pairs and the current in the external circuit can become large. Because one ion pair creates more pairs, which then create more again, this is called an **avalanche effect** (Figure 8.9b).

The source of ionising radiation is placed inside the chamber (though some radiation can also be detected from sources outside). For alpha sources, most of the alpha particle energy will be absorbed in ionisation events within the chamber. With a beta source, however, much beta radiation will reach the chamber walls before losing all of its energy. Nearly all gamma radiation will escape from the chamber before causing enough ionisation to be measured. An ionisation chamber is therefore most suited to measuring alpha activity.

7 Draw a sketch to illustrate an ion-pair 'avalanche effect'.

8 a Explain how a single alpha particle in an ionisation chamber can be responsible for the movement of large numbers of electrons in an external circuit.
b Why must the applied potential difference be within the 'plateau' range?

9 Why does an ionisation chamber show little response to gamma radiation?

The spark detector and modern particle detectors

A **spark detector** deliberately takes advantage of the avalanche effect described above. It uses a high potential difference applied between electrodes which are close together.

In a simple spark detector (Figure 8.10a) one electrode will be a thin wire (and not a metal plate, since alpha particles cannot penetrate through the metal to the space between the electrodes). A **spark chamber** uses the same principle but is more elaborate – it uses layered grids as electrodes so that it is possible to trace the path of an ionising particle (Figure 8.10b).

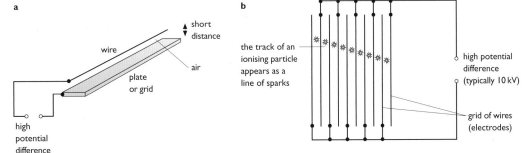

Figure 8.10
a A simple spark detector and **b** the principle of a spark chamber.

Figure 8.11
ALEPH is a multi-layered modern detector at CERN in Geneva. It uses a variety of techniques to track particles and measure their energies. Most of the ionisation that makes this possible takes place within solid material, not air.

A single alpha particle passing between the electrodes produces a dense trail of ion pairs. These accelerate, causing sufficient further ionisation to create a visible and audible spark. The associated burst of current can be detected electrically, or the trail of sparks photographed.

Modern detectors (Figure 8.11) are used to track very energetic particles and can be based on ionisation that takes place in solid material.

10 List three similarities and three differences between ionisation chambers and spark detectors.

The Geiger–Müller tube

Alpha radiation cannot penetrate the metal walls of an ordinary ionisation chamber, so an ionisation chamber is only useful when the alpha source is inside it. A Geiger–Müller tube or **G–M tube** *can* detect alpha radiation from a source in the environment outside itself (Figure 8.12).

A G–M tube's structure is not unlike that of an ionisation chamber; a conducting tube acts as one electrode and a central rod or wire as the other, but the G–M tube has a thin window at one end for radiation to enter from outside. The window must be as thin as possible, otherwise ionising particles – alpha particles, in particular – would be unable to penetrate into the tube.

The G–M tube uses the avalanche effect to amplify the consequences of a single ionisation event. It contains enough gas to make sure that there are enough collisions to make a strong 'avalanche'. An applied potential of 1 to 2 kV provides the necessary acceleration. Each avalanche produces a small pulse of current that is big enough to trigger a count in the external circuit.

A G–M tube is effective at counting beta particles that enter through the window, and for counting alpha particles provided that the window is very thin, and does not absorb the particles before they enter the tube. Gamma ray photons cause very little ionisation of the gas in the tube, but enough to sometimes create a pulse. The tube and its counting circuit only register a small proportion of the gamma photons that pass through it.

Figure 8.12
The Geiger–Müller tube and its circuitry detect bursts of current due to ionisation events and the resulting ion-pair avalanches.

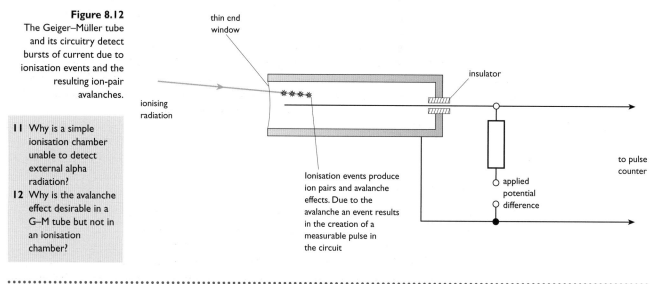

thin end window

insulator

ionising radiation

Ionisation events produce ion pairs and avalanche effects. Due to the avalanche an event results in the creation of a measurable pulse in the circuit

applied potential difference

to pulse counter

11 Why is a simple ionisation chamber unable to detect external alpha radiation?

12 Why is the avalanche effect desirable in a G–M tube but not in an ionisation chamber?

• **Extra skills task** Communication and Information Technology

Neutron radiation is important in the nuclear industry. Find out the properties of this type of radiation, and details of its uses and detection. Is it ionising radiation? Present this information as a poster, including diagrams, and use word processing and desktop publishing skills where appropriate.

 **9** # Nuclei

THE BIG QUESTIONS

● Processes of radioactive decay are detectable by the changes that take place to samples of substances and by the ionising radiation that is emitted. But what are the corresponding changes that take place inside the nuclei of atoms?
● How do we measure the changes that take place in samples of radioactive material?

KEY VOCABULARY

activity   atomic number   background count rate   becquerel   conservation rule count rate   daughter nucleus   decay chain   decay constant   isotope mass spectrometer   natural radiation   neutron   nucleon   nucleon number parent nucleus   population   proton   proton number   radioactive decay radioactive material   radioisotope

BACKGROUND

A **radioactive material** is one which emits radiation that has enough energy to ionise the material through which it passes. The discovery of radioactivity, and the discovery of the nucleus that followed, takes our minds to a new scale of matter – down to objects that are about 10^{-15} metres across.

A world where sizes are measured in multiples of 10^{-15} metres is a world that is strange to us. It is also hard to picture the effects of radiation that emerges from such a very small world into our human-sized world. Many people are suspicious of this ionising radiation. For example, though most scientists would say that food irradiation does us no harm, a lot of people are still uneasy about technologies that make use of radioactive phenomena.

● Changing nuclei

Uranium is an element found in the rocks of the Earth. An element is a material that has characteristic chemical properties and cannot be chemically changed into a simpler material. In places where there is a lot of uranium in the ground, a gas called radon, another element, seeps up continuously from the rocks into buildings. If the rocks are examined they are found to contain some radon, but not nearly enough to explain the continuous upward flow of the gas. The radon seems to be produced continuously from the uranium. In fact it is produced indirectly, by way of intermediate stages. Atom by atom, uranium changes and eventually the atoms of uranium become atoms of radon. The radon atoms also change, and become atoms of other elements. We can express the changes in a simple form:

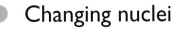
$$\text{uranium} \rightarrow \begin{array}{c}\text{other elements in}\\\text{intermediate stages}\end{array} \rightarrow \text{radon} \rightarrow \text{other elements}$$

The only way that one element can change into another is through changes to the atomic nuclei.

Stable and unstable nuclei

Most of the nuclei in the material in and around you are stable. Most of them were built up from smaller nuclei. That happened in stars that blazed and churned and died before the Sun and its planets came into being. The nuclei have been around for a very long time. They seem to be stable.

However, a small proportion of the nuclei in your surroundings (and in your body) are not so stable. They can change their internal structure and emit energetic, ionising radiation in the process. It's a process called **radioactive decay**. Any one of these unstable nuclei may decay today. Or it may not. The moment of radioactive decay of a particular nucleus cannot be predicted – decay is a random process.

1 'Decay is a random process.' Explain the words *decay* and *random*.

There are plenty of examples of radioactive decays. Some carbon nuclei are radioactive, and such a nucleus can change or decay into a nucleus of nitrogen. We can write this change, in its simplest possible form, as:

carbon → nitrogen

The carbon nucleus is then described as the **parent nucleus**, and the nitrogen nucleus as the **daughter nucleus**. It should be stressed that this only happens for carbon nuclei of a particular structure (see page 289).

Such a change is a once-only occurrence for each carbon nucleus, and may occur after a short time or after thousands or millions of years or longer. Then there will be a sudden readjustment of the nuclear structure and a squirt of radiation into the world outside. The nucleus spends the rest of eternity as a stable nitrogen nucleus.

For a nucleus of some parent materials, however, the newly readjusted nucleus – the daughter nucleus – is also unstable. In this case a particular nucleus may go through a series or chain of changes before finally becoming a stable nucleus. This is a **decay chain**. (You can read more about radioactive decay chains in Chapter 28.)

Very unstable nuclei are not likely to last very long before they decay. For that reason very unstable nuclei are relatively uncommon in our natural environment.

Natural radiation and background count rate

Figure 9.1
Everyone's body contains matter with nuclei that are unstable – we are all radioactive.

There is enough radioactive material in your body for a very large number of nuclei to decay within your body every second. You emit radiation into your surroundings; and your surroundings (the floor and walls, tables and pens, your food and drink, and even the air) emit radiation into you. This is nothing new. Emission of radiation by radioactive nuclei has happened for as long as there have been nuclei. It is **natural radiation**. If you set up a Geiger–Müller tube and a counter it will record such radiation. The count it makes is called the **background count rate**.

If you now place a source of increased radiation (such as a brazil nut, a mantle from a gas lamp or a sample of specially isolated material) in front of the G–M tube then the count rate will increase above the background level. The **count rate** is simply the number of events that the apparatus records in every second:

count rate = number of events counted per second

This quantity has units 'per second' or s^{-1}. In the context of radioactivity, the s^{-1} is called a **becquerel**, or Bq.

2 What is the difference, in terms of base units of the SI system, between one becquerel and one hertz?

3 What is the dimension of count rate?

Measuring activity

The becquerel is used as a unit not just for the counts made by a G–M tube system, but also for any rate of radioactive events. It can be used, for example, to measure the total number of decays per second taking place in a body of material, such as in your body or in a cubic metre of air. This is called the **activity** of the material.

The total activity of the air in a typical house is 2500 Bq. This means 2500 nuclei decay in the air in the house each second. It can often be more useful, however, to measure activity per unit volume. The activity per unit volume of the air around you might be as low as 20 Bq m^{-3}. This means that 20 nuclei undergo radioactive decays every second in each cubic metre. If you are in Cornwall the activity per unit volume of the air might be 200 Bq m^{-3} or more due to the presence of the radioactive gas radon, which seeps up from the granite rocks below. The radon can be trapped inside buildings, so good ventilation, often with special fans, reduces the risk from the radiation.

4 Explain, as fully as you can, what makes the background count rate higher than normal in an unventilated Cornish building.

The activity of a population of nuclei

Imagine a sample of identical atoms. Suppose that the number of atoms (and hence the number of nuclei) in the sample is 10^{18} – this is the **population**, N.

$$N = 10^{18}$$

(10^{18} might seem like a large number, but the mass of 10^{18} typical 'medium-sized' atoms is only about 0.1 mg. It is a very tiny speck of material.)

If the nuclei are unstable then the material is radioactive. Each second, some nuclei decay. Then they become nuclei with a different structure – nuclei of a new element. Suppose, for the sake of simplicity, that these daughter nuclei are stable.

The population, N, of the parent material is falling. N changes as time t changes. We can write the changes in quantities by adding the Greek letter delta, Δ. If the change in the population is ΔN and the time taken for this change is Δt, then:

average rate of change of parent population during time $\Delta t = \dfrac{\Delta N}{\Delta t}$

The change in the population of parent nuclei in the sample, ΔN, and the rate of change are both negative quantities, since the population is decreasing (Figure 9.2).

Figure 9.2
The change in the parent population with time gives a decay curve.

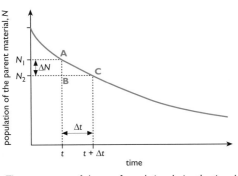

At time t, population $= N_1$

At time $t + \Delta t$, population $= N_2$

During time Δt:

change in population $= \Delta N = N_2 - N_1$

average rate of change of population $= \dfrac{\Delta N}{\Delta t}$

Note that ΔN and $\dfrac{\Delta N}{\Delta t}$ are negative, since N_2 is less than N_1

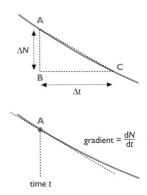

The average rate of change of population during the time Δt, $\Delta N/\Delta t$, is the average gradient of the curve during the time Δt. It can be found from the triangle ABC.

The gradient at the point A, however, is the gradient at a particular instant of time which we have called t. It is not an average gradient but an *instantaneous* gradient – it is equal to the rate of change of population at time t. It is an instantaneous value of the rate of change of population. It can be written as:

instantaneous rate of change of population $= \dfrac{dN}{dt}$

Note that the gradient and the rate of change of population are negative.

Questions 5 to 7 concern the decay of a radioactive parent material into a non-radioactive daughter material.

5 a During radioactive decay of a sample, if daughter nuclei are stable what happens to the population of
i parent nuclei and **ii** daughter nuclei?
b What is the relationship between the rate of change of the parent nuclei population and the rate of change of the population of daughter nuclei?
6 What is the relationship between activity of a sample and the rate of change of the population of daughter nuclei?
7 If the population of the parent material is N_p and the population of the daughter material is N_d, what can you say about $N_p + N_d$ at all times?

Remember that the population is changing because of radioactive decays. The total activity of a sample of material is the total rate at which decay events take place within it. Each decay of a nucleus reduces the population of the parent nuclei by one. So the rate of change of the population is the same size as the activity of the sample:

activity of sample $= -$ rate of change of population

A minus sign is now needed, because although the rate of change of population is negative, the activity of the sample is always positive.

activity of sample at any instant $= -\dfrac{dN}{dt}$

A relationship between population and activity

The decay of nuclei in a sample is a random process. But if nuclei are identical then they all have the same level of instability. They all have the same *probability* of decaying in a fixed amount of time. There are so many nuclei that we can expect them to follow the rules of probability (just as if you tossed a coin two or three times you might see 'heads' each time but if you tossed it 10^{18} times you'd expect very nearly equal numbers of heads and tails).

So the more nuclei we have, the more decays we should expect in each second. If we double the population of nuclei, for example, we'd expect the activity of the sample to double. In fact, we'd always expect activity to change by the same proportion as the size of the population. Activity is proportional to population. We can write this in abbreviated form:

$$\text{activity} \propto N$$

We know that proportionality means that a graph of the quantities concerned will be a straight line passing through the origin. A graph of activity against population is therefore a straight line, and it passes through the origin (Figure 9.3). It has a constant gradient which we will write as λ.

$$\text{activity} = \lambda N$$

λ is called the **decay constant**.

Figure 9.3
Activity is proportional to population.

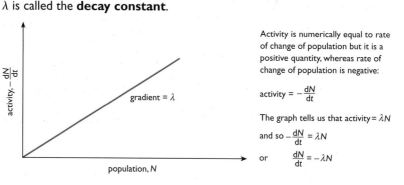

Activity is numerically equal to rate of change of population but it is a positive quantity, whereas rate of change of population is negative:

$$\text{activity} = -\frac{dN}{dt}$$

The graph tells us that activity $= \lambda N$

and so $-\dfrac{dN}{dt} = \lambda N$

or $\dfrac{dN}{dt} = -\lambda N$

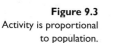

8 If one material has a larger value of λ than another, what differences will you be able to observe in their behaviour?

We will return to the relationship between N and dN/dt in Chapter 28. For the present it is worth noting that the relationship is a proportionality.

Nuclear structure

The proton

Hydrogen atoms are the lightest of all atoms. They are also comparatively easy to ionise completely (to remove all of the negative charge and leave just the nucleus). The charge of a hydrogen atom consists of just one unit of negative charge, carried by its electron, and one unit of positive charge, carried by its nucleus. It loses all of its negative charge when it loses just one electron, and in doing so becomes a positive hydrogen ion. There are no atoms smaller than a hydrogen atom, no atoms with a fewer number of electrons, and no simpler nuclei. The hydrogen nucleus carries a fundamental unit of positive charge, as does a **proton** (Figure 9.4). The hydrogen nucleus *is* a proton.

Figure 9.4
Representations of a proton.

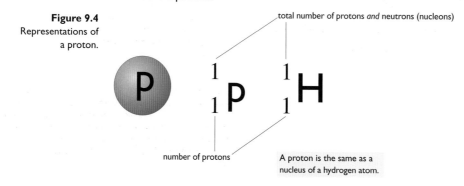

Protons are not too hard to detect. They can be accelerated by electric fields, their pathways can be deflected by magnetic fields, and when they collide with materials they cause strong ionisation effects (Figure 9.5).

Figure 9.5
Fast-moving protons are strongly ionising; like alpha particles, they leave plenty of evidence of themselves. These are tracks created by protons in a bubble chamber. As in a cloud chamber, ionisation stimulates a local change of state – but in the bubble chamber the change of state is from liquid to gas, forming a trail of tiny bubbles in the path of the ionising particle.

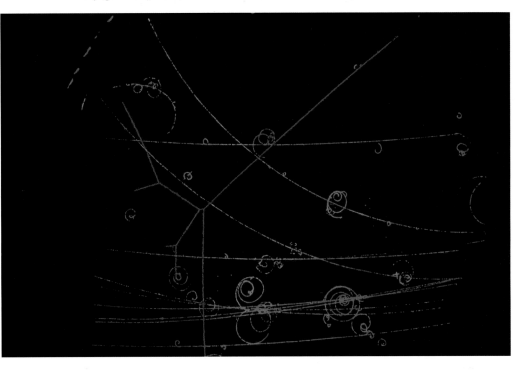

The neutron

All atomic nuclei except hydrogen contain **neutrons** as well as protons. Neutrons carry no electric charge. They do not exert electric force on other particles and so they have little ionising effect. A travelling neutron leaves little direct evidence in its path. Also, neutrons cannot be accelerated by electric or magnetic fields. So they were not detected until 1932, more than 20 years after Rutherford had suggested that alpha-scattering experiments were evidence for the existence of nuclei in atoms. The eventual discovery of neutrons came from the 'knock-on' effect they can have (when they travel fast enough) on protons.

James Chadwick bombarded beryllium nuclei with alpha particles. This produced a radiation that was non-ionising itself, and could only be detected because it could shunt high-energy protons out of a thin layer of paraffin wax (Figure 9.6). The protons had ionising ability and could be detected by Chadwick's detector. The unknown radiation seemed to be a flow of particles with no charge but with mass similar to that of the protons; they were named neutrons.

Figure 9.6
The principle of Chadwick's experiment which led to the discovery of the neutron.

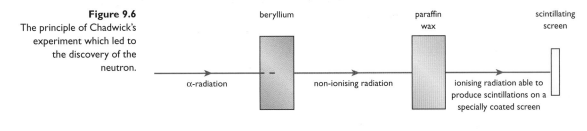

beryllium paraffin wax scintillating screen

α-radiation · non-ionising radiation · ionising radiation able to produce scintillations on a specially coated screen

The ionising radiation that was detected directly consisted of high-energy protons, liberated from the paraffin wax by the bombardment of non-ionising radiation. From consideration of energy and momentum conservation, Chadwick suggested that the non-ionising radiation was a stream of fast-moving particles with no charge but mass very similar to that of protons.

9 Why do fast-moving neutrons cause very much less ionisation than do fast-moving protons?

Vehicles and nucleons

Some people travel in cars, some travel in vans. Cars and vans are vehicles. The word 'vehicle' is less specific but often more convenient than 'car' or 'van'. Likewise it is often more convenient to use the word **nucleons** to describe protons or neutrons or both. A vehicle might be a car or a van. A nucleon is a proton or a neutron.

Isotopes

Every atom of every element is neutral when it contains equal numbers of electrons and protons. The chemical behaviour of an atom is decided by how its electron structure changes – how it loses and regains its neutrality, which depends on how many protons it has. The **proton number**, Z, of an element therefore determines its chemical behaviour.

An atom with eight protons in its nucleus will behave in a certain way. An atom with nine protons behaves very differently. Proton number is so fundamental in determining the chemical identity of an element that it is also called simply its **atomic number**.

Each element has several possible **isotopes** all with the same chemical behaviour, and all with the same number of protons in their nuclei. The difference between isotopes of the same element is in the number of neutrons. All isotopes of an element have the same number of protons but different total numbers of protons and neutrons. They have different total **nucleon number**, A. For most elements, some isotopes are stable while others are radioactive.

There is a standard way of representing nuclear structure, shown in Figure 9.7.

Figure 9.7
Two isotopes of helium, with different nuclear structures.

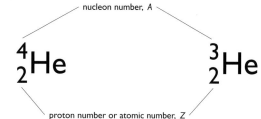

nucleon number, A

proton number or atomic number, Z

10 **a** What are the nucleon and proton numbers of these isotopes of uranium?

$$^{235}_{92}U \quad ^{238}_{92}U \quad ^{239}_{92}U$$

b What is the number of neutrons in each of the isotopes of uranium?

11 Table 9.1 lists details for the isotopes of oxygen. Use a data book to make a similar table for carbon.

12 Add a column to your table showing the number of neutrons in the nuclei of each isotope of carbon.

Table 9.1 lists the isotopes of oxygen and indicates the type of radiation they emit if they are unstable.

Table 9.1
Isotopes of oxygen.

Isotope	Proton number, Z	Nucleon number, A	If unstable, type of radiation emitted	Standard notation
oxygen-13	8	13	beta$^+$ *	$^{13}_{8}O$
oxygen-14	8	14	beta$^+$ and gamma	$^{14}_{8}O$
oxygen-15	8	15	beta$^+$	$^{15}_{8}O$
oxygen-16	8	16	stable	$^{16}_{8}O$
oxygen-17	8	17	stable	$^{17}_{8}O$
oxygen-18	8	18	stable	$^{18}_{8}O$
oxygen-19	8	19	beta$^-$ and gamma	$^{19}_{8}O$
oxygen-20	8	20	beta$^-$ and gamma	$^{20}_{8}O$

*For more information on beta$^+$ radiation see Chapter 28.

Comparing the masses of atoms

All ions (and other charged particles) can be accelerated by electric fields and deflected by magnetic fields. The sizes of the acceleration and deflection depend on the mass and charge of the ions. A **mass spectrometer**, Figure 9.8, uses this to compare and measure the masses.

Figure 9.8
The principle of a mass spectrometer.

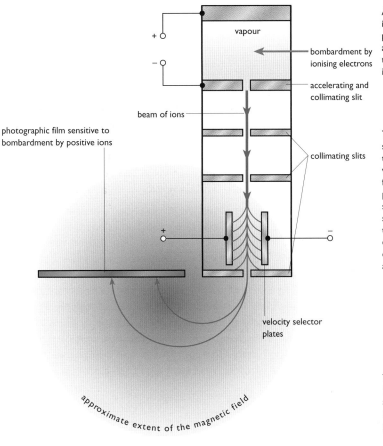

A beam of energetic electrons ionises atoms of a vapour. The positive ions are accelerated by an electric field. In the process they are collimated – focused into a thin beam.

There is some spread in the speeds of the ions. They all pass through a velocity selector, in which electric and magnetic fields tend to deflect their paths in opposite directions, so that only ions with a certain speed emerge and pass through a small slit to the deflection chamber where they experience the magnetic field alone.

The ions hit a photographic plate, from which the radii of their paths can be measured. The mass of the ions can be calculated using the values of the applied electric and magnetic fields.

The magnetic field deflects the ions – now all with the same speed – in circular paths. The radius of their path is proportional to their mass.

The identification of alpha particles

By 1908, 12 years after Henri Becquerel had first noticed the effects of ionising radiation, it was well known that the three principal types of emission had different properties of ionisation and penetration and of behaviour in electric and magnetic fields. Ernest Rutherford and Hans Geiger had found that the charge on an alpha particle was twice as big as the charge on an electron, and of the opposite sign. They also knew the mass, which was very similar to the mass of a helium atom. To check the relationship between alpha radiation and helium, Rutherford and a colleague called Royds then enclosed a source of alpha radiation in a glass tube and, after 6 days, applied a potential difference to electrodes in the tube. They expected that, if helium had collected in the tube, the potential differences would excite the gas and they would be able to detect the characteristic colours (emission spectrum) of excited helium atoms. They saw exactly that.

When Rutherford later discovered the existence of nuclei, it was realised that alpha particles are helium nuclei. This means that each alpha particle is a small cluster of four nucleons – two protons and two neutrons. We can use the standard notation to represent their structure:

$$^4_2\alpha$$

Note that this is identical to the structure of a helium nucleus:

$$^4_2\text{He}$$

Writing nuclear changes

Radioactive decay involves change in the structure of nuclei and emission of radiation. We can use the decay of uranium-238 as an example. Uranium-238 is the commonest isotope of uranium found in rocks in the Earth. It decays by emitting an alpha particle. We can write this as a nuclear change:

$$^{238}_{92}U \rightarrow {}^{234}_{90}Th + {}^{4}_{2}\alpha$$

Uranium-238 is the parent nucleus, and thorium-234 is the daughter nucleus.

Note that the total nucleon number and the total proton number remain the same (238 = 234 + 4 and 92 = 90 + 2). Nucleon number and proton number are conserved. Since these are patterns of behaviour that we see in *all* nuclear changes, we can call them rules or laws: nucleon number obeys a **conservation rule**. Proton number also obeys a conservation rule.

13 a Uranium-234 is another isotope of uranium. It decays by alpha decay to become thorium-230. Write the process as a nuclear change, as in the example of uranium-238.
b Give the nucleon number, proton number and number of neutrons for two isotopes of thorium.

Artificial nuclear changes

Conventional surgery is very messy. It involves sharp instruments, which the surgeon uses to cut into the body to examine, remove or change something that is causing a problem. It is an expensive process, involving a lot of time from people who have been highly trained. It can also be risky to the patient. The use of radioactive isotopes provides a cheaper and safer alternative to many operations.

The penetrating ability of ionising radiation and its detectability allow it to be used for examination of the body's interior without the use of scalpels. Its ionising ability allows it to be used to destroy unhealthy tissue. Hospitals need a supply of radioactive isotopes. Some large hospitals even have their own particle accelerators, or cyclotrons, which produce beams of energetic protons to bombard target material and change the structure of its nuclei. The changes provide the required isotopes. An example of such a process is:

$$^{1}_{1}p + {}^{27}_{13}Al \rightarrow {}^{27}_{14}Si + {}^{1}_{0}n$$

bombarding proton — target nucleus — emitted neutron

An alternative process can produce other isotopes. A target can be bombarded with neutrons rather than protons. The neutrons can be produced indirectly by accelerating ions of deuterium ($^{2}_{1}H$), an isotope of hydrogen, into a beryllium ($^{9}_{4}Be$) target. The energetic neutrons can then in turn bombard a secondary target, such as a sample of sodium. They enter the sodium nuclei, creating nuclei of a new sodium isotope:

$$^{2}_{1}H + {}^{9}_{4}Be \rightarrow {}^{1}_{0}n + {}^{10}_{5}B$$

$$^{1}_{0}n + {}^{23}_{11}Na \rightarrow {}^{24}_{11}Na + gamma$$

The new sodium nucleus is unstable. It is a nucleus of a radioactive isotope or **radioisotope**, which emits beta and gamma radiation.

Alternatively, a nuclear reactor can provide the source of neutrons. A sample of ordinary sodium left inside the shielding of a reactor will absorb neutrons as they emerge from the reactor.

14 What happens to the total number of nucleons in each of the processes in this section?

Counting the costs of the Chernobyl disaster

In 1986 the reactor of the Chernobyl nuclear power station in the Ukraine was allowed to overheat and it exploded. It was a chemical explosion involving, chiefly, the reaction of hydrogen with oxygen. But the reactor contained nuclear fuel and radioactive waste products. About 6 tonnes of this radioactive material was released into the atmosphere and carried across millions of square kilometres of Europe (Figure 9.9). The material had a total activity of 2.6×10^{18} Bq.

In Sweden, workers at a nuclear power station thought they had a disaster themselves when the count of radioactive events on their instruments started to go up. In the hills of Britain, sheep continued to eat the grass. In Saudi Arabia, a shipload of British mutton was declared unfit to eat. The grass had become radioactive, showered with dust from that distant explosion. How could such a modest amount of material, spread over such a vast area of land, have such an impact? It is worth looking at the numbers – see the questions alongside.

15 What would be the activity of
  a 1 tonne
  b 1 kg
  of the radioactive material? (1 tonne = 1000 kg)

16 Estimate, from the map, the total area of the radioactive cloud on 3rd May 1986.

17 It is a great simplification, but suppose that the material was evenly spread over this area.
  a What is the mass of radioactive material, in kilograms, above each square kilometre of land?
  b What is the activity of this material, in becquerel, assuming that the total activity of the released material did not change in the course of its week's travel through the air? (This assumption is another simplification, but it nevertheless gives a meaningful indication of what happened.)
  c Make estimates of the mass and activity of material deposited on a square metre of North Wales hill pasture.
  d Would you expect this to have a significant effect on the activity of the body of a grazing sheep? Explain your answer.

18 The radioactive material contained some very unstable nuclei. Why would the activity of the material actually decrease significantly in the course of the week?

19 Describe the differences between chemical changes to material and nuclear changes.

Figure 9.9
The estimated extent of the radioactive cloud on 3rd May 1986, a week after the explosion at Chernobyl. On that day it rained in western Britain, and the rain brought radioactive material down with it.

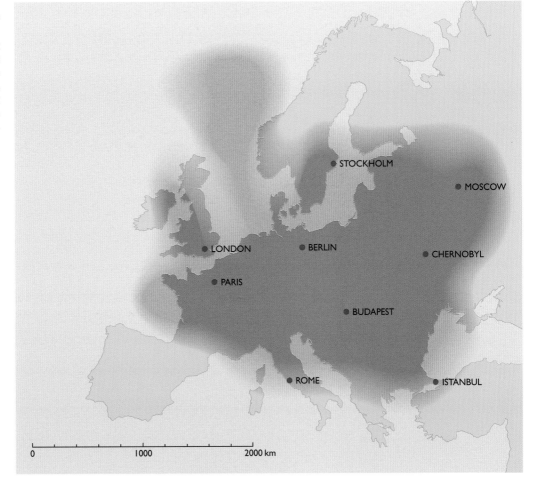

● **Extra skills task**   Communication and Information Technology

We are exposed to risks every day. Decisions are regularly made by individuals and by organisations as to what poses an acceptable risk, and which risks are unacceptable. These decisions are complex and will vary according to priorities. For example, for many people, the everyday convenience of car travel outweighs the relatively high risk of being involved in an accident. For some of these people, air travel is considered unacceptably risky due to the high profile of air accidents, even though, in statistical terms, it is far safer than car travel. The Chernobyl disaster concentrated many minds on the safety, perceived and actual, of nuclear power.

Research the subject of risk thoroughly from different sources, including the Internet, to obtain accurate facts and figures. Contexts to consider might include risk associated with nuclear power stations, air and road travel, asteroid collisions, eating shellfish or pre-cooked foods, smoking and household activities such as mowing grass.

Consider whether you regard the various risks as acceptable or not. Take part in a group discussion. You should make clear and relevant contributions as well as listening and responding to other students.

III
MOTION AND FORCE

10 Variables of motion

THE BIG QUESTION ● What techniques can we use to analyse and predict motion?

KEY VOCABULARY approximation   average speed   calculus   displacement   friction   idealisation   infinitesimal   instantaneous speed   resistive medium   scalar quantity   speed   vector quantity   vector representation   velocity

BACKGROUND Calculations provide extremely reliable predictions of motion such as a ride at an adventure park or even the journey of a space probe to Saturn. This reliability has been impressing people for several centuries, and has seemed to show that physics is powerful in use and meaningful in what it says about the world.

Figure 10.1
Physics provides the ability to predict motion.

● Motion in an ideal world

Imagine an ideal world that does not have the complications that you find in our real world. Everything in an ideal world is reduced to its simplest level. All motion is like motion in deep space – frictionless motion in a straight line. In an ideal world a marble rolls and rolls without resistance; a car drifts effortlessly along the motorway with no need for continuous work by its engine. Physics is indeed simpler in an ideal world.

● Motion in the real world

Motion through air and water is possible, but it involves continuously pushing material out of the way. The motion meets resistance. Air and water are **resistive media**.

Motion of one surface across another also meets resistance. Real surfaces are not perfectly smooth; the surfaces interact and resist the motion. Such **friction** between surfaces is common in the motion of our Earth-bound experience.

In the roll of a real marble across a real table there is resistance to the motion. There is friction between the surfaces of the marble and the table. There is air to be displaced. The marble slows down. However, the rolling motion means that friction between surfaces is quite low. And if the marble's speed is small then air resistance doesn't play a very significant part in affecting its speed. The marble behaves, approximately, like a marble travelling in space. It can reach the end of the table before it has slowed very much.

Motion in an ideal world is so much easier to deal with than motion that involves resistance, that often it is worthwhile to treat real-world motion as ideal-world, or resistance-free, motion. We simplify or **idealise** the situation (Figure 10.2). This gives answers that are **approximately** true. Then we have to make a decision whether an approximation is good enough at describing and predicting real-world behaviour. Depending on the situation, it often is.

Figure 10.2
For a skater, there is little resistance to motion; the motion can be quite reliably idealised.

I Which of the following approximate most and least closely to ideal motion:
a a submarine moving under water
b a satellite in orbit around the Earth
c a skier?

Distance and displacement – a scalar and a vector

We measure distance along lines. The lines can be straight or curved. We can measure the distance straight across a circle – its diameter – and we can measure the distance around the circle – its circumference.

Displacement is not the same thing as distance. Displacement measures a change in position in a specified direction.

Figure 10.3
Distance travelled and displacement are not the same.

When a skater has travelled part of the way around a circle the groove in the ice shows the distance he or she has covered. But the skater's displacement provides a simple comparison of the new position and the initial position – as a straight line between the two points, shown by drawing an arrow (Figure 10.3). This arrow is called a **vector representation** of the displacement.

Distance, however, is a **scalar quantity**; it has magnitude (size) but no special direction. We do not use arrows to represent scalar quantities. Displacement is a **vector quantity**; it has direction as well as magnitude. Though one is a scalar quantity and one is a vector quantity, the unit of both distance and displacement is the metre.

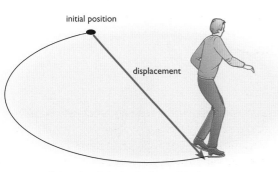

initial position

displacement

distance travelled = length of curved track

● Speed

Speed is rate of change of distance. This can be written in the simplest possible terms as:

$$\text{speed} = \frac{\text{change in distance travelled}}{\text{the time for the change}}$$

Speed, like distance, is a scalar quantity. Its unit is the metre per second, m s^{-1}.

Average and instantaneous speed

If a car journey of 180 km takes 3 hours then we can substitute these values into the simple speed formula, to give speed $= x/t = 180/3 = 60 \text{ km h}^{-1}$. This is the **average speed** for the journey. But the reading on the car speedometer may rarely show 60 km h^{-1}. It varies instant by instant; it shows the **instantaneous speed**.

Average and instantaneous speed can be the same, but very often they are not. If you cycle to the local shop then your speed will vary throughout the journey. You will have many different instantaneous speeds, which could be zero when you stop at a junction or perhaps as much as, say, 20 m s^{-1}. You will have only one average speed for the complete journey, which is the total distance, x, divided by the total time, t:

$$\text{average speed for a complete journey} = \frac{x}{t}$$

For a portion of the journey for which the *change* in distance is Δx and the *change* in time is Δt, we can say:

$$\text{average speed for a portion of a journey} = \frac{\Delta x}{\Delta t}$$

> **2** If, on a journey, a car starts and finishes in towns, but travels some of the way in between on a motorway, what would be typical maximum and minimum instantaneous speeds if the average speed were 60 km h^{-1}?
>
> **3 a** What is the average speed for a 135 km journey that takes 2 hours? Give your answer in m s^{-1}. (1 km = 1000 m; 1 hour = 3600 s)
> **b** Suggest whether the journey might have been made by bicycle.

● Velocity and speed

The velocity of a body has a direction. In a ball game, such as tennis, the direction is all-important. The velocity of a tennis ball can be represented by an arrow which can show its size (drawn to scale) and its direction. Velocity is an example of a vector quantity.

Velocity is defined as rate of change of displacement of a body. Using s for displacement,

$$\text{average velocity for a journey or part of a journey} = \frac{\Delta s}{\Delta t}$$

where Δs and Δt are corresponding changes in displacement and time.

Velocity and speed have the same unit, the metre per second or m s^{-1}. The difference between them is that velocity is a vector and speed is a scalar. Once again, the skater can illustrate the difference. A skater can move gently round and round the ice-rink at a constant speed. But the velocity has to change because the direction has to change (Figure 10.4).

For simple straight line motion the distinction between velocity and speed is less important.

> **4** The skater in Figure 10.4 has constant speed but changing velocity. Is it possible to have constant velocity but changing speed?
> **5** Write down the definitions of speed and velocity and highlight the difference.
> **6** Does a satellite in circular orbit around the Earth have constant speed, constant velocity, both or neither?

Figure 10.4
A skater can have steady speed, but if direction changes then velocity changes.

Motion pictures

We like our information to be visual. We can translate quantitative information such as tables of data into graphs that give us an instant sense of the information. Graphs have stories to tell.

Graphs can represent motion. A graph can show, for example, how displacement varies with time (Figure 10.5a). Time, as usual, is treated as the input or independent variable and measured along the *x*-axis. Displacement is the output or dependent variable, shown on the *y*-axis.

Figure 10.5
Displacement–time graphs.

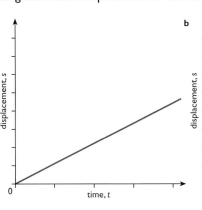
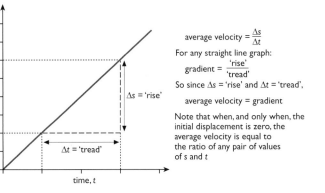

$$\text{average velocity} = \frac{\Delta s}{\Delta t}$$

For any straight line graph:
$$\text{gradient} = \frac{\text{'rise'}}{\text{'tread'}}$$
So since $\Delta s = $ 'rise' and $\Delta t = $ 'tread',

average velocity = gradient

Note that when, and only when, the initial displacement is zero, the average velocity is equal to the ratio of any pair of values of *s* and *t*

A graph tells a story. This graph provides information about a particular journey. A body has no displacement at time 0 (zero). From time 0, displacement increases by equal amounts in equal periods of time.

A different graph tells a different story. For this journey, initial displacement is again zero, but displacement increases more rapidly. This is a journey with a higher velocity.

> **7** On the same suitably numbered axes, sketch displacement–time graphs for bodies moving at 2 m s^{-1} and 5 m s^{-1}.
> **8** A particular displacement–time graph represents 1 s of time by 10 mm and 1 m of displacement by 5 mm.
> **a** Sketch the axes.
> **b** If corresponding values of tread (time) and rise (displacement) are shown at 20 mm and 25 mm respectively, what are the corresponding values of time and displacement?
> **c** If the graph is a straight line and the displacement is zero at time zero, what is the velocity?

A straight line graph has a constant gradient. We can calculate the gradient from the 'rise' and 'tread' of any right-angled triangle drawn with the line of the graph as its hypotenuse (Figure 10.5b). The units of 'rise' are the units of the *y*-axis. The units of 'tread' are the units of the *x*-axis.

A body moving faster produces a steeper displacement–time graph. In fact the value of the gradient is equal to the magnitude of the velocity. For a straight line graph, the velocity is constant; the average velocity is the same as the instantaneous velocity at all times.

Velocity in the lab

An air track provides an excellent approximation to resistance-free motion. Jets of air hold the vehicle away from the track, so that friction is all but absent. Passage of the vehicle along the track can be detected and recorded using light beams that, when interrupted, trigger the start and stop of computer timing devices (Figure 10.6). This arrangement provides information on the change in displacement of the vehicle during a recorded interval of time. That is, it provides values of Δs and Δt for the portion of the journey between the two triggering devices. It allows us to calculate average velocity for that part of the journey. However, it is not capable of giving a continuous record of instantaneous speeds throughout the vehicle's motion. The best we can do is use a series of light beams and timers, so that we can work out average speeds for each short portion of the journey of the vehicle along the track.

Figure 10.6
Timing gates can record the time taken, and the rule on the track shows the distance. From these measurements, average velocity for the journey between the gates can be calculated.

Average and instantaneous values on displacement–time graphs

We could use an air track system to plot a graph of displacement against time. (Note that since motion on the air track is in a straight line, displacement and distance are numerically equal, and so speed and velocity are also numerically equal.) We would know the time for the journey between the two detectors and we would know the total change in displacement. Suppose that for a change in displacement (distance between the detectors) of 1.2 m the time taken is 0.8 s. The average velocity is given by:

$$\frac{\Delta s}{\Delta t} = \frac{1.2}{0.8} = 1.5 \text{ m s}^{-1}$$

We have enough information to plot a graph with two points, Figure 10.7.

Figure 10.7
We can plot points (0, 0) and (0.8, 1.2).

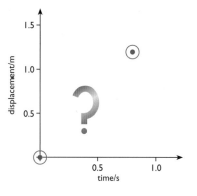

The detecting apparatus tells us nothing about the instantaneous speeds of the vehicle between the two times. We could assume that the vehicle had a steady velocity, covering equal distances in each unit of time. That assumption would let us join the points on the graph, Figure 10.8a. Alternatively, we might guess that the vehicle had slowed down. That would mean that in the earlier stages it covered bigger distances in each unit of time. This would produce a different kind of graph, Figure 10.8b.

Figure 10.8
Different assumptions about the instantaneous velocities between the two known points produce different shapes of graph.

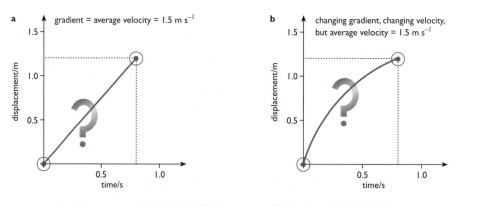

The line between the two points has a gradient of 1.5 m s^{-1}. This represents the average velocity. We do not know anything about instantaneous speeds between the two points.

This graph is based on the assumption that the vehicle slows down, so it has a gradient that decreases. The average velocity is still 1.5 m s^{-1} for the 1.2 m journey.

In Figure 10.8b the velocity is not constant so the gradient is not constant. However, the only truly reliable information we have is of the average velocity of the vehicle. We can only make assumptions about instantaneous velocities.

Using more light beams and timing devices along the length of the track we could collect more information, and plot a graph with several points, Figure 10.9.

Figure 10.9
Now we can plot more points and can see the general shape of the curve much more reliably.

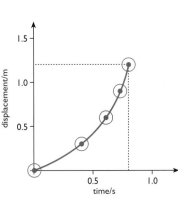

By increasing the number of positions (displacements) at which we measure time of travel, we obtain more complete information about the journey. The average velocity is still 1.5 m s^{-1} for the whole 1.2 m journey, but now we can see that instantaneous velocities are increasing. (We are still, however, making assumptions that velocity changes smoothly between our measurements.)

This takes us rather closer to knowing the instantaneous velocities of the vehicle. We can measure the velocity at a particular time by looking at the graph at that time and then drawing a tangent (Figure 10.10a). The gradient of the curve at a particular point is equal to the gradient of the tangent to the curve at that point, calculated by measuring rise and tread, Δs and Δt, as before.

Figure 10.10
We can find the gradient of the curve at any point to find the value of the instantaneous velocity at that time.

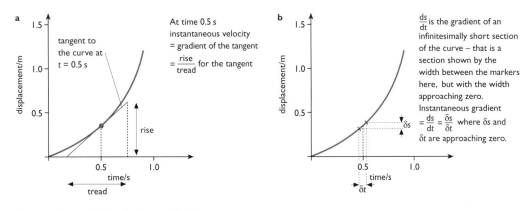

A tangent is a straight line that touches, but does not intersect, a curve at a chosen point.

The single point on the curve represents an instant of time and the matching change in displacement. An 'instant' is an **infinitesimal** time – a time so short that it is approaching zero in duration.

A *small* length of time Δt can be written as δt. A corresponding small displacement is δs. The velocity of a vehicle averaged over time δt is then $\delta s/\delta t$. If δs and δt are so small that they are *infinitesimal* then this ratio is written as ds/dt (Figure 10.10b). This is the *instantaneous* velocity – the velocity at a particular time or instant.

$$\text{average velocity for a portion of the journey} = \frac{\Delta s}{\Delta t}$$

$$\text{average velocity over a very short time} = \frac{\delta s}{\delta t}$$

$$\text{instantaneous velocity} = \frac{ds}{dt} \text{ which is the value of } \frac{\delta s}{\delta t} \text{ when } \delta s \text{ and } \delta t \text{ are infinitesimal}$$

$$= \text{gradient of the tangent that touches the curve at the chosen instant}$$

The mathematics of infinitesimal quantities allows us to deal with change in a way that the ordinary maths of addition, subtraction, multiplication and division cannot. This maths was invented more than 300 years ago, and the development of quantitative science would not have been possible without it. It is called **calculus**.

Figure 10.11 summarises this section on average and instantaneous velocities.

Figure 10.11
In a 100 m race, an athlete has one average velocity and an infinite number of instantaneous velocities. The average velocity is the gradient of the straight line graph. We can work out any instantaneous velocity by measuring the gradient at one point of the curve. To measure this we draw a tangent at the chosen point.

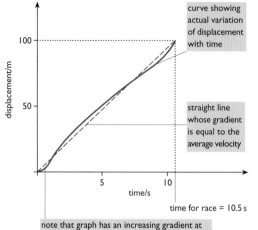

curve showing actual variation of displacement with time

straight line whose gradient is equal to the average velocity

time for race = 10.5 s

note that graph has an increasing gradient at the start – instantaneous velocity is increasing as the sprinter leaves the starting blocks

9 What does *infinitesimal* mean?
10 How can you measure the gradient at a chosen point on a curve?
11 Sketch a displacement–time graph for a straight line journey from your sofa to your TV. Use the graph to illustrate the difference between average and instantaneous velocities, as in Figure 10.11.

● **Extra skills task** Application of Number and Information Technology

Use a datalogging computer system to collect displacement–time information during a motion experiment. You could use a falling object or a trolley which accelerates along a linear air track.

Now use the associated software or a spreadsheet package to analyse the data and produce a displacement–time graph. The gradient of this graph can be used to obtain a value for the instantaneous velocity at any time.

How reliable is this value as an indication of *true* instantaneous velocity?

 11 Newton's force

THE BIG QUESTION	● What is force?
KEY VOCABULARY	acceleration balanced forces deceleration newton Newton's First Law Newton's Third Law reaction force unbalanced force uniform acceleration
BACKGROUND	About 600 years ago Europeans learnt about explosives, a technology that had been used for bombs and grenades in China for some time. Powerful landowners discovered that their castles offered little protection against cannonballs. Warfare changed, and powerful people and their communities had to find ways to do deals with others instead of simply retreating to their individual castle strongholds. Commerce grew.

Those with power became interested in the study of motion of these cannonballs (Figure 11.1). The science of their motion, called ballistics, developed. It used mathematics to provide reliable predictions of the behaviour of projectiles such as cannonballs and shot. The mathematics of military technology, of commerce and of science grew together.

Figure 11.1
Before Galileo, Newton and others had developed the concept of force as the cause of change in motion (acceleration), prediction of the motion of projectiles was unreliable. This illustration of a cannonball's trajectory is by Daniel Santbech, 1561.

● Force before Newton

Some ancient Greek writings, of 2000 and more years ago, held the idea that the natural state of a body was rest. When Europe then fell into chaos and almost universal poverty after the decay of the Roman Empire, around AD400, ideas such as this survived amongst Arabic scholars. Eventually, the Catholic Church in Europe learnt about these ideas and incorporated them into their theology. To challenge the old ideas was to challenge the authority of the Church. Since the Church was not above torture and execution, challenging its authority was not a good idea.

Another key idea from the ancient books was that the Earth was the centre of the Universe. But, in the 16th century Copernicus suggested that there was a different way of explaining how the Sun, the Moon, the planets and the stars appear to move across the sky. He put the Sun at the centre of a system of planets, with the stars outside this system. Wisely, perhaps, Copernicus waited until he was already dying before he published his challenging new idea of the revolutions of the Earth and planets.

Galileo Galilei, who lived from 1564 to 1642, combined a concern for his own safety with a strong belief in new ideas that might have seemed arrogant to people around him. With the new observations that were possible using the telescope that he made for himself, he argued in favour of Copernicus' Sun-centred system. He also had the audacity to challenge other key philosophies. These included ideas about motion, as described in the following paragraphs. Caught up in the politics of his times, he suffered for this, and died under house-arrest.

The ancient books claimed that the natural state of a body was rest and that the natural order was for bodies to resist motion. People noted that moving bodies – from cannonballs to carthorses – always returned to rest. But Galileo watched the ceaseless swing of a pendulum. He thought about the motion of planets and moons through space. He concluded that it was not rest that was normal, but steady motion. 'Rest' was just a particular example of steady motion – steady motion that had a particular value, which was zero.

Galileo's view that steady motion was the natural condition of bodies defied the common sense that was based on cannonballs and carthorses, but not the common sense based on pendulums and planets. While the ancient books said that velocity was unusual and rest was not, Galileo claimed that *acceleration* was unusual and constant velocity was not. It was a big step to take, and it was a whole generation later when Newton took up the idea and wrote it down in formal language:

All bodies remain at rest or in constant straight-line motion unless an unbalanced external force acts on them.

This also tied down the concept of force. **Unbalanced force** is what causes motion to change – it is what causes acceleration (see the next section). This concept is the foundation-stone of Newtonian physics and is called **Newton's First Law**.

A system of **balanced forces** cancel each other out so that they have no net effect.

1 A cannonball is fired along a barrel and flies through the air until it hits a castle wall. Describe the forces acting on it at the various stages of its motion.
2 Explain why the illustration at the start of this chapter gives a false image of the trajectory of a cannonball.
3 From the text of this section, what was Galileo's main contribution to the development of ideas about motion?
4 A bowling ball rolls along the ground and stops. Common sense might lead to the conclusion that the natural state of bowling balls is a state of rest. This would mean that they do not obey Newton's First Law.
   a How would a bowling ball behave if similarly released in space?
   b What causes the illusion that a bowling ball disobeys Newton's First Law?

Acceleration

Motion at constant velocity is the natural state of a body, and can only be changed by the action of an external force, claimed Galileo and Newton. Indeed, we now know that a spacecraft travels at constant velocity – that is, with constant speed and constant direction – unless it experiences a force by firing its rocket motors, by the gravitational pull of a star, planet or moon, or by collision. Any body that is experiencing change in its velocity – in magnitude, direction or both – is said to be accelerating.

Acceleration is defined as the rate of change of velocity. This can be written in its simplest form as:

$$\text{acceleration} = \frac{\text{change in velocity}}{\text{corresponding time change}}$$

Change in velocity is measured in metres per second, m s^{-1}, and change in time is measured in seconds, s. So the unit of acceleration is m s^{-2}. The quantities that we use to describe motion are summarised in Table 11.1.

Table 11.1

Quantity	Unit	Dimension
time	s	[T]
displacement	m	[L]
velocity	m s^{-1}	[L][T]$^{-1}$
acceleration	m s^{-2}	[L][T]$^{-2}$

Acceleration, like velocity, is a vector quantity.

Instantaneous, average and uniform acceleration

Note that we can apply the word acceleration to any motion that involves change in velocity. That includes motion with decreasing velocity as well as motion with increasing velocity. (It also includes motion for which direction is changing, but in this chapter we are interested in straight line motion.) A change in velocity is a difference between two velocities – the final and the initial velocity.

To know the acceleration at *one* particular instant, the final and initial velocities must be the velocities at the start and the finish of that instant. These two velocities are separated by an infinitesimal time. The change in velocity must then also be infinitesimal:

acceleration at an instant = instantaneous acceleration

$$\approx \frac{\delta v}{\delta t} \text{ where } \delta v \text{ and } \delta t \text{ are very small changes in velocity and time}$$

$$= \frac{dv}{dt} \text{ when } \delta v \text{ and } \delta t \text{ are infinitesimal}$$

(Compare this with ideas about instantaneous velocity on page 87.)

We can also calculate the *average* acceleration over an extended period of time, Δt:

$$\text{average acceleration over time } \Delta t = \frac{\Delta v}{\Delta t} = \frac{\text{final velocity} - \text{initial velocity}}{\Delta t}$$

Note that Δ to indicate change in general, δ for small changes and calculus notation using d for infinitesimal changes are applied in the same way throughout this book.

It is common to use v and u for final and initial velocities, respectively. (Initial velocity happens before final velocity, and u comes before v in the alphabet.) We can write the total time for the velocity change as t. Then we can say:

$$\text{average acceleration} = \frac{v - u}{t}$$

Note that if v is bigger than u then velocity has increased and the change in velocity is positive. But if velocity has decreased then v is smaller than u and the change in velocity is negative. In this case the acceleration must be negative. Negative acceleration can be called **deceleration**.

Acceleration may be **uniform** (not changing). Ignoring the effects of air resistance, this is true for a falling object. The average acceleration for a complete fall is then the same as the acceleration at all times during the fall. That is, for uniform acceleration, average and instantaneous values are the same:

$$a = \frac{dv}{dt} = \frac{\Delta v}{\Delta t} = \frac{v - u}{t} \text{ for uniform acceleration}$$

5 a What is the difference between average acceleration and instantaneous acceleration?
b Under what circumstances are they the same?

6 a A body accelerates uniformly from rest ($u = 0$) to 10 m s^{-1} in 4 s. What is its acceleration?
b What is its acceleration in the next 4 s, if velocity falls uniformly from 10 m s^{-1} to 2 m s^{-1}?

Force and acceleration

From Newton's First Law, which includes the idea that a body does not accelerate unless an unbalanced force acts on it, we can go on to say that:

A body *always* accelerates when an unbalanced force acts on it.

If force and acceleration are so strongly linked then it is important to have a mathematical relationship between them. Newton (Figure 11.2) was certainly aware of the usefulness of a mathematical relationship. The relationship that we now use is:

$$F = ma$$

F is force, m is mass (a scalar quantity) and a is acceleration. Force and acceleration are quantities with direction – they are vector quantities. One **newton** (1 N) is the force that gives an acceleration of 1 m s^{-2} to a mass of 1 kg.

The relationship $F = ma$ is derived from Newton's Second Law (see Chapter 14, page 120).

7 In the formula $F = ma$, which quantities are vectors and which are scalars?

8 **a** The relationship $F = ma$ can be considered to be a proportionality relating force to acceleration provided that the mass of the body is constant. Is mass constant for:

i a rocket being launched and ejecting burning fuel very rapidly from its rocket motors

ii an arrow propelled from a bow?

b i Sketch the shape of the graph showing the relationship between force and acceleration for the arrow.

ii What would the graph look like for a more massive arrow?

9 Acceleration has dimensions $[L][T]^{-2}$, and mass has dimension $[M]$. What are the dimensions of force?

10 What force is needed to give a mass of 80 kg an acceleration of 3 m s^{-2}?

11 What acceleration will be experienced by a body of mass 4.5 kg when a force of 27 N acts on it?

Figure 11.2
Isaac Newton established the link between force and acceleration.

Isaac Newton is not remembered for common sense. He forgot to eat meals. He wrote rude letters to his friends. Out of curiosity, he stuck his finger between his eyeball and the surrounding bone (and was lucky not to blind himself). Perhaps it was his lack of common sense that helped him to work with new ideas, and to establish the rules that we can still use today to predict motion very precisely. Newton discovered the absolute link between force and acceleration, as described in the equation $F = ma$, which can still sometimes seem to defy 'common sense'.

● Internal and external force

Figure 11.3
External force causes acceleration.

The tennis ball and the planet are accelerating, due to the unbalanced external forces that are acting on them. But note that the forces acting on the ball and the planet are not the only forces acting in each system. The racquet exerts a force on the ball, but the ball also exerts a force on the racquet, and these two forces are equal in size. Likewise, the Sun's gravitational pull on the planet is matched exactly by the planet's pull on the Sun. In both cases, it is the body with the smaller mass that experiences the bigger acceleration.

Newton's First Law mentions 'external' force. You can't actually pull yourself up by your bootstraps. That would involve one part of a body exerting a force on another. The force would be internal and would produce no acceleration of your whole body. Only a force that acts between you and the world outside you – an external force – can give acceleration to your whole body.

All acceleration requires external force. Consider some very different accelerating bodies, Figures 11.3 and 11.4.

Newton postulated that:

It is impossible for one body to exert a force on another without experiencing an equal and opposite force, called a **reaction force**.

This is a statement of **Newton's Third Law**.

12 When you start walking, what force accelerates you forwards?

13 **a** Explain why it is unwise to use normal walking style to step off a small unmoored boat on to a jetty.

b Why is there less of a problem if the boat is large? (Remember that $F = ma$.)

Figure 11.4
The reaction from the bale of straw provides the external force needed for the cat to leap.

This cat is accelerating forwards. That means it must be experiencing a forwards force. The cat exerts a backwards force on the bale of straw. The bale exerts an equal and opposite force on the cat. It is this forwards force provided by the bale, the reaction force, that is the external force needed to produce the acceleration. The bale's acceleration is very small because its mass is much bigger than the cat's mass.

Action and reaction

Newton's Third Law predicts that forces exist in pairs. A rocket experiences upwards force as it accelerates from its launch pad; this force exactly matches the downwards force experienced by the burning gases which accelerate downwards from the rocket motor. The two forces can be called the 'action' and 'reaction'; they are equal in size and opposite in direction (Figure 11.5).

Figure 11.5
Action and reaction.

force on rocket

force on gases

Figure 11.6
Experiencing a reaction.

Dance partners exert forces on each other (Figure 11.6). If the woman pulls the man by the arm, the man experiences the force. The woman experiences an equal and opposite force. As stated by Newton's Third Law, it is impossible to exert a force without experiencing an equal and opposite force.

When one dancer pushes or pulls the other, he or she experiences equal and opposite force. However, though the forces have the same size, the accelerations of the dancers will only be equal if their masses are equal. For the same force, the smaller dancer will experience the greater acceleration, as predicted by $a = F/m$. Skilful dancers take advantage of this inequality of accelerations to enhance their dance routines.

14 Two ice-skaters stand still, facing each other. One pushes the other but neither falls over. Explain what happens if
a the skaters have the same mass
b one skater is significantly more massive than the other.

● **Comprehension and application**

Lost in space

The motion of a spacecraft in deep space, far from bodies which would exert gravitational force on it, is the simplest type of motion to analyse. There is no resistance to motion. The only external forces are those generated by the spacecraft's rockets.

Suppose that the spacecraft is carrying people and that the rocket motors have broken down. The astronauts need to change its motion. They can push as hard as they like on the walls inside the spacecraft, but this will not produce acceleration. These forces are internal, since the astronauts are effectively part of the body they are trying to influence. The astronauts might try putting on their spacesuits and going outside the spacecraft. They could hold on to the craft with one hand and push with the other. Again nothing useful would happen – the forces being exerted are still just one part of a body pushing on another part. They are still internal forces.

However, if one astronaut wants to make a brave gesture of self-sacrifice, then he or she can push on the spacecraft and let go. The force ceases to act, of course, once there is no contact. But while there is contact the spacecraft and the astronaut experience equal and opposite forces[a]. They both experience acceleration, that of the astronaut being larger[b]. Once contact is lost, acceleration ceases and both bodies continue to move apart with the constant velocity they acquired[c]. The astronauts remaining with the spacecraft can wave a sad and grateful goodbye to their colleague. [a, b, c – see question 16.]

Perhaps a more satisfactory solution would be to repair the rocket motors. Instead of leaving a trail of stranded astronauts behind, the spacecraft can then fire little bursts of gas in the opposite direction to the required acceleration. The force experienced by this exhaust gas is exactly matched by the force experienced by the spacecraft itself. Equal and opposite forces are at work. The mass of the gas is quite small compared with the mass of the spacecraft, so to compensate for this it has to be accelerated as much as possible. In a rocket, this high acceleration of gas is achieved by a chemical reaction, say between hydrogen and oxygen in the rocket fuel.

Should the astronauts decide to visit the atmosphere-free planet Zorg, then the spacecraft will need to have plenty of fuel. As it eases itself to the surface it will need to use its rockets to exert an upwards force close to its own Zorgian weight, to avoid a crash-landing. On take-off it will need to exert an upwards force that is bigger than its own weight, so that the force not only supports the weight of the craft but also provides some upwards acceleration.

15 a Explain how it is possible for a spacecraft to travel at high speed without using fuel.
b When must it use fuel?

16 Which of Newton's Laws of Motion relates most closely to each of the sentences marked *a*, *b* and *c*?

17 Suppose that the astronauts develop a cunning plan. One of them pushes on the outside of the spacecraft and then lets go, successfully causing the spacecraft to accelerate. But the astronaut is now linked to the spacecraft by a lifeline so as not to be left behind. Why won't this overall process work?

18 Explain why the exhaust gases from the spacecraft's rockets must reach a high speed relative to the spacecraft. (Refer to Newton's Third Law and to $F = ma$ to explain this.)

19 a On approaching the surface of the planet Zorg the spacecraft uses its motors to achieve a soft landing. During landing, the acceleration of the spacecraft is zero. What does this say about the net force acting?
b At what other times during its travels through space does the spacecraft experience zero acceleration?

20 a During the early stages of the vertical take-off from Zorg the mass of the spacecraft and its fuel is 6 tonnes (1 tonne = 1000 kg). At the same time, the force provided by the rocket motors is 10 kN greater than the weight (1 kN = 1000 N). Remembering to work in SI units, what is the upwards acceleration?
b Explain why, in this particular situation (using fuel), acceleration increases even if force remains constant.

● **Extra skills task**

Information Technology and Application of Number

Use a spreadsheet to examine the formula $F = ma$. Use typical values that might apply to a spacecraft taking off from a planetary surface to draw and print scattergraphs (point graphs) for:

a constant force, acceleration (*y*-axis) against mass (*x*-axis)
b constant mass, acceleration (*y*-axis) against force (*x*-axis)
c constant acceleration, force (*y*-axis) against mass (*x*-axis).

Draw lines to fit patterns of points.
Which graphs show proportionality and which show inverse proportionality?
Why is acceleration usually plotted as the output variable?

12 Working with vectors

THE BIG QUESTIONS	● How can we add quantities that have direction as well as size? ● How can we add velocities that have different causes? ● How can we find the effect of two forces acting on a body?
KEY VOCABULARY	Cartesian convention co-linear vectors drag lift mutually perpendicular components normal reaction resolution resultant thrust
BACKGROUND	We can see three dimensions of space in the world around us, and some of the quantities that we measure act in particular directions in this space. These are vector quantities. Other quantities, scalars, have no particular direction. The amount of money you have with you is an example of a scalar quantity. You can count your money in a straightforward way. Addition of two scalar quantities is a simple sum, for example £0.20 + £1.30 = £1.50. But two vector quantities can have different directions, and when we add them together we must take account of these directions. That takes us into mathematics with special rules, which are explained in this chapter.

● Motion and force vectors

Velocity and acceleration

1 Which of the following are vectors and which are scalars?

- speed
- amount of money
- velocity
- acceleration

An athlete running around a track can have a constant speed but has a changing direction, and so has a changing velocity. We can use vectors to represent velocity before and after change, and we can also use a vector to represent the change (Figure 12.1).

Acceleration is rate of change of velocity. A body which is changing direction has a changing velocity, so it must be accelerating even if its speed is constant. The direction of the change in velocity is the same as the direction of the acceleration averaged over the same period of time.

Figure 12.1
Velocity vectors for an athlete.

2 A satellite in orbit around the Earth has constant speed. It is, however, accelerating.
a Explain how it is possible to have constant speed but to be accelerating. (Think about the difference between speed and velocity.)
b Draw a sketch to show its velocity at two points in its orbit.

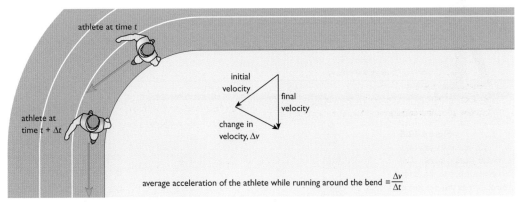

athlete at time t

athlete at time $t + \Delta t$

initial velocity

final velocity

change in velocity, Δv

average acceleration of the athlete while running around the bend $= \dfrac{\Delta v}{\Delta t}$

Force

An unbalanced external force acting on a body changes its velocity, and velocity is a vector quantity. Force, also, is a vector quantity. The direction of a force, just like the direction of velocity, matters (Figure 12.2).

Figure 12.2
Direction of force matters.

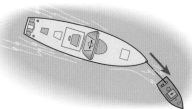

Force, acceleration and velocity vectors

A vector representing acceleration has the same direction as the vector representing the associated *change* in velocity. It does not necessarily have the same direction, at a given instant, as the velocity vector itself.

Force and acceleration are related by the simple equation $F = ma$. If mass stays constant then, whatever the force does, the acceleration changes by an identical proportion. Also, the net force acting on a body and its acceleration are always in the same direction (Figure 12.3).

Figure 12.3
From the direction of the acceleration of a body it is possible to tell the direction of the net force acting on it.

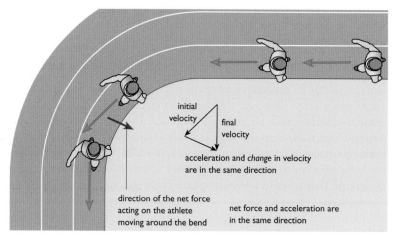

For a body moving and staying in a straight line, velocity and acceleration (if there is any acceleration) are in the same direction. But a body that is changing its direction has acceleration and velocity that are not in the same direction.

3 Consider the motion of a body such as a tennis ball. During a game, are the vectors for the velocity of the ball and the net force acting on it **a** never, **b** sometimes, **c** always in the same direction? Explain.

initial velocity

final velocity

acceleration and *change* in velocity are in the same direction

direction of the net force acting on the athlete moving around the bend

net force and acceleration are in the same direction

Addition of co-linear vectors

Figure 12.4
Addition of co-linear velocities.

The same rules of addition apply to all vectors, including velocity, acceleration and force vectors. Vectors that lie along the same straight line – **co-linear vectors** – can be added and subtracted much like scalar quantities.

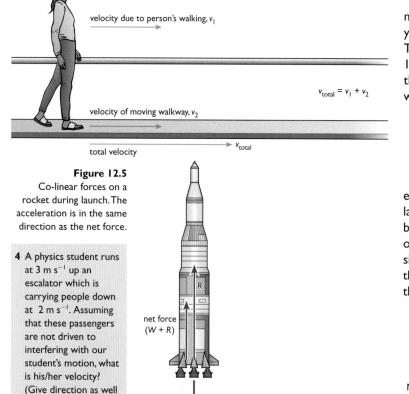

velocity due to person's walking, v_1

$v_{total} = v_1 + v_2$

velocity of moving walkway, v_2

total velocity

v_{total}

For example, suppose that you are on a moving walkway which moves at 1 m s^{-1}, and you also walk in the same direction at 1 m s^{-1}. Then your total velocity will be 2 m s^{-1} (Figure 12.4). If, for some reason, you decide to walk in the opposite direction then your total velocity will be zero.

Figure 12.5
Co-linear forces on a rocket during launch. The acceleration is in the same direction as the net force.

4 A physics student runs at 3 m s^{-1} up an escalator which is carrying people down at 2 m s^{-1}. Assuming that these passengers are not driven to interfering with our student's motion, what is his/her velocity? (Give direction as well as magnitude.)

R

net force $(W + R)$

weight, W net force $= (W + R) = ma$

As a result of burning fuel, a rocket experiences an upward force R. This must be larger than its weight W if the net force is to be upwards (Figure 12.5). The two forces acting on the launching rocket act along the same single straight line – they are co-linear – but they are unbalanced, or not in equilibrium, so that upwards acceleration takes place.

Note that $(W + R) < R$ because W has a negative value. (The $<$ sign means 'is less than'.)

Where forces act along a single straight line then the net or **resultant** force can be found by simple addition. This can be done by scale drawing or by calculation. For the rocket,

$$W \neq -R$$
$$\text{net force} = W + R \neq 0$$

The rocket's acceleration, from $F = ma$, is given by

$$\text{acceleration} = \frac{F}{m} = \frac{(W + R)}{m}$$

Note that we give force in one direction a positive value and force in the opposite direction a negative value. We are free to decide which is which, but having made a decision we need to stick to it. For forces acting vertically it is usual to say that the upward force is positive and the downward force is negative. For horizontal forces we usually say that forces acting towards the right are positive, and those to the left are negative (Figure 12.6). There is no physical reason for making this choice, but it is a good idea to have a single system that everybody uses – a convention. The habit of using positive for upwards and rightwards is called the **Cartesian convention** (after the 17th century French philosopher–mathematician, René Descartes). It is the Cartesian convention that we use when plotting graphs.

Figure 12.6
Co-linear forces
a in opposition and
b in constructive combination.

a

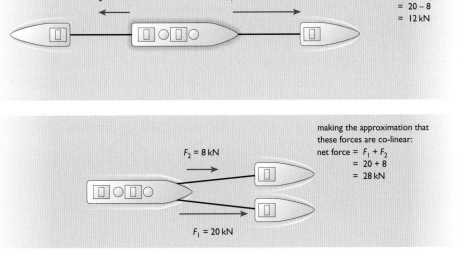

$F_2 = -8\,\text{kN}$ $F_1 = 20\,\text{kN}$

net force $= F_1 + F_2$
$= 20 - 8$
$= 12\,\text{kN}$

If co-linear forces act in opposite directions then they act against each other. The net or resultant force can be found by addition, but one of the forces has a negative value.

b

making the approximation that these forces are co-linear:

net force $= F_1 + F_2$
$= 20 + 8$
$= 28\,\text{kN}$

$F_2 = 8\,\text{kN}$

$F_1 = 20\,\text{kN}$

The resultant of co-linear forces acting together, in the same direction, is a simple addition sum.

5 What is the net force of a 10 N force and an 8 N force when
 a acting in the same direction
 b acting in opposite directions?
6 What is the resultant of each of the combinations of vectors in Figure 12.7?

a   **b**

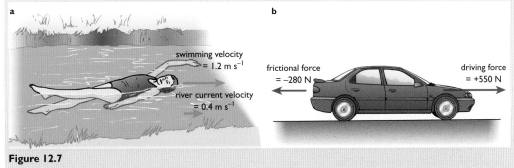

swimming velocity
= 1.2 m s⁻¹

river current velocity
= 0.4 m s⁻¹

frictional force
= –280 N

driving force
= +550 N

Figure 12.7

7 For a rocket with a mass of 100 tonnes and a weight of 1000 kN, what upwards force is needed to produce an upwards acceleration of 20 m s⁻²? (1 tonne = 1000 kg)

Addition of perpendicular vectors

Imagine a rat swimming across a river. The motion of the rat has two independent or separate causes – the rat's own efforts and the water current. If the rat's swimming effort is directed across the river while the current carries it down the river then they act at right angles to each other. The two velocities are **mutually perpendicular components** of the actual velocity.

If the sizes of the two components of the rat's velocity are known, then it is possible to work out the size and direction of the combination or resultant, either by scale drawing or by calculation. By either method, the vectors are added together to find the actual velocity.

'Normal' addition only works for scalars and for co-linear vectors where only one dimension matters. The number of eggs in a nest is a scalar quantity; one egg plus one egg is the same as two eggs. In one dimension, one metre plus one metre is equal to two metres. Working out the resultant of two mutually perpendicular vectors is addition in two dimensions. Not surprisingly, two-dimensional calculation is more complex (Figure 12.8).

Figure 12.8
Finding the resultant of two mutually perpendicular velocities.

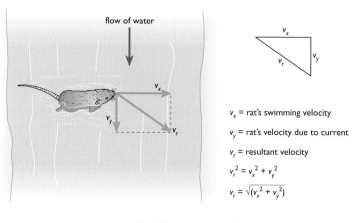

v_x = rat's swimming velocity

v_y = rat's velocity due to current

v_r = resultant velocity

$$v_r^2 = v_x^2 + v_y^2$$

$$v_r = \sqrt{(v_x^2 + v_y^2)}$$

The resultant of two mutually perpendicular velocities can be thought of as forming the diagonal of the rectangle or the hypotenuse of the right-angled triangle that the velocities make. This means we can use Pythagoras' theorem to find the size of the resultant:

$$v_r^2 = v_x^2 + v_y^2$$

where v_x and v_y are the two perpendicular velocities and v_r is their resultant.

In the same way we can find the resultant of any two mutually perpendicular forces, either by scale drawing or by calculation (Figure 12.9).

Figure 12.9
Finding the resultant of two mutually perpendicular forces.

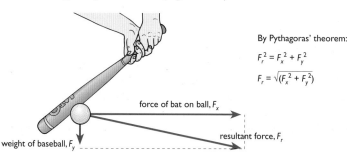

By Pythagoras' theorem:

$$F_r^2 = F_x^2 + F_y^2$$

$$F_r = \sqrt{(F_x^2 + F_y^2)}$$

force of bat on ball, F_x

weight of baseball, F_y

resultant force, F_r

Addition of normal reaction and frictional force

A wall can affect a bouncing ice-hockey puck by two separate processes or mechanisms. There is a force on the puck that is perpendicular to the wall, due to repulsion between wall-particles and puck-particles. This is called a **normal reaction** force, and you can feel a similar force by hitting a table with your fist or just by resting your hand on the table. In the latter case the normal reaction simply balances the weight of your hand. Sitting on a chair or standing on a floor, you experience a normal reaction that balances your body weight.

The wall can also exert frictional force on the puck, just as the chair or the floor can exert frictional force on you. Frictional force is a force that opposes the sliding of surfaces, and it acts in a direction parallel to the surfaces.

The two forces on a bouncing ice-hockey puck – normal reaction and frictional force – act at the same point but in directions that are perpendicular to each other. To find the net force or resultant we need to add the vectors (Figure 12.10).

Figure 12.10
Adding normal reaction
and frictional force.

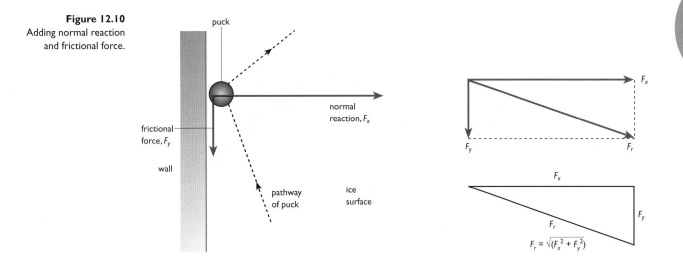

8 Alongside, what operation has been carried out to change the first equation into the second equation?
9 A person walks across the deck of a ship from one side to the other at 2 m s^{-1}, while the ship has a forward velocity of 10 m s^{-1}. Show the size and direction of the person's resultant velocity on a vector diagram.
10 What is the resultant of a 3 N frictional force and a 4 N normal reaction force acting at right angles on an ice-hockey puck?
11 In what circumstances can you add vectors just as if they were scalar quantities?

It helps to be able to refer to their two directions in shortened form. We could say that one force acts (say, the normal reaction) in the *x*-direction and the other (the frictional force) in the *y*-direction. We could call the forces F_x and F_y. The combined effect of the two forces – their resultant – is shown by the hypotenuse of the right-angled triangle they make. We can call the resultant F_r and make use of Pythagoras' theorem to calculate it:

$$F_r^2 = F_x^2 + F_y^2$$

or

$$F_r = \sqrt{F_x^2 + F_y^2}$$

Resolution of vectors

The rat swimming across the river (Figure 12.8) has a forward component to its motion and a lateral or sideways component. By looking at its actual motion relative to the surrounding landscape, and knowing the size and direction of the actual net velocity, it is possible to work out these two mutually perpendicular components of the velocity (Figure 12.11). This can be thought of as the reverse of addition of mutually perpendicular vectors and is called **resolution** of the velocity. Just as for vector addition, a vector can be resolved either by scale drawing or by calculation.

Figure 12.11
Resolution of velocity in two mutually perpendicular directions. Mutually perpendicular components of a vector are a pair of vectors at right angles to each other that have the same effect as the vector.

Each of these pairs is a legitimate analysis of the velocity vector. The first of them, where the components lie in the *x*- and *y*-directions, is the most likely to be useful.

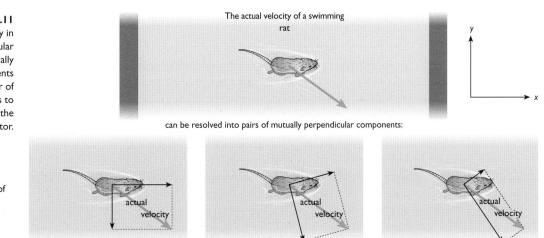

Figure 12.12
We can resolve a vector in *any* two mutually perpendicular directions, such as northerly and easterly.

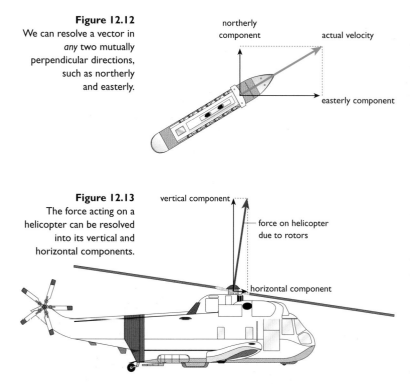

northerly component

actual velocity

easterly component

For a moving object we can resolve actual velocity, if we know its size and direction, into any two mutually perpendicular directions. We could choose to find the northerly component of a ship's motion and the easterly component, for example (Figure 12.12).

Just as we can resolve velocity vectors, we can also resolve a force vector in two mutually perpendicular directions. Think about a helicopter, for example. The force due to its rotors can provide lift (to balance gravity or even to accelerate it upwards) and forward thrust (to overcome drag forces and sometimes to accelerate it forwards). This is achieved by tilting the plane of the rotors, so that the net force on the helicopter has both vertical and horizontal components. From the angle of the tilt we can see the direction of the actual force acting (Figure 12.13). Different tilts produce different combinations of forward and upward force. To work out the relative values of these components we resolve the force vertically and horizontally.

Figure 12.13
The force acting on a helicopter can be resolved into its vertical and horizontal components.

vertical component

force on helicopter due to rotors

horizontal component

A quick rule for working out components

We use the rules of trigonometry to work out the values of mutually perpendicular components. Figure 12.14 shows some examples.

Note that the perpendicular component that is 'next to' (or adjacent to) the named angle is always equal to $F \cos \theta$. The component that is 'opposite' the named angle is always $F \sin \theta$.

Figure 12.14
Calculating the values of perpendicular components of a force.

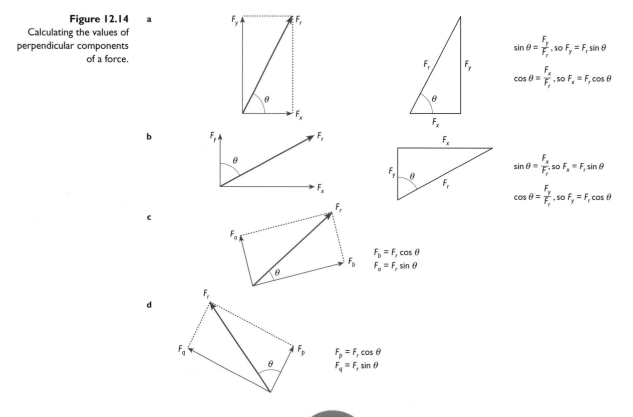

a

$\sin \theta = \dfrac{F_y}{F_r}$, so $F_y = F_r \sin \theta$

$\cos \theta = \dfrac{F_x}{F_r}$, so $F_x = F_r \cos \theta$

b

$\sin \theta = \dfrac{F_x}{F_r}$, so $F_x = F_r \sin \theta$

$\cos \theta = \dfrac{F_y}{F_r}$, so $F_y = F_r \cos \theta$

c

$F_b = F_r \cos \theta$
$F_a = F_r \sin \theta$

d

$F_p = F_r \cos \theta$
$F_q = F_r \sin \theta$

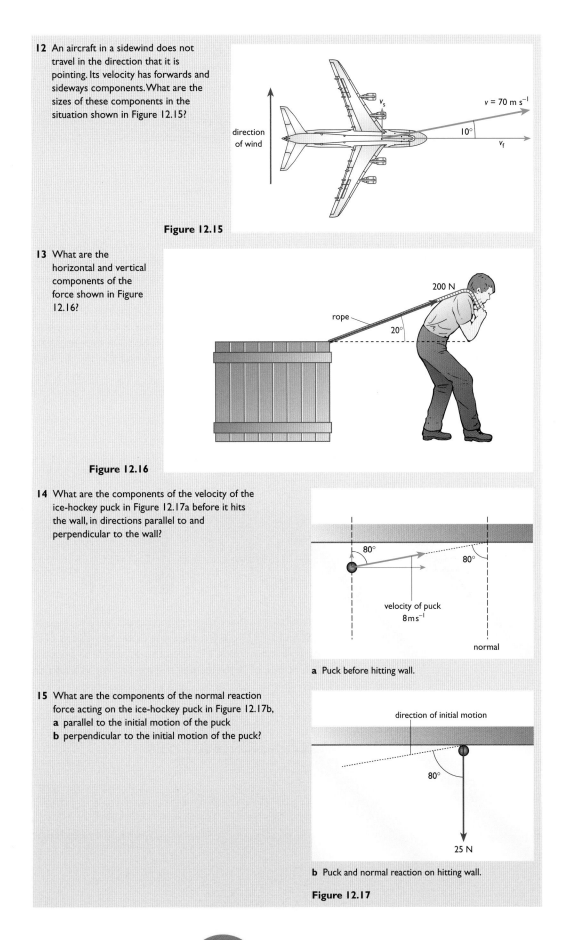

12 An aircraft in a sidewind does not travel in the direction that it is pointing. Its velocity has forwards and sideways components. What are the sizes of these components in the situation shown in Figure 12.15?

Figure 12.15

13 What are the horizontal and vertical components of the force shown in Figure 12.16?

Figure 12.16

14 What are the components of the velocity of the ice-hockey puck in Figure 12.17a before it hits the wall, in directions parallel to and perpendicular to the wall?

a Puck before hitting wall.

15 What are the components of the normal reaction force acting on the ice-hockey puck in Figure 12.17b,
   a parallel to the initial motion of the puck
   b perpendicular to the initial motion of the puck?

b Puck and normal reaction on hitting wall.

Figure 12.17

Addition of non-perpendicular vectors

The resultant of any pair of forces is equal to the diagonal of the parallelogram they make (Figure 12.18). (This is true of mutually perpendicular forces, for which the parallelogram happens to be a rectangle.)

Figure 12.18
Adding two non-perpendicular forces.

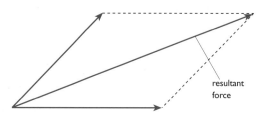

resultant
force

A ferry boat crossing a river backwards and forwards may find it best to have its landing points directly opposite each other. But, if it uses its engines to try to move perpendicularly to the opposite bank then the river will sweep it downstream. Crossing by crossing, the ferry will get closer to the sea. So the ferry powers itself towards a point upstream of its landing (Figure 12.19a). We can add together the two independent velocities of the ferry by completing a parallelogram. The resultant is shown by the diagonal of the parallelogram, and we can find a value for this either by scale drawing or by calculation. The scale-drawing approach requires great care to make sure that the angles, as well as the arrow lengths, in the drawing are the right size.

Figure 12.19
The velocity components of a river ferry.

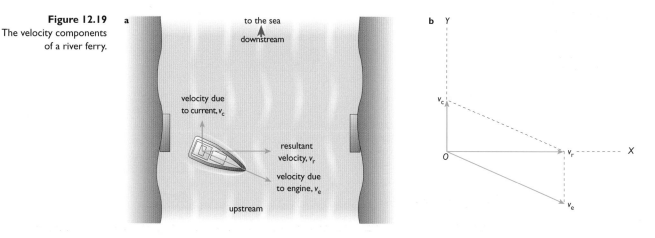

To calculate the resultant we have to work in stages. The first stage is to create two axes, which we can call *OX* and *OY*. (These are the same as the directions of the *x* and *y* axes on a graph, using the Cartesian convention.) To keep this as simple as possible, one of the axes can be in the same direction as one of the velocities (Figure 12.19b). We also need a shortened way of naming the velocities – we will call them v_c and v_e, for velocity due to current and velocity due to engine.

v_e does not act along *OX* or along *OY*, but partly in both directions. We can calculate the contribution that v_e makes in each direction. That is, we resolve v_e along *OX* and *OY* to find its two mutually perpendicular components (Figure 12.20).

Figure 12.20
v_e resolved in the directions *OX* and *OY*.

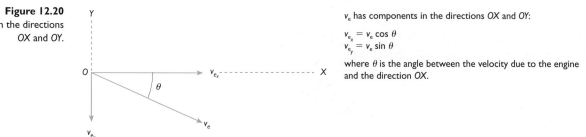

v_e has components in the directions *OX* and *OY*:

$$v_{e_x} = v_e \cos \theta$$
$$v_{e_y} = v_e \sin \theta$$

where θ is the angle between the velocity due to the engine and the direction *OX*.

With this done we can add these components to v_c:

net velocity acting in direction $OX = v_{e_x} = v_e \cos \theta$

net velocity acting in direction $OY = v_c - v_{e_y} = v_c - v_e \sin \theta$

To reach its landing point across the river, the ferry needs a resultant velocity in the direction OX. It can achieve this by choosing the angle θ so that the net velocity in the direction OY is zero (Figure 12.21).

Figure 12.21
The velocity components along *OX* and *OY*.

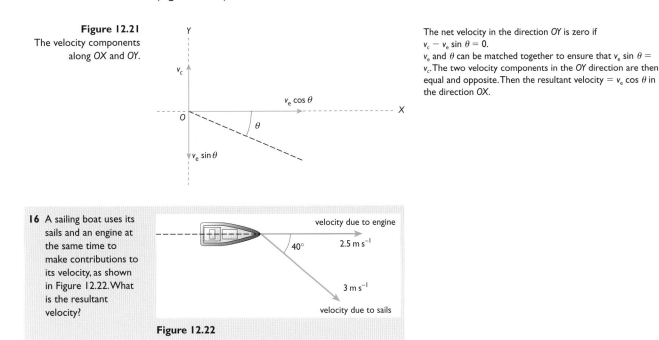

The net velocity in the direction OY is zero if
$v_c - v_e \sin \theta = 0$.
v_e and θ can be matched together to ensure that $v_e \sin \theta = v_c$. The two velocity components in the OY direction are then equal and opposite. Then the resultant velocity $= v_e \cos \theta$ in the direction OX.

16 A sailing boat uses its sails and an engine at the same time to make contributions to its velocity, as shown in Figure 12.22. What is the resultant velocity?

Figure 12.22

Flight vectors

For a bird or a plane to travel at constant velocity, the forces acting on it must total zero. To analyse this it is useful to think separately about the horizontal and vertical forces.

The horizontal forces are the resistive forces due to the air, called **drag** forces, and the forward **thrust** created by the efforts of the bird's wings or the plane's engines. For constant horizontal velocity these forces must be equal and opposite. The net horizontal force must be zero.

The vertical forces are the gravitational force, or weight, and the **lift** created by the flow of air across the wings. For constant vertical velocity, weight and lift must be in balance. The net vertical force must be zero.

Figure 12.23
The forces on a bird in flight.

17 What will start to happen to a bird for which
a lift > weight, thrust = drag
b lift > weight, thrust < drag?

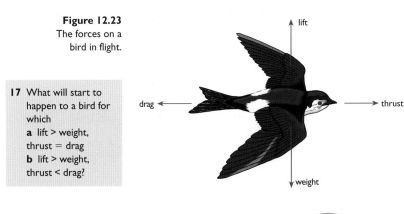

● **Comprehension and application**

Helicopter flight

A helicopter provides a fine illustration of Newton's Third Law in action. Air experiences a force, and also exerts a force. The helicopter is responsible for the force on the air, and the air returns an equal and opposite force. The helicopter and the air are equal partners.

For a helicopter with a single set of main rotor blades, these provide both the force of lift and the force of thrust. Lift is the upward force the aircraft experiences to balance and sometimes overcome the force of gravity, and thrust is the force that allows it to balance or overcome the forces of resistance that oppose its forward motion.

Each rotor blade acts as a wing that exerts force on the air flowing across its surfaces. As with any wing, the net effect is to exert a force on the air that is in the opposite direction to the required force on the aircraft. This only happens to a sufficient extent when the flow of air across the wing surfaces is fast enough. An aeroplane, such as a passenger jet, has to move its whole body in order to achieve these speeds, and so it cannot create enough lift to become airborne until it is moving at high speed. This is why jet planes need long runways. A helicopter on the other hand creates the necessary speed of airflow by moving its 'wings' or rotor blades. They spin at up to 300 revolutions per minute.

Any point you might choose on a rotor is moving in circular motion. Because of this circular motion, one end of a rotor is moving forwards and the other end is moving backwards relative to the helicopter body. The two sides, or blades, of the rotor are called the advancing blade and the retreating blade (Figure 12.24). These backward and forward motions of the ends of the two blades take place whether a helicopter is on the ground or moving forwards at maximum speed.

Figure 12.24
Velocity vectors for
a a stationary helicopter
and **b** a helicopter in
forward motion.

Figure 12.25
Force vectors for a
helicopter **a** hovering,
b accelerating forwards,
and **c** turning to the left.

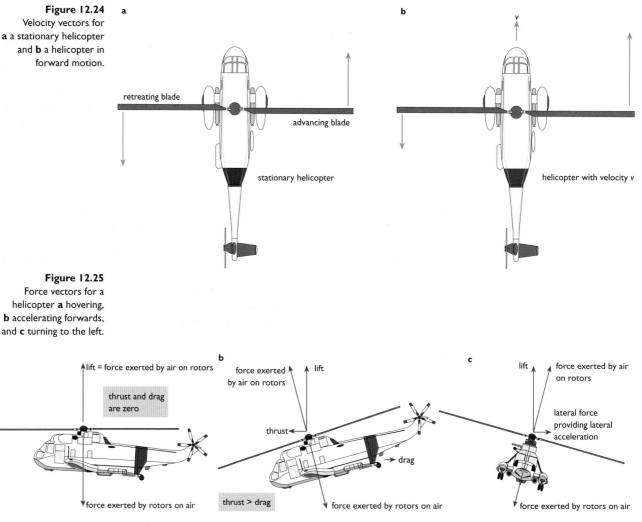

The advancing blade moves through the air more quickly than the retreating blade, so it experiences greater lift. This would make the aircraft topple in flight if it were not compensated for by varying the tilt of the blades to reduce lift on the advancing blade and increase lift on the retreating blade.

In fact, it is by tilting the blades that a helicopter pilot achieves control of the aircraft. To provide thrust, the blades are tilted so that lift is increased as they travel through the rear semi-circle of their motion, and then tilted so that lift is reduced when they come to the front. That tips the whole aircraft, including the plane of the rotors, forwards, giving a horizontal component to the force it experiences (Figure 12.25b).

To steer to the left, the blades are tilted to reduce lift on the left and increase it on the right. This again tips the whole helicopter and provides a horizontal component to the force, but now the force is perpendicular to the direction of travel and not parallel to it (Figure 12.25c).

Finally, there is a problem of spinning. The rotors and the helicopter body are connected by the engine. The rotors have considerably less mass, so they spin at high speed. But the helicopter body is in a low-friction environment and it is not a fixed body. There is therefore a tendency for the engine to make the whole body spin in the opposite direction to the rotors. A tail rotor provides a balancing turning effect.

The engineering that is required to achieve stable flight of a helicopter is complex, but the advantages of manoeuvrability provided by rapidly moving 'wings' are considerable.

18 Why do 'fixed wing' aircraft, such as jet airliners, take off into the wind when this is possible?

19 Draw sketches to show the forces of lift, thrust *and the opposing forces* that are acting on a helicopter in windless air, when
   a travelling at a fixed height and horizontally forwards at steady velocity
   b rising at steady velocity (remember $F = ma$) and travelling horizontally forwards at steady velocity
   c accelerating upwards and travelling horizontally forwards at steady velocity.

20 The speed of the end of a rotor blade relative to the helicopter body is given by
$$\text{speed} = \frac{\text{distance}}{\text{time}} = \frac{2\pi r}{t}$$
where r is the length of each rotor blade and t is the time for each revolution. Calculate
   a the time taken, t, in seconds, for one revolution at a rate of 300 revolutions per minute
   b the speed of the end of the retreating rotor blade relative to the helicopter body if the length of the rotor blade is 4.5 m
   c the speed of the end of the retreating rotor blade relative to the ground when the helicopter is moving at 360 km h^{-1}
   d the speed of the end of the advancing rotor blade relative to the ground when the helicopter is moving at 360 km h^{-1}.

21 Explain fully, in your own words, with the aid of sketch diagram(s), how a helicopter achieves thrust.

22 Calculate the thrust acting on a helicopter, which has a weight of 25 kN, when the plane of its rotors is tipped forwards at an angle of 5° to the horizontal. (Note that 25 kN is the vertical component of the total force provided by the helicopter–air interaction. You will find that a sketch diagram is very helpful!)

●●

● **Extra skills task**   Information Technology

Use a spreadsheet package to produce a mathematical model that can predict the motion of a helicopter for different values of air resistance, thrust, gravitational and lift forces.

A typical weight might be 25 N and a typical range of air resistance for a helicopter with speeds up to 360 km h^{-1} is 0 to 2 kN. Your spreadsheet model should add the vertical forces, and then the horizontal forces, in separate columns.

You will then need to use the formula for resultant vectors to make predictions about net force and hence the direction of the helicopter's acceleration.

Your spreadsheet should allow input in defined cells, prompt the user, and place suitable limits on the input values.

13 Force and motion in gravitational fields

- We are stuck in the gravitational field of a large body, so how can we understand universal rules that apply whether or not a body is in a gravitational field?
- We are stuck in a resistive medium (air), so how can we understand universal rules that apply whether or not a body is in a resistive medium?

KEY VOCABULARY

acceleration due to gravity ballistics field lines (or lines of force) free fall gravitational field strength gravitational mass inertial mass terminal velocity weight

BACKGROUND

Figure 13.1
Spheres within spheres – before Copernicus Europeans saw the Earth as the centre of the Universe.

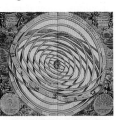

Before the scientific revolution that began with Copernicus, Europeans believed that the Earth was fixed in place but full of imperfections, while revolving around the Earth were perfect heavenly bodies (Figure 13.1). People were sure that the behaviour of earthly things and of bodies in the sky obeyed different rules. Over the centuries, science has changed how we see the world (Figures 13.2, 13.3).

Figure 13.2
Some of the philosophers and scientists who helped to change our view of the world.

Nicholas Copernicus (1473–1543) was a Polish monk. He suggested that the apparent motions of the planets in the skies could be explained very neatly by assuming that the planets and the Earth were all in orbit around the Sun.

Giordano Bruno (1548–1600), a priest, preached in support of Copernicus' ideas about revolutions and was subsequently burnt to death.

Galileo Galilei (1564–1642) used telescopes, a new technology. He discovered that the Moon was not a perfect sphere but has mountains, just like the Earth. He discovered that Jupiter was not a sphere on its own, but had moons. Night by night he tracked the positions of the moons, and could see that they were in orbit around Jupiter. It seemed that everything in the skies did not rotate around the Earth after all.

Johannes Kepler (1571–1630) used data on the motions of the planets across the skies to establish laws that described their orbits, and the orbit of the Earth, around the Sun. The new science had reduced the status of the Earth from being at the centre of the Universe to being just one in a family of planets.

Figure 13.3
'A philosopher lecturing at the orrery', by Joseph Wright of Derby.

This painting, painted more than 100 years after the deaths of Kepler and Galileo and 50 or so years after Newton's death, shows some affluent citizens of Derby listening in awe as a travelling lecturer uses a mechanical model of the solar system, an 'orrery', to describe the motions of planets in the vastness of space. These people were sky-literate – they had no gas or electric lighting, so that the magnificence of space presented itself to them on every cloudless night.

Newton, the apple and the planets

Isaac Newton once had the humility to write:

'*If I have seen further it is by standing on the shoulders of giants.*'

Copernicus, Bruno, Galileo and Kepler were some of the giants he had in mind.

Newton had created the concept of force as we know it, as the direct cause of acceleration. We now show this force–acceleration relationship by a simple equation, $F = ma$. These ideas are contained in Newton's First and Second Laws (see pages 90 and 120). The Third Law (page 92) carries on from here, and says that force is always mutual – when a force acts on a body then the body exerts an equal and opposite force in return.

But the fixing of the concept of force was not Newton's only achievement. It was Newton who realised that the force of gravity – part of everyday experience on Earth – is the same as the force that holds moons and planets in their orbits. Perhaps it was, as the story goes, an apple falling in a Lincolnshire orchard that disturbed Newton's contemplation of the ideas of Copernicus, Galileo and Kepler. In showing that apples and planets are subject to the same laws, Newton continued the work that they had started. He unified earthly gravity and heavenly motion. Nothing has been the same since.

Two definitions of mass

We have already seen that mass can be defined as a measure of a body's resistance to acceleration, as described by the formula $m = F/a$. This is the body's **inertial mass**.

But mass does not just determine the body's resistance to acceleration. Any two bodies with mass interact – they exert gravitational force on each other. The mass of a body determines the extent to which it exerts and experiences gravitational force. It seems we need another distinct definition of mass. **Gravitational mass** quantifies a body's gravitational behaviour – the extent to which it exerts and experiences gravitational force.

These two masses, though they have different definitions, are numerically identical. Newtonian physics has no explanation for this 'equivalence', and treats it as nothing more than coincidence. This is a fundamental weakness in Newtonian physics that is solved by Einstein's General Theory of Relativity.

The mass of a body, inertial or gravitational, changes when it gains or loses material. The mass of an astronaut or a hammer is the same when they are on the Moon as when they are on Earth, provided there is no loss or gain of material. Mass (or, strictly, 'rest mass') does not vary from place to place.

The Sun is a body with mass. It has a (large!) quantity of material. It is meaningless to talk about the direction of this quantity. We cannot draw arrows to show which way the Sun's mass points. Mass is a scalar quantity.

1 Describe ways in which a visit to the dentist could potentially result in small changes to your personal mass.
2 What effects could an increase in your personal mass have on
  a how easy it is to start and stop on a bicycle
  b how much you compress the air in the tyres when you sit on the bicycle?
3 **DISCUSS**
  a Imagine a Universe that consists of a solitary body. Would the body have
  i inertial mass
  ii gravitational mass?
  b Imagine a more sophisticated Universe, with just two identical bodies that are initially stationary some distance apart. What happens next? Describe the roles of the gravitational and inertial masses of the bodies in what happens.

Gravity and mass

All bodies with mass exert gravitational force on one another; this seems to be true throughout the Universe. Most places in the Universe are a long way from any large star or planet, and so in such places we would expect human-like bodies to experience little gravitational force. But we live on the surface of a body that is big enough to hold our own bodies to it. Few of us expect to ever be able to escape from the Earth's gravitational pull. We are so used to this pull that we can sometimes forget that it is a local effect.

Weight is the name we give to the gravitational force acting on a particular body. Weight varies from place to place – it is a locally varying quantity and not a universally fixed one.

The mass of a body is measured in kilograms, and its weight in newtons. A simple investigation can be carried out to determine the relationship between the mass of a body and the gravitational force it experiences and exerts when it is in a particular place. A newtonmeter can be used to measure the weights of known masses. The easiest place to carry out the investigation is at the Earth's surface. The results reveal a simple proportionality, which means that the gradient of a graph of weight against mass is constant and we can write:

weight = gradient × mass

If we were able to repeat the investigation at different heights above the Earth's surface, we'd find that the gradient of the graph decreased as we moved further outwards (Figure 13.4).

The gradient provides a measure of the ratio of the force of gravity to the mass of the body. It is called **gravitational field strength**.

$$\text{gravitational field strength} = \frac{\text{weight}}{\text{mass}}$$

$$g = \frac{W}{m}$$

or

$$W = mg$$

Gravitational field strength has the unit N kg^{-1}. At the surface of the Earth its value is 9.81 N kg^{-1}, though there are small variations from place to place. Gravitational field strength, like force itself, is a vector quantity. At the surface of the Earth it acts downwards, towards the centre of the Earth.

Figure 13.4
At any one place, weight is proportional to mass. The gradient of the line is the gravitational field strength at that place. Graphs plotted at different heights above the Earth's surface have different gradients.

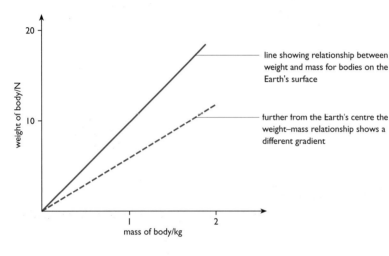

line showing relationship between weight and mass for bodies on the Earth's surface

further from the Earth's centre the weight–mass relationship shows a different gradient

4 As you've grown, what has happened to
  a your mass
  b your weight?
5 How could you change your weight without changing your mass?
6 Is it possible to change your mass without changing your weight? How?

Gravitational field

Gravitational field is said to exist in a space in which gravity acts on any mass. You, the Earth and the Sun are all responsible for gravitational field in the space around you (though your field is relatively feeble). Mass placed in this space will experience force. It helps to have ways to visualise this abstract notion of field.

Visual representation of gravitational field

Starting with the concept of force, we can talk about the force that *would* act on a body *if* a body were at a particular place in the field. We don't have to specify what would happen to a particular body, just the size and direction of the force that would act on each kilogram – the force that would act per unit mass. This is the gravitational field strength at that point.

One way to illustrate gravitational field around a body such as the Earth would be to show a forest of arrows, each arrow acting as a vector representation of the field strength at a different place (Figure 13.5).

Figure 13.5
Every point in a gravitational field has a field strength. No body is needed at the point – gravitational field strength is a property of a point in the field, not of a particular body.

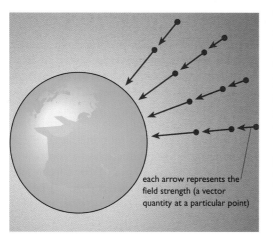

each arrow represents the field strength (a vector quantity at a particular point)

7 Sketch the pattern of field lines around the Earth.
8 Is gravitational field strength a vector or a scalar quantity?

The 'forest of arrows' diagram can be tidied if we draw continuous lines with arrows on them. Each line then shows the direction of the field strength at each point along it. These are **lines of force** or **field lines**. They no longer show the strength of the field by the lengths of arrows, but the closeness of the lines gives an indication of strength.

The most useful visual models are simple. So it is usually better to draw a field pattern in just two dimensions. Field lines are useful for thinking about complex gravitational fields, such as the Earth–Moon system.

Acceleration due to gravity

We have seen how the weight of a body at a particular point in space is related to its mass, with the local value of gravitational field strength as the constant of proportionality:

$$W = mg$$

We also know that there is a universal relationship linking force and mass:

$$F = ma$$

where a is the acceleration of the mass m due to force F. Weight is a force. So we can write:

$$W = ma$$

This means that g and a are interchangeable. The gravitational field strength and the **acceleration due to gravity** of a body at a particular point in space are numerically identical. At the Earth's surface:

gravitational field strength $g = 9.81$ N kg^{-1}
acceleration due to gravity $a = 9.81$ m s^{-2}

Dimensional analysis shows that units N kg^{-1} and m s^{-2} are dimensionally identical. So the abbreviation g is often used for acceleration due to gravity, as well as for gravitational field strength.

Gravitational field strength is a property of a point in space, not a property of a particular body. So acceleration due to gravity is also a property of a point in space and not a property of a particular body.

Galileo demonstrated, to the surprise of everyone else at the time, that the acceleration of a body due to the Earth's gravity was not a property of a particular body with a particular mass. He dropped bodies of unequal mass (from the leaning tower of Pisa, the story goes) and both fell together, with identical accelerations (of 9.81 m s^{-2}). Astronaut Buzz Aldrin repeated the demonstration on the surface of the Moon, with a hammer and a feather. Both fell with identical accelerations (of 1.6 m s^{-2}, the value of g being smaller at the Moon's surface since the Moon is less massive than the Earth). Buzz Aldrin was also demonstrating motion without air resistance, something that Galileo could only imagine.

9 State the dimensions which match the following units:
a kg   e m s^{-2}
b m   f N
c s   g N kg^{-1}
d m s^{-1}
10 What is the value of the gravitational field strength, in N kg^{-1}, at the Moon's surface? (Refer to the text alongside.)
11 If Galileo and the leaning tower of Pisa were on the Moon, how would the result of Galileo's demonstration have been different?

Falling with and without resistive force

Imagine a planet identical in size (volume and mass) to the Earth, in orbit around a distant star. This planet, planet X, has gravitational field strength close to the surface of about 10 N kg^{-1}, but it has no atmosphere. Imagine also that you work for the Galactic Space Agency, and your job is to plan a mission to land a capsule on the surface of planet X. The capsule will be released from a larger spacecraft that will hover above the planet.

Straight away you can rule out the prospect of using parachutes. They depend on a resistive medium, such as air, to work. On the plus side, the motion is relatively simple. Your maths is going to be easier than it would be on Earth. Under the action of a constant force, the capsule experiences a constant acceleration (of about 10 m s^{-2}). For a freely falling body, velocity increases uniformly right up until the moment of impact (Figure 13.6).

Figure 13.6
Falling without resistive force.

Close to the surface of planet X the acceleration of the capsule is constant (the acceleration due to gravity) so the velocity increases uniformly.

On the Earth the presence of the atmosphere makes a very big difference to what happens. A body released from a hovering craft into the atmosphere (or the craft itself if it switches off its motors) behaves as it would on planet X only at the very start of its fall. At time zero (the moment of release) it has not yet acquired any velocity, so there is no resistance to its motion. Just for that instant, the atmosphere has yet to have any resistive effect. So at time zero the only force acting is gravitational force. The body has acceleration to match (about 10 m s^{-2}), and its velocity increases above zero. But as soon as the body acquires velocity it starts to experience resistance due to the air. That is, a force develops that acts in the opposite direction to the velocity. This resistive force opposes the force of gravity and reduces the total or net force acting:

net force = sum of forces acting
        = gravitational force + resistive force

$$F_t = F_g + F_r$$

or

$$F_t = W + F_r$$

Direction matters, and if we say that the downwards force W has a positive value then the upwards resistive force F_r must have a negative value. The net force F_t is less than the gravitational force alone. But the net force is still downwards, so the body still accelerates downwards. Velocity continues to increase.

The resistive force grows until it is equal to the downwards gravitational force. When that is achieved the net force is zero. Acceleration must also then reach zero. The body continues to fall with no net force, no acceleration, and constant velocity (Figure 13.7). This constant velocity is called its **terminal velocity**.

Figure 13.7
Falling with resistive force.

The resistive force increases with velocity. When the weight of the capsule and the resistive force are balanced, the acceleration is zero and the capsule falls with constant velocity.

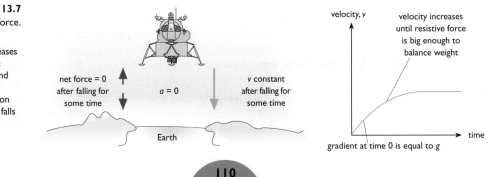

110

Free fall

The capsule released above planet X goes into **free fall**. Its acceleration, throughout its fall, is related to its weight by the simple formula $W = ma$, with no interference by any resistive force. If you were silly enough to be inside the capsule, you would also be in free fall. You and the capsule would be falling together. The walls of the capsule would not exert any forces on you unless you throw yourself at them or push yourself off them. The floor would not support your weight. Suppose that the capsule had no windows. There would be no way – no experiment that you could perform – to find out whether or not you were moving or whether you were floating in free space. Such a free fall experience is indistinguishable from actual weightlessness (until you hit the planet's surface!).

12 a Explain why, in the Earth's atmosphere, net force on a falling body dropped with zero velocity decreases until it is zero.
  b Describe the matching acceleration.
  c Describe the matching velocity.
13 a Describe the motion of a falling body which experiences zero net force.
  b Sketch a velocity–time graph for its motion.

Roller-coaster vectors

For a body falling vertically, acceleration, force and velocity all act or take place along the single straight line of the fall. The vector quantities are co-linear. This is not generally the case; consider an adventure park ride, Figure 13.8.

A car on a curved track has velocity parallel to the track. But many of its *changes* in velocity during the ride are not parallel to the track, because the net force causing the acceleration is not parallel to the track.

Figure 13.8
Net force and acceleration are always co-linear, but velocity doesn't have to be; *change* in velocity is co-linear with acceleration.

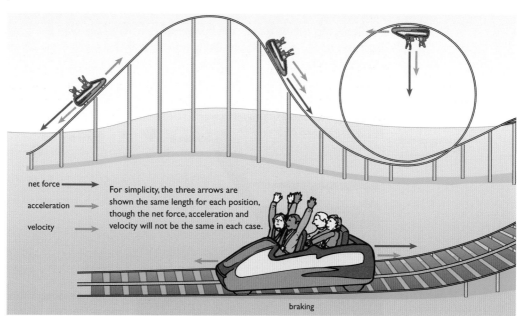

net force ⟶
acceleration ⟶
velocity ⟶

For simplicity, the three arrows are shown the same length for each position, though the net force, acceleration and velocity will not be the same in each case.

braking

14 What factor determines the relative sizes of the force and acceleration of the roller-coaster?

Ballistics

Ballistics is the name given to the study of the motion of projectiles such as cannonballs. It is a study that had huge military and political importance in the 17th century. Before then, theory claimed that cannonballs flew in straight lines until they ran out of 'force', and then they fell straight to the ground (see Figure 11.1). Galileo's new methods of mathematical analysis made it possible to plot a more accurate course for a cannonball.

Galileo's breakthrough was the realisation that the flight of a cannonball could be analysed by considering its horizontal and vertical motion separately. The trick was to take the motion apart into horizontal and vertical components, Figure 13.9a, analyse these and then put them back together again. It is easiest to do this by first assuming that air resistance plays a negligible part in what happens.

Figure 13.9
Independent analysis of the
horizontal and vertical
components of motion
gives more reliable
predictions of a cannonball
trajectory. Better maths
provided an advantage
in warfare!

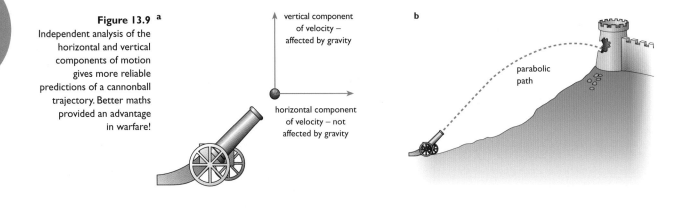

15 What is the
difference between
the real and idealised
trajectories of a
cannonball?

The result of the mathematics is then that the cannonball will follow a parabolic path (Figure 13.9b). It can make further predictions, for example that the maximum range over flat ground is achieved by firing the cannon at an angle of 45° to the horizontal. Of course, air resistance has a significant effect on the flight of the cannonball, providing a reminder that simplified calculations are useful but do not tell the complete story.

Projectile motion that is initially horizontal

We can use the technique of separate analysis of horizontal and vertical motion to predict what will happen if a cannonball, or any other object, is projected horizontally, as it might be from the side of a ship.

Neglecting air resistance, the horizontal velocity of the object is constant. The vertical velocity increases under gravity. The resulting velocity makes the projectile travel in a parabolic path (Figure 13.10).

Figure 13.10
The velocity of the
projectile has a constant
horizontal component and
an increasing vertical
component.

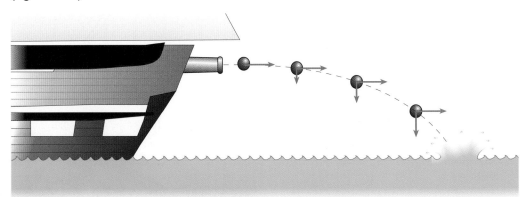

Figure 13.11
Observation of projectile
motion using 'multiflash'
strobe photography.

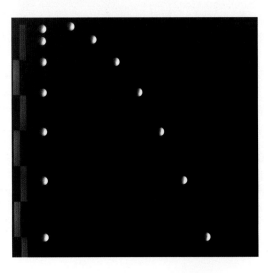

Strobe observations (Figure 13.11) confirm the prediction made by the analysis of vectors that the horizontal motion and the vertical motion of projectiles can be considered independently.

16 An arrow is fired
horizontally across a
large flat field. Another
arrow is dropped
from the same height
at the same instant.
Which would you
expect to hit the
ground first, or would
you expect them to
land together? Explain.

These two bodies were released at the same time, one dropped and one projected horizontally. The image shows clearly that the two bodies are at the same height at each instant – the horizontal component of the motion has no effect on the vertical motion.

Circular orbit

Imagine that you are a window cleaner working on the world's tallest building. Having lost interest in job security and the safety of people on the ground, you decide to experiment. You throw your bucket horizontally away from the building. The bucket moves out and away from you, while at the same time accelerating downwards.

Suppose now that the building is so tall that it sticks out of the top of the atmosphere so that there is no air resistance or wind to play any part. Even at the top of the Earth's atmosphere, gravitational field strength is not much less than it is at the surface. However hard you throw the bucket, it accelerates downwards towards the centre of the Earth at about 10 m s^{-2}. But the harder you throw it horizontally, the faster it travels away from you (Figure 13.12). It is possible, if you could throw hard enough, to launch the bucket into space so that it moves away from the Earth for ever. The Earth's gravity will curve the path of the bucket, but the curvature becomes less as the Earth becomes more and more distant and its gravitational influence fades.

Figure 13.12
Bucket trajectories!

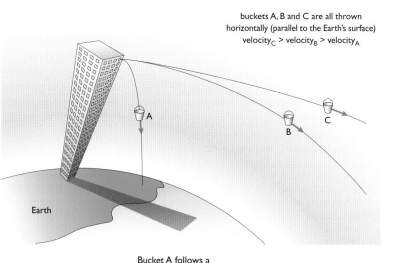

buckets A, B and C are all thrown horizontally (parallel to the Earth's surface)
velocity$_C$ > velocity$_B$ > velocity$_A$

Bucket C has a very large initial horizontal speed. Though its path is curved by the effect of the Earth's gravity, it moves away from the Earth.

Bucket B has just the right initial horizontal speed for the curvature of its path to match the curvature of the Earth. It remains at a constant distance from the Earth – in orbit.

Bucket A follows a parabolic path to the ground.

Earth

17 An apple falls to Earth. Why doesn't the Moon apparently do the same?

18 a For a body in orbit, which vector is in a different direction from the other two – velocity, acceleration or force?
   b Describe the direction of the 'odd one out' relative to the direction of the other two.

If you throw it at just the right speed, it is possible to put the bucket into permanent orbit. It is then forever accelerating 'towards' the Earth, but has such high speed that, the Earth being round and not flat, it never gets any closer to the surface. Speed is a constant but velocity changes due to the Earth's pull. It changes such that it is always parallel to the Earth's surface.

● Comprehension and application

Dancing in the sky

Kitsou Dubois (Figure 13.13) is a dancer with ambition. She wanted to know what it would be like to dance in space – to be truly weightless. She hasn't managed to persuade anybody to let her travel in space – not yet at least. But she has done the next best thing. She persuaded the French air force to take her on an astronaut's training flight.

The aircraft rises and falls. It uses its engines to reach high speed and begins to rise, and then switches off the thrust of its engines and dives. Its motion is then parabolic – that is, it is the same shape as the path of a cannonball. While describing this parabolic path, the aircraft and its contents behave very much like cannonballs do. On such a path, a cannonball, an aircraft and Kitsou are all subject to gravitational force and have a downwards acceleration to match. This is (again ignoring resistive effects, which for Kitsou inside the aircraft are non-existent because the air and everything else in the cabin moves with her) equal to 9.81 m s^{-2}. They also, of course, have velocity due to their motion at the start of the parabola.

Figure 13.13
Kitsou Dubois dancing in effective weightlessness, thanks to the French air force.

When Kitsou is acting as such a human cannonball there is no force acting on her other than the force of gravity. The floor and walls of the aircraft exert no force on her, allowing gravity to take full effect and accelerate her downwards. But since the floor and walls exert no force on her, unless she bumps into them, she floats freely in the cabin. She feels as if she is weightless.

When the aircraft comes out of the final part of the parabola – that is, when it comes out of its dive – then the floor of the cabin will exert force on her again. This upward force must become bigger than her weight, in order to decelerate her. She needs to take care that the force isn't too big or she could get badly hurt.

So, as the aircraft flies through one parabolic flight path after another, Kitsou changes from feeling weightless and accelerating downwards to feeling a large force accelerating her upwards. It takes a strong stomach to dance in the sky.

19 Kitsou may feel weightless, but is she? Explain.

20 Draw Kitsou's parabolic path and mark six points along the path. Draw vectors to show Kitsou's velocity, acceleration and force experienced at each of these six points. For velocity, show the horizontal and vertical components.

21 Describe examples, for the benefit of an imaginary audience who know little physics, of situations in which acceleration and velocity are
a in the same direction
b in opposite directions
c in neither the same nor opposite directions.
Use Kitsou's motion as an example where appropriate.

22 Kitsou's cabin was windowless. Would it have been possible for her to sense any difference between her experience, during parabolic flight, and the experience she would have had if she had
a been in a similar cabin that had travelled to deep space and out of the Earth's gravitational field
b been in a similar cabin that fell vertically towards the Earth
c been in a similar cabin in orbit?

23 a Draw a single sketch to show
i Kitsou's weight (as a vector)
ii an upward force bigger than her weight exerted by the aircraft floor on Kitsou
iii the net force.
b Why must the upward force be bigger than her weight at the end of a parabolic period of flight?

● **Extra skills task** ## Communication and Information Technology

Many spacecraft have been sent into orbit or on paths that will take them far into outer space. One major problem to overcome is the gravitational pull of the Sun.

Cassini is one such space probe recently sent to investigate Saturn (the Cassini–Huygens mission). The kinetic energy of Venus, Earth, Saturn and Jupiter were used to sling the spacecraft on its way, conserving fuel and accelerating it rapidly towards its final destination.

Use as many sources as you can to research how designers for this, and other, projects have accelerated space vehicles using planets' motion and gravitational fields to help them attain their final goals. Sources may include the Internet, CD ROMs, libraries and scientific journals. You should produce a written report of approximately four sides in length, including diagrams and other images, which clearly describes the physics behind these feats of planning. To make your report clearer, use headings, sub-headings and highlighting in an appropriate way.

14 Mass, motion and momentum

THE BIG QUESTION

● Bodies, from pairs of particles to entire galaxies, interact. How can we analyse the effects of these interactions on motion of the bodies?

KEY VOCABULARY

conserved (quantities)   elastic collisions   explosion   impulse   inelastic collisions
momentum   Newton's Second Law   principle of conservation of momentum

BACKGROUND

Figure 14.1
A simple collision between particles.

Figure 14.1 shows the effects of a collision in a world of particles smaller than the atom. In fact the collision is between an alpha particle and a hydrogen nucleus. There may not be direct contact in such a process, but the particles involved come close enough together to repel and change each other's motion. Collisions take place on all other scales – between molecules of gas, between dancers on a dance floor, between galaxies.

Mass

Mass, like time, is one of the few words in physics that is difficult to define. In Newtonian physics there are not one but two definitions of mass: inertial mass and gravitational mass (see Chapter 13). In this chapter we are dealing with motion, so it is inertial mass that we are concerned with. Inertial mass provides a measure of a body's resistance to change in its motion (resistance to acceleration).

Two kinds of collision

Collisions between free sub-atomic particles, such as an alpha particle and a nucleus, are usually relatively simple. The two particles approach each other, exert electrical force on each other, and their motion is changed. The two then have new velocities. Nothing else changes – there is no screeching of brakes, no crunch of metal, nothing changes shape, no heating of material takes place. The two particles have kinetic energy (see Chapter 15) before the collision and, though one may gain velocity while the other loses it, their combined or total kinetic energy after the collision is just the same as it was before.

Collisions in which the total kinetic energy is **conserved** (unchanged by the process taking place) are called **elastic collisions**.

A collision between two cars is very different. During the collision shapes change – there is deformation of material. Once this has finished the temperature of the materials involved (including surrounding materials, such as the air and the tar on the road) will have risen a little. Energy has been transferred from the kinetic energy of the cars to internal energy (see Chapter 15) of the material they are made of. The total kinetic energy of the two cars after the collision is less than their total kinetic energy before.

Collisions in which the total kinetic energy changes are called **inelastic collisions**. Inelastic collisions are harder to predict, because there's more going on. There are more variables, such as the quantities of internal energy gained by the colliding materials, which influence the outcome. Collisions in the human-sized world, involving larger bodies of material than particles, are always inelastic (Figure 14.2).

However, some human-sized collisions approximate closely enough to simple elastic processes for it to be worthwhile to suppose that they are elastic (Figure 14.3). A head-on collision between swinging pendulum bobs that do not deform also approximates to an elastic collision.

1 What would you expect to find if you were to swing two pendulum bobs into each other many times and then measure their temperature?

2 Two ice-hockey players collide. Is this an elastic or an inelastic collision?

Figure 14.2
An inelastic collision
on a large scale.

Here two spiral galaxies
are colliding in the
constellation of Corvus.
 A galaxy is not a simple
one-particle body but has
an internal structure made
up of millions of stars. The
kinetic energy of this
collision has triggered the
creation of over 1000
bright, young star clusters
(white/blue).
 The total kinetic energy
of the pair of colliding
galaxies is changed by the
collision. Such collisions are
inelastic.

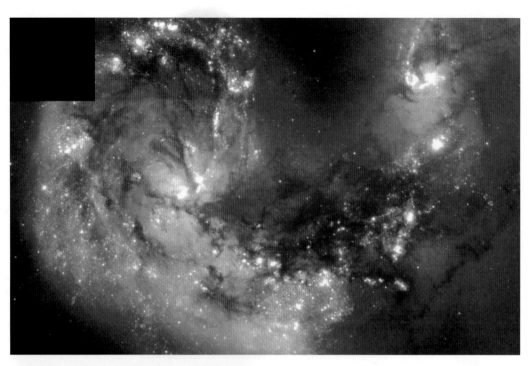

Figure 14.3
An approximation to an
elastic collision in the
human-sized world.

Elastic particle collisions are relatively simple.
Snookerball–snookerball collisions approximate to this, but
snooker players complicate the situation by introducing
more variables, such as the spin of the balls and the
interaction of the travelling balls with the surface of the
table. A useful general technique in physics is to consider
what happens in the simple or 'ideal' case, such as an elastic
collision, and then consider why and how actual behaviour
might not match the ideal.

Momentum and its conservation

The usefulness of the variables of velocity and mass in analysing collisions is so strong that it is
worth combining them into a single quantity, and giving a name to this new quantity we've
created. The product of mass and velocity (that is, mass multiplied by velocity) is called
momentum. Expressed as an equation:

momentum, p = mass $\times$ velocity

With mass measured in kilograms and velocity measured in metres per second, the unit of
momentum is kilogram metres per second, or $kg\ m\ s^{-1}$. Although momentum is a useful quantity,
the unit $kg\ m\ s^{-1}$ has never been given a special name.
 The particular usefulness of momentum in analysing collisions is due to an empirical
(observed) finding:

 In *all* collisions, the total momentum of the bodies before the collision is the same as that
 after the collision.

This statement is the **principle of conservation of momentum**. It allows quantitative analysis
of collisions. We can write an equation:

 total momentum before interaction = total momentum after interaction

or

$$p_1 = p_2$$

3 What does the word
conservation mean?
4 Why don't we use m
as an abbreviation of
momentum?

Momentum conservation in inelastic and elastic collisions

Momentum is conserved in all collisions. Kinetic energy is conserved only in elastic collisions. For *inelastic* collisions, all that we know is:

total momentum before collision = total momentum after collision

For *elastic* collisions, the two conservation laws together give us two equations which are useful for predicting the outcome of collisions:

total momentum before collision = total momentum after collision

total kinetic energy before collision = total kinetic energy after collision

Figure 14.4
Analysing **a** inelastic and **b** approximately elastic collisions.

a

b
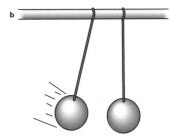

Collisions in the human-scale world are inelastic. Momentum is conserved but total kinetic energy decreases while total internal energy (of the colliding bodies and then of their surrounding materials) increases.

total momentum before = total momentum after
total kinetic energy before > total kinetic energy after
total internal energy before < total internal energy after

A collision between metal balls of a 'Newton's cradle' is *approximately* elastic.

total momentum before ≈ total momentum after
total kinetic energy before ≈ total kinetic energy after
total internal energy before ≈ total internal energy after

Inelastic collisions between bodies that stick together after colliding

For inelastic collisions we can write only one equation – the equation based on conservation of momentum. If, however, the bodies stick together after their collision then they have the same velocity. This simplifies the mathematics, as shown in the following example.

Example 1 Two trucks are moving in the same direction along the track of an adventure park ride. One has a velocity of 8 m s^{-1} and the other, with twice the mass, has a velocity of 6 m s^{-1}. They collide and link together. What is their new speed?

initial momentum of first truck $= mv_1 \ \ = m \times 8 \ \ = 8m$
initial momentum of second truck $= 2mv_2 = 2m \times 6 = 12m$

(We don't know the masses, but we do know that one is twice as big as the other, and for the moment we can just write the masses as m and $2m$.)

final momentum of linked trucks $= (m + 2m) \times v = 3mv$

The trucks are now acting as a single vehicle, mass $(m + 2m)$ and new velocity v.

total momentum before = total momentum after
$$8m + 12m = 3mv$$

Mass, m, appears in all three terms of the equation, so can be 'cancelled'. (Cancelling here is a short-hand method for dividing both sides of the equation by m.)

$$8 + 12 = 3v$$
$$v = \frac{8 + 12}{3}$$
$$= 6.7 \text{ m s}^{-1}$$

5 What will be the final velocity of the linked trucks in Example I if the faster truck has a mass of 90 kg and the slower one has a mass of 50 kg?

6 The collision between the trucks that stick together in Example I is inelastic. What can be said about the total kinetic energy before and after the collision?

7 Collisions between two protons are usually elastic.
  a Write down two equations to show conservation of quantities in an elastic collision between two protons.
  b Explain why a collision between larger bodies, such as tennis balls, can never be perfectly elastic.

8 Explain the effect of the custard pie on its thrower in terms of Newton's Third Law (see Chapter 11).

Note that it is acceptable not to include the units of quantities in the intermediate steps in a calculation, but the appropriate unit *must* be given as part of the answer unless the quantity is dimensionless (see page 14) and has no unit.

Momentum conservation in collisions between bodies of very different masses

A custard pie provides a nice example of a body that can have significant momentum but which makes inelastic collisions. Deformation of the pie, and perhaps of its target, take place. Temperatures rise. Kinetic energy is not conserved.

If you are hit by a custard pie it loses momentum, but you don't seem to gain it to compensate for this. It might seem that total momentum after the collision is less than that before. However, if you were resting on a frictionless surface when hit, you would find that you do gain obvious momentum – the impact of the custard pie would cause you to slide. But in most cases you are effectively connected to the ground by a force of friction that resists such sliding. Unless you slide, you can only move if the whole of the ground moves with you. Your effective mass is not just your personal mass but also the mass of the ground you are standing on. The mass is huge compared with that of the custard pie, and so any gain in velocity of the you/ground system is very tiny.

Momentum has direction

Momentum is a vector quantity. This should not be surprising, because it is the product of a scalar, mass, and a vector, velocity.

Think about two identical bodies – they could be galaxies, cars, skaters, cannonballs or protons – moving towards each other along a single straight line so that they collide 'head on'. In each case, the value of the total momentum before the collision depends on the directions of the momenta of each body. The momenta of the two bodies have opposite signs (Figure 14.5).

Figure 14.5
Momentum has direction.

total momentum before collision $= p_1 + p_2$

If we say that p_1 is positive then we have to say that p_2 is a negative quantity. If p_1 and p_2 are the same size then

total momentum before collision $= 0$

- -

Example 2 The two adventure park trucks of Example 1, still of mass m and $2m$, are now moving towards each other with velocities 8 m s^{-1} and 6 m s^{-1} and again they link up after their collision. What is the new velocity?

$$\text{initial momentum of first truck} = mv_1 = m \times 8 = 8m$$
$$\text{initial momentum of second truck} = 2mv_2 = 2m \times (-6) = -12m$$
$$\text{final momentum of linked trucks} = 3m \times v$$

(The trucks are now acting as a single vehicle, mass $3m$ and velocity v.)

$$\text{total momentum before} = \text{total momentum after}$$
$$8m + (-12m) = 3mv$$

9 a Draw a sketch using arrows to show the various velocities before and after this collision.
  b Explain why the final velocity is shown as negative.
  c Draw a sketch using arrows to show the various momenta before and after the collision.
  d If the faster truck had a mass of 90 kg and the slower truck had a mass of 50 kg, what would the new final velocity be?

Mass, m, appears in all three terms of the equation, and can be 'cancelled':

$$8 - 12 = 3v$$
$$v = \frac{8 - 12}{3} = \frac{-4}{3}$$
$$= -1.3 \text{ m s}^{-1}$$

- -

Explosions

Imagine two ice dancers standing still and close together, who then push on each other and move apart across the ice. We wouldn't, in everyday language, say that the dancers have exploded. However, we need a name to describe this kind of interaction that, like collisions, involves exchange of momentum between bodies. Such a process, in which bodies are initially stationary relative to each other is, in physics, called an **explosion**. The total momentum acquired as a result of an explosion must be zero, since momentum is conserved in all processes.

The same word, explosion, is used to describe what happens when a number of particles which are all initially stationary move apart thanks to a sudden transfer of energy, perhaps from the potential energy of chemical bonds. However, such a chemically-driven explosion is complex in that there are many particles of different sizes and they spread into three dimensions. The motion of the two skaters is along a single straight line – it is one-dimensional, and so is much easier to think about.

For the study of one-dimensional explosions, dynamics trolleys are fitted with sprung plungers (Figure 14.6). The release of a plunger pushes a pair of trolleys apart.

Figure 14.6
How a sprung dynamics trolley works.

rod is struck, e.g. with a hammer, to free the plunger from the lip

lip holds plunger in place

spring

plunger

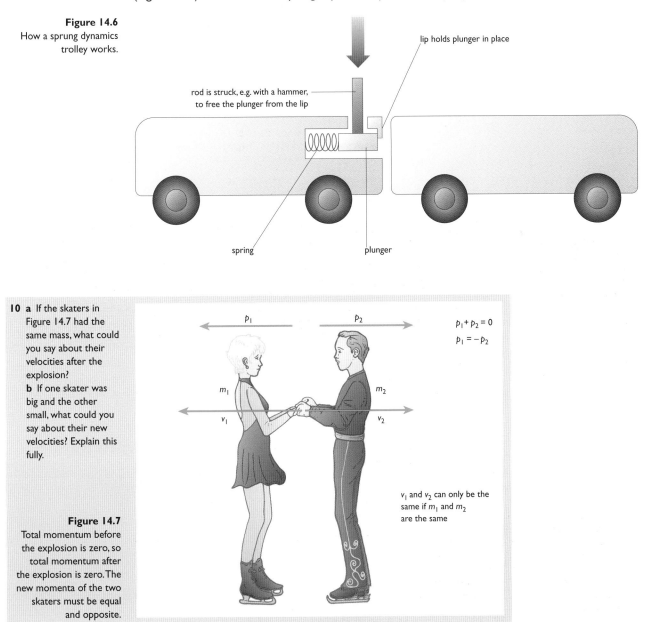

10 a If the skaters in Figure 14.7 had the same mass, what could you say about their velocities after the explosion?
b If one skater was big and the other small, what could you say about their new velocities? Explain this fully.

p_1 p_2

$p_1 + p_2 = 0$
$p_1 = -p_2$

m_1 m_2

v_1 v_2

v_1 and v_2 can only be the same if m_1 and m_2 are the same

Figure 14.7
Total momentum before the explosion is zero, so total momentum after the explosion is zero. The new momenta of the two skaters must be equal and opposite.

Momentum: throwing and receiving

A clown who throws a custard pie gives momentum to it. He is unlikely to have been on a frictionless surface when he did so. But if he had been, say, on ice, then momentum conservation would have been visible in his sliding in the opposite direction to the throw. The tendency is for the thrower to move in exactly the opposite direction to the person the pie hits (Figure 14.8).

Figure 14.8
Throwing and receiving,
a with and
b without friction.

a

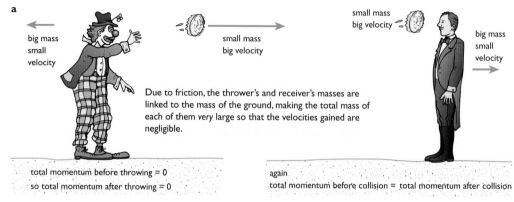

big mass small velocity

small mass big velocity

small mass big velocity

big mass small velocity

Due to friction, the thrower's and receiver's masses are linked to the mass of the ground, making the total mass of each of them *very* large so that the velocities gained are negligible.

total momentum before throwing = 0
so total momentum after throwing = 0

again
total momentum before collision = total momentum after collision

b

In the absence of friction it is a different story. The velocities gained by two freely floating astronauts who throw and receive an object the size of a custard pie are small but not negligible. The net effect of the exchange of the object is that the astronauts move away from each other.

Impulse

The **impulse** that a body experiences during an interaction with another is the change in its momentum.

$$\text{impulse} = \Delta p = \Delta(mv) = \text{final momentum} - \text{initial momentum}$$

A ball of mass m that flies at a wall at $90°$ to its surface at velocity v has an initial momentum of mv. Suppose that it is an elastic collision and the ball bounces off the wall with the same speed, but now with velocity $-v$. The final momentum is $-mv$.

$$\text{impulse} = \text{final momentum} - \text{initial momentum} = -mv - mv = -2mv$$

11 If a stone has an initial velocity of 20 m s^{-1} and a mass of 0.02 kg, what is the impulse on hitting a wall and bouncing off with a velocity of 2 m s^{-1}?

Momentum and Newton's Second Law

In his First Law of Motion (see page 90), Newton provided a clear statement of the need for a force for a change in motion to take place. He effectively defined force as the cause of change in motion.

In his Second Law of Motion, Newton made a direct link between force and other quantities. In modern language we express **Newton's Second Law** as:

force is proportional to rate of change of momentum

This can be written in mathematical symbols:

$$F \propto \frac{\mathrm{d}(mv)}{\mathrm{d}t}$$

Note that *average* force $\propto \dfrac{\Delta(mv)}{\Delta t}$, but the relationship, force $\propto \dfrac{\mathrm{d}(mv)}{\mathrm{d}t}$, does not relate just to average force, but to any force at any instant.

For a body of constant mass, change in momentum is entirely due to the change in velocity, so we can write d(*mv*) as *m* dv. So,

$$F \propto m \frac{dv}{dt}$$

Any proportionality relationship can be written as an equation by inserting a constant of proportionality:

$$F = km \frac{dv}{dt} \quad \text{where } k \text{ is the constant of proportionality}$$

It makes sense to define the unit for force so that the mathematical relationship is as simple as possible. We can also call the unit what we like, as long as we all agree. The internationally agreed name is the newton, N, and we define the newton so that $k = 1$. This gives us:

$$F = m \frac{dv}{dt} \quad \text{where force is measured in newtons and other quantities in their usual SI units}$$

As we have seen, dv/dt = acceleration, *a*, so

$$F = ma$$

This equation is of central importance in physics. In this book we have already used it in Chapters 11 and 13. We have now seen how it is derived from Newton's Second Law.

> **12** Explain why $\Delta(mv) = m\Delta v$ is true for a caravan being towed from rest by a car but not (quite) true for the car and its load (including the contents of its fuel tank).

● ●

● **Comprehension and application**

The asteroid threat

Most (if not all) scientists now believe that a collision of an asteroid with the Earth played at least a strong part in the climatic and other changes that made the dinosaurs extinct. There is evidence of this event in layers of dust in rock strata around the world. The material of the asteroid became part of the Earth. That was 64 million years ago, and almost 64 million years before the appearance of the modern human species.

> **13** Explain why a collision between the Earth and a meteor or an asteroid is inelastic.
>
> **14** The circumference of the Earth's orbit around the Sun is about 10^{12} m. The journey takes 1 year. The mass of the Earth is about 6×10^{24} kg. A large asteroid might have a radius of 5 km.
> **a** What is the distance travelled by the Earth in 6 months? What is its displacement? Draw a sketch to illustrate the difference between distance and displacement.
> **b** What is the Earth's orbital speed, in m s^{-1}? Why is it true to say that this speed is constant but the velocity is constantly changing?
> **c** What is the Earth's momentum due to its orbital motion? What momentum will it have 6 months from now? Does this disobey the principle of conservation of momentum?
> **d** What is the volume, in m³, of the large asteroid assuming that it is spherical (as given by $V = \frac{4}{3}\pi r^3$)? If it has an average density of 2500 kg m^{-3}, what is its mass?
>
> **15** Imagine that an asteroid is travelling in a straight line and straight for a collision with the Earth. Sketch displacement–time graphs for the asteroid's travel,
> **a** assuming that its speed is constant
> **b** more realistically, allowing (in general terms) for the gravitational attraction of the Earth.
>
> **16** Suppose that the asteroid is moving in the opposite direction to the Earth, relative to the Sun. In fact this is very unlikely – an asteroid is, like the Earth, in orbit around the Sun, and planets and other bodies orbit in the same direction, though asteroids can have irregular orbits. However, it makes an interesting 'worst case scenario' to consider a head-on collision. Suppose the asteroid has the same speed as the Earth, and its mass is as you calculated in question 14d.
> **a** What is its momentum?
> **b** What is the total momentum of the Earth–asteroid system?
> **c** By what percentage will the collision change the Earth's momentum?

Most asteroids orbit the Sun, making a belt of rocks between the orbits of Mars and Jupiter. Most of them seem to behave in the same way as any other orbiting body, like the planets themselves, and they remain in a steady orbit. Some asteroids, safely in the asteroid belt beyond the orbit of Mars, can be detected from Earth as individual bodies, and have been given names. *Ceres* is the biggest at about 1000 km across. But some seem to have been thrown out of this pathway by some cataclysmic event long ago, and they follow more unusual routes. Some of them cross the path of the Earth's orbit.

In November 1937 the asteroid *Hermes* passed the Earth at a distance of 800 000 km. There have also been actual collisions in recent times – an area of Siberian forest was devastated in 1911.

If you look up at a clear night-time sky it usually won't be many minutes before you see a meteor burning up in the Earth's upper atmosphere. Most of these meteors are just grains of dust and they don't penetrate far into the atmosphere, but some are larger and can reach the surface – the rocks that are occasionally found are called meteorites.

Asteroids are much bigger than meteors. Are we due for another big collision? That is hard to answer, largely because most asteroids are not big enough to see until they are quite close to the Earth, so it's hard to tell when or if there is anything coming our way. In any case, there is really not much we can do to protect ourselves from an impact.

● **Extra skills task** ## Communication and Information Technology

The possibility of an asteroid–Earth collision is real but it is no greater than it has been throughout human history. It has been said that space scientists, afraid that politicians will stop providing funding for their work, have been keen to draw attention to the possibility – to save their jobs.

In a group, each person should take one of the following roles:

1 scientist keen to emphasise the asteroid threat in order to protect his/her job
2 scientist who is keen that decisions should be based on a careful assessment of the risk
3 politician keen to cut government spending on research as much as possible
4 politician keen to react to public opinion in making decisions
5 journalist who works for a popular sensationalist newspaper
6 journalist who works for a newspaper which prides itself on 'intelligent' assessment.

The first four should use the Internet (and other resources if possible) to research the asteroid threat, and should each prepare a short submission (about one page) to promote their position. They should present their submissions to the group. The journalists should take notes and write short articles for their newspapers about the asteroid threat. These articles should be word processed and presented back to the group at a later date.

IV
ENERGY

15 Energy for working and heating

THE BIG QUESTIONS
- What is energy?
- Why is it such an important concept?

KEY VOCABULARY
conservation   dissipation   efficiency   Einstein's mass–energy equation
elastic strain energy   electrical potential energy   electronvolt   energy
gravitational potential energy   internal energy   joule   kinetic energy
mass–energy   potential energy   principle of conservation of energy
principle of conservation of mass   principle of conservation of mass–energy
violation (of principle of conservation)   watt   work

BACKGROUND
All changes – from motions and interactions of sub-atomic particles to changing your socks – involve energy transfer from place to place, from system to system. If there were no energy transfer there could be no change. The Universe would be in a state of inactivity.

Figure 15.1
There are plenty of energy transfers from place to place, from system to system, going on here – from Sun to plants, from body to bicycle to air.

Doing and not doing work

Work is defined in physics as the product of force and the displacement that takes place in the direction of the force while the force is acting. Work can only be done when both force and displacement are involved.

Figure 15.2
In which of these situations is work being done?

a

The tennis player is exerting a force. The post is deformed very slightly but it does not move. It simply provides a force (a reaction) to oppose the push of the player's body. The player is doing no work on the post.

force

b

If we neglect the small force of air resistance then we can say that the ball is experiencing no net force and is moving at constant velocity. Where there is displacement but no force, no work is being done. The ball is doing no work, and no work is being done on the ball.

c

force

While the ball is in contact with the racquet it is experiencing force. It is also displaced (it moves in a particular direction) while under the action of this force. The racquet is doing work on the ball. (Since the racquet experiences force and displacement we can also say that the ball does work on the racquet.)

I In which of the situations in Figure 15.3 is work being done on the weights? Explain your answers.

a

lifting the weights

b

holding the weights high

c

in a lift moving upwards at steady speed

Figure 15.3

At its simplest, we can express the relationship that follows from the definition of work as follows:

work = force × displacement under the action of the force and in the same direction

$$W = Fx$$

The unit of work could be correctly called a newton metre. However, work is such a useful quantity that the newton metre has been given a name of its own. A newton metre is called a **joule**, or J for short.

What if the force and the displacement are not in the same direction?

Force and displacement are both vector quantities. When they are in the same direction then the work done is simply their product, $W = Fx$. (Note that though work is the product of two vectors, it is not itself a vector quantity but a scalar. Work has no direction.) To calculate the work done when force and displacement act in *different* directions we need to consider their parallel components.

The force applied to a crate to pull it across a floor, for example, might not be parallel to the crate's displacement. Force and displacement are then not in the same direction (Figure 15.4).

Figure 15.4
Here the applied force is at an angle to the crate's displacement.

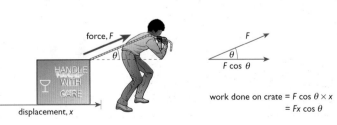

force, F

θ

displacement, x

F

θ

F cos θ

work done on crate = F cos θ × x
= Fx cos θ

2 What is the work done by the force exerted by each tug boat in Figure 15.5?

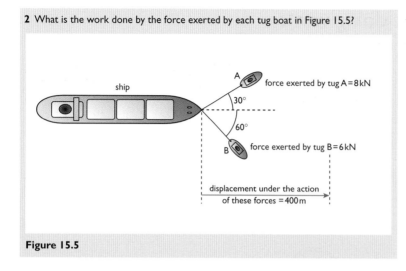

ship

A — force exerted by tug A = 8 kN

30°

60°

B — force exerted by tug B = 6 kN

displacement under the action
of these forces = 400 m

Figure 15.5

For the crate of Figure 15.4 and similar situations, we need to think about parallel components of the force and displacement. We can either find the component of force that is parallel to the actual displacement, or we can find the component of the displacement that is parallel to the actual force.

The component of the force which is perpendicular to the displacement is balanced by the force of gravity and has no other effect. It is not doing work. Only the component of force that acts in the same direction as the displacement can be thought of as doing work.

If the angle between the force F and the displacement is θ, the component of F parallel to the displacement is $F \cos \theta$ (see page 100).

Work done by the action of a constant force

The force needed to overcome friction and push a crate across the floor at a constant velocity is approximately constant; we can call the average force F_1. Due to friction, F_1 might have to be quite large. The displacement of the crate can be called x, so that

$$\text{work done on the crate to overcome friction} = F_1 x$$

If the person moving the crate has the sense to put rollers under it then he or she can reduce the average force needed, and reduce the amount of work that has to be done. If we call the new reduced average force F_2, then

$$\text{work done on the crate to overcome friction, with rollers} = F_2 x$$

We can create useful visual representations of the two situations. Figure 15.6 shows graphs of force (y-axis) plotted against displacement (x-axis).

Figure 15.6
A greater force for the same displacement means that more work must be done. Work done = 'area under graph'.

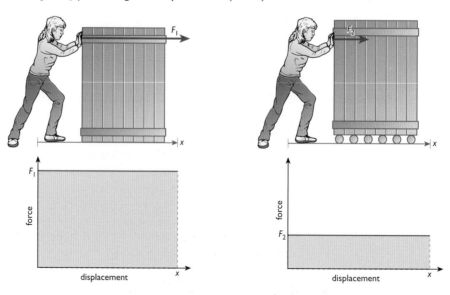

3 Sketch graphs to represent a small force and a large force doing *equal* amounts of work.

Notice that both graphs produce simple rectangles. The areas of these rectangles are called the 'areas under the graphs'. The height of each rectangle is the force, and the width is the total displacement of the crate while the force is acting. The area of each rectangle is the force multiplied by the displacement. This, of course, is equal to the work done.

Work done by a varying force

Stretching a spring

The force required to extend a spring is not constant. The larger the extension, the bigger the force. In fact (provided that you don't pull too far and permanently distort the spring) the force is proportional to the extension. Extension is the same thing as the displacement of the moving end of the spring. The proportionality produces a graph which is a straight line, passing through the origin (Figure 15.7).

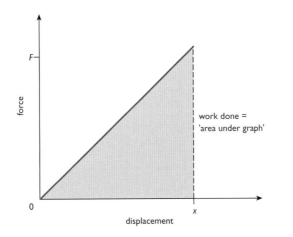

Figure 15.7
Force–displacement graph for the extension of a spring.

The total amount of work done is again the product of the average force and the displacement. The average force is not quite so simple as when applying an approximately constant force to push a crate against friction. Fortunately, though, the situation has not become too complicated. The force increases steadily so we can use just two values of force to work out its average – the first (initial) and the final values of force:

$$\text{average force for the spring extension} = \frac{\text{initial force} + \text{final force}}{2}$$

Since the initial force is zero, this simplifies further:

$$\text{average force for the spring extension} = \frac{\text{final force}}{2}$$

So the work done on the spring is given by:

$$\text{work done} = \text{average force} \times \text{displacement}$$
$$= \frac{\text{final force} \times \text{displacement}}{2}$$
$$= \frac{Fx}{2}$$

The final force is the height of the triangle made by the force–displacement graph. The displacement is the base of the triangle, so notice that

$$\text{work done} = \frac{\text{height of triangle} \times \text{base}}{2}$$

and that the right-hand side is identical to the standard formula for working out the area of a triangle. So, the work done is again equal to the area under the graph.

More complex systems

A tennis racquet and tennis ball interact in a complex way. The force on the ball rises from zero as the moving ball hits the racquet. The shape of the graph is not so easily described as for a simple spring.

Figure 15.8
Possible
force–displacement
graph for a tennis ball as it
hits the racquet.

The tops of the narrow slices very closely follow the curving shape of the graph. The narrower the slices, the more smoothly and precisely they follow the curve.

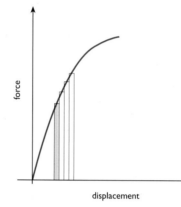

Nevertheless, we can think about a small portion of the graph – a vertical slice for which change in displacement is small and force can be considered approximately constant (Figure 15.8). For one slice we find, again, that the work done is equal to its area, and this is true for every single slice of the graph. It is therefore approximately true for the whole graph, and the narrower the slices the more precise this approximation becomes. So the statement that:

work done = area under the force–displacement graph

is not only true for constant forces and for forces that are proportional to displacement (as for the spring), but is found to be true for all shapes of graph.

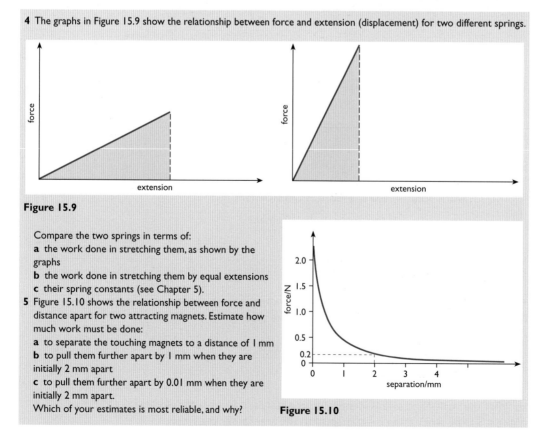

4 The graphs in Figure 15.9 show the relationship between force and extension (displacement) for two different springs.

Figure 15.9

Compare the two springs in terms of:
a the work done in stretching them, as shown by the graphs
b the work done in stretching them by equal extensions
c their spring constants (see Chapter 5).
5 Figure 15.10 shows the relationship between force and distance apart for two attracting magnets. Estimate how much work must be done:
a to separate the touching magnets to a distance of 1 mm
b to pull them further apart by 1 mm when they are initially 2 mm apart
c to pull them further apart by 0.01 mm when they are initially 2 mm apart.
Which of your estimates is most reliable, and why?

Figure 15.10

The ability to do work

A demolition ball, like any moving body, can do work (exert a force that acts over a distance) on other bodies with which it collides.

An arrow in a stretched bow is not yet moving, but the arrow-and-bow system already has the ability to do work due to the tension in the string. This tension is a force which is only temporarily balanced by the pull of the fingers.

Likewise, water in a mountain reservoir is subject to a gravitational force which is temporarily balanced by the support of the ground beneath it. The water will fall as soon as that becomes possible, and the moving water will have the ability to do work. The ability to do work exists before the motion begins.

A potato, when it reacts chemically with oxygen, can allow materials to do work. The material of the potato is fuel. Burning the fuel (even slowly as in a human body) can, for example, set material in motion. The principle is the same for the combustion of any other fuel. Petrol can, by reacting with oxygen, set pistons in motion. A fuel, in the presence of oxygen, has the ability to do work.

The ability to do work is called **energy**. The amount of energy that a system possesses is equal to the total amount of work that it can potentially (under ideal circumstances) do. Energy, just like work, is measured in joules.

Figure 15.11
Each of these systems possesses the ability to do work.

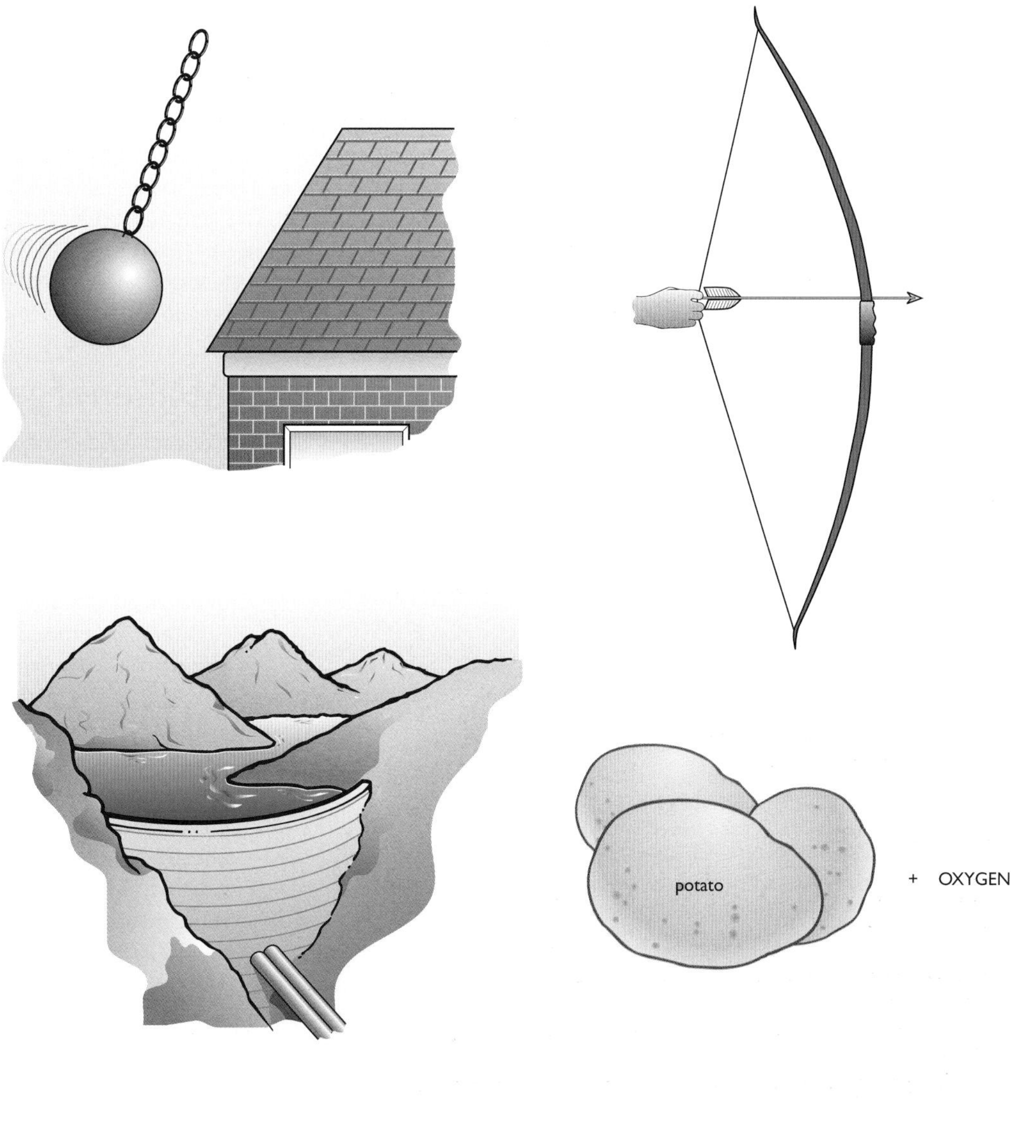

potato   +   OXYGEN

Energy for working and heating

Figure 15.12
This diagram shows the relative values of total available energy and the energies transferred by working and by heating in a car engine.

Doing work is a process of transferring energy. Combustion of petrol in a car engine is an example of an energy transfer process and of a process of doing work – exerting force and displacing pistons. The processes inside the engine transfer energy mechanically to the pistons. But energy transfer by thermal processes, or 'heating', also takes place. Thermal transfer of energy to a passive body of material, such as a lump of metal or a container of water, can result in an increase in temperature or a change of state, or both.

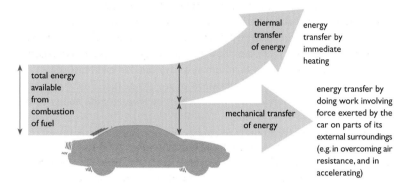

In all real energy transfer processes the total initial amount of energy is not transferred only by doing work. Some energy is always transferred to heat the system and/or its surroundings. An energy transfer diagram can use arrows whose widths show the relative amounts of the available energy that is transferred by working (mechanical transfer) and heating (thermal transfer) in a system, such as a car engine (Figure 15.12).

Energy carried by light

Electromagnetic radiation from the Sun can allow work to be done. It heats the atmosphere, not just raising its temperature but creating differences in temperature that lead to movements of the air that are sometimes violent, as in hurricanes and tornadoes. On a molecular level, it is the light of the Sun that makes photosynthesis possible. The light and other electromagnetic radiations transfer energy. Almost all the energy available to us – from foods, fuels and renewable resources like winds and tides – is provided by the Sun. (Nuclear and geothermal energy resources are the exceptions.) All the work your body will ever do is made possible by the Sun.

Introducing kinetic and potential energy

6 A swinging demolition ball has the ability to do 1 kJ of work on a wall with which it collides. How much kinetic energy does it have just before the collision?

7 Table 15.1 shows energy values associated with some different processes.

Table 15.1

Process	Energy/joules
energy transferred by combustion of 1 ml of petrol	38 000
energy transferred by average human metabolism in 1 s	150
kinetic energy of the Earth in orbit around the Sun	2.65×10^{33}
energy of a photon (quantised energy carried by electromagnetic radiation) from a sodium street light	3×10^{-18}

a What is the range of energies covered by the data in the table?
b Write each value of energy in megajoules, MJ.
c How much energy is made available by the combustion of 1 litre of petrol?
d How many street light photons would you have to absorb to fuel your metabolism for 1 second? What changes to the human body would be necessary for this to be possible?
e There are approximately 10^{10} people on Earth. How many people could live for 1 second on the kinetic energy of the orbiting Earth? By what factor is this bigger than the population of the Earth? For how long could the Earth's population survive by making use of this kinetic energy? If this kinetic energy could be made available for our use in this way, what would happen to the Earth?

We can make some fundamental classification of mechanical energy that a system possesses. The energy of a body due to its motion is called its **kinetic energy**. The energy of a system of material due to the future action of unbalanced force is its **potential energy**.

Water in a mountain reservoir has **gravitational potential energy**. For the bow and arrow and for a fuel, the forces involved are not gravitational but forces acting between particles at a molecular level; these forces, however subtle and complex, are electrical forces. At a molecular level, the energies involved are therefore **electrical potential energies**. The falling water, the released arrow and the particles in a burning fuel all involve loss of potential energy and gain in kinetic energy.

Energy storage as potential and kinetic energy

If the water in a mountain reservoir can be controlled, as in a hydro-electric power system, then energy can be made available when it's wanted. In the meantime, energy is 'stored' as gravitational potential energy.

Energy can be stored not only as potential energy but also as kinetic energy. A flywheel is a rotating body with significant mass. It can gain kinetic energy and spin faster, and then make this energy available later by slowing down. This is useful in a car engine, where the explosions in the cylinders do their work in sudden bursts. The rotation of the crankshaft is made smoother by the flywheel that is attached to it. Speeding up after each explosion and slowing down in between explosions still takes place, but to a much reduced extent because the large mass of the flywheel can take significant energy into storage at each explosion and release it again between explosions.

Fuels

Most naturally occurring chemical reactions make energy available to the world outside, heating the surroundings and allowing work to be done. In the case of a fuel in controlled combustion, the energy can be made available when it's wanted. In the meantime the energy is 'stored'.

The analogy between the gravitational and electrical situations is useful. Since we can picture water in reservoirs in the hills, the gravitational system makes a good model to help us to understand the electrical system. The change in electrical potential energy that takes place in a combustion process (a chemical reaction) can be compared to the change in gravitational potential energy that occurs when water flows from one mountain reservoir to another one lower down (Figure 15.13).

Figure 15.13
The release of stored energy.

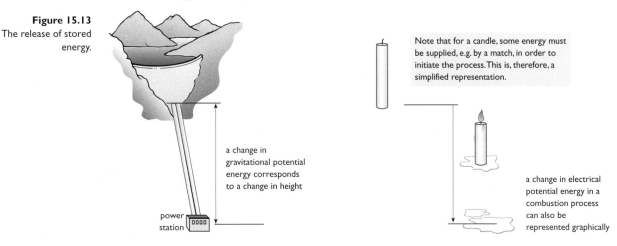

a change in gravitational potential energy corresponds to a change in height

power station

Note that for a candle, some energy must be supplied, e.g. by a match, in order to initiate the process. This is, therefore, a simplified representation.

a change in electrical potential energy in a combustion process can also be represented graphically

Energy storage by a spring

Work is done in stretching a spring. As a result, the spring is capable of doing this same amount of work when it is released. For as long as it is kept in an extended state the spring is a system with an ability to do work. While it is extended we can think of the spring as an energy storage system.

energy stored = amount of work that can be done on release
= amount of work done in first stretching the spring

$$= \frac{Fx}{2} \quad \text{(see page 127)}$$

8 Calculate the energy stored by a spring which stretches 2 cm when a force of 3 N is applied. (Remember to use consistent SI units.)

The energy stored by a spring is potential energy and is related to forces between particles in the spring. It is also called **elastic strain energy**. Note that the formula above applies to any system for which the displacement (or extension) is proportional to the applied force. It therefore applies to wires and cables as well as to springs (provided that the limit of proportionality is not exceeded).

Internal energy of bodies

In the early 1840s, James Joule got married, and he and his wife went to the green meadows, fresh mountain air and gushing rivers of the Swiss Alps for their honeymoon. At that time, rail travel was a very new possibility. Steam engines, however, had existed for some time. They involved hot steam turning into not-so-hot water, and work being done – that was the technology, but there was no complete scientific theory to explain the processes.

James Joule took his scientific instruments with him to Switzerland. It was the gushing rivers that he wanted to investigate. He measured the temperature of the water at the top and at the bottom of a waterfall and detected a small temperature difference in the water (Figure 15.14). The energy of the water – whether its potential energy at the top of its fall or the kinetic energy at the bottom – though apparently 'lost', did not cease to exist but had an effect on the water. It was evidence that heating could be achieved by transfer of potential and kinetic energy.

Figure 15.14
The mechanical energy of the water falling had gone into the material of the water itself – becoming internal energy – and the effect of this was to raise the temperature. (Note that the internal energy of a body is related to, but not the same as, its temperature.) Joule showed that there was a direct relationship between the mechanical energy made available by the fall of the water and the rise in temperature.

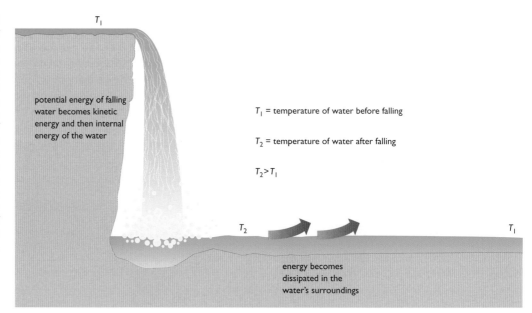

T_1

potential energy of falling water becomes kinetic energy and then internal energy of the water

T_1 = temperature of water before falling

T_2 = temperature of water after falling

$T_2 > T_1$

T_2

T_1

energy becomes dissipated in the water's surroundings

Joule went on to show a direct relationship between work done and heating effect. His work led to the notion of **internal energy** of material. Now, with our theory of the particle nature of matter, we can picture this internal energy of a body as the total kinetic and potential energy of its particles. We are, of course, completely unable to measure the kinetic and potential energies of individual particles, but we can measure changes in the internal energy of a body of material. Changes in internal energy are usually shown as ΔU. Such changes can be associated with changes in temperature, changes in state, and chemical changes.

9 a The temperature of a bag of sugar is increased. What happens to its total internal energy?
b A bag of sugar is burned (reacting with oxygen and producing waste materials). What happens to the total internal energy of the materials during this reaction?
c What other kind of change could alter the sugar's total internal energy?

Energy transfers and dissipation

James Joule made measurements on a physical system which was transferring energy. Initially, the water had potential energy. The energy became its kinetic energy, and then internal energy due to the increased agitation of its particles. As the water moved away from the foot of the waterfall, interaction between water particles and those of the surrounding materials, such as air and rock, took place. The surrounding particles became more agitated, and the water particles gradually slightly less so; energy was transferred between particles. Not far downstream from the waterfall any increase in temperature of the water above the initial temperature T_1 became too small to measure. The energy had transferred into the large quantity of surrounding materials. The energy had been **dissipated**.

10 The energy that can be made available by combustion of a 1 kg bag of sugar is 17 MJ.
  a What is the maximum amount of work that could be done using the sugar as fuel?
  b Why will the actual amount of work be less than this?
  c To climb stairs you need to experience an average upwards force equal to your own weight (e.g. 750 N). To what height (displacement) could you climb if the energy available is 17 MJ?
11 A hammer hits a nail repeatedly and the nail sinks further into a block of wood. The temperatures of the hammer and nail increase.
  a Draw an energy transfer diagram for this process.
  b The hammer and nail are left 'to cool down'. What happens to
  i their internal energy
  ii the internal energy of the materials in their surroundings?
  c Use your answer to **b ii** to explain the term 'dissipation of energy'.
12 **DISCUSS**
  'At the end of a journey, all of the energy made available to a car and its contents by combustion of fuel has been dissipated.'
  Is this always true?

Energy is a scalar quantity. It has no direction. We can't illustrate quantities of energy by using the kind of arrows we might use for vectors such as velocity and force. However, we can show energy transfers from system to system, with arrowheads to show the transfer, and the thickness of the arrow providing an indication of the relative quantities of energy involved (Figure 15.15).

All energy transfer processes can be represented by diagrams like these.

Figure 15.15 Diagrams showing the dissipation of energy in different transfer processes.

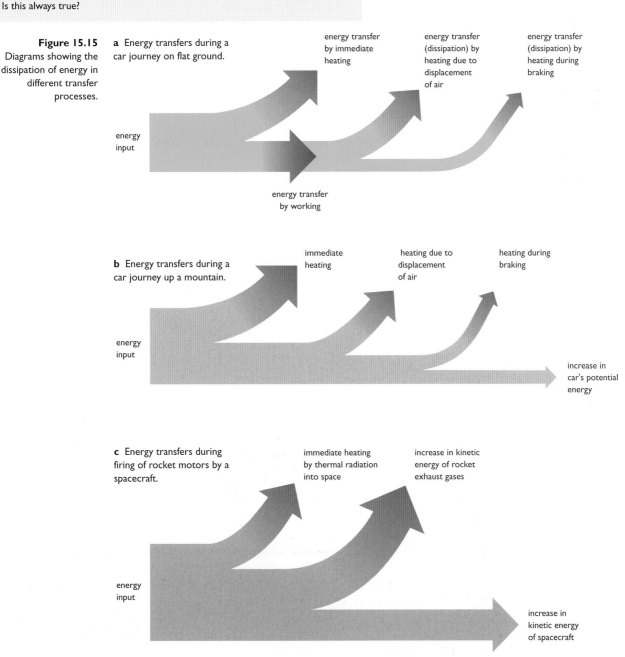

Energy conservation

In all the energy transfers shown in the previous section the total amount of energy is unchanged. That is, energy is conserved. This behaviour is so ever-present that it is called the **principle of conservation of energy**. An alternative statement of this principle is that energy cannot be created or destroyed, but only transferred between systems or locations.

13 When a bag of sugar burns there is a change in total internal energy of the materials. Is this consistent with the principle of conservation of energy?

Power as rate of transfer of energy

Power is a quantity that is closely related to energy, and in many circumstances it is a more useful measure of processes. Power is the rate of transfer of energy. (It can also be said that power is the rate of doing work, since doing work involves energy transfer.) We can write this definition of power as an equation:

$$\text{power}, P = \text{rate of transfer of energy}$$

$$\text{average } P = \frac{\Delta E}{\Delta t}$$

Note that in the special case for which power is constant, that is, when energy is transferred at a steady rate, average power is equal to instantaneous power, and then we can say that

$$\text{instantaneous } P = \frac{\Delta E}{\Delta t} \quad \text{where } P \text{ is constant}$$

However, this applies to the special case only. It is more generally true that

$$\text{instantaneous } P = \frac{dE}{dt}$$

The SI unit of power is the **watt**, W, named after the engineer James Watt (Figure 15.16). One watt is equivalent to a rate of transfer of energy of one joule per second. Table 15.2 shows typical power inputs for some energy transfer systems.

Table 15.2
Typical power inputs.

	Power input/W	Energy input in one hour/J
coal-fired power station	5×10^9	1.8×10^{13}
car	1×10^5	3.6×10^8
human	150	5.4×10^5

Figure 15.16
James Watt (1736–1819).

James Watt was a practical engineer with a good sense of marketing. Not only did he develop a steam engine with increased efficiency, but he sold it to owners of the new factories by comparing its work rate, or power, with that of horses. He described his engines in terms of their 'horsepower'.

14 Calculate the average power involved in using a 20 N maximum (final) force to stretch a spring by 4.4 cm in 0.6 s.

15 a Calculate the average power involved in exploding a bag of sugar (with 17 MJ of available energy) in 0.01 s. (Such a rapid change is possible if oxygen can be supplied quickly enough to the sugar, perhaps by using powdered sugar in an oxygen-rich environment.)
b Sketch graphs of energy against time to show constant and varying power, and use the graphs to show the distinction between
$\frac{\Delta E}{\Delta t}$ and $\frac{dE}{dt}$

Efficiency

Natural processes of change transfer energy. We also live in a world full of 'human-made' energy transfer devices – from watches to the engines of supertankers. All energy transfers involve some dissipation of energy. This means that the energy supplied, such as from a battery or fuel, does not all provide work or heating (or lighting) that we find useful. We have to supply more energy than we can actually take advantage of. Useful energy output is less than energy input. If this were not the case then we would use much less fuel, for example to generate electricity, and so we would save money and reduce pollution. Engineers, therefore, are always keen to maximise the **efficiency** of energy transfer processes. That is, they try to maximise the ratio between 'useful' energy output and the energy input that takes place over the same time. As a matter of convention, they express this ratio as a percentage:

$$\text{efficiency} = \frac{\text{useful energy output}}{\text{energy input during the same time}} \times 100\,\%$$

$$= \frac{\Delta E_2}{\Delta E_1} \times 100\,\%$$

where ΔE_1 and ΔE_2 are energy input and energy output.

Efficiency is a ratio, expressed as a percentage, of two quantities with the same dimensions. It is therefore a dimensionless quantity and has no units.

When power is unchanging, as we will assume here, then energy is related to power by $\Delta E = P\,\Delta t$. So, for a constant power input P_1 and a constant power output P_2,

$$\Delta E_1 = P_1\,\Delta t \qquad \text{and} \qquad \Delta E_2 = P_2\,\Delta t$$

So

$$\text{efficiency} = \frac{P_2\Delta t}{P_1\Delta t} \times 100\,\%$$

$$= \frac{P_2}{P_1} \times 100\,\%$$

$$= \frac{\text{useful power output}}{\text{power input}} \times 100\,\%$$

Engineers often prefer to use this definition of efficiency because it removes the need to specify that the two quantities in the ratio must relate to identical periods of time.

16 A power station has an efficiency of 35%.
a If its power output is 100 MW, what must be its power input?
b What is the source of this power?
c How much energy does the power station, operating at 100 MW, make available to the electrical grid system (its energy output) in 1 hour?
d In the same hour, how much energy is made available to the power station from the burning of fuel or other source (the energy input)?
e In any period of time, only 35% of the input energy is transferred to the electrical grid system. What happens to the rest of the energy?

Mass, energy and units in particle physics

The total mass of the products of a chemical reaction, as far as we can usually expect to measure, is the same as the total mass of the reactants – total mass does not measurably change. We say that chemical reactions **conserve** mass. This can be called the **principle of conservation of mass**.

Chemical reactions involve atoms. The nuclei of atoms are much, much smaller than the whole atoms, and at the nuclear level a very different sort of process takes place. In these processes, such as radioactive emission or the fusion reactions in the Sun, the mass of material after the change is measurably less than that before. Some of the mass of the original material has ceased to exist, and the principle of conservation of mass has been broken or **violated**.

The product particles from such a nuclear reaction tend to fly violently apart, and some energy can also be carried away by electromagnetic radiation. The particles and the radiation together have more energy than was present before the change. The principle of conservation of energy has also, apparently, been violated. However, if we think of mass and energy as interchangeable versions of the same thing, a single **mass–energy**, then the increase in energy matches the decrease in mass and we can still have a conservation rule that is not violated by either chemical or nuclear changes. This is the **principle of conservation of mass–energy**, which states that total mass–energy is the same before and after all change processes.

Energy seems to be able to exist as mass, or, to put it the other way, mass seems to be energy that is localised. However, we normally measure mass in kilograms and energy in joules, treating them as very different quantities. Using these units we can convert quantities of mass into quantities of energy using the formula:

$$E = mc^2$$

where c is the speed of light in a vacuum, 3×10^8 m s^{-1}. The formula is **Einstein's mass–energy equation**.

If the history of physics had been different, we'd have a different system of units, and if the equivalence of mass and energy had been known about sooner then it is possible that we would now be using a system of units in which energy and mass have the same unit. In such a system of units we would say that $E = m =$ mass–energy and $c = 1$ unit of speed. We might want to give a name to this unit of speed; borrowing from science fiction we could call it a 'warp', so that $c = 1$ warp. Then we could talk about 5 warp or 'warp factor 5' as five times the speed of light. However, coming back down to Earth, since the speed of light is so much faster than the ordinary speeds that we can directly experience, we'd have to use very small numbers for measuring the speeds of bodies as slow as snails.

Particle physicists deal with small particles which each have quite small quantities of energy, so they find the joule too big to be a convenient unit of energy. They measure energy in **electronvolts**, or eV; 1 eV is equivalent to 1.6×10^{-19} J. It is the energy transferred to an electron by a potential difference of 1 V. Such a potential difference does 1 eV of work on the electron and (if there are no other forces acting) provides 1 eV of kinetic energy to the electron.

They are also working with mass–energy conversions as a matter of everyday routine. So they use the electronvolt as a unit of mass as well as energy. That saves them having to use $E = mc^2$ and perform unnecessary calculations. If the energy wrapped in a proton's mass could, for example, be made available as kinetic energy, then the amount of energy available would be 9.38×10^8 eV, or 938 MeV (Table 15.3). In the process, the mass of the proton would cease to exist. 938 MeV is the mass–energy of the proton.

17 a If all of the mass of a 1 kg bag of sugar could be made available as useful energy, how much energy would this be?
  b Repeat **a** for the mass of a proton.
18 a What would be the advantage(s) and disadvantage(s) of abandoning separate concepts of mass and energy and using only the concept of mass–energy?
  b Suppose that such a new system was adopted. Then we would no longer need to use the kilogram or the joule as units, but we would need a new unit of mass–energy. Provide advice to the SI committee in choosing a name for this new unit.
  c Particle physicists use the MeV as their unit of mass–energy. What is the equivalent amount of
  i mass in kilograms
  ii energy in joules?

Table 15.3

Name of particle	Mass–energy/million electronvolts, MeV	Mass/kg
electron	0.511	9.11×10^{-31}
proton	938	1.67×10^{-27}

Note that 1 eV is equivalent to a mass of 1.78×10^{-36} kg.

● Extra skills task ## Information Technology, Application of Number and Communication

The table shows some typical efficiencies of various devices:

Device	Efficiency / %	Device	Efficiency / %
early steam engine	1	diesel engine	40
filament lamp	5	domestic boiler	80
fluorescent tube	20	electric motor	80
car engine (petrol)	30	torch battery	90
coal-fired power station	35	electric heater	100

Use a computer to create a visual display of the data.

The calorific value of coal is 35 MJ kg^{-1}. Calculate the useful energy outputs of an early steam engine and a coal-fired power station for each tonne of coal burned.

Explain the wide difference between the quoted efficiencies of a filament lamp and an electric heater.

16 Changes in potential and kinetic energy

THE BIG QUESTION
- How can we use ideas of potential and kinetic energy to interpret the behaviour of mechanical systems?

KEY VOCABULARY
equilibrium separation

BACKGROUND

The world is complex, but some systems within it are surprisingly simple. Think about the motions of planets, moons and comets, for example. Though their orbits are not simply circular, they are simple enough for it to be possible to predict their motions for centuries to come. Energy transfers take place away from a planet in orbit very slowly indeed.

The main leakage of energy away from the Earth's daily spin is a result of heating effects of the tides of the Earth's oceans. These are very slowly making the days longer, but you won't notice the difference in your lifetime.

The Earth is just a little closer to the Sun at some times of year. It's not enough to make a significant difference to the climate, but it is enough for the Earth to speed up and slow down as it goes around the Sun. As it moves slightly closer to the Sun, the Earth loses potential energy and this becomes kinetic energy. Then as it moves back a little further from the Sun, the potential energy increases and the kinetic energy decreases. We all go along for the ride, but without careful observations of the changing positions of the stars we never notice the difference.

Calculation of changes in gravitational potential energy

Imagine going to work on the tenth floor of a building. Each morning, you take the lift. The lift does work on your body, and your body can be said to gain potential energy. Your body now has extra ability to do work because of its position in the gravitational field. One way to illustrate this ability would be to leap out of the window, convert the potential to kinetic energy, and do work by exerting a force over a (short) distance on the roof of a car parked below.

The potential energy that you gain from your ride in the lift is equal to the work that the lift did in lifting you. 'Doing work' is a process of energy transfer, in this case from the lift system to your body–Earth system:

potential energy gained = work done on body by lift
= average force exerted by lift × change in height
$$\Delta E = F_{av} \times \Delta h$$

The 'deltas' here, again, indicate changes in quantities. The average force does not change and applies to the whole of the upwards journey.

The average force is equal to your weight. (The actual force is rather larger than this when the lift accelerates upwards at the start of the journey, and rather smaller when it decelerates as the lift reaches the tenth floor. You might be particularly conscious of the inertia of your breakfast during these accelerations. Average force for the whole journey remains equal to weight.)

Your weight is related to your body's mass by the conversion factor of 9.81 N kg^{-1} that applies on and around the Earth's surface. This is the gravitational field strength at the Earth's surface, generally abbreviated simply as g. That is, $F_{av} = mg$.

So

$$\Delta E = mg\, \Delta h$$

I a Estimate your weight.

b If the height difference between adjacent floors of a building is 2.7 m, how much potential energy will you gain in rising
i one floor   **ii** eight floors?
c What is the ratio of these potential energy changes?

2 Still thinking about your changes in potential energy as you travel up and down a tall building, sketch graphs of your changes in potential energy gained, against
a your height above ground level, in metres
b total number of floors through which you rise.
What information is carried by the gradients of these two graphs?

3 A reservoir holds 10^9 kg of water. What is the total change in potential energy of the water if the dam bursts and the water falls through a height of 10 m? Is this change positive or negative? (Use $g = 10$ N kg^{-1}.)

4 The relationship between gravitational field strength, g, and distance from the centre of the Earth, r, is given by $g = \dfrac{4 \times 10^{13}}{r^2}$ where r is in metres.

What is the ratio of the value of g at the Earth's surface (where $r = 6.37 \times 10^6$ m) and at the height of orbit of a space shuttle (300 km higher, where $r = 6.67 \times 10^6$ m)? (To find their ratio, call these values $g_{surface}$ and g_{orbit}.)

Note that the value of g varies with locality in space. It changes as your height above the Earth changes. However, the value of g on the tenth floor of a building is almost identical to that at ground level. In using the formula $\Delta E = mg\,\Delta h$ we assume that g is constant. We can use the formula for all gravitational potential energy changes that take place close to the Earth's surface. However, when calculating the energy changes experienced by a spacecraft travelling far from Earth, for which Δh is large and g cannot be assumed to be constant, the formula will not provide reliable answers. It is a locally useful, and not a universally true, formula.

Calculation of changes in kinetic energy

On your journey to your ten-floor office building, your bus accelerates away from the bus stop. Your seat exerts a force on you so that you also accelerate. Your body experiences displacement while it experiences the force, and if the bus is travelling in a straight line then force and displacement are in the same direction. The force and your displacement under the action of the force relate to the work done on your body:

$$\text{work done} = \text{average force} \times \text{displacement}$$

The effect of the work done on your body is to give you kinetic energy:

$$\text{work done} = \text{kinetic energy gained}$$

So

$$\text{kinetic energy gained} = \text{average force} \times \text{displacement}$$
$$\Delta E = F_{av} \times \Delta x$$

Average force produces acceleration, according to $F_{av} = ma$, where m is mass (which we can usually assume to be constant), and a is average acceleration, so

$$\Delta E = ma\,\Delta x$$

In Chapter 31 you will see that, for a body starting from rest, $a\,\Delta x$ is equal to $v^2/2$, where v is the velocity acquired during the acceleration. So, for a body starting from rest, kinetic energy gained is given by

$$\Delta E = \frac{mv^2}{2} = \tfrac{1}{2}mv^2$$

5 a What is the gain in kinetic energy of a car of mass 10^3 kg that accelerates from rest to 20 m s^{-1}?
b What is its change of kinetic energy if it
i accelerates from 20 m s^{-1} to 25 m s^{-1}
ii decelerates from 20 m s^{-1} to rest?

6 Use a computer to help you to plot a graph of kinetic energy against speed for a car of mass 1×10^3 kg.

7 Which has more kinetic energy, a truck of mass 3×10^3 kg with a speed of 10 m s^{-1} or a car of mass 1×10^3 kg with a speed of 30 m s^{-1}?

8 A pile driver is a heavy weight that is allowed to fall on a post to drive it into the ground.
a If a pile driver has a mass of 500 kg and is allowed to fall through a height of 1.2 m, and if all its potential energy becomes kinetic energy just before it hits the post, how much kinetic energy does it gain?
b On hitting the post, it drives it 0.07 m into the ground. If all of the energy of the pile driver does work on the post, what is the average force on the post?

Since the body started from rest and had no initial kinetic energy, the amount of kinetic energy it gained, ΔE, is equal to its total final kinetic energy, E. That is,

$$\Delta E = \text{final kinetic energy} - \text{initial kinetic energy}$$
$$\Delta E = E - 0$$

Final kinetic energy is therefore related to mass and velocity by

$$E = \tfrac{1}{2}mv^2$$

Note that this final kinetic energy, E, is not just a change in kinetic energy but is the actual value of total kinetic energy of a body of mass m that has velocity v.

Potential and kinetic energy of a pendulum

The bob of a pendulum rises and falls. Its potential energy increases and decreases, and so does its kinetic energy. However, a pendulum is a mechanical system that must obey the principle of conservation of energy. Total energy is considered constant if the influence of air resistance and friction, which could cause dissipation of energy, are small enough to be negligible for all practical purposes. Potential energy is continually changing to kinetic energy, and back again (Figure 16.1).

Figure 16.1
Changes in potential and kinetic energy of a simple pendulum.

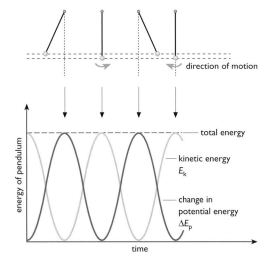

The pendulum continuously loses and gains potential energy during its swing, having minimum potential energy when at its lowest point (where horizontal displacement is zero) and maximum potential energy at its highest point (where horizontal displacement is maximum). The increase in potential energy above its minimum when it is at a particular point can be written as ΔE_p.

The pendulum also gains and loses kinetic energy, with this being maximum at the bob's lowest point and minimum at the bob's highest point. The kinetic energy at a particular point can be written as E_k.

The minimum kinetic energy is zero – the bob has zero velocity for an instant when at its maximum displacement. The minimum potential energy, however, is not zero. This is explained in *Further Advanced Physics*.

$$\Delta E_p + E_k = \text{total pendulum energy}$$

The pendulum continuously exchanges potential and kinetic energy.

Potential and kinetic energy of a mass on a spring

A mass suspended on a spring oscillates, just as a pendulum bob does. Its changes in kinetic and potential energy follow the same patterns as for the pendulum. (For more information on the 'simple harmonic motion' of the pendulum and the mass on a spring, see *Further Advanced Physics*.)

The mass–spring system is a little more complicated in that its changes in potential energy have two origins. One of these is gravitational, in that the mass is moving up and down in a gravitational field. The other is fundamentally electrical, in that the spring's extension varies (and so the separation of its constituent particles varies). The associated energy is sometimes called strain potential energy. However, overall potential energy changes still follow the same pattern as for the pendulum, and graphs of the changes in potential and kinetic energy have the same shapes (Figure 16.2).

Figure 16.2
Changes in potential and kinetic energy of a mass on a spring.

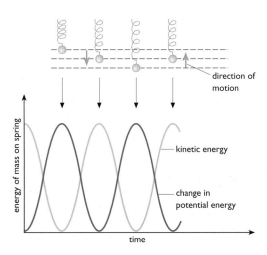

Note that potential energy is minimum not when the mass is at its lowest point, but when it is at the position where it would settle to rest. Its potential energy when displaced from rest is a combination of gravitational potential energy and strain potential energy (due to extension and compression of the spring).

Potential and kinetic energy of an atom in a solid

A simple but useful model of the forces between particles in a solid is a spring model. The particles are simply thought of as being masses connected by springs of very small mass. This is at its simplest when we think about just two masses connected by a massless spring (Figure 16.3).

Suppose that such a simple system floats in space, so that we don't have to worry about gravitational forces. This is a reasonable simplification, since we are establishing a model of the behaviour of particles, and the forces of gravity acting on individual particles are very small compared with the electrical forces. When the spring is neither extended nor compressed the masses settle at a particular distance apart. This is their **equilibrium separation**. When the masses move closer together, the spring becomes compressed and the force between them is effectively repulsive (Figure 16.4a). When they move further apart the spring provides an effective attractive force. Thus the spring model behaves like a real pair of particles.

When the spring is either compressed or extended then a force is applied to the masses and parallel displacement is involved. Work is done. The system gains potential energy (Figure 16.4b). If the compressed or extended spring is released, this potential energy can then become kinetic energy. The mass–spring system oscillates with a total energy that is equal to the work done in compressing or extending the spring. If no energy leaks away from it then it will oscillate for ever, continuously exchanging potential and kinetic energy but with constant total energy.

9 A mass on a stretched spring is released and potential energy becomes kinetic energy. Is this consistent with the principle of conservation of energy?

10 At what point in their oscillation do a pair of particles have maximum kinetic energy?

11 For a pair of masses connected by a spring:
   a What is the force when the spring is neither extended nor compressed? (Assume the spring is responsible for all of the forces acting on the system.)
   b What happens to the force as the spring is extended? Show this as a graph.
   c How does the graph show the work that has been done?
   d The mass–spring system has gained energy as a result of work being done on it and the spring being stretched. Classify this energy.
   e What happens to the classification of this energy when the spring is released, assuming that the system keeps the same total energy and none is transferred to the surrounding environment in any way?
   f Suppose that, after this release, the system is isolated from the world around it and can never lose energy. Describe
   i the motion of the masses
   ii the total energy
   iii the potential energy
   iv the kinetic energy.

Figure 16.4
Graphical representations of force and potential energy changes for the mass–spring system. Note that the graphs look different, but they represent the same event.

a Force–separation graph, using the convention of showing repulsive force as positive.

b Change in potential energy with separation.

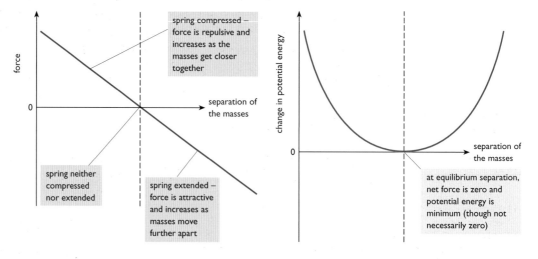

spring compressed – force is repulsive and increases as the masses get closer together

spring neither compressed nor extended

spring extended – force is attractive and increases as masses move further apart

at equilibrium separation, net force is zero and potential energy is minimum (though not necessarily zero)

Potential and kinetic energies of Pluto–Charon

● **Comprehension and application**

The planet Pluto and its moon Charon (Figure 16.5) have a combined mass of about 1.5×10^{22} kg, and they rotate around their common centre of mass so that they can be thought of as a single body in orbit around the Sun. Their orbit is unusual compared to that of other planets: firstly, the other planets rotate in approximately the same plane, but the plane of the Pluto–Charon orbit lies at an angle of about 17° to this; secondly, the other planets have orbits that are slightly elliptical, but not far from circular, whereas the Pluto–Charon orbit is strongly elliptical. So, even though they are usually the furthest recognised planets from the Sun, they are sometimes closer to the Sun than Neptune is. Pluto–Charon vary in their distance from the Sun from a minimum of 4.4×10^9 km to a maximum of 7.3×10^9 km.

Figure 16.5
Pluto–Charon and the distant Sun, too far away to be bright but close enough to trap Pluto and Charon in its gravitational field.

Pluto–Charon orbit the Sun every 247.7 Earth years – none of us would celebrate a first birthday if we lived on their surfaces. As they orbit, the variation in their distance from the Sun means that they experience big changes in gravitational potential energy. When they are furthest away from the Sun (the aphelion of their orbit) they have much more potential energy than when they are closest to the Sun (perihelion of their orbit). In this process they do not gain or lose total energy. Instead, their kinetic energy changes. When they are far from the Sun, the potential energy is higher and kinetic energy is lower, so, in the course of a Pluto–Charon year, the bodies experience considerable change in orbiting speed.

We cannot calculate changes in planetary potential energy by using an equation of the form $\Delta E_p = mg\Delta h$, where g is gravitational field strength due to the Sun's gravitational field. The value of g varies strongly with distance; the formula $\Delta E_p = mg\Delta h$ is an adequate approximation only when Δh is small. But for Pluto–Charon, Δh is very large indeed. The change in potential energy experienced by a body initially at distance r from the Sun and increasing this by an amount Δr is:

$$\Delta E_p = k\,\frac{\Delta r}{r(r + \Delta r)}$$

where k is equal to 2.0×10^{42} N m². This is a very large number because the bodies involved are large.

12 What are the values of r and Δr for Pluto–Charon?
13 Calculate the changes in potential energy of Pluto–Charon that take place during one orbit.
14 What corresponding changes in kinetic energy take place?
15 Estimate the distance travelled by Pluto–Charon during one orbit of the Sun. Use this distance to estimate the average speed. Then use this to estimate maximum and minimum speeds.
16 Explain fully why the formula $\Delta E_p = mg\Delta h$ cannot be applied to Pluto–Charon. When can it be applied?
17 Make a sketch of the orbit of Pluto–Charon. Draw bar charts to show the relative sizes of potential energy and kinetic energy at their nearest point to the Sun (perihelion) and their furthest point (aphelion).
   Explain how the sizes of these bars vary during one orbit.

V
ELECTRICITY

17 Electric charge

THE BIG QUESTION
● What models have people created to describe their ideas of the behaviour of matter that underlies the electrical phenomena that we can see?

KEY VOCABULARY
charge (*noun*)   charge (*verb*)   charge carriers   conventional current   coulomb
discharge (*verb*)   electric field lines   electric field strength   electrode   electrolysis
Fleming's left-hand rule   magnetic flux density   negative charge   net charge
neutral   positive charge   specific charge   tesla

BACKGROUND

Figure 17.1
Michael Faraday
(1791–1867).

We've come a long way in the last 200 years or so, since the first experiments on electric charge, current and electromagnetism. Perhaps no other technology of the last two centuries has transformed our lives more than electrical technology. Certainly whole industries are based on electricity, and none of us would like to manage without using it. The early experimenters – people like Michael Faraday (Figure 17.1) – changed the world.

Faraday was not the greatest of mathematicians, but he was an innovative and enthusiastic experimenter, an exceptionally creative visual thinker and a passionate explainer. His work on charge, circuits and fields established ways of thinking that we all still use today.

● Charge

You only have to drop a plastic ruler to be aware of the effects of gravity. You see the acceleration of the ruler. Electric force is not quite so obvious, but you can make the ruler exert and experience electric force by rubbing it with a cloth. The force can be attractive – the ruler can lift small pieces of paper or bend a gentle flow of water from a tap. It can also be repulsive – two rulers rubbed with the same cloth will repel each other. The ability to exert and experience electric force is called **charge**. A plastic ruler can be **charged** by friction – it gains charge when rubbed. It can also lose its charge again – it can be **discharged**. Such a charge on an object that's big enough to see and feel, such as a ruler, is not permanent. Note that the rubbing also causes the cloth to become charged.

There are some similarities between electric and gravitational force. Both can act at a distance. For both, the strength of the force between two objects decreases as they get further apart. Both act between bodies that have a named property: gravitational force acts between bodies with mass, electric force acts between bodies with charge.

However, straight away we see a clear difference between gravitational and electric force. Gravitational force is always attractive, while electric force can also be repulsive. The conclusion to this is that while there is only one kind of mass, there are two types of charge. The names chosen for these two types of charge deliberately suggest opposite behaviour – **positive charge** and **negative charge**.

Charge and particles

CHARGE AND PARTICLES

Figure 17.2
In a bubble chamber (see Chapter 29), charged particles leave trails of bubbles. The track here of an electron is bent because the bubble chamber is surrounded by a magnetic field, and moving charged particles can experience magnetic force.

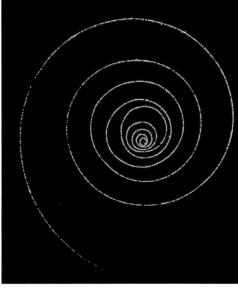

Our knowledge of atomic structure (see Chapter 7) tells us about the role of electric force between sub-atomic particles, that is between positively charged protons in nuclei and negatively charged electrons in orbit. Neutrons play no direct part in such electrical activity – they are electrically **neutral**. All protons are positively charged at all times, electrons are always negatively charged, and all neutrons remain uncharged. A particle always conserves its charge.

Figure 17.3
In terms of mass, there are two possibilities; particles either have positive mass or no mass at all. But electrically speaking there are three possible kinds of particles: those with positive charge, those with negative charge, and neutral particles with no charge at all.

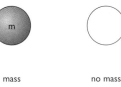

mass   no mass

 positive charge   negative charge   neutral (no charge)

The total mass of any body, like a ruler, is the sum of the masses of its particles. The charge on a ruler is also due to particles, but this is a little more complicated because some of its particles have negative charge and some have positive charge. Balance and imbalance between these negative and positive charges decide whether or not the ruler has a **net charge**.

So, what can we say about the ruler and the cloth used to rub it? When they are uncharged (before rubbing), then either:

- they are made up of a mixture of these kinds of particles, with the total positive and negative charges in balance, or
- they are made up of particles with no charge.

The second hypothesis cannot survive the observations made after rubbing the ruler. If the ruler and the cloth contained only particles with no charge, and if all charge is carried by particles which cannot change their charge, then the ruler and the cloth could not become charged by rubbing.

The cloth and the ruler normally have a balance of positively and negatively charged particles. Rubbing upsets this balance, by transferring charged particles from one surface to the other. A plastic ruler becomes positively charged (Figure 17.4).

1 Why don't we need to take the trouble to talk about positive and negative mass?

Figure 17.4
When the ruler is electrically neutral and shows no electrical behaviour, its positive and negative particles are in balance and it has no net charge. When rubbed, the positive and negative charges are no longer in balance.

neutral   positively charged – due to a deficit of negative charge

Once charged the ruler can be returned to neutrality, or discharged, by restoring the balance.

Phenomena and models

Observation of phenomena and testing of ideas through experiment led, over several generations, to our present-day theory of particles as charge carriers. We make a lot of use of visual models, each one consistent with the general theory, to interpret the phenomena.

Figures 17.5 to 17.10 illustrate the richness of particle theory in providing fundamental descriptions of how the world behaves. General ideas of movement of positive and negative charge provide the beginnings of a theory (see text on pink background). But ideas about electrons and ions as carriers of charge provide explanations or descriptions of mechanisms in terms of the fundamental nature of matter (see text on mauve background).

Figure 17.5
Charging by friction.

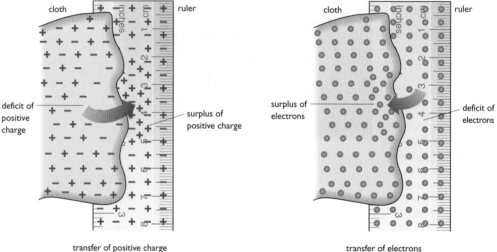

Charge can be transferred between bodies, such as a ruler and a cloth. A simple model of charge treats it as a quantity, indicated by + and − signs. Before present-day particle theory, this was the only reasonable way to think about charge.

Particle theory can provide a more detailed model, by considering that the transfer of charge is due to a transfer of charged particles. When a cloth and ruler rub together there is a transfer of negatively charged electrons.

Figure 17.6
The gold-leaf electroscope.

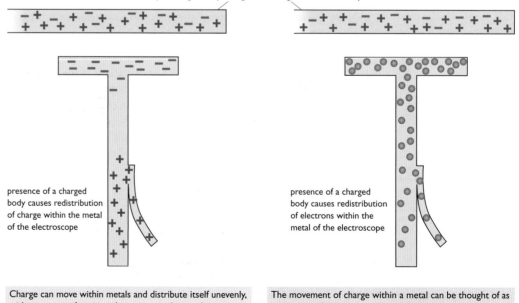

Charge can move within metals and distribute itself unevenly, with an excess of positive charge in one place and negative charge in another. One part of a metal body can then visibly repel another, as in the gold-leaf electroscope.

The movement of charge within a metal can be thought of as movement of electrons, so that there is an excess of electrons in one place and a shortage in another.

Figure 17.7
Discharge of a positively charged metal body.

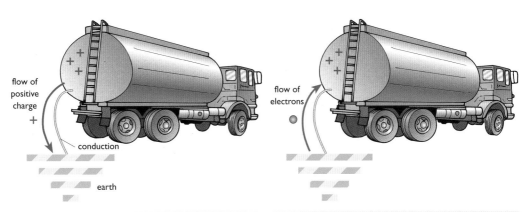

A positively charged metal body becomes neutral when connected to 'earth' by a conductor. ('Earth' is a very large neutral body.) Charge can be thought of as flowing along the conductor. The direction of flow of positive charge is the same as the direction of **conventional current** (see page 148).

In terms of particles, the discharge or neutralisation of the positively charged metal is thought of in terms of flow of electrons on to it. The negative electrons neutralise the positive charge.

Figure 17.8
A current in a liquid, resulting in chemical change: **electrolysis**.

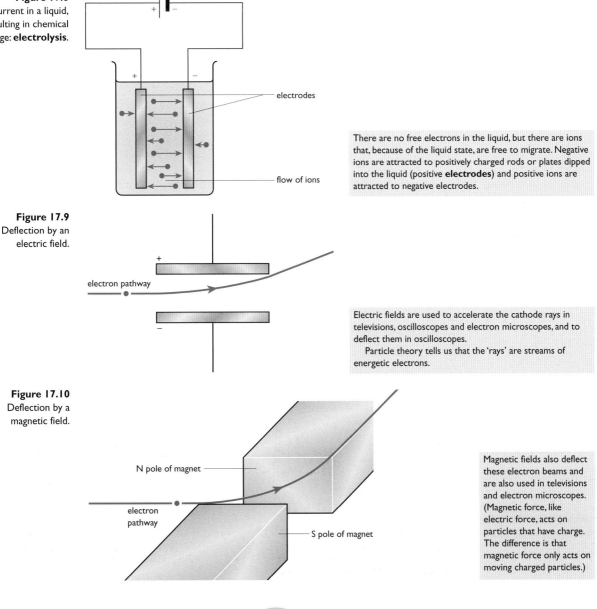

There are no free electrons in the liquid, but there are ions that, because of the liquid state, are free to migrate. Negative ions are attracted to positively charged rods or plates dipped into the liquid (positive **electrodes**) and positive ions are attracted to negative electrodes.

Figure 17.9
Deflection by an electric field.

Electric fields are used to accelerate the cathode rays in televisions, oscilloscopes and electron microscopes, and to deflect them in oscilloscopes.
  Particle theory tells us that the 'rays' are streams of energetic electrons.

Figure 17.10
Deflection by a magnetic field.

Magnetic fields also deflect these electron beams and are also used in televisions and electron microscopes. (Magnetic force, like electric force, acts on particles that have charge. The difference is that magnetic force only acts on moving charged particles.)

Conventional current

Figure 17.11
Conventional current and
electron flow are in
opposite directions.

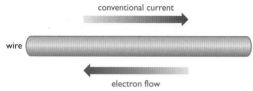

conventional current

wire

electron flow

2 What particles are transferred between a ruler and a cloth when they are rubbed together?
3 If the ruler becomes positively charged, in which direction did particles transfer? What particle movements will discharge the ruler?
4 Explain how electrons in a neutral gold-leaf electroscope can become unevenly distributed.
5 In electrolysis, which particles flow in the *same* direction as conventional current?

Conventional current and electron flow are in opposite directions. This is an unfortunate accident, which happened because a lot of work was done on electricity before electron theory was developed. By the time it was realised that when there is an electric current in a metal wire the charge carriers are electrons, the habit had already developed of supposing that flow was from the positive to the negative terminal of a cell (the direction in which positive charges would flow). Since the flow in a conductor is beyond the reach of our senses, it doesn't actually matter for practical purposes which way we say the flow goes, as long as we are clear about what we mean. Throughout this book we will show conventional current with red arrowheads and electron flow with blue arrowheads, as in Figure 17.11.

Note that in a solution where there are ions that are free to move, some ions have positive charge and some have negative charge. A solution can therefore contain both positive and negative **charge carriers**. In a solution that is connected via electrodes to a battery, the negative charge carriers flow away from the negative towards the positive terminal of a battery, which is in the opposite direction to conventional current. Positive charge carriers flow away from the positive towards the negative terminal – in the same direction as conventional current.

Measuring charge

6 How many electrons are needed for their total charge to equal −1 C?
7 A charged piece of plastic holds 8.55×10^{16} protons and 8.54×10^{16} electrons. What is the excess of the number of protons over the number of electrons? What is the charge on the plastic, in coulombs?
8 What is the charge, in coulombs, of a kilogram of electrons?

The charged particles in materials each have very small mass and very small charge. For mass, the SI unit is the kilogram. One kilogram of carbon contains about 10^{25} atoms. For charge, the SI unit is the **coulomb** (C). One coulomb of charge is the same as the charge on approximately 10^{18} carbon nuclei:

$$\text{charge on one proton} \quad = 1.6 \times 10^{-19}\,\text{C}$$
$$\text{charge on one electron, } e = -1.6 \times 10^{-19}\,\text{C}$$

The ratio of the charge to the mass of an electron is called the electron's **specific charge**. It is a very large number:

$$\frac{e}{m} = -1.7 \times 10^{11}\,\text{C kg}^{-1}$$

The minus sign is present because electrons have negative charge. The large number suggests that electrons have a lot of charge packed into a small mass. (The mass of an electron is a very small quantity, just 9.1×10^{-31} kg.)

Charged particles and electric field

Figure 17.12
Some of Faraday's
equipment for investigating
magnetic fields and
electromagnetism.

The region of space around a charged object, in which force is experienced by another charged object, is called its electric field. A ruler charged by friction has an electric field around it, but it is

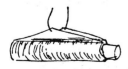

drawing of a coil from
Faraday's laboratory
notebook of 1831

relatively weak. A stronger electric field can exist around metal bodies because these can be given larger deficits or surpluses of electrons.

The concept of the field line or line of force, created by Michael Faraday to help him to think about magnetic fields (Figure 17.12), can also be used to illustrate gravitational and electric fields (Figure 17.13). **Electric field lines** show the direction of the force that will act on a small ('point'-sized) positive electric charge. The spacing of the field lines gives an indication of the electric field strength – see *Further Advanced Physics*.

Figure 17.13
Field lines (in two dimensions): gravitational and electric fields compared.

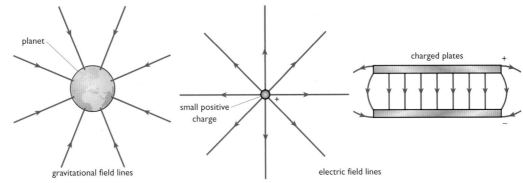

planet

gravitational field lines

small positive charge

electric field lines

charged plates

In thinking about gravitational fields we use a quantity that provides information about the field itself rather than about any particular body that happens to be in the field. This useful quantity is called gravitational field strength, g. The value of g at a point 1 m in front of you is $9.81\,N\,kg^{-1}$. The point 1 m behind you has the same value. Gravitational field strength of a point is defined as the force that would act on a unit of mass placed at that point (see page 108).

We can argue in exactly the same way for an electric field, and define the **electric field strength** of a point as the force that would act per unit charge at that point. Electric field strength is usually abbreviated as E, and in SI units it is measured in $N\,C^{-1}$.

$$E = \text{force that can act per unit charge} = \frac{F}{q}$$

If we know the electric field strength of a point and know the value of a charge then we can predict the force that will act on that charge at that point.

9 Why is the deflection of electrons in an electric field in the opposite direction to the arrows on the field lines?

Charged particles and magnetic field

Bodies with mass experience gravitational force, and bodies with charge experience electric force. Bodies with charge *and* velocity experience magnetic force. So, we can create a quantity for magnetic field, similar to the field strengths that we already have for gravity and electricity. This is called the **magnetic flux density** and it is abbreviated as B. It is analogous to gravitational and electric field strengths, and just as for g and E it is a property of a point in the field; its definition follows a similar pattern: it is the force that would act at a point per unit charge with unit velocity.

$$B = \left(\begin{array}{l}\text{force that can act per unit charge with unit velocity} \\ \text{when velocity and field lines are mutually perpendicular}\end{array}\right) = \frac{F}{qv}$$

An acceptable SI unit of B would be the $N\,C^{-1}\,m^{-1}\,s$, but this is too long and inconvenient. So this unit is given its own name, the **tesla**, T:

$$1\,T = 1\,N\,C^{-1}\,m^{-1}\,s$$

If we know the magnetic flux density at a point then we can predict the force that will act on a particle of known charge and known velocity *perpendicular to the field lines* when at that point:

$$F = B\,q\,v$$

Just as we can draw field lines to provide visual representations of electric field, we can do the same for magnetic field. However, magnetic field lines show the direction of the force that will potentially act on a small North pole placed in the field, and this is not the same as the force acting on a moving charge. The direction of the force on a moving charged particle is perpendicular

10 a If an electron in a beam has a mass of 9.1×10^{-31} kg, a charge of -1.6×10^{-19} C and is moving with velocity of 5×10^7 m s^{-1} in a region where

$g = 9.81$ N kg^{-1}, $E = 1 \times 10^3$ N C^{-1}, $B = 8 \times 10^{-4}$ T

and the field lines and electron velocity are all mutually perpendicular, then calculate the forces due to each of the fields, and the resulting accelerations.

b Make sketches of the field lines in each case, ensuring that arrows are added appropriately. Use these arrows to then show the direction of the force that will act on an electron at a point in the field.

both to the motion of the particle and to the magnetic field lines. The force acting on a positive charge moving perpendicularly to the local field lines can be determined using **Fleming's left-hand rule** (Figure 17.14). Care must be taken, therefore, when dealing with electrons – their negative charge means that the force they experience is opposite to that experienced by positive charges.

Figure 17.14
Magnetic fields and forces on positive and negative charges (left-hand rule).

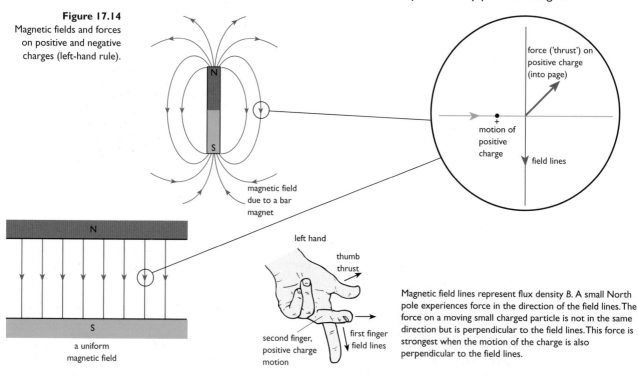

Magnetic field lines represent flux density B. A small North pole experiences force in the direction of the field lines. The force on a moving small charged particle is not in the same direction but is perpendicular to the field lines. This force is strongest when the motion of the charge is also perpendicular to the field lines.

Electron beams

The process of emission of electrons from a heated negatively charged metal surface – thermionic emission (see page 59) – makes the television and the electron microscope possible.

In both the TV and the electron microscope, thermionic emission takes place from a cathode which is heated by a nearby heating filament. The negative charge of the cathode repels the electrons, and a positive charge on an anode attracts them. They therefore accelerate from cathode to anode. If the latter has a narrow hole, then only a narrow beam of electrons can pass through the hole (Figure 17.15), and these fast electrons can be further 'steered' by using either more electric fields or magnetic fields.

Figure 17.15
Creating a narrow beam of electrons.

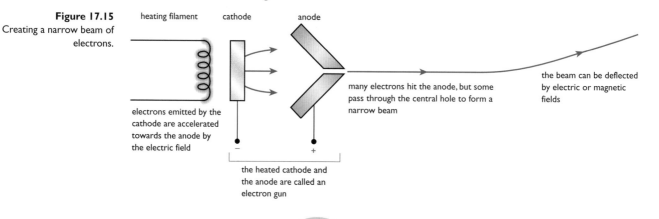

In a colour TV tube there are three electron beams (to produce the three primary colours red, blue and green – see page 23). These are made to scan the screen rapidly by deflecting them with varying magnetic fields (Figure 17.16). You can read about the electron microscope in Chapter 26, page 270.

Figure 17.16
Electron beams in a TV set.

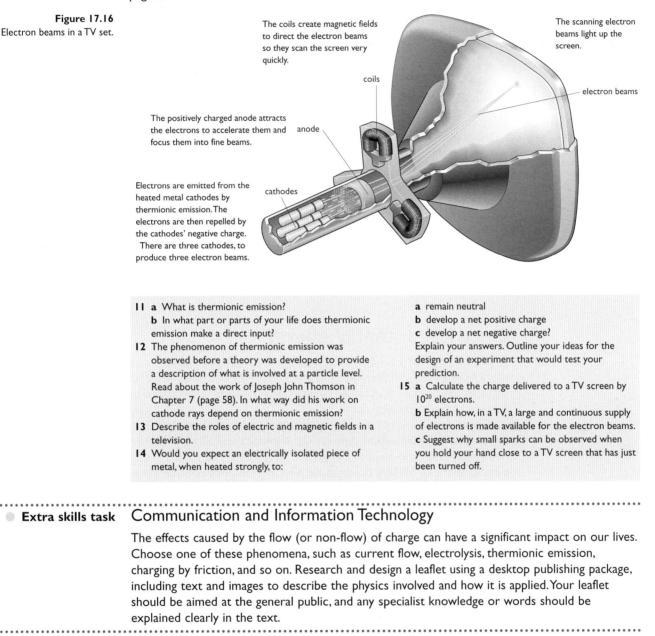

The coils create magnetic fields to direct the electron beams so they scan the screen very quickly.

The scanning electron beams light up the screen.

coils

electron beams

The positively charged anode attracts the electrons to accelerate them and focus them into fine beams.

anode

Electrons are emitted from the heated metal cathodes by thermionic emission. The electrons are then repelled by the cathodes' negative charge. There are three cathodes, to produce three electron beams.

cathodes

11 **a** What is thermionic emission?
   b In what part or parts of your life does thermionic emission make a direct input?
12 The phenomenon of thermionic emission was observed before a theory was developed to provide a description of what is involved at a particle level. Read about the work of Joseph John Thomson in Chapter 7 (page 58). In what way did his work on cathode rays depend on thermionic emission?
13 Describe the roles of electric and magnetic fields in a television.
14 Would you expect an electrically isolated piece of metal, when heated strongly, to:

a remain neutral
b develop a net positive charge
c develop a net negative charge?
Explain your answers. Outline your ideas for the design of an experiment that would test your prediction.
15 **a** Calculate the charge delivered to a TV screen by 10^{20} electrons.
   b Explain how, in a TV, a large and continuous supply of electrons is made available for the electron beams.
   c Suggest why small sparks can be observed when you hold your hand close to a TV screen that has just been turned off.

● **Extra skills task** Communication and Information Technology

The effects caused by the flow (or non-flow) of charge can have a significant impact on our lives. Choose one of these phenomena, such as current flow, electrolysis, thermionic emission, charging by friction, and so on. Research and design a leaflet using a desktop publishing package, including text and images to describe the physics involved and how it is applied. Your leaflet should be aimed at the general public, and any specialist knowledge or words should be explained clearly in the text.

18 Electrical potential energy, potential and potential difference

THE BIG QUESTIONS
- How can we analyse the changes in potential energy experienced by charged bodies due to the action of electric force?
- How do potential energy changes experienced by particles (or idealised point-sized charged bodies) in simple imaginary systems relate to energy transfers in real circuits?

KEY VOCABULARY electrical potential energy   e.m.f. (electromotive force)   lost volts   potential   potential difference   unit charge   volt

BACKGROUND Within a hydrogen atom, electric force exists between just two particles – an electron and a proton. For larger atoms more particles are involved and their electric interaction becomes more complex. Such interactions give rise to a huge area of study, which we call chemistry. The study of 'electricity' in physics is mostly concerned with the movement of electrons that are free to move inside metals. In this chapter we start by thinking about a simple two-particle system – something rather like a hydrogen atom – and move on to electric circuits, which are systems that contain vast numbers of migrating particles.

Electrical potential energy

When the height of a fixed-mass body above the Earth changes then its gravitational potential energy changes. In just the same way, when two charged bodies move further apart or closer together, their electrical potential energy changes. In both gravitational and electrical cases, force is acting and work is being done.

A system made up of two small (point-sized) charged bodies and nothing else is a simple system that provides a good start for our thinking about electrical behaviour. Suppose that our two bodies both have positive charge, and to start with they are an infinite distance apart. Also suppose that one of them is, somehow, held in a fixed position and the other is free to move in space that is otherwise empty. At infinite separation they have no influence on each other – they are outside each other's fields. The electrical potential energy of the system is zero.

We can now move the mobile charge closer to the fixed one. Repulsive force acts between them. We (or some other agency) have to push to overcome the electric force. We have to do work to bring the charges together. If at some point we let go of the mobile charge it will accelerate away. It will return to infinite separation. In doing work, we give the system potential energy (Figure 18.1) which the system will transform to kinetic energy if we let go.

The **electrical potential energy** of a two-body system, at separation x, is equal to the work that must be done to bring the two bodies from infinite separation:

$$\left(\begin{array}{c} \text{electrical potential energy of a system} \\ \text{of two charged bodies at separation } x \end{array} \right) = \left(\begin{array}{c} \text{work done in bringing them from} \\ \text{infinite separation to separation } x \end{array} \right)$$

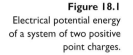

Figure 18.1
Electrical potential energy of a system of two positive point charges.

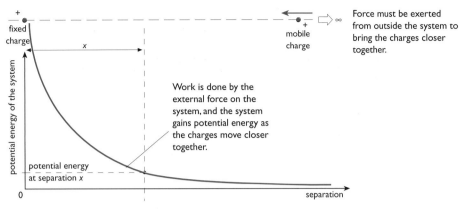

Force must be exerted from outside the system to bring the charges closer together.

Work is done by the external force on the system, and the system gains potential energy as the charges move closer together.

If the two bodies have *opposite* charge then they attract. *We* (acting as the external agency) don't have to do work to bring them closer together – the system is capable of doing work on *us*. We can just relax and enjoy an exhilarating ride. This means effectively that we have to do a *negative* amount of work to bring them together, and the potential energy of the system decreases (Figure 18.2).

Figure 18.2
Electrical potential energy of a system of a positive point charge and a negative point charge.

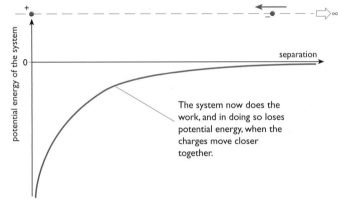

The system itself provides the attractive force that acts on the mobile charge.

The system now does the work, and in doing so loses potential energy, when the charges move closer together.

This is similar to what happens in a gravitational field. The force between massive bodies is attractive. As they move closer together (or 'downwards' if one of the bodies is the Earth), potential energy is lost. As they move further apart ('upwards') the system gains potential energy.

The system of the two opposite charges (or the two massive bodies) has zero potential energy when they are an infinite distance apart. It then loses potential energy as they move together. If a system starts with nothing and then loses energy, it must end up with a negative quantity. The potential energy of an attracting two-body system is therefore negative.

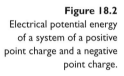

1 Two particles with opposite charge are moved further apart.
  a Is it necessary to do work on the system to achieve this?
  b Explain whether this process increases or decreases the potential energy of the system.
  c At what separation does the system have zero potential energy?
2 In a two-particle system, when the two charged particles move closer together, in what circumstances does the potential energy
  a increase
  b decrease?
3 Explain why the potential energy of an attracting two-body system is negative.

Electrical potential

A system of two charged bodies has potential energy. The amount of potential energy depends on the amounts and type of charge each body has and on their distance apart. But suppose that we want to consider one fixed body and its electric field on its own. It will then be useful to be able to predict what would happen not when a particular mobile charge happens to come along, but what would *potentially* happen for *any* charged body. The way to do this is to use the concept of **unit charge**. In SI units, unit charge is 1 coulomb.

The work we have to do to push a unit charge from infinite separation (that is, outside the field) to separation x is called the electrical **potential** of point x:

$$(\text{electrical potential of point } x) = \left(\begin{array}{l} \text{work done per unit charge in bringing} \\ \text{charge from outside the field to point } x \end{array} \right)$$

In abbreviated form:

$$V = \frac{W}{q}$$

4 What is the potential of a point, P, in a field if 10 J of work must be done to bring a charge of 2 C from outside the field?

5 By doing 24 J of work, how much charge can be brought from a point outside a field to a point, P, in the field that has a potential of 6 V?

6 Write down as many differences as you can between *potential* and *potential energy*.

Since the unit of energy is the joule and the unit of charge is the coulomb, the unit of potential is a 'joule per coulomb'. This can be abbreviated to J C^{-1}. However, potential is a very useful concept, and so the joule per coulomb is given its own name, the **volt**, V. Electrical potential involves unit charge and not some particular charged body whose charge we may or may not know, so it is a more generally useful concept than potential energy. Potential energy is a property of a particular system whereas potential is a property of a *point* in an electric field (Figures 18.3, 18.4).

Figure 18.3
Potential energy is a property of a system whereas potential is a property of a point.

a particular system of charges has potential energy

all points in an electric field have potential

Figure 18.4
The potential of a point depends on the charge creating the field.

If the fixed charge creating the field is negative, then the work done on charge $+q$ will be negative and so the potential will be negative.

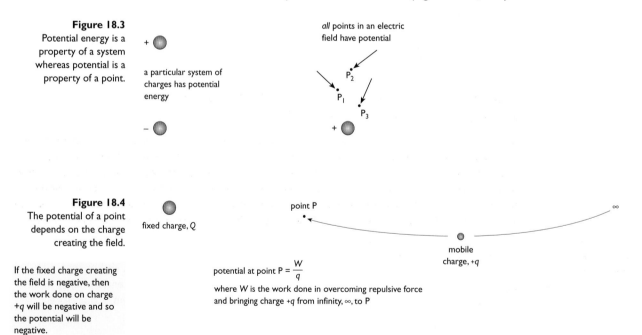

fixed charge, Q

point P

mobile charge, $+q$

potential at point P $= \dfrac{W}{q}$

where W is the work done in overcoming repulsive force and bringing charge $+q$ from infinity, ∞, to P

7 A mobile positive charge of 0.1 C gains kinetic energy of 2 J when attracted from an infinite distance away to a point, P, in the field of a fixed charge.
a Is the fixed charge positive or negative?
b How much work is done in overcoming repulsive force?
c What is the potential of point P?

These ideas are developed in *Further Advanced Physics*. Our purpose here is not to study potential in detail but to provide a basis for understanding energy transfers and potential difference in circuits.

Potential difference

8 When do potential and potential difference have the same value?

9 What is the potential difference between two points if 5 J of work must be done to transfer 0.2 C of charge between them?

10 How much work must be done to overcome repulsive force and transfer a charge of 0.1 C through a potential difference of 2 V?

Two points in a field may have the same potential. This means that it takes the same amount of work to bring a unit charge from infinity to either point. It also means that no net work has to be done to move the unit charge from one point to the other.

Two other points may have different potentials. There is then a **potential difference** between them. The potential difference between two points is the work that must be done per unit positive charge to move charge between them.

$$\left(\begin{array}{l}\text{electrical potential difference} \\ \text{between two points}\end{array}\right) = \left(\begin{array}{l}\text{work done per unit positive charge} \\ \text{in moving charge between them}\end{array}\right)$$

$$\Delta V = \frac{\Delta W}{q}$$

Electrical potential difference exists between points in circuits in which charge is moved and work is done. Just as for electrical potential, its unit is the volt. A voltmeter can be connected to the two points in a circuit whose difference in potential we want to know (Figure 18.5).

Figure 18.5
A voltmeter measures the difference in potential between two points, A and B.

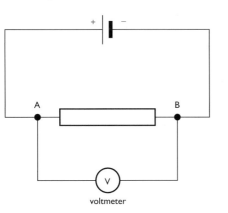

Modelling potential difference between the terminals of a battery*

Each point in a circuit has a potential. Some points may be at the same potential as others. Some pairs of points will have a potential difference between them. Certainly there will be a potential difference between the two terminals of the battery*, or 'across' the battery. The potential of the positive terminal of the battery is higher than that at the negative terminal (because we would have to do more work to bring a unit of positive charge from infinity to the positive terminal, due to repulsion). We can represent this difference as a difference in potential levels (Figure 18.6).

Figure 18.6
There is a difference in potential between the terminals of a battery.

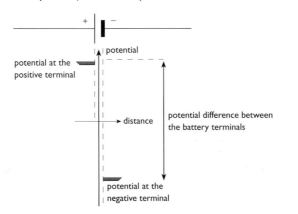

* Strictly, a battery is a number of electric cells working together. Here, as in everyday language, we will use the term battery also to mean a single cell.

A free particle with positive charge will experience force and move from higher to lower potential. In a circuit that is connected to the two terminals of a battery, it will move away from the positive terminal towards the negative terminal. A free particle with negative charge, such as an electron, will move in the opposite direction, from lower to higher potential. We can represent potential levels and electron flow in a simple circuit as shown in Figure 18.7.

Figure 18.7
In this simple circuit the potential difference between the battery terminals (A and X) is the same size as the potential difference between points on each side of the lamp (B and Y).

If there is no resistance between two points in a circuit then no work needs to be done to move charge between them. So there is no difference in potential between them.
   This means that A and B are points at the same potential. X and Y also have equal potentials.

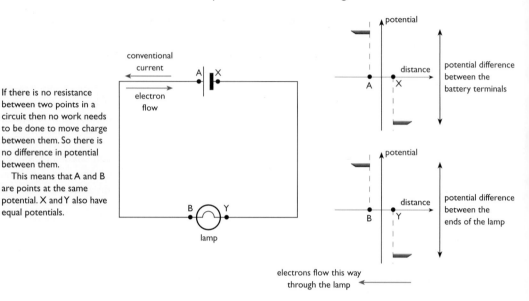

Energy transfers in circuits

A battery can be thought of as transferring energy to electrons. The amount of energy transferred to electrons in the circuit is given by:

$$\Delta W = \Delta V q$$

where ΔV is the battery potential difference and q is the total amount of charge transferred.
   The electrons do not only experience the electric force that results from the battery's potential difference. As they move through the metal of the external circuit they experience interactions with each other and with metal ions. These interactions resist their motion. The electrons must do work to overcome this resistance. In doing this work, they transfer energy to the metal of the circuit, causing heating. They may also do work on electrons in other wires, by way of magnetic interaction, as in a transformer, or they may enable mechanical work to be done, as in an electric motor.

Figure 18.8
Energy transfers in circuit components.

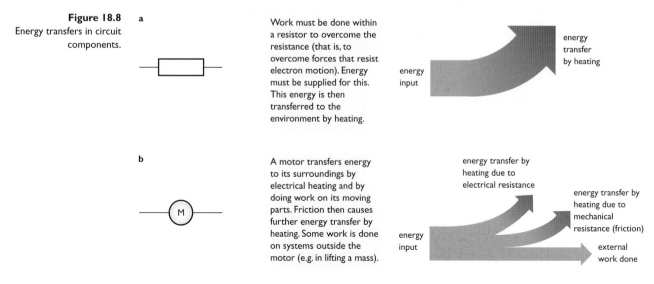

a   Work must be done within a resistor to overcome the resistance (that is, to overcome forces that resist electron motion). Energy must be supplied for this. This energy is then transferred to the environment by heating.

b   A motor transfers energy to its surroundings by electrical heating and by doing work on its moving parts. Friction then causes further energy transfer by heating. Some work is done on systems outside the motor (e.g. in lifting a mass).

Potential difference and e.m.f. of a battery

Batteries are usually the energy sources in d.c. circuits. The behaviour of a battery can be measured in terms of the energy it can supply to each coulomb of charge that is moved around an external circuit – that is, in terms of potential difference across its terminals:

$$\text{potential difference across terminals} = \frac{\text{energy transferred to the external circuit}}{\text{charge moved}}$$

Batteries can also be measured in terms of **electromotive force, e.m.f.** This is not a force at all but, like potential and potential difference, is a ratio of work (or energy) to charge. So, like potential and potential difference, it is measured in volts. It takes account of the work that must be done per unit charge to move charge through the battery itself as well as through the external circuit. This work causes heating within the battery, and so the energy is dissipated (Figure 18.9).

$$\text{e.m.f. of cell} = \frac{\text{energy transferred to the external circuit} + \text{energy dissipated within the battery}}{\text{charge moved}}$$

11 During a particular period of time a battery with a potential difference of 1.25 V between its terminals moves 0.01 C of charge through an external circuit.
a How much work does the battery do during this period?
b How much work does it do on each electron? (1 C is the charge carried by about 6.3×10^{18} electrons.)

Figure 18.9
Total energy transfer from a battery, taking account of the work that must be done to move charge through the battery as well as through the external circuit.

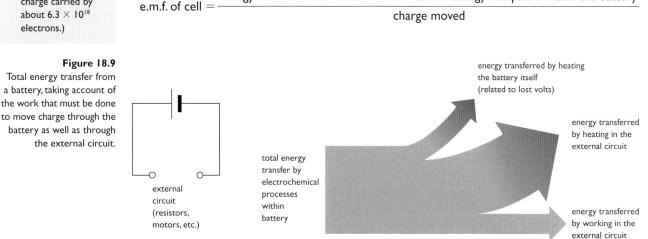

external circuit (resistors, motors, etc.)

total energy transfer by electrochemical processes within battery

energy transferred by heating the battery itself (related to lost volts)

energy transferred by heating in the external circuit

energy transferred by working in the external circuit

12 What is the unit of e.m.f.?
13 A battery quoted as a 1.5 V battery provides a terminal potential difference of 1.2 V when powering a torch. What is the value of lost volts?
14 A battery is connected to a variable resistor. This allows the resistance of the external circuit to be steadily reduced, so that the circuit current, including the current in the battery, increases. What would you expect to happen to
a the rate of dissipation of energy within the battery due to the increased current
b the lost volts?
15 Components in a car circuit are connected in parallel, all to the 12 V supply. Explain why high power devices need a faster rate of transfer of charge through them.

The value written, and measured in volts, on the side of a battery is its e.m.f. The potential difference that the battery can actually make available to drive current in an external circuit is always less than the quoted e.m.f. The difference between battery e.m.f. and terminal potential difference is called the **lost volts**. The existence of lost volts is due to work that must be done to move charge through the battery itself:

$$\text{e.m.f.} - \left(\begin{array}{c}\text{potential difference} \\ \text{across terminals}\end{array}\right) = \text{lost volts}$$

19 Current

THE BIG
QUESTIONS

● How can we tell when a wire is carrying a current?
● How can we measure the current in a wire?
● How can we induce current in a wire?

KEY
VOCABULARY

alternating current, a.c.   amp (or ampere)   calibration   current
direct current, d.c.   electromagnetic induction   full scale deflection
moving coil meter   multiplier   range (of meter)   relay   residual current device
restoring moment   'root mean square' voltage   shunt   transformer

BACKGROUND

For nearly all of human history, nobody knew anything about electric current. This is not surprising. The currents that we now treat as part of everyday existence must be carefully induced or generated. Even then, we can look at a wire, made especially for the purpose of carrying current, and be unable to tell whether or not it is carrying a current. We can only detect current by its effects. It took a long time to build a theory of what happens inside current-carrying wires.

In the early part of the 19th century a few scientists had begun to establish a concept of flow. People like Charles Augustin de Coulomb, Alessandro Volta and Georg Simon Ohm developed ideas of charge and potential. Then for several decades investigation of electrical phenomena was focused on magnetic effects, which were investigated by Hans Christian Oersted, André Marie Ampère and Michael Faraday. Some of these names have been preserved in the names of SI units.

Faraday's work, in particular, led in two directions. First, it led to developments in the theory of fields and of the travel of light itself, and to the work of James Clerk Maxwell and Albert Einstein in areas of physics that at first might seem very different. Secondly, it established rules of electromagnetism that eventually made commercial generation of electricity a possibility. By the end of the 19th century, electric lamps were lighting streets and motors were doing work in vehicles and other machinery. Vacuum cleaners, washing machines and other labour-saving devices lay in the not too distant future, and would change the way that people lived.

● A definition of current

Electric current is a rate of flow of charge. The **current** at a point in a circuit is the quantity of charge that flows past that point in each second. We can write:

$$\text{current} = \text{rate of flow of charge}$$

$$\text{average current during a period of time} = \frac{\text{charge that flows}}{\text{time for it to flow}}$$

Using I for current and q for charge (since it is clearly a good idea to avoid confusion by using C for either), then we can use Δq for change in charge (the charge that flows past the point) and Δt for the corresponding time. Average current during time Δt is given by:

$$\text{average } I = \frac{\Delta q}{\Delta t}$$

If Δq and Δt become very small, we call them δq and δt; they can then provide a value of average current during the short time δt:

$$\text{average current during a short period of time } \delta t = \frac{\delta q}{t}$$

1 **a** If there is a steady current of 0.4 A through a point in a circuit for 1 minute, how much charge has been moved through the point?
   b If the current is a result of electron flow, express the answer to **a** in terms of number of electrons. (The charge on one electron = e = −1.6 × 10⁻¹⁹ C. In this context the minus sign is not physically important, and can be ignored.)

2 Each electron carries a very small charge and, when a current flows, drifts quite slowly. How is it possible for a current as large as, say, 1 A to flow?

3 Sketch graphs for which
   a $dq/dt = \Delta q/\Delta t = $ constant
   b $dq/dt \neq \Delta q/\Delta t$.
   What can you say about the current in each case?

If δt and δq become infinitesimal (vanishingly small) then the calculated value of current is the current at just one instant:

$$\text{instantaneous current} = \frac{dq}{dt}$$

$\dfrac{dq}{dt}$ is the value of $\dfrac{\delta q}{\delta t}$ where δq and δt are infinitesimal.

The unit of current is the **ampere**, or **amp**, abbreviated to A. 1 A is equivalent to 1 C s⁻¹.

Figure 19.1
A value of current can be found from the gradient of a graph of charge against time.

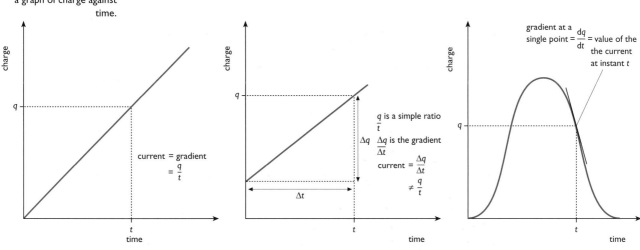

The heating effect of current

In Chapter 17 we dealt with simple systems of two charged particles which had no interaction with anything else. Working in such a way is useful for establishing some basic ideas, but in the 'real world' particles interact with others. In a metal wire, electrons interact with metal ions and with other electrons. In solutions, where ions are the mobile charge carriers, these ions interact with other particles in the solution. Both in the wire and in the solution, these interactions transfer energy from particle to particle. Energy is transferred away from the coherent flow that constitutes an electric current, and the material is heated. Since energy is transferred away from the current, if the current is to keep flowing then the energy must be replaced. In a d.c. circuit it is the battery that provides the continuous source of energy to replace that which is lost by heating.

Calculating the energy transfer

We know that the potential difference between two points is equal to the work that must be done to move one unit of charge (1 coulomb, in SI units) between them. If the two points are the terminals of a battery in a circuit then it is the battery that does the work to transfer charge from terminal to terminal around the circuit. The work done is equal to the energy transferred to the external circuit. (The external circuit is the circuit from terminal to terminal, excluding the interior of the battery itself.) In a simple circuit of, say, a battery and a lamp, the energy is transferred entirely by heating. We can write:

$$\text{potential difference between battery terminals} = \frac{\text{work done (or energy transferred)}}{\text{charge moved around circuit}}$$

$$\Delta V = \frac{\Delta W}{q}$$

It is normal practice when dealing with circuits to simply use V for potential difference and W for work done, so:

$$V = \frac{W}{q}$$

and

$$W = Vq$$

The work done is equal to the energy transferred to the external circuit by heating. Can we relate this to the current? We know that the rate of flow of charge is:

$$I = \frac{dq}{dt}$$

Over a period of time t, if the current is steady then it is related to the total charge transferred by:

$$I = \frac{q}{t}$$

which means that:

$$q = It$$

So, going back to the work done (= energy transferred), and replacing q with It,

$$W = VIt$$

Knowledge of this relationship allows us to make measurements of the energy transferred by a circuit or part of a circuit in a certain time. An ammeter and a voltmeter are used to measure current and voltage, and the period for which the current flows is timed (Figure 19.2).

4 Calculate the energy transferred in an external circuit in a car to which a potential difference of 12 V is applied, if it carries a current of 1.2 A for 20 minutes.

5 How long will it take to transfer 1 kJ of energy in the external circuit of a small torch, using a potential difference of 1.3 V and a current of 0.05 A?

Figure 19.2
The energy transferred by the heater in this circuit is given by $W = VIt$.

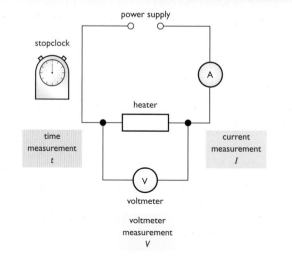

The magnetic effect of current

Energy is transferred from a current to the materials of a circuit and then into its surroundings due to heating. But there is another very important way in which energy is transferred from current and circuit into systems that lie outside. This is by way of the magnetic field that exists around moving charged bodies.

Note that both electric force and magnetic force act on charged particles, but while all charged particles exert and experience electric force, only *moving* charged particles exert and experience magnetic force. An electric field exists around any charged particle; a magnetic field exists around any moving charged particle.

Electrons in atoms are moving (including spinning). They therefore have magnetic fields around them. In most materials all of these magnetic fields cancel each other out. But in a few substances, like iron, every atom acts as a tiny magnet which can be aligned with all the other atom-sized magnets. So that instead of cancelling each other's magnetism, they combine together to produce magnetic effects that can be detected in the human-sized world (Figure 19.3a).

Figure 19.3
Magnetism that has origins in the micro-world of particles creates a magnetic field that is observable in the human-sized world in two ways: **a** in magnetised materials such as iron, and **b** in current-carrying wires.

Combination of the effects of large numbers of atom-sized magnets, as in an iron or steel magnet, is not the only way in which we can observe a magnetic field. The travel of electrons in huge numbers in the same direction in a wire, making an electric current, also results in combination of very many tiny magnetic fields (Figure 19.3b). This is the basis of electromagnets, in which wires are shaped into coils (with many loops or turns to further enhance the observable magnetic effects). Most electromagnets also have iron cores. The magnetic field due to the current causes temporary alignment of the atoms within the iron core, and this alignment further reinforces the coil's magnetic field.

a   bar magnet

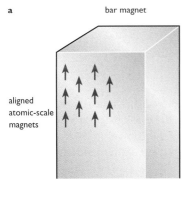

aligned atomic-scale magnets

b   current-carrying wire

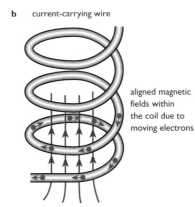

aligned magnetic fields within the coil due to moving electrons

We have already seen in Chapter 17 that the force acting on a charge q when moving at velocity v in a direction perpendicular to the field lines in a magnetic field of flux density B is given by:

$$F = Bqv$$

In a wire carrying a steady current there is a continuous flow of charge, and the total charge transferred over a time t can be calculated from

$$q = It$$

If in this same time t each charge-carrying particle moves a distance l along the wire, then they have velocity v which is given by

$$v = \frac{l}{t}$$

So, the force acting on all of these charge carriers together, which is the force acting on the length l of wire is

$$F = BIt\,\frac{l}{t}$$
$$= BIl$$

Note that this force F is perpendicular to both velocity v and flux density B. The direction of the force acting on a *positive* charge carrier is given by the left-hand rule (see page 150). Conventional current is considered to be a flow of positive charge carriers, from the positive to the negative terminals of a cell or battery. So the left-hand rule can be applied, where the second finger provides the direction either of the motion of the positive charge, or the direction of the conventional current. (For the force on the electrons, the second finger of the left hand is pointing in the wrong direction.)

Note also that where v and B are not perpendicular to each other, the force is less than that given by the formula. The formula $F = Bqv = BIl$ gives the maximum force that can act.

Devices that use the magnetic effect to transfer energy

A current-carrying wire can do work on other systems by means of magnetic force. At its simplest, a current-carrying wire can do work on an external magnet. This is what happens in a **relay**. A relay is a switch that allows use of a low-current circuit, such as a circuit containing delicate electronics, to switch on a high-current circuit. A coil forms part of the low-current circuit, and when its field is strong enough it attracts an iron armature which closes the switch in the high-current circuit (Figure 19.4).

Figure 19.4
A simple relay.

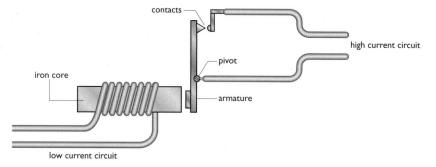

A loudspeaker also transfers energy from a coil, but here it is the magnet that remains fixed while the coil itself is allowed to move (Figure 19.5). Nevertheless, the work that is done in moving the coil is made available from the circuit by the magnetic effect of the current in the coil. Work is then done on the surrounding air, and a sound wave carries energy away from the loudspeaker.

Figure 19.5
The principle of a simple loudspeaker.

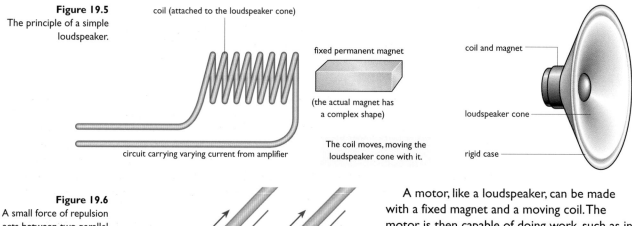

Figure 19.6
A small force of repulsion acts between two parallel conductors.

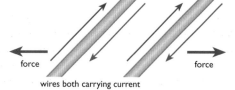

A motor, like a loudspeaker, can be made with a fixed magnet and a moving coil. The motor is then capable of doing work, such as in lifting a weight. Energy will also be required to replace that which is transferred in doing work to overcome frictional forces.

No net force acts between a wire carrying a current and a wire that is not. The passive wire has no magnetic field and neither exerts nor experiences magnetic force. However, if *both* wires carry current then a force will act between them (Figure 19.6). Most practical electric motors take advantage of the force between two current-carrying coils.

6 Where telephone signals are carried by electric current in wires it is possible to detect the current, and its signal, from outside the wire and so 'tap' a telephone call. What effect of the current is being detected?

7 A copper wire in which no current flows is full of electrons moving freely. Explain why the wire shows no observable magnetic behaviour.

8 To what system is energy transferred directly, by the work done due to the magnetic effect of current in a coil, in
  a a relay
  b a loudspeaker
  c a motor spinning freely
  d a motor lifting a weight?

Using the magnetic effect for measurement of current

Current is defined as rate of flow of charge. However, it is not possible to make measurements, inside a wire, of flow of charge. To measure current in a working circuit we must measure it indirectly, by measuring the effects it has.

We could make measurements of the heating effect of a current. We could measure the amount of energy transferred by heating, and also the time this heating takes and the relevant potential difference. Then we could transform the relationship

$$W = VIt \quad \text{into} \quad I = \frac{W}{Vt}$$

but this is hardly a practicable way of measuring the current, instant by instant, in a circuit.

It makes more sense to use the magnetic effect. In the **moving coil meter**, the current to be measured passes through a coil. The current in the coil interacts with a permanent magnet, producing rotation of the coil. A pointer can be attached to the coil so that readings can be made from a scale (Figure 19.7).

Figure 19.7
The principle of a moving coil meter.

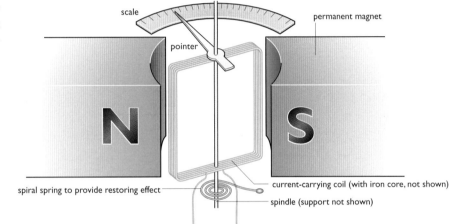

The bigger the current in the coil, the stronger its tendency to rotate. The coil may spin quite quickly for a small current and very quickly for a larger current. The speed of rotation of the coil does not provide a useful practical indication of the size of the current. However, if a spring is attached to the coil so as to oppose its turning, then the bigger the current the more the magnetic effect will tend to overcome this opposition (sometimes called a **restoring moment**). The coil and its attached pointer will come to rest when the magnetic turning effect balances the spring's opposing or restoring turning effect. The pointer position when such 'rest' is achieved is used as an indication of current.

The pointer turns through an angle and the scale is marked or graduated in an angular fashion, but it is necessary to **calibrate** the scale so that current can be read directly from it. Calibration can be achieved by passing a known current through the coil, and marking the corresponding pointer position, as well as its position when no current flows. If it is then assumed that the angle of rotation of the coil is proportional to the current, the scale can be marked accordingly (Figure 19.8).

Figure 19.8
Calibration of an ammeter.

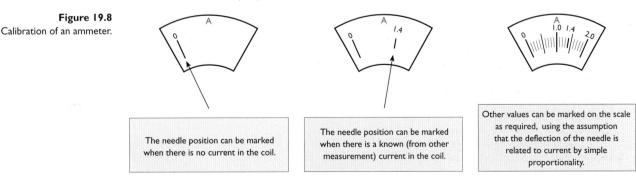

Problems and solutions in measuring current

A moving coil meter acting as an ammeter uses the angular displacement of a pointer to provide a measurement of current. The current to be measured must pass through the coil. It is inevitable that the coil will have resistance, so it will have an influence on the size of the current. This is undesirable, but we have to accept that a moving coil meter is a measuring instrument that affects the value of what it is measuring. (We'd be surprised if this happened with a simpler measuring instrument, such as a ruler.) In cases where the resistance of the circuit is much bigger than the resistance of the coil in the meter then the influence of the meter is comparatively small. For current measurement, it is therefore desirable to use a coil with small resistance.

A particular meter will have a particular current that makes the pointer move to the end of the scale – that is, produces **full scale deflection**. If such a deflection corresponds to, say 0.1 A, then the meter may only be used for measuring currents of 0 to 0.1 A. This is the meter's **range**. The effective range of a meter can be increased, so that it can read up to, say 1.0 A, by allowing only a proportion of the total circuit current to pass through the coil. To do this, the rest of the current is made to 'by-pass' the coil. The 'by-pass' connected in parallel with the meter coil is called a **shunt** (Figure 19.9). The resistance of the shunt must be carefully chosen, so that known proportions of the current pass through the shunt and the coil.

Figure 19.9
The use of a shunt to alter the range of a meter.

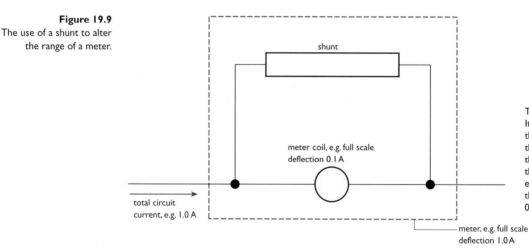

shunt

meter coil, e.g. full scale deflection 0.1 A

total circuit current, e.g. 1.0 A

The shunt acts as a by-pass. Its resistance is matched to that of the meter coil so that a fixed proportion of the total current passes through the coil. For example, 0.1 A may pass through the coil for every 0.9 A through the shunt.

meter, e.g. full scale deflection 1.0 A

Comparing measurement of current and potential difference

An ammeter is connected into a circuit so that a current flows through it. It should have a low resistance to minimise its influence on the current to be measured.

The same moving coil meter can be adapted for use as a voltmeter. The principle of operation is still based on the magnetic force acting on a coil in which a current flows. However, since the current in the coil depends on the potential difference that is applied to it, the deflection of the pointer provides an indirect measurement of potential difference. Of course, the meter must be re-calibrated so that its scale is marked in volts.

9 Choose one or more words for each space.
   A moving coil meter uses the angular ....
   of a pointer as an indication of the magnetic
   .... acting on a coil, which in turn is an
   indication of the .... in the coil.

There is a major difference in the ways in which a meter is connected into a circuit to measure current and to measure potential difference. In the latter case it must be connected between or 'across' two points in a circuit. That is, it is connected in parallel with a part of the circuit. There will be some circuit resistance between these two points. If the meter were to have a low resistance compared with that part of the circuit with which it is in parallel, then it would take a large proportion of the current and have a strong influence on the behaviour of the circuit which it is meant to be measuring. So, when a moving coil meter is used as a voltmeter, resistance must be added to it. A resistor connected in series with a meter coil in this way is called a **multiplier** (Figure 19.10).

Figure 19.10
The addition of a multiplier to increase the meter's resistance.

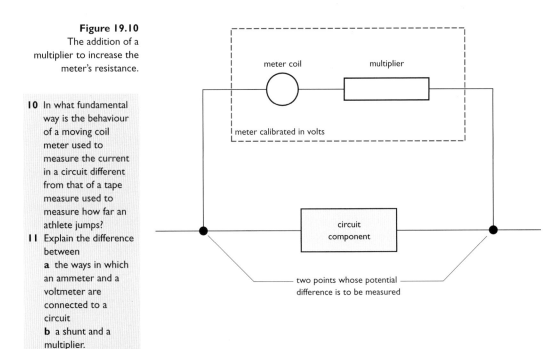

10 In what fundamental way is the behaviour of a moving coil meter used to measure the current in a circuit different from that of a tape measure used to measure how far an athlete jumps?

11 Explain the difference between
a the ways in which an ammeter and a voltmeter are connected to a circuit
b a shunt and a multiplier.

Inducing mass movement of electrons in metals

A metal can be thought of as an array of positively charged bodies in fixed positions, surrounded by a 'sea' or a 'gas' of electrons that can travel throughout the metal. The positively charged bodies are positive ions – the nuclei of the metal atoms surrounded by some electrons that are not free to wander. There are not enough of these 'attached' electrons to make the ions neutral. But the metal overall has no charge; the total number of 'free' and 'attached' electrons is enough to balance the positive charge of the nuclei and achieve neutrality.

In any metal object, there are huge numbers of 'free' electrons. They are in constant motion. This motion is normally random, like the random thermal motion of molecules in a liquid or a gas. However, an electric or a changing magnetic field can induce the electrons to move in a non-random way. They do not lose their random motion, but the field causes all the free electrons to have the same drifting motion superimposed on the random motion. It is this drifting motion that results in current in a wire.

Drift or flow of electrons produced by an electric field is away from repulsive negative charge and towards attractive positive charge, or from lower to higher electric potential.

Drift or flow of electrons along a wire due to a magnetic field is also very important; generation of such an effect is called **electromagnetic induction**. The induced motion of the electrons only takes place when there is relative motion of the electrons and the field. This can happen either when the electrons are moving because the whole wire is moving (Figure 19.11), or when the magnetic field is moving or changing. The former effect is applied in a simple generator, the latter in a transformer.

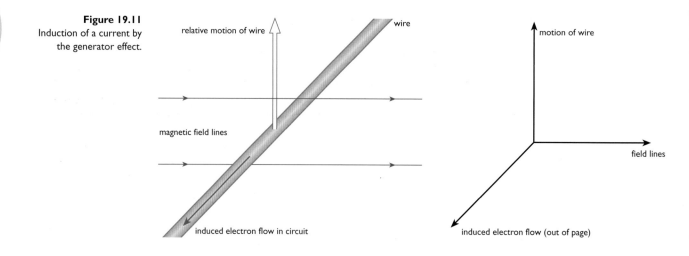

Figure 19.11
Induction of a current by the generator effect.

relative motion of wire

wire

magnetic field lines

induced electron flow in circuit

motion of wire

field lines

induced electron flow (out of page)

A **transformer** uses two coils (Figure 19.12), usually both wrapped around the same iron core which strengthens the magnetic field that is created by current in one of the coils, called the primary coil. The two coils, the primary and the secondary coil, are insulated from each other and are parts of two completely separate circuits. It is not surprising that when there is a steady current in the primary coil then nothing happens in the secondary coil. What is surprising at first is that when the current in the primary coil changes then an e.m.f. is induced across the secondary coil which can result in a current in the secondary circuit. So, a varying current in one circuit induces current in a second circuit that is not connected electrically but only by magnetic effects. A transformer transfers energy from circuit to circuit (Figure 19.13).

Figure 19.12
Current can also be induced by the transformer effect.

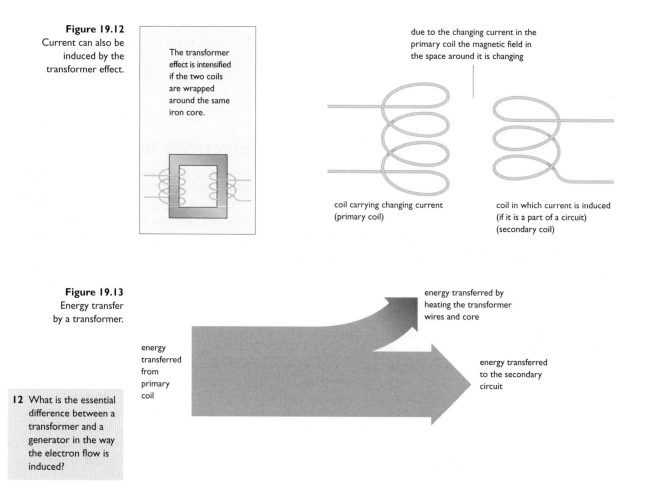

The transformer effect is intensified if the two coils are wrapped around the same iron core.

due to the changing current in the primary coil the magnetic field in the space around it is changing

coil carrying changing current (primary coil)

coil in which current is induced (if it is a part of a circuit) (secondary coil)

Figure 19.13
Energy transfer by a transformer.

energy transferred from primary coil

energy transferred by heating the transformer wires and core

energy transferred to the secondary circuit

12 What is the essential difference between a transformer and a generator in the way the electron flow is induced?

Direct and alternating currents

Chemical processes inside an electric cell or battery create an e.m.f., and so a potential difference between its terminals. When the terminals are connected by a conducting loop – a circuit – then charge moves and work is done on it. The source of mobile charge (electrons) is the battery's negative terminal, which is supplied with electrons by the chemical processes, and repels these into the circuit. The battery's positive terminal is an electron 'sink' – electrons flow into it to be combined with positive ions inside the battery (Figure 19.14). The terminal potential difference doesn't change, and the direction of the flow of electrons around the circuit is constant. This is simple **direct current, d.c.**

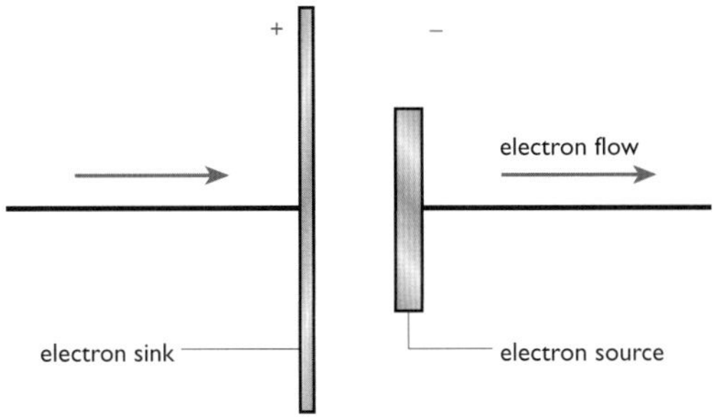

Figure 19.14
A battery (a cell or group of cells) provides an electron source and an electron sink.

electron flow

electron sink

electron source

In a simple generator, a coil rotates in a magnetic field. One side of the coil passes first in one direction through the field, and then in the other direction (Figure 19.15). The induced flow of charge also changes direction. This is **alternating current, a.c.**

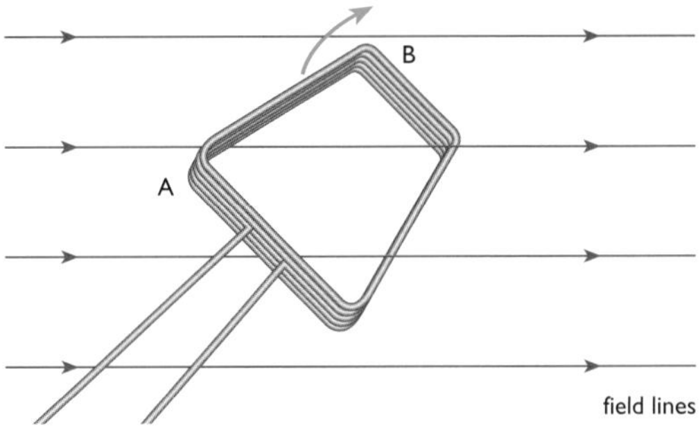

Figure 19.15
Motion of a simple generator coil through a magnetic field.

B

A

field lines

Here, side AB of the coil moves upwards through the field and then downwards. The direction of the current induced in it also reverses.

In a generator, the wire in which the electromagnetic induction takes place lies in an environment in which an external magnetic field is changing. The change is produced by *relative movement* of the wire and the 'source' of the magnetic field. The source of the field may be a permanent magnet or an electromagnet.

In a transformer, the secondary coil in which electromagnetic induction takes place also lies in an environment in which an external magnetic field is changing. The change is produced by variation in the magnetic field that is produced by a fixed source. This fixed source is another coil – an electromagnet – and in order for its magnetic field to vary it must carry a *varying current*. This may be a.c. or varying d.c. (Figure 19.16).

13 Describe the difference between a d.c. circuit and an a.c. circuit in terms of movement of electrons.
14 Explain why current is induced in the secondary circuit of a transformer when its primary circuit is driven by a battery *only* very briefly when the circuit is being connected and disconnected, and not while the current remains continuously connected.

Figure 19.16
d.c. has a single direction; a.c. changes direction. Varying d.c. has a varying size of current but the flow is always in the same direction.

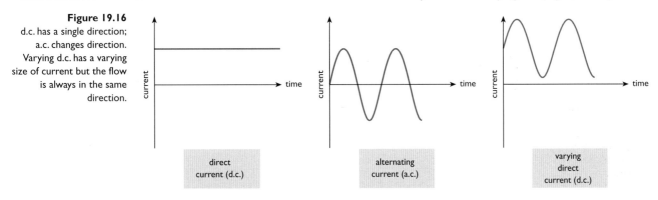

direct current (d.c.)

alternating current (a.c.)

varying direct current (d.c.)

Comprehension and application

Mains supply and the residual current device

Study the diagrams and text panels in Figures 19.17 to 19.22, then answer questions 15 to 18.

Figure 19.17
Electrical supply wires.

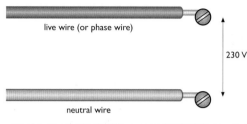

There will be a current between the live and the neutral wire, passing through whatever connects them.

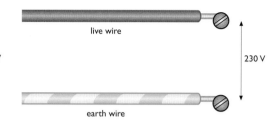

There would also be a current between the live and the earth wire if they became connected.

Figure 19.18
Alternating current, a.c., and the corresponding alternating potential difference between live and neutral wires.

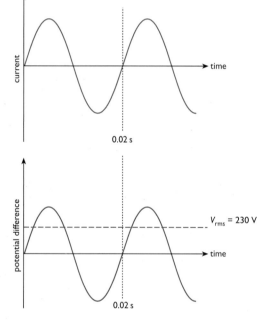

The mains supply is alternating.

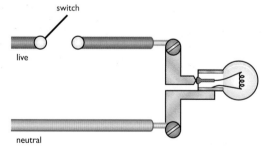

V_{rms} is the '**root mean square**' voltage, and is equal to the constant d.c. potential difference that would provide the same rate of energy transfer as the a.c. potential difference supplied. In the European Union the value of V_{rms} is 230 V and the a.c. frequency is 50 Hz.

Figure 19.19
Switches and fuses are always connected into the live wire.

A switch can isolate a lamp or other appliance from the live wire.

If the live and the neutral wire are connected by a low resistance, a very large current can flow. A fuse acts as a weak point in the circuit. If the current exceeds a set value – the fuse rating – the fuse wire melts and breaks the circuit, acting as a switch. If the fuse did not melt, any appliance at X could be damaged by the high current.

Figure 19.20
The danger of
electrocution.

If the connection between live and neutral is through a human body then a *VERY* dangerous current can flow. The effect on the body is made worse by the fact that it is alternating current, as this causes repeated contraction of the muscles.

The same effects take place if a person who is connected to earth then connects to the live wire.
- An alternating current of around 10 mA at 50 Hz through the fingers causes muscle contractions, which make it impossible for the person to let go of the live wire they have gripped.
- Current across the skin can cause burning. The skin has a thin coating of ionic liquid (salt solution) that has a relatively low resistance.
- Currents of tens of milliamps can cause heart fibrillation – uncoordinated action of the heart muscle which overrides its regular beating and leads to death.

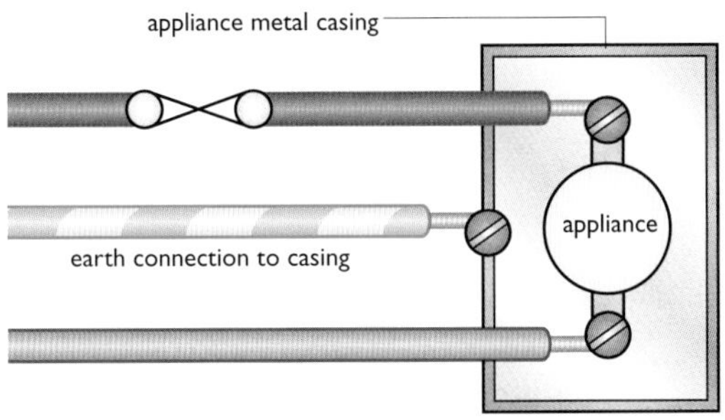

Figure 19.21
The danger of a faulty
appliance.

appliance metal casing

earth connection to casing

appliance

It is possible that the metal casing holding an appliance can accidentally become connected to the live wire. If the appliance is earthed, the earth wire then behaves like the neutral wire, and a large current passes which can melt the fuse wire, isolating the appliance from the supply.

The fuse wire melts before any wires in the appliance melt, so it protects the appliance from damage. However, a significant current can flow before the fuse melts, so a fuse provides only partial protection to people. A **residual current device** (RCD) is designed for extra safety – to protect people rather than appliances.

Figure 19.22
A residual current device.

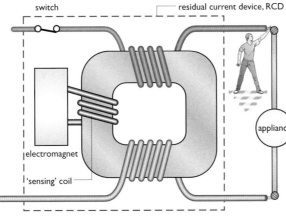

switch

residual current device, RCD

electromagnet

'sensing' coil

appliance

In normal circumstances the currents in the live and neutral wires are equal. In a residual current device these currents create equal and opposite magnetic fields in coils wrapped around an iron core.

Should accidental earthing (for example through a human body) occur, then the current in the live wire is increased, but that in the neutral wire is decreased.

The two magnetic fields created by the coils in the residual current device are no longer equal. A net current is induced in a third coil, which effectively 'senses' the magnetic field inequality. This net current triggers the opening of a switch by the action of an electromagnet.

15 a The potential difference between mains live and neutral wires is often zero and varies between +325 and −325 V. Show this information on a graph.
b Use the graph to show the d.c. potential difference that provides the same rate of energy transfer. Show the value of this potential difference.

16 a Calculate the energy transferred by a domestic mains appliance that operates at 3.5 A when used for
i 1 minute
ii 1 hour.

b Calculate the energy transfer taking place in a human muscle when connected to the mains and carrying a current of 10 mA for a time of 0.1 s.
c What is the effect of such a current?
d Why is it possible for current across skin to be very much higher than 10 mA?
e What is the effect of such a current?

17 Describe the role of electromagnetic induction in the action of a residual current device.

18 'A fuse protects appliances and wiring, but a residual current device protects people.' Explain this statement.

● **Extra skills task**   Communication and Information Technology

It is said that King William IV visited Faraday while he was researching electromagnetic induction, but was underwhelmed. At this point in time, Faraday was carrying out 'pure' research; that is, finding out about things that interested him. When asked by the clearly unimpressed King what was the use of his fiddling around, Faraday replied that the results of his work were so important that they would be taxable in years to come.

Obviously, we now realise how far-sighted Faraday was. Unfortunately, nowadays, funding for pure research is hard to come by. More money is available for projects with clear-cut commercial aims.

Take part in a group discussion, which covers the issues that face governments and private investors. Consider questions such as: 'Is it right that research should pay for itself directly, and only be financed if it is very likely to lead to a marketable product? Is it wrong to pay scientists for 'playing' in laboratories, or should we encourage them, knowing that often there will be commercially valuable 'spin-offs' from work which is carried out simply to explore the way nature works?'

Before the discussion, carry out research to allow you to make clear relevant contributions. Give examples, such as the relationship between the Internet and particle physics, or the development of new materials as a result of space science. Listen to the other contributors, and develop the various points and ideas.

20

Resistance of circuit components

THE BIG QUESTIONS

- How can we predict what will happen when particular components are connected into circuits?
- What different kinds of behaviour do different components show?

KEY VOCABULARY

base   base current   characteristic curve   collector current   current amplification
depletion layer   doping   forward bias   intercept   internal resistance
n-type (semiconductor)   ohm   ohmic conductor   Ohm's Law
p-type (semiconductor)   p-n junction   positive feedback   potential gradient
resistance   reverse bias   semiconductor   semiconductor diode
slide potentiometer   switching   thermal runaway   thermistor   transistor

BACKGROUND

We can manipulate resistance to make circuits serve useful purposes. First we need to know the rules of how resistance influences circuit behaviour. With such knowledge, electricity is under our control.

Figure 20.1
Knowledge is power to make life better. These farmers in the Sudan can channel solar energy to provide the energy to drive a water pump.

Resistance and energy transfer

1 Rough or bumpy horizontal surfaces resist the motion of a skate-boarder, due to friction. Sloping ground can compensate for this to make it possible for the skate-boarder to travel across such surfaces without slowing down and without making any extra personal effort.
a Where does the energy come from to compensate for friction and allow a skate-boarder to move effortlessly at a constant speed down a slope?
b In an electric circuit, how do we 'compensate' for resistance? Where does the energy come from for this?

The action of a resistor in a circuit is to oppose the flow of charge; this is comparable to the way in which mechanical friction opposes motion. In both cases, work must be done to overcome the opposition, and the end result of this work is heating of the materials and their surroundings. The energy is dissipated. In a d.c. circuit, the battery is usually the source of the energy that is needed. Its potential difference allows work to be done on charged particles.

A definition of resistance

The **resistance** of a component is defined as the ratio of the potential difference between its ends to the current that passes through it:

$$\text{resistance} = \frac{\text{potential difference}}{\text{current}} \quad \text{or} \quad R = \frac{V}{I}$$

The unit of resistance is the volt per amp, which is more often called an **ohm**, Ω. A large voltage relative to a small current produces a large ratio, and denotes a high resistance. If, on the other hand, the ratio is small, then a relatively small potential difference results in a large current; resistance is low.

2 What happens to the ratio V/I if:
   a V increases but I stays the same
   b I increases but V stays the same
   c I decreases but V stays the same
   d V and I increase by the same proportion
   e V and I increase by the same amount. (Warning: trick question!)
3 What is the resistance of a wire that requires a potential difference of 6 V to drive a current of 0.2 A?
4 What potential difference must be applied to a 16 Ω resistor to drive a current of 0.25 A through it?

Calculating energy transfer by a resistor

We know that the energy transferred by a current, I, flowing due to potential difference, V, for time t, is given by the formula:

$$W = VIt$$

(see page 160).

Resistors transfer energy from circuit to surroundings, and heat their surroundings. If we want to see how the amount of the energy transfer is related to resistance, we can use $R = V/I$, rearranged to $V = IR$. We can then substitute IR for V:

$$W = IRIt = I^2Rt$$

Power dissipation by a resistor

A resistor transfers energy from the circuit to its surroundings, heating them. The energy is dissipated. The rate at which the resistor transfers energy to its surroundings is its power:

$$\text{average power} = \text{average rate of transfer of energy} = \frac{\text{change in work}}{\text{change in time}} = \frac{\Delta W}{\Delta T}$$

$$\text{instantaneous power} = \frac{dW}{dt}$$

The unit of power is the watt, which is equivalent to the joule per second: $1\,W = 1\,Js^{-1}$ (see page 134).

When energy is transferred at a steady rate starting at time $t = 0$, then the power is constant. We can then calculate power from the total energy transferred, W, over a period of time, t:

$$\text{power (when constant)} = \frac{\text{total energy transferred}}{\text{time taken}} = \frac{W}{t}$$

$$= \frac{VIt}{t}$$

$$= VI$$

Thus the power of a domestic appliance is the product of the applied voltage (230 V when connected to the mains) and the current. Since domestic mains voltage has a fixed value, and the operating power of an appliance is often known (from a label on the appliance), we can calculate the current required by the appliance:

$$I = \frac{P}{V}$$

where P is power.

This can be used to determine the value (current rating) of a fuse that should be used with the appliance. For example, for a 400 W hairdryer:

$$I = \frac{P}{V} = \frac{400}{230} = 1.74 \, A$$

A 1 A fuse would melt with this appliance, while a 13 A fuse would allow a current much larger than the normal operating current, which could damage the appliance wiring. Therefore, a 3 A fuse is most suitable.

Note that since $W = VIt = I^2Rt$, then constant (or average) power $= VI = I^2R$.

The kilowatt-hour

5 How much energy is transferred by a heater running with an applied potential difference of 24 V and a current of 0.8 A for
  a 10 seconds
  b 10 minutes?
6 a What is the resistance of the heater in question 5?
  b What is its power when the applied potential difference is 24 V?
7 How much energy would the same heater transfer if it carried a current of 1.6 A for
  a 30 seconds
  b 2 minutes?
8 For how long would the heater have to carry a current of 2.0 A in order to transfer
  a 600 J of energy
  b 200 kJ of energy
  c 1 kWh of energy?
9 a Take the formula $W = VIt$ and use $I = V/R$ to eliminate I, and so find a new formula relating energy transfer to potential difference, resistance and time.
  b Use this new formula to calculate the energy transferred when a potential difference of 230 V is applied to a 100 Ω kettle heater element for 5 minutes.
  c What happens to this energy?
10 Sketch graphs of energy transfer against time to show constant and varying power. Use the graphs to show that $\frac{W}{t}$ can be, but is not always, equal to $\frac{dW}{dt}$.
11 What is the operating current of a 2.4 kW heater when connected to the domestic mains?
12 Sketch a graph of power against current for a resistor of constant resistance.

Electricity bills are sent out every 3 months. A lot of energy transfer takes place in the average home in a 3-month period, so the joule is too small a unit to be convenient. One solution is to measure domestic energy 'consumption' in megajoules, MJ. However, it is easier to use the kilowatt-hour, since this is defined in terms of typical use of a household appliance. The kilowatt-hour, kWh, is the energy transferred by a 1 kW appliance in 1 hour. 1 kW is 1000 W and 1 hour is 3600 seconds, so

$$\begin{aligned} 1 \, kWh = \ & \text{energy transfer by a } 1000 \, W \\ & \text{appliance in } 3600 \text{ seconds} \end{aligned}$$

Since, for constant power,

$$P = \frac{W}{t}$$
$$\begin{aligned} W &= P \times t \\ &= 1000 \times 3600 \\ &= 3\,600\,000 \, J \end{aligned}$$

So $1 \, kWh = 3\,600\,000 \, J = 3.6 \, MJ$

Potential gradient across a resistor

A length of uniform wire has the same resistance in each millimetre of length. That is, a uniform wire has uniform resistance per unit length. We can test this by connecting a voltmeter to two points along the wire, using crocodile clips as contacts (Figure 20.2). We can choose any two points. Suppose, for example, we place the contacts 100 mm apart. We can move the contacts to as many different pairs of points on the wire as we wish. If the pairs of points are all 100 mm apart then we will see the same voltmeter reading in all cases. If we increase the distance above 100 mm, then the voltmeter reading will increase. If the distance is less than 100 mm then the potential difference between the two points will be smaller.

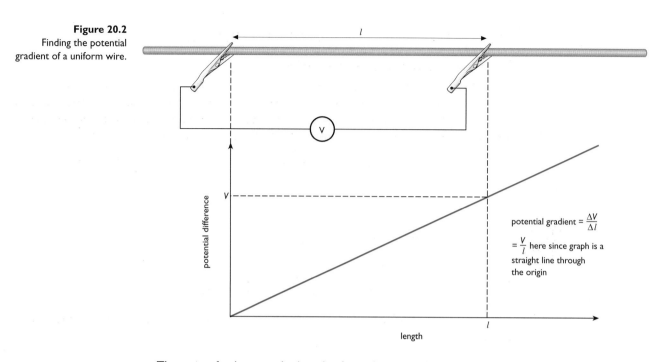

Figure 20.2
Finding the potential gradient of a uniform wire.

potential gradient $= \dfrac{\Delta V}{\Delta l}$

$= \dfrac{V}{l}$ here since graph is a straight line through the origin

The ratio of voltage to the length of wire between the voltmeter connections is called **potential gradient**:

$$\text{potential gradient} = \frac{\text{potential difference}}{\text{distance between voltmeter connections}}$$

$$= \frac{\Delta V}{\Delta l}$$

The unit of potential difference, ΔV, is the volt and the unit of distance, Δl, is the metre. So potential gradient has the unit volt per metre, V m^{-1}.

We can make use of this when we want to vary the potential difference applied to a particular component, using a fixed voltage supply such as a battery. We apply the battery potential difference to a length of uniform resistance wire. Then simply by sliding the two contacts attached to the wire, further apart or closer together, we make use of larger or smaller portions of the total battery potential difference. This arrangement is sometimes called a **slide potentiometer** (Figure 20.3a), and provides the way in which the voltage applied to very many components is controlled, such as volume, brightness and contrast controls in a TV. For these, the wire may be coiled, and the coil bent into an arc of a circle, so that adjustments are made by turning a contact to a different point on the coil (Figure 20.3b).

13 **a** Sketch a graph of potential difference against distance between voltmeter connections on a metal wire.
  b What is the name of the gradient of the graph?
14 Sketch a circuit that you could use to vary the potential difference applied to a lamp.

Figure 20.3
a A simple slide potentiometer circuit, and
b the inside view of a commercial potentiometer.

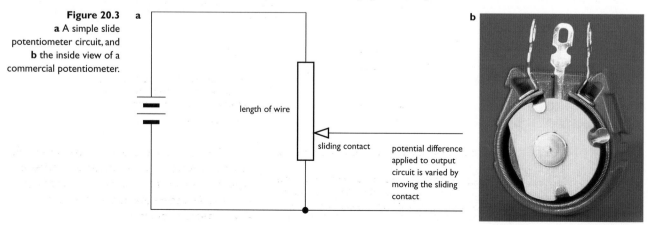

length of wire

sliding contact

potential difference applied to output circuit is varied by moving the sliding contact

The relationship between current and applied potential difference

The relationship $R = V/I$ provides a definition of resistance and is true at all times. For a component for which the ratio V/I varies, resistance also varies. A component for which the ratio V/I is constant has a constant resistance. The constant ratio means that a graph of current against potential difference is then a straight line passing through the origin (Figure 20.4).

Figure 20.4
I–V graph for a component with constant resistance.

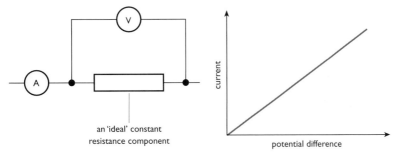

The graph has a constant gradient, equal to I/V. This means that the gradient is the inverse, or reciprocal, of the resistance:

$$\text{gradient} = \frac{I}{V} = \frac{1}{R}$$

The plotted quantities have a proportional relationship – the current is proportional to the potential difference.

The statement that current is proportional to potential difference is known as **Ohm's Law**. It applies only to circuit components that have current–potential difference graphs that are straight lines passing through the origin; these components are called **ohmic conductors**. Metal wires are ohmic conductors provided that their temperature (and tension, shape and density) do not change. Ohmic behaviour is idealised behaviour. It has the benefit of simplicity, but real metal components are only ohmic under carefully controlled conditions, and non-metal components are rarely ohmic at all.

15 Show, by step by step rearrangement, that the relationship $I/V = 1/R$ agrees with the relationship $R = V/I$.

Metal wires

A metal wire is normally ohmic (that is, has constant resistance, a constant current–voltage ratio and a straight line current–voltage graph) until the heating effect of the current becomes significant. So the behaviour of a metal wire approximates to that of an 'ideal' ohmic conductor when it is cool. However, its resistance increases with its temperature, and the current–voltage line becomes curved (Figure 20.5).

Figure 20.5
I–V graph for a metal wire.

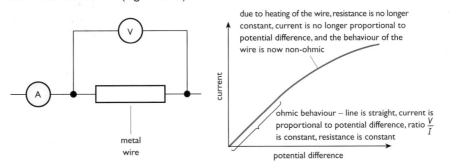

16 Explain whether you would expect the filament of a lamp to be an ohmic conductor when in use.
17 Sketch, using the same voltage scale, graphs of current against voltage and resistance against voltage for a metal wire. (Your second graph should show what happens to the resistance when the behaviour of the wire becomes strongly non-ohmic.)

The shape of a current–voltage curve tells us a good deal about the component. Different types of component have different shapes of curve, and so these are sometimes called **characteristic curves**.

Thermistors

A device made from **semiconductor** material has a high resistance when cool but much lower resistance when warmer. Some semiconductors, like silicon or carbon in the form of graphite, are moderately good conductors at room temperature. Carbon in the form of diamond is also a semiconductor, but is a poor conductor at room temperature and only conducts well when it is very hot.

A cool semiconductor contains few free electrons within its crystal structure. As the material is heated then more electrons gain enough energy to become free of individual atoms. The material becomes a much better conductor. This means that the resistance of a semiconductor decreases rapidly as its temperature increases.

Note that temperature has an opposite effect on the resistance of a semiconductor to that of a metal wire. Another difference is that the resistance of a semiconductor is much more sensitive to changes in temperature (Figure 20.6).

Figure 20.6
Variation in resistance of **a** a typical thermistor with temperature, and **b** a typical metal wire over the same temperature range.

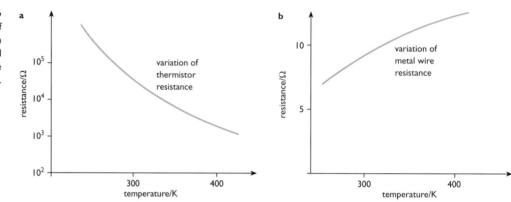

Note the logarithmic scale – equal lengths of scale correspond to increases in resistance by equal *multiples*. This allows the scale to show a wide range of values.

There is no need to use a special scale here because the variation in resistance is much smaller.

A **thermistor** is a piece of semiconductor material that is used in circuits for its strongly temperature-dependent resistance. It can be used to make the behaviour of a circuit depend on temperature. The circuit then 'senses' changes in temperature. Such circuits can be used to control heating or ventilating systems (Figure 20.7).

Figure 20.7
The heater circuits for this baby incubator switch on and off automatically, depending on the temperature inside. A circuit containing a thermistor responds to temperature changes.

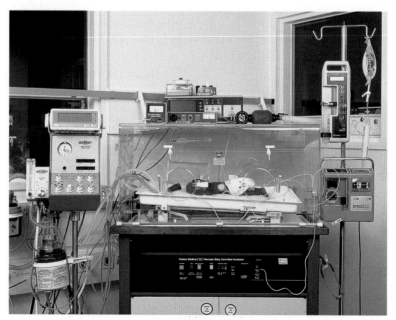

Figure 20.8
Positive feedback – an increase in current causes further temperature rise and further increase in current. Positive feedback results in instability and this can lead to thermal runaway.

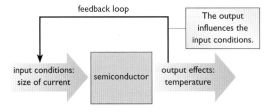

feedback loop

The output influences the input conditions.

input conditions: size of current → semiconductor → output effects: temperature

Current in the semiconductor itself causes heating which reduces its resistance. So an increase in current is itself the cause of a further increase in current. This is an example of **positive feedback** (Figure 20.8), and if uncontrolled is called **thermal runaway**.

18 Why does a thermistor respond differently from a metal wire to an increase in temperature? How is this revealed in the shape of the characteristic curve shown in Figure 20.9?
19 Which of the following has a direct relationship and which has an inverse relationship:
    a thermistor temperature and the current in it, for a fixed potential difference
    b thermistor temperature and its resistance?

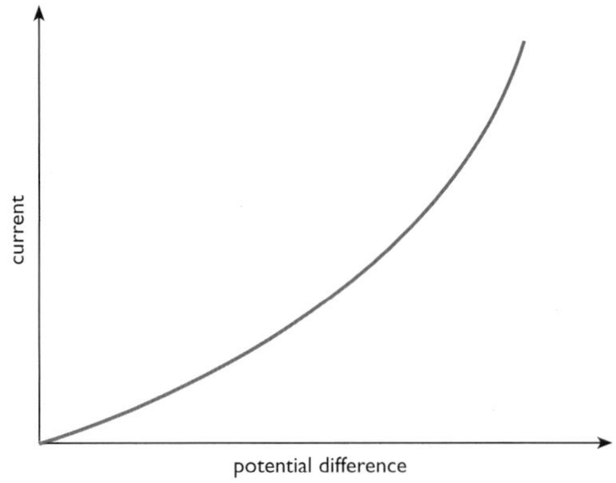

Figure 20.9
A characteristic curve for a thermistor.

current

potential difference

Semiconductor diodes

20 An *ideal* diode would have infinite resistance for all negative or reverse potential differences, and zero resistance for all positive or forward potential differences. Sketch the characteristic curve for such a device.

A **semiconductor diode** consists of two pieces of semiconductor. When a potential difference is applied in one direction a current can flow across the boundary between these two pieces. When the potential difference is reversed, no current can flow.

A semiconductor diode is analogous to a valve in a heart or other mechanical pump. It allows one-way flow only. In electrical terms, it has low resistance when a potential difference is applied in the 'forward' direction, and very high resistance when a potential difference is applied in the 'reverse' direction (Figure 20.11).

Figure 20.10
Diodes, resistors, capacitors and microchips in a circuit.

A microchip is, in simplified terms, a collection of transistors. A transistor is similar to a diode but with three rather than two layers of semiconductor material. This difference makes its behaviour very different (see page 181).

Figure 20.11
A characteristic curve for a semiconductor diode.

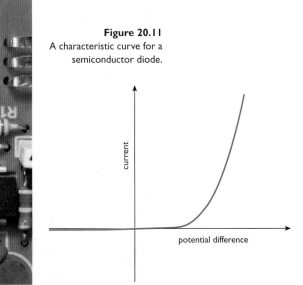

current

potential difference

Internal resistance of a battery

A battery is connected in series with the circuit to which it provides energy. The size of the current in the battery is the same as that in the external circuit. Because of current in the battery, energy is dissipated within it. We can say that during any period of time:

$$\text{total energy transfer by the battery} = \left(\begin{array}{c}\text{energy transferred}\\\text{to external circuit}\end{array}\right) + \left(\begin{array}{c}\text{energy dissipated}\\\text{within the battery}\end{array}\right)$$

We could abbreviate this to:

$$W_T = W_X + W_B$$

We could then divide this by the charge, q, that is transferred around the circuit during the same period of time:

$$\frac{W_T}{q} = \frac{W_X + W_B}{q}$$

$$\frac{W_T}{q} = \frac{W_X}{q} + \frac{W_B}{q}$$

These are quantities of energy divided by charge. That is, they are measured in volts – they are voltages:

$$\frac{W_T}{q} = \text{e.m.f. of battery, } \varepsilon$$

$$\frac{W_X}{q} = \text{potential difference across battery terminals, } V$$

$$\frac{W_B}{q} = \text{'lost volts'}$$

So,

$$\varepsilon = V + \text{lost volts}$$

This agrees with the formula given on page 157.

If V is the potential difference across the battery then it must also be the potential difference across the external circuit. It is related to the resistance of the external circuit by:

$$V = IR$$

The value of the lost volts does not depend on the resistance of the external circuit but on the effective resistance of the battery itself, which is called the battery's **internal resistance**, r. Since the same current, I, exists in the battery as in the external circuit,

$$\text{lost volts} = Ir$$

So we can say that:

$$\varepsilon = V + \text{lost volts}$$
$$= V + Ir = IR + Ir = I(R + r)$$

We can rearrange $\varepsilon = V + Ir$ to make V the subject:

$$V = \varepsilon - Ir$$

It is interesting and important to see what happens to the potential difference that a battery makes available to a circuit when different current demands are made on it. We can control the current by connecting a battery to a variable resistor. Current is then the input variable and the battery potential difference is the output variable, and these can be plotted on a graph. The formula $V = \varepsilon - Ir$ predicts that this graph will have a positive **intercept** of ε. An intercept is the value of a plotted variable when its partner variable equals zero. When current is zero, then $V = \varepsilon$ = the intercept. The intercept is shown on a graph as a point at which the line crosses an axis. The formula also predicts that the graph will be a straight line with a negative gradient equal in value to the internal resistance of the battery, r (Figure 20.12).

21 What is the internal resistance of a battery which has an e.m.f. of 1.5 V and drives a current of 0.2 A through an external resistance of 3.5 Ω?

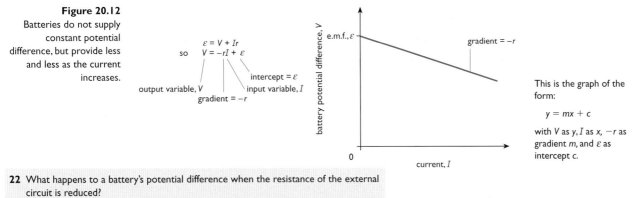

Figure 20.12
Batteries do not supply constant potential difference, but provide less and less as the current increases.

22 What happens to a battery's potential difference when the resistance of the external circuit is reduced?

23 **a** Sketch a graph to show the relationship between circuit current and terminal potential difference for a battery with an e.m.f. of 6 V and an internal resistance of 0.3 Ω.
b At what value of current is the terminal potential difference
i equal to the e.m.f. **ii** 60% of e.m.f?

Car batteries

Car batteries have the benefit of being able to be 'revived', or recharged, time after time. They are made up of cells called secondary cells or accumulators. When they drive current, a chemical reaction takes place and they become discharged. They are recharged when the chemical reaction is reversed. During discharge, a battery makes energy available to an external circuit; this is needed, for example, to start the car by means of the starter motor. Once the car is running, its alternator generates the electricity and the battery is then not usually needed. In fact, the alternator actually recharges the battery.

A car battery has an e.m.f. of a little over 12 V, and its terminal potential difference when driving a current is even smaller. But, a starter motor is a fairly high power device, and so, to produce this power the battery must be capable of driving a high current at its low voltage. However, the battery must also have a low internal resistance or this would lead to a high value of 'lost volts' and a low value of battery potential difference, as explained below.

average starter motor power $= IV$ (where V is the battery terminal potential difference)
for a starter motor, the current I is typically as much as 80 A
if the battery e.m.f. is 12 V and its internal resistance is 0.05 Ω, then

$$\varepsilon = I(R + r) = V + Ir$$
$$V = \varepsilon - Ir$$
$$= 12 - (80 \times 0.05)$$
$$= 12 - 4 = 8\,V$$

So, clearly, it is desirable that the battery should have low internal resistance in order to provide as much terminal potential difference as possible.

For a partly discharged or 'flat' battery, the e.m.f. is little changed relative to the state of full charge. However, the internal resistance increases, so, when the starter motor demands a high current, the terminal potential difference of the battery drops very significantly. Reduced power to other circuits in the car, resulting in temporary dimming of the lights for example, is evidence of this fall in terminal potential difference. If the discharge is more severe, the starter motor receives insufficient power to turn, and the car will not start.

● **Comprehension and application**

Variable resistance in semiconductor components

A semiconductor such as carbon (graphite), silicon or germanium has, at room temperature, a limited number of free charge carriers. An increase in temperature provides the energy needed to increase the number of charge carriers that are free to move within the material. This is the principle of the thermistor (page 176), which is a resistor whose value is very dependent on its temperature.

It is also possible to change the number of charge carriers in a semiconductor by slightly altering its composition. This process of introducing impurity is called **doping**. Some impurities have the effect of increasing the number of free electrons – they increase the number of negative charge carriers in the material. The material is then called an **n-type** semiconductor, where 'n' stands for negative.

Other impurities increase the number of free spaces within the crystal structure that are available to electrons. An electron can move into such a free space, but when it does it leaves a free space behind it. The spaces are usually called 'holes'. As electrons drop in and out of holes, the holes move around the material, acting like moving positive charges. Though it is electrons that physically move, it becomes easier to think of this process as the movement of positive holes. A material doped to create mobile holes is called a **p-type** semiconductor, where 'p' stands for positive.

Interesting things happen when n-type and p-type materials are in contact. In any collection of mobile particles, diffusion takes place due to random movement. At a **p-n junction** (Figure 20.13a) electrons diffuse randomly across the junction. There is a net movement of electrons from the n-type material, where there are more free electrons to begin with. Some randomly wandering electrons fill holes in the nearby p-type material. The result is that there are fewer free electrons on the 'n' side of the junction and fewer free holes on the 'p' side. A layer of material has been naturally depleted of its free charge carriers. This is called a **depletion layer**.

If a potential difference is applied to the junction in one direction – **reverse bias** – then it makes this depletion worse (Figure 20.13b) and the material does not carry current. It has a very high resistance. But if the potential difference is applied in the opposite direction – **forward bias** – then free charge carriers are forced back into the depletion layer (Figure 20.13c) and the material has low resistance. It may carry a large current. So a block of semiconductor with a p-n junction acts as a one-way device; such a device is called a diode. Figure 20.13d shows its characteristic behaviour.

Figure 20.13
A semiconductor diode formed from a p-n junction, and its behaviour.

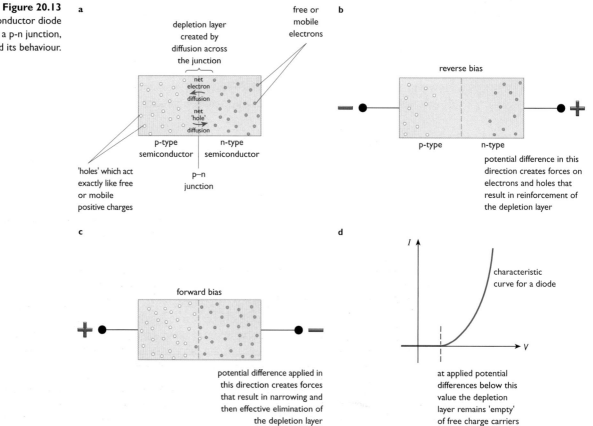

A whole new world of electronic possibilities is created by making a sandwich of n-type or p-type materials acting as 'bread' with a very thin 'filling' of the opposite type of material. The whole of the 'filling' and the 'bread' either side is depleted of charge carriers. The 'filling' is called the **base**, and the whole sandwich is a **transistor**. By introducing charge carriers directly into the base by means of a third wire to the transistor, we can very greatly reduce its resistance. A small flow of these charge carriers – the **base current** – makes a very big difference to the effective resistance of the transistor. This creates possibilities of **current amplification** – using a small current to control a bigger one, so that the bigger one copies the pattern or signal carried by the smaller one. The larger current is called **collector current** (Figure 20.14).

Figure 20.14
Connections to a transistor, and its behaviour as a current amplifier.

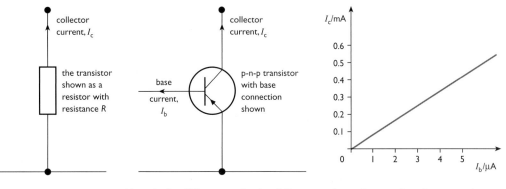

Note that I_b is 100 times smaller than I_c. Variation in the small current I_b produces a matching variation in the larger current I_c, hence varying current patterns, or signals, are amplified.

There are also possibilities of electronic **switching** – using one voltage, called the input voltage, to change the effective transistor resistance from high to low or from low to high. Such switching is what makes computers possible (Figure 20.15).

Figure 20.15
The transistor's behaviour as a switch.

The transistor's resistance flips from high to low as the potential difference applied to the base (V_{input}) increases. This is the basis of computer switching.

24 Why does semiconductor resistance decrease with temperature?

25 Six people sit on six chairs in a doctor's waiting room. The end person gets up to see the doctor. The next person moves into the space that is now available, and each of the other four people also move along one place.
   a Where is the space after they have done so?
   b How many people physically moved?
   c Why might it be easier to think about such movement as movement of the space?
   d What does this analogy have to do with conduction in p-type semiconductors?

26 What would you expect to happen to the rate of diffusion across a p-n junction when its temperature increases?

27 When does a transistor have
   a high resistance
   b low resistance?

28 a What is the value of the ratio I_c/I_b for the graph shown in Figure 20.14?
   b Assuming that the potential difference applied to the transistor stays the same, sketch a graph to show the dependence of resistance, shown in the diagram as R, on base current.

29 Suppose that, for switching purposes, the transistor is connected in series with another resistor, R_L, which is constant (see Figure 20.15). What happens to the resistance of the transistor as V_{input} increases?

● **Extra skills task** Information Technology and Application of Number

Investigate the changes in resistance with temperature for a thermistor and a light bulb using a data-logging program with suitable sensors to collect the data. Analyse the data using a spreadsheet program to produce graphs that display the trends seen in the experiment. If you can obtain a straight-line graph (which can be described by the equation $y = mx + c$) you will have revealed a relationship that is 'linear', and if $c = 0$ the relationship is direct proportionality. Otherwise, try to explain the results you have obtained.

VI
THEMES OF PHYSICS

21 Words of physics

KEY VOCABULARY   conservation law   extrapolate   falsification
                     hypothesis   law   model   theory

Knowing the words of physics

Very many students complain that it's the maths in physics that gives them a headache. But every year, examiners find that the real problems are in how students use *words* to explain ideas. It's something that takes practice.

Whatever subject you study, it will have its own habits of language. In history or law you may have to argue the point, and be persuasive. In physics, the priority is to describe and explain in terms of theory. When talking or writing about a particular topic, always imagine that the person you are communicating with is an intelligent and questioning person but who happens not to know much about that topic.

The first step in being able to use words and pictures to explain ideas is to know what the words mean. The Physics Specification is packed with words that each have a very specific meaning – words such as radiation, diffraction, tension, proportionality, model, energy, energy level, internal energy, ionisation, instantaneous, elastic, plastic, force, power, rate of change, potential, potential difference, potential gradient ... Some of them sound similar to each other. Many of them, such as 'energy', are used in everyday language in a less precise way. Some, like 'plastic', can be used in everyday language to mean something different.

Make it your first challenge – your first target – to know what the words of physics mean. Make sure that, if asked, you could write an accurate and complete definition of every one of them that you come across. You could do this parrot-fashion to start with; you'll find that understanding follows later. You will then have the framework on which to build your understanding. Neglect the words, and you'll struggle all the way through.

If you know and understand, for example, that resistance is a property of a circuit component and is defined as 'the ratio of potential difference applied to it, to the current in it', then you've already understood a lot. Even the maths gets easier, and we'll come back to that in the next chapter.

1 Obtain a copy of your examination specification. Identify every word that has a specific meaning in physics. You could use highlighter pens, or simply underline or circle words, to pick out
   a the words that you should and do already understand well enough to write a definition
   b the words that you should understand well enough to write a definition, but about which you are unsure
   c the words from parts of the specification that you haven't yet covered.
2 Which terms are defined by the following:
   a the product of force and parallel displacement under the action of the force
   b the ratio of potential difference applied to a component, to the current in it
   c rate of transfer of energy or of doing work
   d rate of change of velocity
   e the potential difference between two points when 1 J of work is done in transferring 1 C of charge between them
   f the force required to accelerate 1 kg at 1 m s^{-2}
   g a frequency of 1 cycle per second
   h rate of radioactive decay or detection at 1 event per second?
3 The terms in question 2 can be grouped into just two types – what are these? Group the terms into these two types.
4 The following are names of phenomena. Describe each phenomenon using sketches where they are helpful:
   a diffuse reflection
   b diffraction
   c anomalous expansion of water
   d ionisation
   e dissipation.
5 These are processes or operations used in physics. Give their names:
   a ignoring the effects of complex influences
   b finding the values of mutually perpendicular components of a single force.
6 Sort these nouns into those with units and those without:
   ray   density   spring constant   strain   Young's modulus   pressure
   charge   circumference   thrust   conservation   ion   lost volts   field line
   semiconductor
7 Do gradients and intercepts of graphs have units,
   a always
   b sometimes
   c never?
   Explain this, giving examples.

8 A sketch diagram is a powerful and speedy way of communicating. This makes it very useful in the examination situation. It should be drawn with care but without using a ruler, and it should be labelled. Large sketches are usually clearer because they have plenty of space for clear labels.

Use *sketch* diagrams to illustrate the following:

a a parallel beam (of light)

b the law of reflection

c refraction

d subtraction of colours

e the Vernier scale

f Brownian motion

g the photoelectric effect

h a G–M tube

i proportionality

j inverse proportionality

k the resultant of two mutually perpendicular forces (approximately to scale)

l energy transfer by a person riding a bicycle

m transformer

n potential gradient.

9 (This is a major task, both in terms of the effort it will take and its importance to you.) Prepare a personal glossary of all the words of physics that you come across. Give it some personal style if you wish. Give a defining formula where appropriate (such as $R = V/I$ for resistance). Use sketch diagrams wherever possible.

Laws of physics

In physics, a **law** is an observed pattern of behaviour. There's the Law of Reflection, Hooke's Law, Newton's First, Second and Third Laws of Motion and Ohm's Law. These consist of more words and you need to be able to write a short statement for each one.

Laws are patterns of behaviour that we've repeatedly seen so far. There is no absolute reason why a mirror shouldn't send light in a new and exciting direction; it's just that all of the mirrors we've ever carefully observed reflect light such that angles of incidence and reflection are the same. There is no *proof* that all mirrors will obey this law at all times, but we'd get a surprise if we didn't find it happening.

Many laws apply only in certain conditions – Ohm's Law, for example, applies to a metal wire provided it isn't heated by the current in it and its temperature doesn't change.

10 State

a Hooke's Law

b Ohm's Law

c Newton's Third Law

d the Law of Reflection.

11 Put the laws listed in question 10 into the following classifications:

• often disobeyed

• rarely disobeyed

• has never been known to be disobeyed.

Conservation laws

Conservation laws (or 'principles') are patterns of behaviour that are worth a special mention because they are so useful. If we know that the total energy, for example, is going to be the same after a process as before, then we have a useful aid to predicting the process. The following are some quantities that are conserved:

• energy (other than in nuclear changes)

• mass (other than in nuclear changes)

• mass–energy (always)

• charge (of a particle)

• current in a series circuit

• momentum.

12 Sketch an energy transfer diagram to illustrate an example of the principle of conservation of energy.

13 Do the following violate the principle of conservation of mass?

a A rusty iron nail has more mass than it had when it was shiny.

b An astronaut has less weight when 100 km above the ground than when on the ground.

c The various particles that exist after a radioactive decay have less mass than the particles that existed before.

d Explain c in terms of mass–energy.

14 When you comb your hair the comb and hair become charged. How is this consistent with the conservation of charge?

15 Why is current not the same at all points in a circuit if part of it is connected in parallel? (Sketch a diagram to help you with this explanation.)

Theories

A **theory** is a set of descriptions of processes that lie behind our observations. We have a theory of evolution that provides descriptions of the processes by which so many species of living thing can have developed. It provides suggestions of plausible mechanisms for how things have become the way they are. In terms of hard logic, the theory of evolution provides no *proof* of anything, but it is the best theory we've got for providing us with a source of descriptions of invisible underlying processes for the way the living world is. Such descriptions are often called 'explanations'. We also have particle theory which provides a rich source of explanations for the behaviour of so much of the physical world – electrical, thermal and chemical behaviour.

Both the theory of evolution and particle theory are good theories because they provide so much explanation of mechanisms that lie behind what we see. They are also good scientific theories because they are capable of being **falsified**, or found to be false. That's important because it's what makes scientific theory special. Scientific theories are forever being tested, and just one observation could be enough for them to fail the test and be proved wrong. Scientists would then have to accept, perhaps after many of them had debated the importance of the observation, that the theory was no longer a reliable source of explanation for all observations. The theory would be dumped. When we say that scientific theories are falsifiable we mean that they are not accepted as unquestionable truth but *can* be dumped.

The theories of evolution and the particle nature of matter are constantly being put to the test in jungles and in labs across the world. Any scientist who can make an observation, which other scientists can check, that shows an existing theory to be false may become famous (Figure 21.1).

Figure 21.1
Scientists who became famous through falsification of theories.

Thomas Young's observations of the diffraction and interference of light helped to falsify the 'corpuscular theory' (a particle theory of light) that had previously been used as a source of explanation for reflection and refraction.

Ernest Rutherford's conclusions from observations of the pathways of alpha particles through a thin leaf of gold falsified the theory of atoms of uniform density, which had previously been used as a source of explanations for the behaviour of matter.

16 DISCUSS

Can you think of any observation that would have the power to falsify the particle theory of matter? You'll need to be imaginative.

17 Say which of the following are scientific statements. For those which are scientific statements, explain how you might try to falsify them.

a 'Only my own mind exists. There is no world outside my mind. Everything that happens, happens only in my mind.'

b '1 + 1 = 2'

c 'Fossils are the remains of living things, turned to stone.'

d 'Atoms contain negatively charged electrons and positively charged nuclei.'

e 'Light is a kind of wave.'

f 'Light travels in a way that is like the travel of waves.'

g 'There is no such thing as absolute truth.'

If someone says, 'I saw a stork leave a baby under a gooseberry bush last week and I've developed a theory that this is how all babies arrive', then they will be putting forward a theory that we know will not survive. It establishes a general rule from very limited observation and any such theory should be subject to particularly careful examination. We can use the theory to make a **hypothesis** – a testable prediction. If all babies come from beneath gooseberry bushes then we can predict that if we chop down all gooseberry bushes, or just look in a part of the world where there are no gooseberry bushes, then we should find that the supply of babies comes to a stop. When this hypothesis has been found to be incorrect, the gooseberry bush theory of babies has been falsified, and we need a new theory.

Theories that cannot, by their nature, be falsified by observation are not scientific theories. If someone says, 'God exists', then this does not provide a scientific theory because we cannot test it in a particular way. Each of us makes a personal decision about this statement. The question of the existence of God is not a scientific question.

Different kinds of models

A theory is intended to provide descriptions, that agree with *all* observations, of hidden mechanisms that shape the world we see. Where the descriptions successfully agree with all observations then we can describe the theory as *complete*. A single new observation can show that a theory that was previously thought to be complete is in fact incomplete and therefore not the best possible way of thinking about the world. The single observation 'falsifies' the theory. A **model**, however, does not have to pretend to provide a complete picture, but only a *useful* one. A model is a representation of reality. When we use a model we are not always saying 'this agrees with *all* observations', but we are saying 'it's a bit like this, and this is a useful way of thinking about it'. There are different kinds of model, as discussed on the following pages.

Mathematical models

Any formula is a mathematical model. A formula 'mimics' reality. To take a fairly simple example, the mathematics of the formula:

$$I = I_1 + I_2$$

mimics the behaviour of current in a circuit, where I is total circuit current and I_1 and I_2 are the currents in each of a pair of parallel resistors (Figure 21.2). It is because we know that the formula copies reality so closely that we can use it to make predictions about behaviour in circuits that haven't yet been built. That, to say the least, is useful. However, a formula is not reality – it models reality. Mathematics is stunningly reliable at modelling reality.

Figure 21.2
A representation of reality.

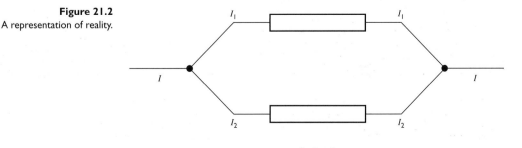

$$I = I_1 + I_2$$

Visual models

Think about an atom. When you do, pictures may form in your mind. There may be a picture of a simple ball, or you may picture a central nucleus with electrons in orbit, and the electrons might be dots or small balls or fuzzy clouds. That's fine – these are all models of atoms and electrons that can be useful. That's all a model has to be – useful. It doesn't have to be true. In fact the idea of a true *picture* of an atom or an electron is nonsense. Particles so small are fundamentally invisible. They don't and never can 'look like' anything. Yet we keep on drawing pictures to help us to understand atoms. It seems that we're hooked on pictures. If they help, why not? It's just important to remember that they are only pictures and not the real thing.

A circuit diagram, as in Figure 21.2, is a useful visual model. A graph is both a mathematical model and a visual model, and it is an extremely useful representation of reality.

Physical models

A toy car is a representation of reality. It's a physical model. A 'ball and stick' model of a molecule, made out of real balls and real sticks, is a physical model. Such physical models are very useful to chemists.

A two-dimensional picture of a car is as much a model as is a three-dimensional toy car. A 2D model may seem less useful, but its advantage is that it can exist on paper or on a computer screen.

Computer models

A simple graph is a model of how two variables interact in the real world. Now think about the Earth's climate. Variables that can make a difference include rate of burning fossil fuels, rate of emission of methane from flatulent cows, rate of burning of rainforest, amounts of cloud reflecting sunlight, growth of living things in the oceans, destruction of peat bogs, etc. Many climate variables are interrelated, and not necessarily by simple proportionality. People have the intelligence to try to work out which variables matter and what the relationships between them are, and to set the rules for a computer to follow, then the computer does the mathematical donkey-work. A computer needs a good model built into its software, and good data, or the predictions it provides will have nothing to do with reality. A computer that is capable of modelling the climate – mimicking reality so that it can be used to predict what reality might do next – needs to be big and fast.

Dealers in stocks and shares make decisions about financial deals using computer models. They use mathematical formulae to try to predict what will happen next. They use graphs and they try to **extrapolate** – use what they know from existing data to make suppositions about what might happen next (Figure 21.4). Stock market dealers predict future behaviour with only past data to go on. If their predictions are good they make money, and if they are bad they lose it.

Figure 21.3 (above) Making predictions with the help of computer models.

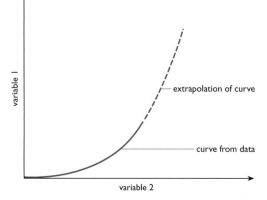

Figure 21.4 Extrapolation is one way of predicting behaviour.

Figure 21.5
Using different models.

18 Give examples of
   a a mathematical model
   b a visual model
   c a physical model.
19 Sketch three different visual models of an atom that you might find in a modern physics or chemistry textbook. In what circumstances would you use each of the three different models?
20 Sketch a spring being extended by a force. Sketch a graph of the extension (y-axis) against force (x-axis). When is each of these representations of reality more useful than the other?
21 Suggest why, despite the use of large and fast computers, scientists are still unable to predict future climate with a high level of certainty.
22 What is the difference between a theory and a model?
23 Does the Universe exist only inside your mind?

Scientists use whatever models help them to think and to predict behaviour – physical models, mathematical models, computer models and visual models (Figure 21.5). Very often they use different models of the same reality – such as a physical 'ball and stick' model of a molecule, and a computer 'space filling' model of the same thing. One of the chief new skills that you will gain from studying science is a sophisticated ability to use models.

● **Extra skills task**   Information Technology and Communication

Choose one scientific model of some aspects of the physical world. Possible models to choose include ray or wave models of light, mathematical models of structures or motion, or visual models of electrons.
  Research it thoroughly using sources from computers, libraries, journals and other literature.
  Prepare an extended essay including diagrams, which explains the model, the theories associated with it, and how it has helped the scientific community to understand that aspect of the physical world.

22 Quantities and relationships in physics

KEY VOCABULARY

accuracy   base units   dependent   derived units   dimension
dimensional analysis   dimensionless quantity   error   error bar   exponential
Gaia model   holism   homogeneity   independent   inverse proportionality
inverse square   line of best fit   precision   proportionality   reciprocal
reductionism   significant figures   sinusoidal   subject (of equation)   uncertainty
variable

● Keeping it simple

'Simple' and 'easy' do not mean the same thing. The opposite of easy is 'difficult'. The opposite of simple is 'complicated'. Physics is a simple subject, or at least it looks at a complicated world and seeks simple patterns. At the heart of physics is the process of examining **variables** – measurable quantities that can take different values – just two at a time.

The real world, as in the example of global climate, is packed with interrelated variables. It's both complicated and very, very difficult. It's so difficult that despite the work of thousands of scientists around the world, we still have no idea of how humans are affecting the climate and what life, as a result, will be like in 50 years' time. Dealing with two variables at a time is relatively very simple (Figure 22.1).

Figure 22.1
Standard graphs have just two axes – one for each variable. Some relationships between variables are simpler than others, but relationships between pairs of variables can be shown as shapes of lines.

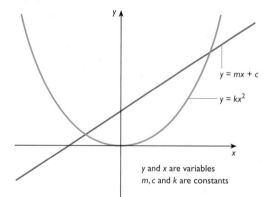

$y = mx + c$

$y = kx^2$

y and *x* are variables
m, *c* and *k* are constants

The process of looking at the world in the simplest possible way – often looking at just small parts of a system because the whole thing is too complicated – is called **reductionism**. It is a powerful way of working, but we need to be careful. For one thing, there is a danger of over-simplification, or over-idealisation. (Note that for a physicist, 'simple' and 'ideal' can often mean much the same thing.) For example, we might, for the sake of simplicity, ignore air resistance or friction in analysing motion. Sometimes that works, and sometimes it doesn't.

Secondly, the way that parts of a system work when they are apart from each other may be different from the way they all work together. In 'reducing' a system to its parts we can miss the really important features of the system. (A single ant doesn't tell you very much about an ant colony.) Reductionism is a powerful way to improve understanding of the world, in a simple or one-bit-at-a-time approach, but it is not the only way to understand the world.

The process of looking at a system, such as the Earth's ecosystem, as a whole generally provides less precise understanding of individual relationships, but it can still provide useful new ways of thinking. For example, scientist James Lovelock worked for NASA in the 1960s, examining planets of the Solar System for signs of life. He was looking at Venus and Mars as complete systems, and so he started thinking about the Earth in the same way. This has encouraged other scientists to examine, in new ways, the interrelationships that support life on Earth, and to discover new connections between variables. This view of the world as a whole is called the **Gaia model**. The process of examining any system as a whole entity, in contrast to reductionism, is called **holism**.

Dependence and independence

What is the simplest possible relationship between two variables? Before answering that we need to remember that two variables may have no relationship at all – they could be **independent**.

Suppose that we investigate the dependence on temperature of a number of different types of behaviour. We would find that many properties of materials, such as density, refractive index and resistivity, vary with temperature. They are **dependent** on temperature. But we would also find other behaviour that is independent of temperature.

For example, we could measure the activity of a sample of radioactive material. If it is a small sample we might notice that its activity varies from measurement to measurement, but such variation would not be influenced by its temperature.

Or we could measure the acceleration of a body of fixed mass when a constant force is applied. We know that $F = ma$ and therefore that $a = F/m$. Nothing else appears in this equation. If F and m are fixed during a series of measurements, then a will also stay the same. Change in temperature of the body will make no difference. Acceleration of a body of fixed mass when subject to a fixed force is independent of temperature.

Figure 22.2
The output variables are both independent of temperature.

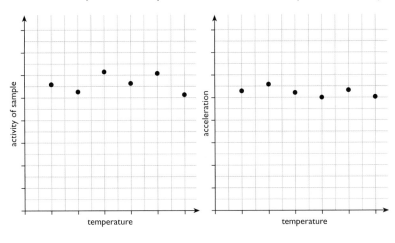

Any variation in our measurements of acceleration is due to the limitations of our experimental measurements. (Such limitations are sometimes called experimental **error**, though this term is unfortunate in that it implies something is the fault of the experimenter rather than there being a natural limitation of the measuring process; a better word is **uncertainty**.)

We could plot graphs of activity of the radioactive sample and acceleration of the body against temperature (Figure 22.2).

In an obsessive search for patterns in the world, we could try plotting activity of the radioactive sample at a particular temperature against acceleration of the mass m at the same temperature, using the same data. Now the points will be scattered widely, making no pattern at all. From this we would be forced to conclude that the two variables are independent of each other.

Searching for simple patterns

Figure 22.3
The pattern here shows some dependence between the variables.

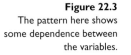

Using an everyday example, we could plot the distance travelled by a bus on its round trip from bus station back to bus station, against time. To do this, we could note its position, and so its distance from its starting point, at regular time intervals. As usual, a graph provides a very useful visual record from which it is relatively easy to see any general pattern (Figure 22.3).

There seems to be a pattern. We could use statistical techniques or simple judgement (which is much quicker, quite effective but not as reliable as the statistics) to draw a curve that follows the pattern. The pattern is enough to tell us that some sections of the journey are slower than others.

Figure 22.4
Graph showing
proportionality.

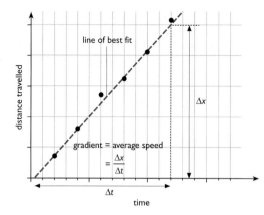

Suppose that the local council then introduces a ban on all cars in the area, meaning that the bus can maintain a much more uniform speed. Now if we make new measurements and plot a new graph, the pattern is more obvious (Figure 22.4). It appears to be a straight line through the origin. Again we can use judgement or statistics to draw the best straight line, or the **line of best fit**, through the points. (Computers can do the donkey-work of the statistics to produce a line of best fit to any set of data.)

The straight line passing through the origin shows that distance travelled is **proportional** to time. That is, whatever the change in time, the distance changes by the same proportion. The gradient is constant, equal to change in distance, Δx, divided by change in time, Δt, and is measured in units of distance per unit of time. That is, it is equal to the average speed.

The gradient of a straight line graph plotted from experimental data is a physical quantity. Its units are decided by the units of the plotted variables (Figure 22.5).

Figure 22.5
Examples of straight
line graphs.

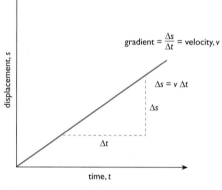

If displacement is measured in metres, m, and time in seconds, s, then the unit of the gradient is m s^{-1}.

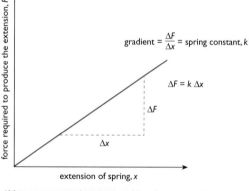

If force is measured in newtons, N, and extension in metres, m, then the unit of the gradient is N m^{-1}.

1 DISCUSS
Which provides the most
a reliable **b** meaningful
way of examining a complex system (such as the human brain) – holism or reductionism?

2 DISCUSS
Components of a car include the engine, the transmission system that connects the engine to the wheels, the wheels themselves, and the electrical system. Components of the Earth include the oceans, the atmosphere, rock and living matter.
a Apart from the obvious differences in size, what are the fundamental differences between a car and a planet in terms of interaction of the components?
b In which case does the reductionist approach work most easily?
c Is a human body more like the car or the planet?
d If you go into hospital for treatment or diagnosis, most of the methods used will have been developed by the reductionist approach. Is it possible to understand the human body and develop effective medical treatment without using a reductionist approach?

3 One politician might claim that, say, crime is caused by unemployment. Another politician might claim that crime has nothing to do with unemployment. Politicians work in a world that is much more complicated than the world of reductionist science.

Suggest a reductionist approach to finding out more about the relationship between crime and unemployment. This will involve investigating the relationship between the variables.

4 What is the significance of the gradient of
a a displacement–time graph
b a velocity–time graph
c a graph of energy transfer against time
d a graph of charge transfer in a circuit against time
e a force–extension graph for a wire
f a stress–strain graph?

5 A quantity y is plotted against time t.
a For what kind of graph is the following true?
$$\frac{dy}{dt} = \frac{y}{t}$$
Sketch such a graph.
b Sketch a graph for which
$$\frac{dy}{dt} \neq \frac{y}{t}$$
Use your graph to show the difference between dy/dt and y/t.
c Give an example of a situation in which the difference between dy/dt and y/t has physical importance. (y could be displacement, velocity, energy transfer or charge transfer, for example.)

Inverse proportionality and the reciprocal

Variables may be **inversely proportional**. That is, whenever one increases then the other decreases by the same proportion. Plotted on a graph a curve results, like that in Figure 22.6.

Figure 22.6
Inverse proportionality.

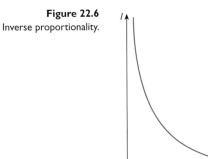

As resistance increases, with a fixed potential difference, current in the circuit decreases by the same proportion.

A curve provides less immediate information than a simple straight line. To check whether a relationship is really an inverse proportionality, we can plot a second graph with the inverse or **reciprocal** of one variable on one of the axes (Figure 22.7).

The reciprocal of variable x is $\dfrac{1}{x}$.

(Your calculator may have a $1/x$ or x^{-1} button. You should find it useful.)

The reciprocal of resistance R is $\dfrac{1}{R}$.

Figure 22.7
Proportionality.

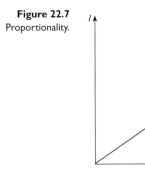

From this graph, we can say with confidence that I is proportional to $1/R$.

Other relationships between variables

As your physics course proceeds you will come across other standard relationships and their matching graphs. These include:

- the **sinusoidal** relationship (Figure 22.8)
- the **inverse square** relationship (Figure 22.9)
- the **exponential** relationship (Figure 22.10).

Figure 22.8
A sinusoidal relationship.

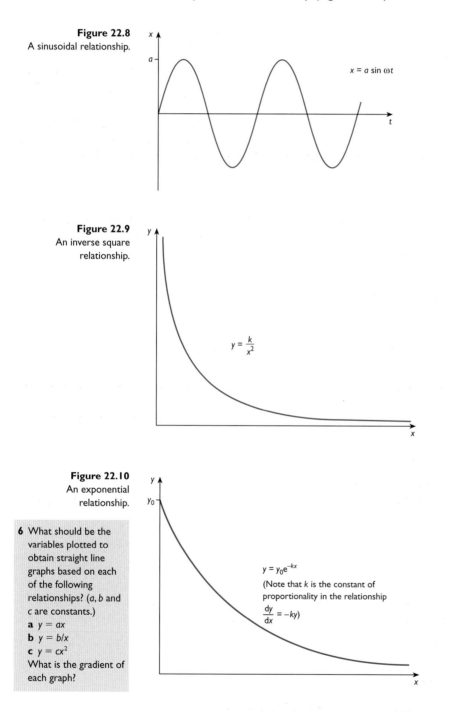

$x = a \sin \omega t$

Figure 22.9
An inverse square relationship.

$y = \dfrac{k}{x^2}$

Figure 22.10
An exponential relationship.

$y = y_0 e^{-kx}$

(Note that k is the constant of proportionality in the relationship $\dfrac{dy}{dx} = -ky$)

6 What should be the variables plotted to obtain straight line graphs based on each of the following relationships? (*a*, *b* and *c* are constants.)

a $y = ax$
b $y = b/x$
c $y = cx^2$

What is the gradient of each graph?

Using formulae

When trying to understand formulae it helps to start in the middle. In the middle of every formula is an equal sign. This makes a statement of equality (Figure 22.11). It is telling you that the two collections of symbols on either side can, and must, be treated as being the same. If they are the same, they are interchangeable. One can replace the other.

Figure 22.11
A formula involves equality.

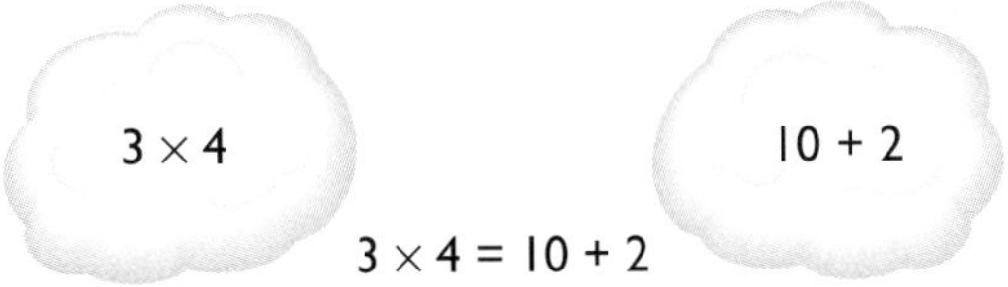

$$3 \times 4$$

$$10 + 2$$

$$3 \times 4 = 10 + 2$$

Where we know that two expressions are interchangeable we can link them by an equal sign.

$$3x = 24$$
$$x = 8$$

This is a statement providing information that $3x$ and 24 are interchangeable, or equal. From this statement we can then work out that x must have a value of 8.

This does not mean that you can't make any changes to the symbols on one side of the equal sign or the other. It's just that there are rules about what you can and cannot change. Note that a quantity written on its own on the left-hand side of the equal sign is called the **subject** of the equation.

Major Rule 1

You can make changes to the symbols on one side of a formula, without changing the other side, provided that the changes are the operations of the processes of multiplication and division, addition and subtraction. For example:

Division

The statement that $\dfrac{20}{4} = 5$ can be changed to $5 = 5$

Multiplication

The statement that $9 \times 4 = 6 \times 6$ can be changed to $36 = 6 \times 6$, or $9 \times 4 = 36$, or $36 = 36$

This may look obvious when done in simple numbers, but you also have to be able to do similar operations using symbols like x and y.

Mathematical statements can include more than one kind of operation and then operations must be done in the right order, such as in:

$$3 + 4 \times 5 = 26 - 9 \div 3$$

Here, carrying out the operations in a different order produces different answers. We perform multiplication and division first, and then addition and subtraction (M and D before A and S). That gives:

$$3 + 20 = 26 - 3$$

and so

$$23 = 23$$

Remember, however, that brackets lump parts of a formula together, and the brackets must then be dealt with first:

$$(3 + 4) \times 5 = (73 - 3) \div 2$$

so that

$$35 = 35$$

So an overall Minor Rule, relating to order of operations, is:

M and D before A and S, but always Brackets first

(Morning and Daylight before Afternoon and Sunset, but always Breakfast first.)

7 Carry out the operations to rewrite the following statements in simpler form:
a $2 + 7 = 1 + 8$
b $14 - 7 = 9 - 2$

8 Try doing the addition and subtraction before the multiplication and division:
$3 + 4 \times 5 = 26 - 9 \div 3$

9 Try ignoring the brackets:
$(3 + 4) \times 5 = (73 - 3) \div 2$

10 If you don't like 'Morning and Daylight before Afternoon and Sunset, but always Breakfast first', try making up your own mnemonic.

Major Rule 2

You can do anything else you like to a formula provided that you do it to both sides. These are some examples.

Starting with

$$36 = 6 \times 6$$

we can:

- add 1 to both sides

$$36 + 1 = (6 \times 6) + 1$$

- or divide both sides by 6

$$\frac{36}{6} = \frac{6 \times 6}{6}$$

- or take the square root of both sides

$$\sqrt{36} = \sqrt{(6 \times 6)}$$

- or take the reciprocal of both sides

$$\frac{1}{36} = \frac{1}{6 \times 6}$$

Starting with

$$\frac{ab}{c} = \frac{x}{y}$$

we can multiply both sides by c:

$$\frac{abc}{c} = \frac{xc}{y}$$

which simplifies to

$$ab = \frac{xc}{y}$$

Starting with

$$\frac{p}{q} = s + t$$

we can multiply both sides by q:

$$p = q(s + t)$$

which is the same as

$$p = qs + qt$$

11 Add and/or subtract from both sides of this equation to find x:
$25.5 - x = 14.5 + x$

12 Rearrange the following to make a the subject. In each case, explain your actions.

a $2a = b - c$

b $2p = a + q$

c $ax = 5y - z$

d $\dfrac{a}{w} = 3v$

e $4c + d = \dfrac{3e}{a}$

f $3(p - a) = 4q$

g $3c = \dfrac{d}{2(a - b)}$

h $\dfrac{a}{3} + \dfrac{b}{4} = 5c$

i $\dfrac{1}{a} + \dfrac{1}{x} = \dfrac{1}{y}$

j $\dfrac{p}{m} + \dfrac{q}{n} = \dfrac{r}{a}$

Major Rule 2 is the ultimate rule that you follow in all rearrangements of formulae.

Base SI units

SI stands for Système International d'Unités, which is French for 'international system of units'. It is essentially the same as the metric system on which it is based. The other system still in use is the Imperial system, based on British units of the foot (for length), the pound (for mass) and the second (for time). This is still in use largely because American industry used the British system and has been slow to change. The problem with using two systems is that mistakes might happen – this has resulted in some disasters, including the failure of a multi-billion dollar space mission to Mars. British industry now uses the metric or SI system. Physicists throughout the world use the SI system.

The SI system is built up from seven **base units** that are defined in terms of observation of the physical world. The quantities associated with base units can be expressed so as to allow them to be considered in terms of **dimensions**. See Table 22.1.

Table 22.1
Base SI units, their associated quantities and dimensions.

Quantity	Base unit	Quantity expressed to allow analysis of dimensions	Definition of unit in terms of observation of the physical world
mass	kilogram, kg	[M]	1 kg is equal to the mass of the international prototype kilogram (a piece of platinum–iridium kept in Paris). Note that because the mass of the international prototype is subject to change due to chemical action on its surface, the definition of the kilogram is to be changed in the future
length	metre, m	[L]	1 metre is the length of the path travelled by light in a vacuum during a time interval of 1/299 792 458 s
time	second, s	[T]	1 second is the duration of 9 192 631 770 periods of the radiation corresponding to the electron transition between the two hyperfine levels of the ground state of the caesium-133 atom
electric current	ampere (or amp), A	[I]	1 ampere is the current in two straight parallel conductors of infinite length and negligible cross-sectional area, placed 1 metre apart in a vacuum, that produces a force of 2×10^{-7} newton per metre of length
thermodynamic temperature	kelvin, K	—	1 kelvin is 1/273.16 of the thermodynamic temperature of the triple point of water
amount of substance	mole, mol	—	1 mole is the amount of substance which contains as many elementary entities (such as atoms, molecules) as there are in 0.012 kg of carbon-12
luminous intensity	candela, cd	—	1 candela is the luminous intensity, in a given direction, of a source of light that emits monochromatic radiation of frequency 540×10^{12} hertz and that has intensity in that direction of 1/683 watt per steradian

Dimensional analysis at A-level is usually limited to situations involving mass, length and time.

It is important that you understand that the definitions of the base units in Table 22.1 are the basis for *all* SI units. However, unless you have a reason to be especially interested in the technical details of these definitions you do not need to remember them.

Derived SI units

All of the other SI units are built up from the base units; they are **derived units**. Some derived units have special names; others do not. See Tables 22.2 and 22.3.

Note that the dimensions of any derived quantity show the powers to which the base quantities must be raised for the derived quantity.

Table 22.2
Some derived SI units without special names.

Quantity	SI unit	Quantity expressed to allow analysis of dimensions
area	m^2	$[L]^2$
volume	m^3	$[L]^3$
speed and velocity	$m\,s^{-1}$	$[L]\,[T]^{-1}$
acceleration	$m\,s^{-2}$	$[L]\,[T]^{-2}$
mass density	$kg\,m^{-3}$	$[M]\,[L]^{-3}$

Table 22.3
Derived SI units with special names.

Quantity	SI unit	Unit in terms of other units	Unit in terms of SI base units	Quantity expressed to allow analysis of dimensions
plane angle	radian, rad	—	—	ratio of lengths, $[L]\,[L]^{-1}$, so dimensionless
solid angle	steradian, sr	—	—	ratio of areas, $[L]^2\,[L]^{-2}$, so dimensionless
frequency	hertz, Hz	—	s^{-1}	$[T]^{-1}$
force	newton, N	—	$kg\,m\,s^{-2}$	$[M]\,[L]\,[T]^{-2}$
pressure	pascal, Pa	$N\,m^{-2}$	$kg\,m^{-1}\,s^{-2}$	$[M]\,[L]^{-1}\,[T]^{-2}$
energy, work	joule, J	$N\,m$	$kg\,m^2\,s^{-2}$	$[M]\,[L]^2\,[T]^{-2}$
power	watt, W	$J\,s^{-1}$	$kg\,m^2\,s^{-3}$	$[M]\,[L]^2\,[T]^{-3}$
charge	coulomb, C	—	$A\,s$	$[I]\,[T]$
electric potential	volt, V	$J\,C^{-1}$	$kg\,m^2\,s^{-3}\,A^{-1}$	$[M]\,[L]^2\,[T]^{-3}\,[I]^{-1}$
capacitance	farad, F	$C\,V^{-1}$	$kg^{-1}\,m^{-2}\,s^4\,A^2$	$[M]^{-1}\,[L]^{-2}\,[T]^4\,[I]^2$
electrical resistance	ohm, Ω	$V\,A^{-1}$	$kg\,m^2\,s^{-3}\,A^{-2}$	$[M]\,[L]^2\,[T]^{-3}\,[I]^{-2}$
electrical conductance	siemens, S	$A\,V^{-1}$	$kg^{-1}\,m^{-2}\,s^3\,A^2$	$[M]^{-1}\,[L]^{-2}\,[T]^3\,[I]^2$
magnetic flux	weber, Wb	$V\,s$	$kg\,m^2\,s^{-2}\,A^{-1}$	$[M]\,[L]^2\,[T]^{-2}\,[I]^{-1}$
magnetic flux density	tesla, T	$Wb\,m^{-2}$	$kg\,s^{-2}\,A^{-1}$	$[M]\,[T]^{-2}\,[I]^{-1}$
inductance	henry, H	$Wb\,A^{-1}$	$kg\,m^2\,s^{-2}\,A^{-2}$	$[M]\,[L]^2\,[T]^{-2}\,[I]^{-2}$
luminous flux	lumen, lm	$cd\,sr$	cd	—
illuminance	lux, lx	—	$cd\,m^{-2}$	—
activity or radioactive count	becquerel, Bq	—	s^{-1}	$[T]^{-1}$
absorbed dose	gray, Gy	$J\,kg^{-1}$	$m^2\,s^{-2}$	$[L]^2\,[T]^{-2}$
dose equivalent	sievert, Sv	$J\,kg^{-1}$	$m^2\,s^{-2}$	$[L]^2\,[T]^{-2}$

Unit prefixes

The SI system incorporates a standardised system of prefixes, such as kilo-. A 'kilo' in front of an SI unit converts it into a unit 1000 times larger; a kilometre is a thousand times larger than a metre. The same rule applies to a kilovolt or a kilonewton, for example.

The prefixes that provide such multiplications, from 10^{-18} to 10^{18}, are given in Table 22.4.

Table 22.4
Unit prefixes.

Factor	Prefix	Symbol	Example
10^{18}	exa	E	exajoule, EJ
10^{15}	peta	P	petaohm, PΩ
10^{12}	tera	T	terabecquerel, TBq
10^{9}	giga	G	gigawatt, GW
10^{6}	mega	M	megahenry, MH
10^{3}	kilo	k	kilogray, kGy
10^{-3}	milli	m	millicoulomb, mC
10^{-6}	micro	μ	microamp, μA
10^{-9}	nano	n	nanokelvin, nK
10^{-12}	pico	p	picofarad, pF
10^{-15}	femto	f	femtometre, fm
10^{-18}	atto	a	attosecond, as

Homogeneity of formulae

$2 + 3 = 5$, but two dogs and three cats can never equal five camels. In any physically meaningful equation the units must be the same on both sides. That is, the equation must be **homogeneous**. Take, for example,

$$\text{force} = \text{mass} \times \text{acceleration} \quad \text{or} \quad F = ma$$

We know the units of the three quantities and they do not immediately appear to be the same on both sides of the equation. But the newton is a derived unit, and it is in fact derived from this equation. The size of the newton has been chosen so that it is the same as a kilogram metre per second squared. This is written into the formal definition of a newton:

1 newton is the force required to give a mass of 1 kg an acceleration of 1 m s^{-2}

We can express the homogeneity of the equation either in terms of units or in terms of dimensions:

$$F = ma$$

In terms of units,

$$\text{kg m s}^{-2} = \text{kg m s}^{-2}$$

In terms of dimensions,

$$[M]\,[L]\,[T]^{-2} = [M]\,[L]\,[T]^{-2}$$

Note that a ratio of two dimensionally identical quantities has no dimensions itself. How many times does an area of 3 m^2 fit into an area of 12 m^2? The question is asking for a ratio:

$$\frac{12 \text{ m}^2}{3 \text{ m}^2}$$

The answer is 4, not 4 m^2. It is a **dimensionless quantity**.

Since a physical equation must be homogeneous, we can use this as a useful check of our equations. Suppose that you are in the exam room and you can't remember the formula relating the period of pendulum to its length. You know that it is either

$$T = 2\pi \sqrt{g/l} \quad \text{or} \quad T = 2\pi \sqrt{l/g}$$

where T is the period, l is the length and g is the gravitational field strength. Only an equation that is homogeneous can be correct. You could perform a **dimensional analysis**. The dimensions are given by:

for T [T]
for 2   it is just a number and has no unit or dimensions
for π it is a ratio of two lengths (circumference and diameter) and has no unit or dimensions
for l [L]
for g this is force per unit mass; the dimensions of force are [M] [L] [T]$^{-2}$ and the dimensions of mass are [M], so the dimensions of g are [L] [T]$^{-2}$

So for the first version of the equation:

$$[T] = \sqrt{[L]\,[T]^{-2}/[L]}$$
$$= \sqrt{[T]^{-2}}$$

which simplifies to

$$[T] = [T]^{-1}$$

which is clearly wrong.

For the second version of the equation:

$$[T] = \sqrt{[L]/([L]\,[T]^{-2})}$$
$$= \sqrt{1/[T]^{-2}}$$
$$= \sqrt{[T]^2}$$

which simplifies to

$$[T] = [T]$$

which is clearly correct.

The second version has homogeneous dimensions.

This does not prove that the second version must be right. (The formula $T = \sqrt{l/g}$ is also homogeneous, for example). However, of the choice between the two formulae with which you began, you have established which one is certainly wrong.

---

13 Which other base unit is used in the definition of the metre?

14 **DISCUSS**
What problems could arise if there were no standard international definition of the metre?

15 Suppose that the SI committee agrees that the SI unit of acceleration should be given its own name.
  a Suggest a name. (Note that most such units are named after scientists from the past, though you do not have to stick to this tradition.)
  b Write a formal definition for this unit.

16 Explain why strain (see Chapter 5 on extension of wires) is a dimensionless quantity.

17 The relative density (or specific gravity) of a substance is the ratio of its density to that of water:

$$\text{relative density of substance} = \frac{\text{density of substance}}{\text{density of water}}$$

Comment on the unit and dimensions of relative density.

18 a Express resistance in terms of its dimensions.
  b R_1, R_2, R_3 and R_4 are all resistances. A student writes:

$$R_1 + R_2 = \frac{R_3}{R_4}$$

Explain how you can quickly say that this must be wrong.

19 a Express in pm:
     1 pm + 100 fm
   b How many times bigger is 1 gigabecquerel than 10 kilobecquerel?

20 a How many ns are there in 1 minute?
   b How long is 1 Es in years?
   c Which is closest to your life expectancy:
     i 1 Ms   ii 1 Gs   iii 1 Ts   iv 1 Es?

Measurement, precision, errors and accuracy

A textbook provides the background of ideas, but it cannot provide practical experience. Physics is, at heart, a practical subject based on observation of the world. A focus of attention is the investigation of quantitative relationships between variables. Measurement of variables is therefore a central activity in physics.

Unfortunately, there is no such thing as a perfect measurement. All that we can achieve is a certain level of **precision**. If we say we know that the resistance of a piece of wire is 2.428 Ω then we are claiming a high level of precision. We are saying that we can tell the difference between 2.428 Ω and either 2.427 Ω or 2.429 Ω. There are four **significant figures**, or digits that contribute to the precision of the stated value, in the value we are giving for resistance. We are claiming that our measurements are precise to 0.001 Ω. We need to be sure that our measurements justify this.

It may be that the best we can do is give a value for the resistance to two significant figures, so we say that the resistance is 2.4 Ω. We might need to be careful about that, and admit that even this less precise figure is subject to error or uncertainty, and that the best we can say with reasonable confidence is that the resistance lies somewhere between 2.3 Ω and 2.5 Ω. We can build this uncertainty into our quoted value of resistance, and say

$$R = (2.4 \pm 0.1)\,\Omega$$

The 0.1 Ω is an uncertainty or an assessment of the maximum error in a measurement. It is sometimes called simply the error in the measurement. Such errors can be important when plotting a graph – they can increase our confidence in drawing a line to fit the plotted points. To make sure that we are not reaching unjustified conclusions, we can draw **error bars** on our graph (Figure 22.12). We then know that the line must lie inside the limits of the error bars.

Precision is sometimes, but not always, valuable. If you want to know how long it will be to your next meal you probably don't need to know to the nearest second or even the nearest minute. A high level of precision has little value here.

Accuracy does not mean the same thing as precision. Accuracy is an indication of the reliability of the measurement. An inaccurate measurement is an unreliable indication of the real value, and further measurement may be able to reveal the inaccuracy. The statement 'it's about two hours to lunch', for example, may be perfectly accurate but it's not very precise. An imprecise measurement can be completely accurate. On the other hand, the statement is inaccurate if it is in fact three hours to lunchtime.

Figure 22.12
Error bars take account of the uncertainty of measurements.

Here the plotted points appear to show a pattern of rising and falling, but it is also possible to fit a straight line inside the error bars. In this case, it is not possible to declare a reliable conclusion about the shape of the line.

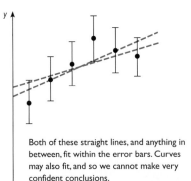

Both of these straight lines, and anything in between, fit within the error bars. Curves may also fit, and so we cannot make very confident conclusions.

Figure 22.13
The time for an athletics race is usually given to an agreed and quite high level of precision – to the nearest 1/100th of a second, or to two decimal places of seconds.

For the hundred metres race, for example, the times for the first, second and third places might be:

9.96 s   9.98 s   9.99 s

People responsible for the measurement are making a claim that their measuring instruments can tell the difference between a time of 9.96 s and 9.97 s. But the instruments cannot distinguish between 9.962 s and 9.964 s. Both are recorded as 9.96 s.

We can provide a quoted value of uncertainty, or possible error, in measurements. For example, a time of 9.96 s could be written as (9.96 s ± 0.01) s, which tells us that the time lies between 9.95 s and 9.97 s, but that we are unable to say any more than this.

21 In investigating a relationship, why is it important to make measurements over a range of input variables that is as large as possible?

22 Express the output variable measurement in Figure 22.14 in writing, giving uncertainty values in ± format.

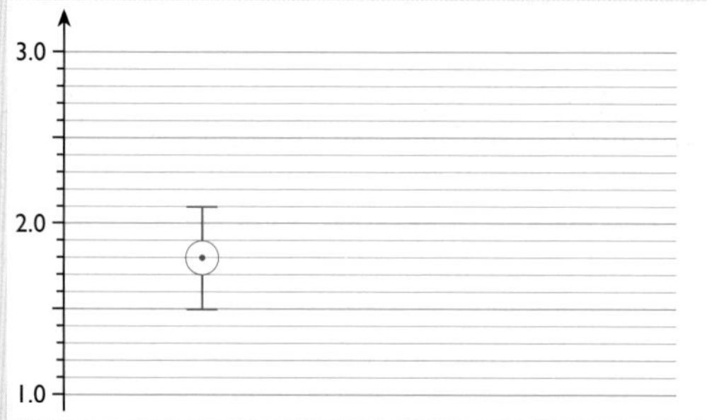

Figure 22.14

23 The value of the current in a circuit is stated as 0.025 A. This has two significant figures.

    a How many significant figures are there in these measurements:

      i 1270 N

      ii 0.000 565 m?

    b Write the measurements in **a** to two significant figures. What effect does this have on the precision of the values given?

    c Why is it important that the precision of a stated value should not be greater than the precision of the measurement?

24 a Each of the following measurements is known to have an uncertainty of 5% of the stated value. Write down the value showing the error. (The error should be rounded to match the number of significant figures in the stated value. Any unjustified claims of precision in the stated value should be removed.)

    For example, a measurement of 2.41 V becomes (2.4 ± 0.1) V.

    i 38.8 mm

    ii 0.05 J

    b Repeat **a**, assuming an uncertainty of 1%.

    c How often, in general terms, are you able to make measurements with an uncertainty as low as 1%?

● **Extra skills task** Communication, Information Technology and Application of Number

Spend a short time, either:

- in a Year 10 maths lesson on algebra, or
- talking to a maths teacher about the problems encountered by Year 10 pupils in becoming competent in algebra.

Prepare a poster or other graphic aid (which might be computer-based) to help Year 10 pupils to deal with manipulation of equations.

VII
LIGHT AND MATTER

23 Refraction of light

THE BIG
QUESTIONS
- How can we analyse the travel of light across boundaries between media?
- How does this analysis help us to
  a understand the world in new ways
  b develop new technologies?

KEY
VOCABULARY

absolute refractive index   angle of incidence   angle of refraction   bit
bit-rate   cladding (of optic fibres)   coherent (optic fibre bundle)
critical angle   endoscope   graduated index fibre   LED (light emitting diode)
monomode fibre   multimode fibre   optical density   partial internal reflection
principle of reversibility   refractive index   replication (of results)
step index fibre   time division multiplexing   total internal reflection   wavefront

BACKGROUND

Figure 23.1
Refraction makes us see
things differently.

Light that is transmitted first through one material, or medium, and then through another has experienced change in the process. The changes seem to alter the geometry of the world (Figure 23.1). This effect is due to refraction at boundaries between transmitting materials, a phenomenon caused by a change in the speed of light. Except where the light crosses a boundary travelling along a normal, refraction also involves a change in direction. We can analyse refraction using rays, waves and mathematical relationships as models. Having done this we can apply our knowledge to any kind of wave but most usefully to light. This has given us not just new ways of seeing but new ways of communicating.

Rays and refraction

Air, glass, perspex and water are all materials that are good at transmitting light. But where light passes through boundaries between such transparent materials it experiences refraction. Using ray representation of the travel of the light we can draw diagrams to analyse all such refraction phenomena accurately (Figure 23.2).

Figure 23.2
Ray analysis of refraction is consistent with the observation that a hidden coin becomes visible when a beaker is filled with water. The presence of the water provides a new pathway for the light.

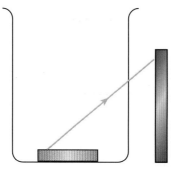
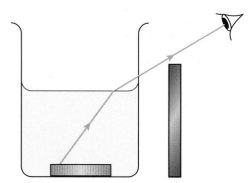

Rays, which are visual representations that we draw, are so useful that it is worth trying to create experimental conditions that imitate them. (This is a reversal of what we are more often trying to do, which is to create models that imitate reality!) A narrow beam of light shining through a block of glass, or a line of pins on either side of the block, help us to create ray diagrams showing refraction. Figure 23.3a shows such attempts, and Figure 23.3b shows the resulting ray diagram, complete with labelled **angle of incidence** and **angle of refraction**.

Figure 23.3
a A glass or perspex block, either rectangular or semi-circular in horizontal cross-section, provides the simplest possible start to the study of refraction.
b A ray diagram can be created to help analyse the refraction.

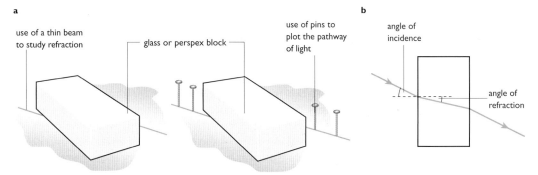

A mathematical model of refraction

From ray diagrams created by experiment, if drawn carefully enough, we can make measurements of the angle of incidence and the angle of refraction at a boundary between two materials such as air and glass. We can investigate the relationship between the angles, varying the angle of incidence and measuring the angles until we can start to look for a mathematical pattern that applies to all of the data.

Unfortunately, inspection of the measurements of angles of incidence and refraction does not reveal a relationship that is anything like so simple as the equality of angles that we find for reflection. We have to look more closely. A scientific calculator gives plenty of scope for the necessary manipulation of numbers. It is only when we start to use the trigonometrical buttons that a pattern appears in the inspected data. We find that we need to look at the sines of the angles. The data, and our manipulation of it, reveal that all pairs of sines of angles of incidence and refraction have the same ratio. Dividing one sine by the other in all cases, within the limits of the accuracy of our measurements, we get the same answer. Expressed as briefly as possible:

$$\frac{\sin i}{\sin r} = \text{a constant}$$

where i is the angle of incidence and r is the angle of refraction (Figure 23.4).

From Chapter 24 onwards, we'll find it more useful to measure angles in radians rather than degrees. However, when measuring angles of incidence and refraction it is simpler to take measurements directly from a protractor, and protractors are most often calibrated in degrees. In this chapter we will continue to use degrees for measuring angles.

Figure 23.4
Inspection of the relative sizes of our measured angles of incidence and refraction at a boundary between air and glass reveals a mathematical relationship:
$\frac{\sin i}{\sin r} = \text{a constant}$.

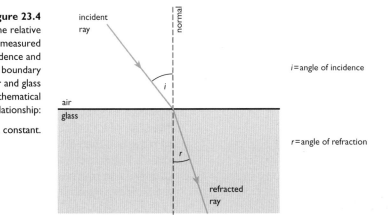

Other boundaries – refractive index

An experimental result is only considered to be reliable if it can be **replicated** – that is if other experimenters can repeat the experiment and get the same result. And indeed, repeating the experiment with blocks of identical glass is found to give the same result, with the same value of the constant, time after time. However, if the blocks are made of different types of glass then the value of the constant is found to be slightly different. With perspex blocks, it is different again. It seems that when the boundary is between air and different substances, we get different results. To express our result in a form that is more truly replicable we need to be careful with our words:

$$\frac{\sin i}{\sin r} = \text{a constant for a specified boundary}$$

The constant is clearly important enough to have its own name. It is called the **refractive index** of the boundary. Strictly the value is constant only for light of a specific wavelength or colour. The value of the refractive index varies very slightly with wavelength. We are familiar with this variation because of dispersion – the separating of white light into colours, as achieved most strongly by a prism. However, in making laboratory measurements on, say, red light and blue light, we would need to be achieving a high level of precision to measure a difference in the values of refractive index.

In mathematics, it saves a lot of effort to abbreviate language, and so the refractive index of, for example, an air-to-glass boundary, can be written as $_an_g$. For a glass-to-water boundary, it is $_gn_w$.

If you are now asked to provide a formal definition of refractive index, you could write something like the one in Figure 23.5.

Figure 23.5
A formal definition of refractive index.

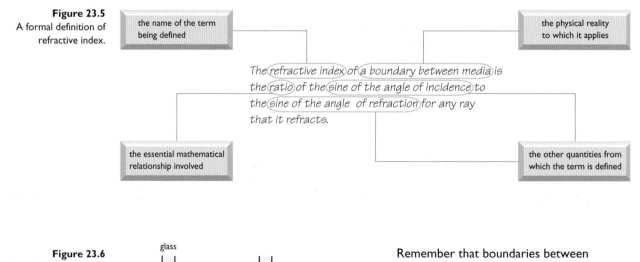

Figure 23.6
The refractive index, and so the change in light direction, is different for each of these boundaries.

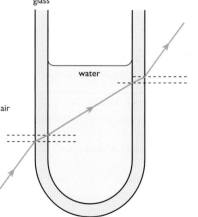

Remember that boundaries between different materials have different refractive indices (Figure 23.6).

The refractive index of a surface relates to the relative sizes of the angle of incidence and the angle of refraction. If the angle of incidence is bigger than the angle of refraction then the boundary's refractive index is more than 1. If the angle of incidence is smaller than the angle of refraction then the refractive index is less than 1. Refractive index is a ratio of the sines of angles and so it has no units – it is a dimensionless quantity.

For light travelling from air into a typical glass block, the refractive index $_an_g$ is more than 1; the angle of incidence is bigger than the angle of refraction. Light travelling in the opposite direction, from glass to air, will be refracted so that the angle of refraction is bigger than the angle of incidence. In fact a ray of light can follow exactly the same pathway, but in exactly the opposite direction. The observation that light can pass one way through an optical system and can also pass in the opposite direction along an identical pathway is called the **principle of reversibility** (Figure 23.7).

Figure 23.7
Light obeys the principle of reversibility – it passes along the same pathway in either direction.

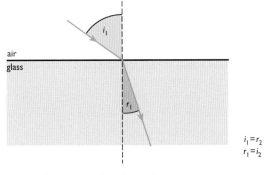

$i_1 = r_2$
$r_1 = i_2$

For the air-to-glass boundary:

$$_an_g = \frac{\sin i_1}{\sin r_1} \quad \text{(i)}$$

For the glass-to-air boundary:

$$_gn_a = \frac{\sin i_2}{\sin r_2}$$

But since $i_1 = r_2$ and $r_1 = i_2$,

$$_gn_a = \frac{\sin r_1}{\sin i_1} \quad \text{(ii)}$$

Note the relationship between the right-hand sides of equations (i) and (ii). One is the inverse of the other. So the left-hand sides must also be inverses of each other:

$$_gn_a = \frac{1}{_an_g}$$

1 **a** For sketches **i** and **ii** in Figure 23.8, calculate the refractive index of the boundary.
   b For sketches **iii** and **iv**, calculate the angle of refraction.
   c For sketches **v** and **vi**, calculate the angle of incidence.
2 Why, in the case of refraction, is a ratio ($\sin i/\sin r$) more useful than a difference ($\sin i - \sin r$)?
3 **a** If $_wn_g = 1.13$ (w = water and g = glass), what is $_gn_w$?
   b Calculate the angle of refraction when light crosses a water-to-glass boundary with an angle of incidence of 60°.

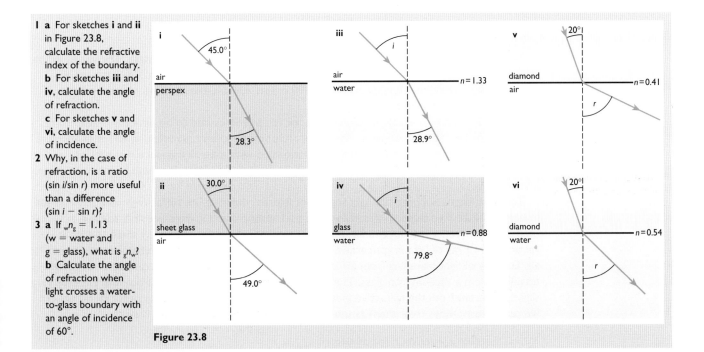

Figure 23.8

Absolute refractive index

With some special effort in setting up experimental arrangements, it's possible to make measurements of refraction as light passes from a vacuum into a light-transmitting substance. The refractive index of such a boundary can be written as:

$$_{vacuum}n_{material} \quad \text{or simply} \quad n_{material} \quad \text{(or, indeed, just } n\text{)}$$

This gives us a useful way of comparing all light-transmitting materials. It is called the **absolute refractive index** of the material.

The formal definition of absolute refractive index (Figure 23.9) is similar to that of the refractive index of a particular boundary given in Figure 23.5, but here we need to specify that absolute refractive index is a property of a material, and that a vacuum lies on the 'incident side' of the boundary. Table 23.1 gives some values for different materials.

Figure 23.9
A formal definition of absolute refractive index.

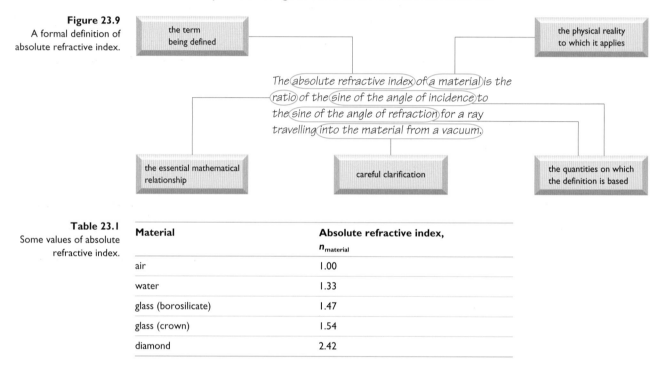

the term being defined

the physical reality to which it applies

The absolute refractive index of a material is the ratio of the sine of the angle of incidence to the sine of the angle of refraction for a ray travelling into the material from a vacuum.

the essential mathematical relationship

careful clarification

the quantities on which the definition is based

Table 23.1
Some values of absolute refractive index.

Material	Absolute refractive index, $n_{material}$
air	1.00
water	1.33
glass (borosilicate)	1.47
glass (crown)	1.54
diamond	2.42

Note that the absolute refractive index of a material varies with wavelength. The values in the table are for wavelengths in the middle of the visible range – about 550 nm. Over the whole range of visible wavelengths, approximately 400 nm to 700 nm, the absolute refractive index of crown glass varies by about ± 0.02.

A material with high absolute refractive index is sometimes said to be optically dense. Thus glass has a higher **optical density** than water. (Optical density is quantified in terms of absolute refractive index. It has no direct and fixed relationship with mass density of a material, as measured in kg m^{-3}. Note that when the word density is used alone, it should be taken to mean mass density.)

To a high degree of precision (six significant figures) the absolute refractive index of air at sea level is 1.000 29. The fact that this is so close to 1 means that if it were possible to have a boundary between a vacuum and air, the degree of refraction would be very small. As far as refraction of a ray of light is concerned, there is only a small difference between a vacuum and air. So a ray of light travelling from air into glass, say, behaves almost identically to a ray of light travelling from a vacuum into glass. The refractive index for a vacuum–glass boundary is only very slightly different from the refractive index for an air–glass boundary, and for practical purposes we can almost always treat them as the same. So

$$\text{absolute refractive index of glass} = n_{glass} \approx {_a}n_g$$

4 For an angle of incidence of 30° an air–glass (borosilicate) surface refracts violet light with an angle of refraction of 19.7° and red light with an angle of refraction of 20.0°. Calculate the refractive indices for the two colours.

5 Use the data in Tables 23.1 and 23.2 to find out whether there is a clear relationship between optical density (as quantified by absolute refractive index) and mass density.

6 a If it were possible to have a vacuum–air boundary, with refractive index 1.000 29, what would be the angles of refraction for angles of incidence of

   i 30°   **ii** 30.0°   **iii** 30.00°   **iv** 30.000°?

  b Describe the pathway of a ray of light across such a boundary.

7 Calculate the angle of refraction when a ray of light makes an angle of incidence of 80° with

  a a vacuum–water boundary (if that were possible)

  b a vacuum–diamond boundary.

Table 23.2

Material	Mass density/kg m^{-3}
air	1.26
water	1000
glass (borosilicate)	2230
glass (crown)	2900
diamond	3510

Real and apparent depth

A pool of water looks shallower than it really is, due to refraction (Figure 23.10). We see the bottom of the pool apparently shifted upwards; similarly a stick dipped into the water looks shorter than it does in air. If the real depth of the water is x and the apparent depth is y, then

$$\tan i = \frac{\text{opp}}{\text{adj}} = \frac{z}{x} \quad \text{and} \quad \tan r = \frac{\text{opp}}{\text{adj}} = \frac{z}{y}$$

where z is the distance between the two normals shown. So

$$\frac{\tan i}{\tan r} = \frac{z/x}{z/y} = \frac{y}{x}$$

Figure 23.10
The rays appear to be diverging from a point above their actual place of origin.

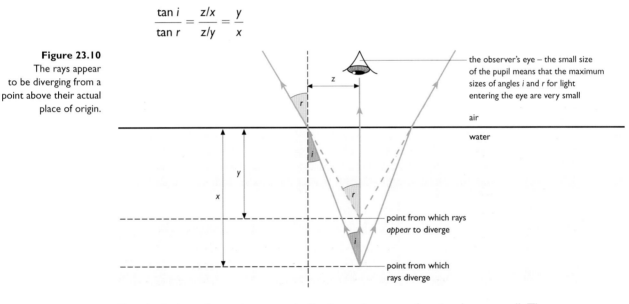

the observer's eye – the small size of the pupil means that the maximum sizes of angles i and r for light entering the eye are very small

air

water

point from which rays *appear* to diverge

point from which rays diverge

If we look down from almost vertically above the water then i and r are small. Then

$$\tan i \approx \sin i \quad \text{and} \quad \tan r \approx \sin r$$

8 Make calculations based on the figures in Table 23.3 to enable you to give your view on whether the approximation

$$n_{\text{water}} = \frac{\text{real depth}}{\text{apparent depth}}$$

is reasonable for the angles given. Note that we know

$$\frac{\tan r}{\tan i} = \frac{\text{real depth}}{\text{apparent depth}}$$

and

$$\frac{\sin r}{\sin i} = n_{\text{water}}$$

Table 23.3

$\sin r/\sin i = n_{\text{water}}$	1.33	1.33	1.33	1.33
r (= angle in the air)	30°	10°	3°	1°

So

$$\frac{\tan i}{\tan r} \approx \frac{\sin i}{\sin r} = \frac{1}{n_{\text{water}}}$$

(i is the angle in the water)

We can therefore say, to an approximation that improves as our direction of viewing becomes more and more vertical (so making i and r smaller), that

$$n_{\text{water}} = \frac{x}{y} = \frac{\text{real depth}}{\text{apparent depth}}$$

Variation of refractive index of a material with its mass density

If two materials have the same absolute refractive index then a ray of light passes from one to the other without a change of direction, but just a small difference in refractive index can make a difference to the ray's pathway. The absolute refractive index of a material varies as its mass density changes. Small changes in ray direction can thus occur within a single material due to variations in mass density.

Local temperature differences can cause small differences in density, leading to refraction. In air or water such local temperature differences are unlikely to be stationary, but will move around due to movement of material by convection currents. So, when you mix hot and cold water in your bath you see moving patterns of light and 'shadow' on the bottom of the bath. Or as you look across a lake on a summer's evening the lights on the far shore twinkle as the intensity of light reaching your eyes fluctuates. And on a hot day you might see mirages – shimmering patterns appearing to be on or near the ground (Figure 23.11).

Figure 23.11
A mirage is an image of the sky seen on or near the ground.

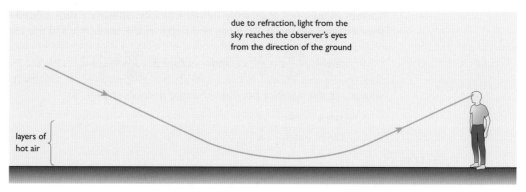

Total internal reflection and the critical angle

As we have seen, the angle of incidence is the angle between a ray striking a surface and the normal at the same point, and the angle of refraction is the angle with the normal made by the ray travelling away from the boundary. In whichever direction the light is travelling, the angle in the optically denser medium is always the smaller one (Figure 23.12).

Figure 23.12
Remember that the angle of incidence is the angle with the normal made by the ray hitting the boundary.

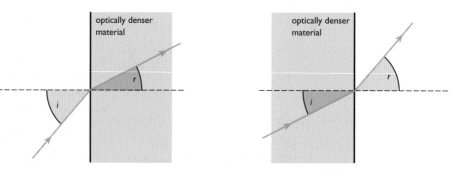

So for light travelling through glass and reaching a boundary with air, for example, the angle of refraction in the air is bigger than the angle of incidence in the glass. If the angle of incidence increases, the angle of refraction also increases. The angle of refraction reaches 90° before the angle of incidence does. The angle in the air cannot be any bigger than 90°. So any further increase in the angle of incidence makes emergence of the light into the air an impossibility. Instead, the light travels back into the optically denser medium. It obeys the law of reflection, and the phenomenon is called **total internal reflection**.

The size of the angle in the denser medium, the angle of incidence, at which the external angle becomes 90° is called the **critical angle**. For angles of incidence less than the critical angle, light is refracted into the less dense medium. For angles of incidence bigger than the critical angle, total internal reflection takes place (Figure 23.13).

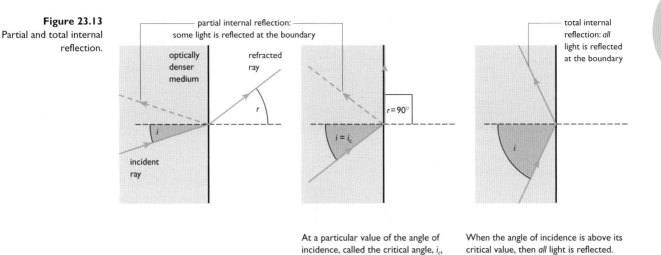

Figure 23.13
Partial and total internal reflection.

At a particular value of the angle of incidence, called the critical angle, i_c, the angle of refraction is 90°.

When the angle of incidence is above its critical value, then *all* light is reflected. This is total internal reflection.

A small proportion of the light reaching a boundary between an optically denser medium and a less dense medium is always reflected by the surface, and obeys the law of reflection. The reflected light remains in the optically denser medium. This is **partial internal reflection**.

Critical angle and refractive index

We know that, for refraction in general between two media, a and b,

$$_a n_b = \frac{\sin i}{\sin r}$$

Total internal reflection can take place when medium a is optically denser than b, for angles i bigger than the critical angle. When i is equal to the critical angle, i_c, then angle r is equal to 90° (Figure 23.14).

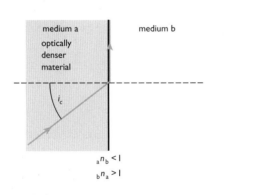

Figure 23.14
When the angle of incidence (the angle in the optically denser medium here) is equal to the critical angle, then the angle of refraction is 90°.

medium a
optically denser material

medium b

i_c

$_a n_b < 1$
$_b n_a > 1$

So we can write

$$_a n_b = \frac{\sin i_c}{\sin 90°}$$

and, since $\sin 90° = 1$,

$$_a n_b = \sin i_c$$

Most commonly, the less dense medium involved in total internal reflection is air, while the denser medium is water, glass, perspex or perhaps diamond. For water, for example,

$$_{water} n_{air} = \sin i_{c\,water}$$

The refractive index for light passing from water to air is the inverse of that for light passing from air to water:

$$_{water} n_{air} = \frac{1}{_{air} n_{water}}$$

so, for water,

$$\sin i_{c\,water} = \frac{1}{_{air} n_{water}}$$

But $_{air} n_{water} \approx _{vacuum} n_{water}$, which is the absolute refractive index of water, n_{water}, so we can say, to a good approximation,

$$\sin i_{c\,water} = \frac{1}{n_{water}}$$

This means that the critical angle, where the external medium is air, is a property of a material. Some values of critical angle are given in Table 23.4.

Table 23.4
Some critical angles and refractive indices.

Material	Absolute refractive index, n_{material}	Critical angle, $i_{c\,\text{material}}$
water	1.33	48.7°
glass (borosilicate)	1.47	42.9°
diamond	2.42	24.4°

Figure 23.15
Totally reflecting prisms in one half of a pair of binoculars. The light undergoes a series of 90° reflections.

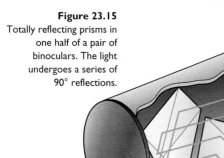

one objective

one eyepiece

The prisms increase the effective distance between the objective and the eyepiece lens, and also compensate for the inversion that is produced by the lens system.

Images produced by total internal reflection are brighter than those produced by mirrors, because much less absorption of light takes place at a totally internally reflecting surface than at a mirror surface. So optical devices such as cameras and binoculars (Figure 23.15) make use of total internal reflection within glass prisms rather than reflection by mirrors. This is made simple by the fact that the critical angle of the glass is less than 45°.

Note that diamond has an exceptionally high refractive index and an exceptionally low critical angle. Total internal reflection within the diamond, and the ultimate escape of light into the air, give diamond its special appearance (Figure 23.16b).

Figure 23.16 a
Glass can never sparkle quite like a diamond.

b

9 Write down the relationship between critical angle and absolute refractive index of a medium, with the latter as the subject of the equation.
10 Comment on the statement:
   'The critical angle is just an angle of incidence, but not *any* angle of incidence.'
11 Draw a sketch to show what happens when light travelling in
   a glass
   b diamond
   strikes a boundary with the air at an angle of incidence of 30°. How does this difference in behaviour affect the difference in appearance of glass and diamond?
12 What is the critical angle for a ray of light travelling in glass (light Ba crown) and striking a boundary with the air? (See Table 23.1.)

Optic fibres

At one time all telephone conversations were carried by analogue electrical signals. Essentially, the continuously varying patterns in the electric current in the wires mimicked patterns in the sound. Now, though the telephone mouthpiece first creates an analogue electrical signal, the transmission of information over long distances is in digital form – as a series of 'ons' and 'offs' either as electrical or, more often, as optical signals.

Optic fibre communication is made possible by total internal reflection. The fibre of glass is long and thin, and light can travel along and hit the surface of the glass at large angles of incidence – much larger than the critical angle, unless the cable is bent too sharply (Figure 23.17).

Figure 23.17
Excessive bending of an optic fibre allows escape of light.

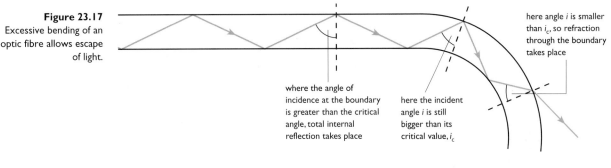

here angle i is smaller than i_c, so refraction through the boundary takes place

where the angle of incidence at the boundary is greater than the critical angle, total internal reflection takes place

here the incident angle i is still bigger than its critical value, i_c

Figure 23.18
Light can escape at an imperfection in the surface of a fibre.

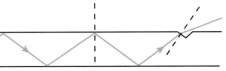

No absorption takes place when light is totally internally reflected by a perfectly smooth surface, and so at each reflection in the optic fibre the light intensity stays the same. However, any scratches in the surface change the local angle of incidence. At such irregularities in the surface, light can emerge (Figure 23.18). For a badly scratched surface, the intensity of light travelling along the fibre is reduced.

Optic fibres are designed so that the total internal reflection does not take place at the glass–air surface. The fibre can be built in one of two ways (Figure 23.19). It may have a central core of high optical density glass which carries the signal, and an outer sheath of optically less dense glass. The total internal reflection takes place at the boundary between these. Such a fibre is a **step index fibre**. Alternatively, the fibre is made with high optical density at the centre, decreasing gradually, with low refractive index material in the outer layers. Such fibres, called **graduated index fibres**, are expensive to make.

13 Imagine a step index fibre made of two types of glass 'the wrong way round' – that is, with the optically less dense glass as the core and the denser material as the sheath. Sketch the paths of some rays of light that enter the end of such a fibre.

14 The critical angle for the boundary between the central core and the outer sheath of a step index fibre is large. Explain why. Also, explain why this doesn't matter.

Figure 23.19
Transmission of light through, and variation in refractive index across, **a** a step index fibre and **b** a graduated index fibre.

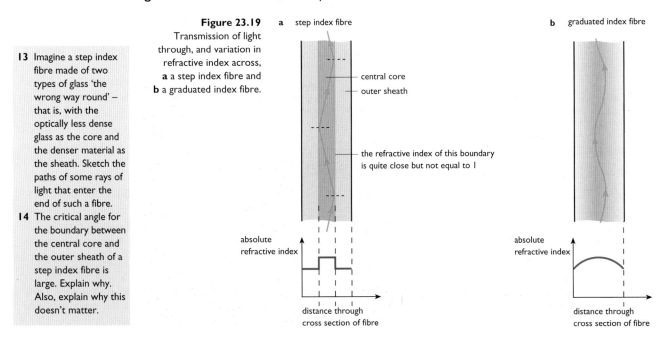

a   step index fibre

b   graduated index fibre

central core

outer sheath

the refractive index of this boundary is quite close but not equal to 1

absolute refractive index

distance through cross section of fibre

absolute refractive index

distance through cross section of fibre

Optic fibre endoscopy

A bundle of optic fibres can transmit clear images, provided that light from a small area of an object is transmitted so that the received light does not overlap with light from other small areas to produce blurring. This requires the use of very fine fibres, and that all of the fibres in the bundle remain in the same relative positions, or are **coherent**, at the two ends of the bundle.

For medical use, such a bundle of fibres is accompanied by a second bundle which carries light down to the site being viewed. This second bundle need not be coherent. There may also be other wires running the length of the bundle, connected to surgical instruments. A device consisting of these bundles and cables together is called a medical **endoscope** (Figure 23.20). It may be inserted into the body by a natural channel, such as the throat, or by a surgically created aperture, thus making 'keyhole surgery' possible.

Figure 23.20
An endoscope.

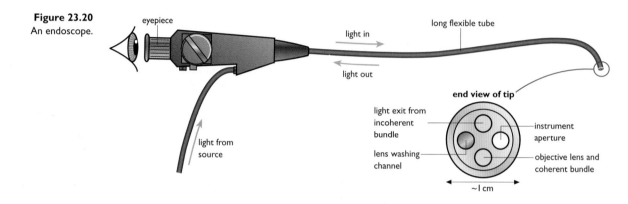

The wave model of refraction

Figure 23.21
Rays and wavefronts are alternative representations of the same behaviour in the real world.

So far we have dealt with refraction in the simplest possible way, in terms of rays, which show direction of travel. It is, however, useful to see what happens when we switch to using waves to model the same behaviour of light. We can represent waves visually by lines that are parallel to the crests and troughs – such lines are called **wavefronts**. The direction of travel of waves is always perpendicular to the wavefront. So the switch between the two models should not be too difficult; see Figure 23.21. Huygens suggested that each point on a wavefront acts as a point of disturbance, and so as a source of secondary wavelets. These wavelets combine together (by superposition) to create the wavefront that travels forwards.

We know that light travels more slowly in optically denser media than in a vacuum or in air. We can picture this in terms of wavefronts. (Note that since rays show only the direction of travel, in themselves they can tell us nothing about the speed of travel. Ray models are inadequate for thinking about the speed of waves.) Consider a wave passing from a vacuum or air into glass. Unless the wave travels normally to the surface the refraction results in a change of direction. The further a point on a wavefront travels through the optically denser medium the further it gets left behind compared with a point on the wavefront that is still in the less dense medium (Figure 23.22).

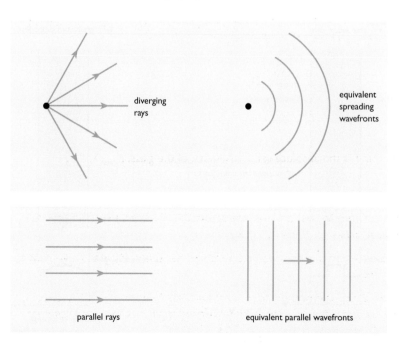

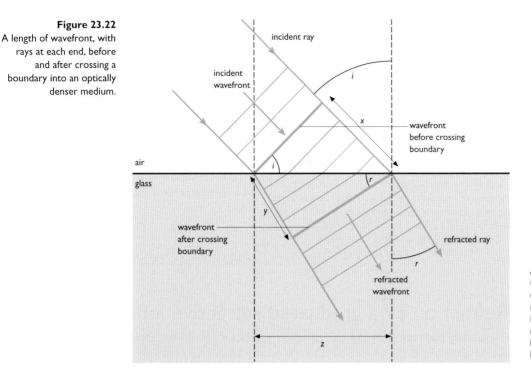

Figure 23.22
A length of wavefront, with rays at each end, before and after crossing a boundary into an optically denser medium.

Analysis of refraction in this way was first carried out by Christiaan Huygens in about 1678. Note that it is not just the speed but also the wavelength of the light that changes at the boundary.

For a length of wavefront that is just about to pass through a boundary from air to glass, its speed is c and its direction of travel is such that it makes an angle i (angle of incidence) with the normal to the boundary. After a certain time, t, this length of wavefront has just crossed the boundary and now has a reduced speed v, and its direction of travel makes an angle r (angle of refraction) with the normal. One end of this portion of wavefront has been travelling through air, distance x, while the other end has been travelling through glass, distance y. We have

$$\frac{x}{t} = c \qquad \text{and} \qquad \frac{y}{t} = v$$

so

$$\frac{c}{v} = \frac{x/t}{y/t} = \frac{x}{y} \qquad\qquad \text{(i)}$$

But note that $\sin i = \text{opp/hyp} = x/z$ and $\sin r = \text{opp/hyp} = y/z$, where z is the length of boundary that has been crossed (see Figure 23.22). This tells us that

$$\frac{\sin i}{\sin r} = \frac{x/z}{y/z} = \frac{x}{y} \qquad\qquad \text{(ii)}$$

Bringing (i) and (ii) together,

$$\frac{\sin i}{\sin r} = \frac{c}{v}$$

We already know that $\sin i/\sin r$ is the absolute refractive index of the glass, n_{glass}, and so

$$n_{\text{glass}} = \frac{c}{v}$$

Measurements of the speed of light in glass are consistent with this. Note that we also have a new, alternative, way to define absolute refractive index:

> The absolute refractive index of a material is the ratio of the speed of light in a vacuum to its speed in the medium.

This definition is in agreement with the one given earlier (page 208). We can use either one. (Remember there is some variation of refractive index with wavelength, but this does not alter the fundamental definitions.)

Refraction, frequency and wavelength

Note that the number of waves leaving one medium per second is the same as the number arriving in the other. Refraction does not affect frequency. However, given that $v = f\lambda$ (see Chapter 2), then if speed changes but frequency does not, wavelength must change. A decrease in speed results in a decrease in wavelength.

If the refractive index of a boundary between two materials is $_1n_2$ and the speed in the first medium is v_1 and the wavelength is λ_1, and in the second medium these become v_2 and λ_2, then

$$_1n_2 = \frac{v_1}{v_2} = \frac{f\lambda_1}{f\lambda_2} = \frac{\lambda_1}{\lambda_2}$$

15 Sketch a diagram showing a series of straight wavefronts travelling from glass into air.

16 Light of wavelength 6.00×10^{-7} m passes from air into a human cornea, which has an absolute refractive index of 1.38. What is its new
   a speed   b wavelength   c frequency?

17 The shortest wavelength in air of visible light is about 3.7×10^{-7} m. Taking the human eye to be a simple ball of water with absolute refractive index of 1.33, what is the wavelength of this light when it reaches the retina?

18 When you are swimming under water does the colour of your swimsuit appear (to you) different from its colour when you are out of the water? Explain this in terms of speed, frequency and wavelength of light in the media through which it must pass to reach your retinas when the swimsuit is in air and when it is in water.

Refraction by a graduated refractive index

A boundary between glass and air is sharp. In graduated index fibres the glass with the highest optical density is at the centre, but there is no sharp boundary between materials of different refractive index. There is a gradual decrease in refractive index. Similarly, in the case of a mirage (see page 210) the boundaries between regions of warmer and cooler air are not sharp. We can understand more easily why this leads to a change in direction of travel if we use the wave model of the light rather than the ray model (Figure 23.23).

Figure 23.23
The bending of wavefronts causing a mirage.

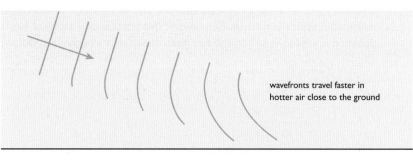

wavefronts travel faster in hotter air close to the ground

We can use this same treatment to think about the behaviour of other types of wave. Waves on the sea, for example, travel faster in deep water than in shallow water. So as a straight wavefront approaches a sloping beach those parts of it that are in shallower water travel more slowly. The shape of the wavefront therefore begins to change. The longer that part of the wavefront travels in shallower water, the further behind it falls compared with the part that is still in deep water. Water normally continues to get shallower and shallower as the wave approaches the shore. The shape of the wavefront changes so much that the waves approaching the water's edge are almost parallel to it (Figure 23.24).

Figure 23.24
Bending of wavefronts at a sloping shoreline.

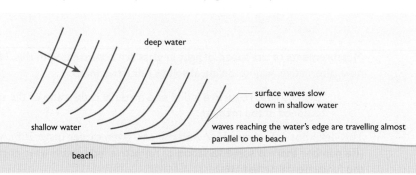

deep water

surface waves slow down in shallow water

shallow water

waves reaching the water's edge are travelling almost parallel to the beach

beach

Shallow water over sandbanks can have interesting effects on wavefronts. These effects are made hard to see because long wavefronts – with continuous crests and troughs – are not often seen on the sea, due to superposition of waves that have been created at different places and different times under different conditions. Waves on the sea tend to be complex. But seen from the air, patterns can be seen (Figure 23.25).

With sound waves, too, refraction occurs. Sound has a higher speed in warmer air and so follows curved pathways when there are layers of air at different temperatures.

Figure 23.25
An area of shallow water such as a sandbank may have an interesting effect on the travel of surface water waves.

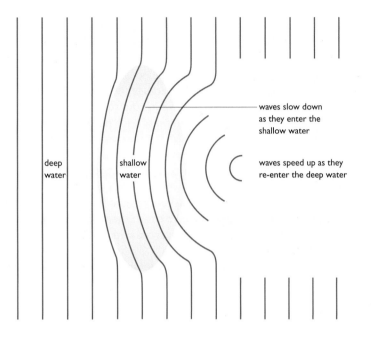

waves slow down as they enter the shallow water

deep water   shallow water

waves speed up as they re-enter the deep water

19 Use wavefronts to sketch the travel of sound waves in still air:
a above hot ground during a summer's day
b above cold ground during the night.

Refraction at curved surfaces

A 'parallel beam' of light is arriving at both lenses in Figure 23.26. Rays and wavefronts are both good models – both show the parallel nature of the incident beam, and each can be used to show different versions of the same reality. We can construct the ray diagram with the knowledge that $\sin i / \sin r = n$. We can construct the wave diagram with the knowledge that $c/v = n$, and see that the slower speed of the light during its travel through the glass produces the curving of the wavefronts.

For careful analysis of refraction by lenses, rays are more useful and the next chapter uses rays to consider the behaviour and applications of lenses.

Figure 23.26
The ray model and the wavefront model of refraction by a convex lens.

20 Sound waves travel at higher speed through helium than through air at the same pressure. Sketch wavefronts passing through a helium-filled balloon.

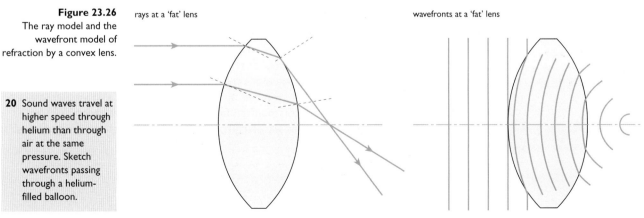

rays at a 'fat' lens

wavefronts at a 'fat' lens

● **Comprehension and application**

Refraction, digital signals and optic fibres

Use of digital signals has advantages:

- the digital signal can be carried by rapid bursts of electric current *or* of light;
- many continuous signals (such as are generated by many telephone conversations) can be sent along a single fibre at the same time, by using very short coded bursts which travel along the fibre in turn. This is called **time division multiplexing**.

Where the signal is carried by light rather than by electric current there is a further advantage: the flashes of light in a digital optic fibre system can travel further without loss of the signal.

A **bit**, or binary digit, of information is projected into the end of an optic fibre as a pulse of light, with intensity rising almost instantaneously to its maximum value and falling just as quickly. Sharp and rapid pulses are desirable, because they allow more information to be transmitted in a given length of time. The speed with which the source of light can be switched on and off is the main limitation on reducing the duration of each pulse, but it is possible to emit pulses lasting less than one tenth of a nanosecond. With such pulses of duration 10^{-10} s, it is possible to send up to 10^{10} bits per second. This value is called the **bit-rate**.

With analogue systems the information is encoded in the shape of the wave (such as in electrical analogues of the sound waves of speech or in AM or FM radio waves). If the shape of the wave is changed during transmission then the information is distorted. With digital pulses it is necessary only that one pulse can be distinguished from the next, and that the absence of a pulse (indicating a binary value of 0) is not blurred with an actual pulse (which indicates a binary value of 1).

One pulse, represented as a plot of intensity against time, looks like a section of a square wave (Figure 23.27), and not like a section of a sine wave. However, the pulse arriving at the other end of the fibre will not have such a sharp profile: the longer the fibre, the less sharp it will be.

Figure 23.27
A sharp pulse, or bit.

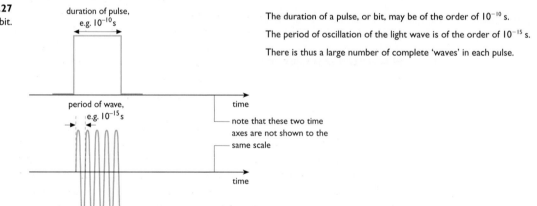

duration of pulse, e.g. 10^{-10} s

period of wave, e.g. 10^{-15} s

time

The duration of a pulse, or bit, may be of the order of 10^{-10} s.

The period of oscillation of the light wave is of the order of 10^{-15} s.

There is thus a large number of complete 'waves' in each pulse.

note that these two time axes are not shown to the same scale

time

If the light used for transmission contains a range of wavelengths then a pulse received at the end of a digital optic fibre will have become broader and rounder, effectively blurring the pulse. Violet light travels more slowly than red light through the glass, so violet light arrives later than red and the time difference between the arrival of the first light to the last is increased (Figure 23.28). This problem is easily solved by using monochromatic light – laser light or light from an **LED (light emitting diode)** that can be switched on and off very quickly. With either lasers or LEDs as sources, the 'light' used for telecommunications is in fact infrared radiation, with a wavelength in the region of 1.5 μm. At this wavelength, absorption of the light by the glass itself is minimised.

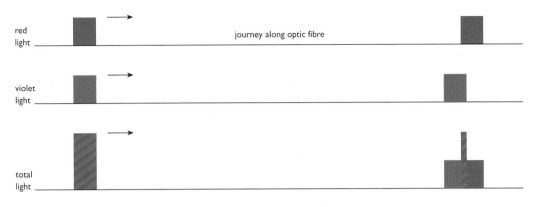

Figure 23.28
The absolute refractive index of the glass is higher for violet light than for red light. A red pulse travels more quickly than a pulse of violet light along the optic fibre.

A second cause of broadening of the profile of the pulse is the different lengths of the different pathways that are available to the light, and hence the different times taken to travel along the fibre. Some light will take the straightest possible route along the fibre, while some might take a much longer route, experiencing many total internal reflections during its journey (Figure 23.29).

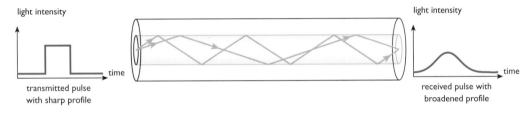

Figure 23.29
Different pathways or modes have different lengths and so take different times. This causes an increase in the total duration of the pulse and a blurring of its start and finish. This is called pulse broadening.

These different journeys are called modes, and fibres in which there are many possible modes are called **multimode fibres**. Over short distances the increased spread in arrival times is not enough to blur one pulse into another, but over long distances it is more of a problem. The effect can be reduced by using a very thin fibre, which approximates to a **monomode fibre** with just one straight pathway. But such fibres are more difficult to manufacture and are more easily snapped. Another, but still expensive, solution is to use graduated index fibres (see page 213), with the material at the centre having the highest optical density, this gradually decreasing towards the outside of the fibre. This cleverly solves the problem, since although there are different lengths of pathway and the shortest is straight along the centre, the light that takes this shortest route travels slowest.

A typical multimode optic fibre is made of a glass core of radius of about 60 μm, surrounded by a layer of glass **cladding** (or sheath) of about the same thickness. This reduces the problem of 'leakage' of light at imperfections of the boundary, since the glass-to-glass boundary can be manufactured to be extremely even and is not then subject to scratches as a glass–air interface would be. Also, by using two types of glass of similar but not identical refractive index, the critical angle can be made quite large. This reduces the amount of zig-zagging that light can do and hence reduces the maximum path length. The difference in these refractive indices is very important, because if they are too similar then it will not be possible to bend the cable very much without leakage of light.

Optic fibres are the highways of the information revolution. Optic fibre cables (Figure 23.30) are lighter and thinner than copper cable. They cannot be 'tapped' illegally because, unlike electric currents which create a magnetic field in the surrounding space, they are not detectable from outside the cable. But most importantly, because of the very high bit-rates that are possible, a cable can carry thousands of telephone conversations at once.

Figure 23.30
Many optic fibres can be fitted into a single cable.

Each fibre within the cable is capable of carrying many separate streams of data at the same time, by time division multiplexing. Each telephone conversation, for example, is broken down into a short 'train' of pulses, and the trains take turns to travel along the fibre (Figure 23.31).

Figure 23.31
Time division multiplexing of digital signals.

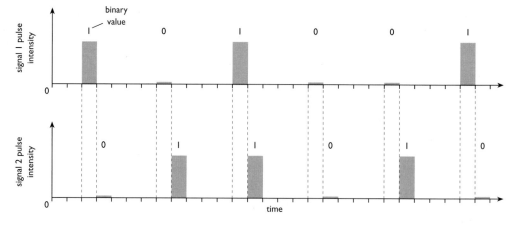

If receiving equipment 'reads' the pulses at set regular time intervals it will receive signal 1.

If further receiving equipment reads the pulses at *different* times, but separated by an equivalent time lapse, it will receive a different signal, signal 2.

Thus two sets of equipment can read different signals from the *same* optic fibre. This is the principle of time division multiplexing.

21 a Given that the absolute refractive indices of the glass used in an optic fibre for red light and violet light are 1.51 and 1.53 respectively, calculate the times taken by red light and violet light to travel 10 km along such a fibre.
    b Show this as a representation of a red pulse and a violet pulse at the start of their journeys and at the end of their journeys, with a labelled and numbered time axis.
    c How is the problem of this separation solved?

22 Imagine a crude optic fibre system that uses fibres with no cladding.
    a Draw a sketch to show a light ray travelling along the axis of a fibre and a light ray that makes repeated reflections from the fibre walls at angles of incidence of 45°.
    b On your sketch, mark the distance, x, along the length of the fibre, between adjacent reflections from the same side of the fibre.

    c In terms of x and the angle 45°, what is the distance travelled by the light between these two reflections?
    d What is the total distance travelled along a fibre 10 km long, in this mode?
    e What is the time taken to travel along the 10 km fibre in this mode, if the speed of light in the cable is $1.95 \times 10^8 \, \text{m s}^{-1}$?
    f What is the time taken for light to travel 10 km along the axis of the fibre? Compare the times for the two journey modes. Draw a sketch to show how this might influence the profile of a pulse.
    g Explain how cladding can greatly reduce this time lag.
    h What other advantage does the cladding give?
    i Explain, with the aid of a sketch diagram, how graduated index fibres solve the problem of pulse broadening.

Examination questions

1 a With the aid of clearly labelled diagrams, explain the terms
   i 'refractive index' (2)
   ii 'critical angle' (2)
   iii 'total internal reflection'. (2)

b Use an annotated diagram to show how the wave theory of light accounts for the refractive bending of light at a boundary between two different media. (2)

c A parallel beam of monochromatic light in water has a frequency 5.0×10^{14} Hz. It passes into air with an angle of incidence at the water/air boundary of 45°. The refractive index of the water is 1.33. Determine
   i the speed of the light beam in the water (2)
   ii the frequency of the beam in the water (2)
   iii the wavelength of the beam in air and in the water (3)
   iv the critical angle. (2)

d A student is swimming under water without wearing goggles. She finds it difficult to focus on nearby underwater objects. Explain why. (3)
International Baccalaureate, Physics, Subsidiary Level, Nov 1997

2 A ray of red light is incident on the face AB of a glass prism ABC as shown in the figure.

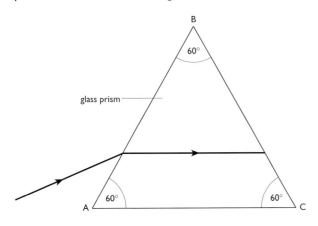

a i On the figure, mark the angle of incidence *i* and the angle of refraction *r* at the face AB.
   ii Define refractive index in terms of *i* and *r*. (2)

b i The refractive index of the glass of the prism for red light is 1.60. The angle of incidence of the ray at the face BC is 30°. Show that the ray will be refracted and not totally internally reflected at face BC.
   ii Without further calculation, sketch on the figure the path of the ray as it emerges from face BC. (3)
OCR, Sciences, Physics, Foundation, Mar 1999

3 a State three properties of the image of an object as seen in a plane mirror. (3)

b When light is incident on an air/glass boundary, some of the light is refracted and some is reflected. The figure shows a ray of light from an object O incident on a thick glass mirror.

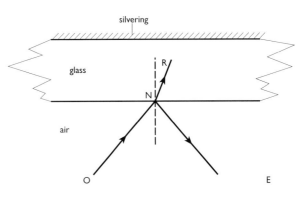

   i On the figure, continue the path of the ray NR until it has emerged into the air again.
   ii Suggest and explain what might be seen by an observer at E. (4)

c The speed of light in glass is 2.03×10^8 m s^{-1} and in water, the speed is 2.26×10^8 m s^{-1}. Calculate, for a ray of light passing from glass into water,
   i the refractive index
   ii the maximum angle of incidence for the ray to be refracted. (4)
OCR, Sciences, Physics, Foundation, June 1999

4 The diagram shows the path of a ray of monochromatic light through a glass prism.

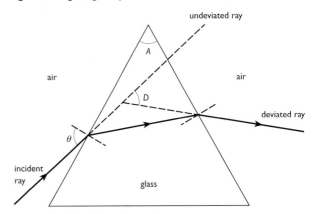

a i Describe how you would measure the angle *A* of the prism to within ±0.5%, i.e. to about a quarter of a degree.
   ii It is suggested that the ray would follow exactly the same path if its direction were reversed.
   Describe how you would test this suggestion experimentally. (6)

b As the angle of incidence θ is varied, the angle of deviation D also varies. When A is equal to 60.0°, a series of corresponding values of θ and D were found experimentally with the following results:

θ/deg	D/deg
32.5	44.8
35.0	42.4
37.5	40.5
40.0	39.0
45.0	37.5
50.0	37.3
55.0	37.9
60.0	39.0
65.0	40.6

i Plot a graph of D against θ.
Write down the value of the minimum deviation D_m and the corresponding value of the angle of incidence.
ii When the deviation is a minimum, the ray passes symmetrically through the prism. The refractive index n of the glass of the prism is then given by

$$n \sin\left(\frac{A}{2}\right) = \sin\left(\frac{A + D_m}{2}\right)$$

where A, here 60.0°, is the angle at the top of the prism.
Use the equation to calculate a value for n. **(10)**
c i When the angle A of the prism is less than about 10° the prism is a thin wedge of glass. The deviation D is then small and is constant over a wide range of values of the angle of incidence.
Use the equation in **b ii**, together with the small angle approximation $\sin x \approx x$ (in radians), to predict how D ($= D_m$) depends on n and A for thin prisms.
ii The deviation produced by a thin prism, though independent of θ, does depend on the refractive index n of the glass. The value of n depends in turn on the wavelength (colour) of the light.
The table below gives the values of n for four colours for a prism with $A = 5.00°$.

colour	n
red	1.520
green	1.526
blue	1.531
violet	1.538

Draw up a table of corresponding values of n and D for this prism. Plot a graph of D against n. **(8)**
Edexcel (London), Physics, Module PH5, June 1999

5 The figure illustrates an experiment where a source is made to emit a single, brief pulse of light into a straight length of step index optic fibre.

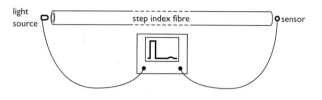

Sensors at each end of the fibre enable the input and output powers of the light pulse to be monitored. The resulting graph of the variation with time of the light power entering and leaving the fibre is shown below.

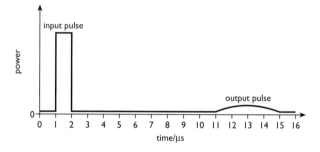

a Explain the following observations.
i The output pulse covers a smaller area on the graph than the input pulse.
ii The output pulse lasts for a longer period of time than the input pulse. **(4)**
b Use the graph to determine the shortest period of time that light takes to travel through this fibre. Explain your answer. **(2)**
c The length of the fibre is 2.0 km. Calculate
i the speed of light in the fibre
ii the refractive index of the core. **(4)**
d Explain why step index fibres are no longer used in the telecommunications industry for the long-distance transmission of high bandwidth signals. **(3)**
OCR, Sciences, Telecommunications, Nov 1999

24 Lenses, images and instruments

THE BIG QUESTIONS

- What do lenses do, and how?
- What do combinations of lenses do, and how?
- How have optical instruments changed the way people understand the world?

KEY VOCABULARY

accommodation   achromatic lens   angular magnification   aperture
astronomical refracting telescope   chromatic aberration
compound microscope   converging lens   diverging lens   eyepiece
focal length   focal plane   Galilean telescope   linear magnification
long sight   magnifying power   normal adjustment   objective
principal axis   principal focus   radian   Rayleigh criterion   real image
'real is positive' convention   reflecting telescope   resolution
resolve (neighbouring points)   short sight   simple microscope
spherical aberration   spherical lens   virtual image

BACKGROUND

On many clear nights you can look up and see Jupiter looking just like a star, if a rather bright one. That's all that people could do until the invention of the telescope; this gave the stunning realisation that Jupiter was not at all like the stars, but had a banded surface, and moons in orbit. 'The moons of Jupiter' is an expression people sometimes use to speak of some new evidence that changes everything. The sight of these moons and their shifting positions night by night were the first evidence to support Copernicus' revolutionary theory that all of the bodies in the sky are not in orbit around the Earth.

Galileo was the bold observer who was the first to watch these moons, in the year 1610. Along with his discovery of the mountains of our own Moon, the moons of Jupiter showed Galileo that heavenly bodies were not perfect spheres all in orbit around the Earth, as the ancient books claimed. Two thousand years of accepted ideas had to be abandoned. The development of the telescope and Galileo's observations happened alongside the move out of the medieval world and into the modern world.

So if you ever get the chance to shiver in a cold dark field and see Jupiter with the help of a telescope, seize the opportunity. The experience will put you in closer touch with your Universe.

Figure 24.1
Jupiter and its moons seen through a telescope.

Fat lenses, thin lenses and the principal focus

We can illustrate the action of a lens by drawing it as a series of prisms (Figure 24.2).

Figure 24.2
The action of
a a converging lens and
b a diverging lens.

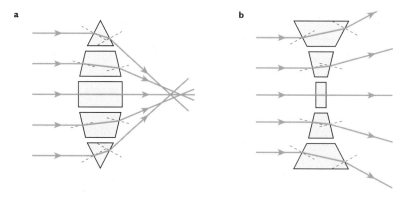

Prisms arranged as in Figure 24.2a change the paths of the rays, which are parallel before refraction, and make them converge. A lens which does this is a **converging lens**. A **diverging lens** (Figure 24.2b) not surprisingly causes parallel rays to diverge.

We know that prisms cause dispersion of light, and this happens with lenses too. The refractive index for violet light is larger than that for red light, so the violet light experiences more bending of its path. With a telescope this produces some separation of colour in the image and consequently some blurring. This effect is called **chromatic aberration**, and it was a problem that astronomers had with telescopes right from the start. Newton struggled to try to get rid of it and had only partial success – but in the process he reached the conclusion that white light is a mixture of the colours. There are now complex lens systems that use layers of different kinds of glass to minimise chromatic aberration. These are called **achromatic lenses**.

Lenses, of course, are not made of blocks or prisms but have continuously curved surfaces. The easiest way to make them is to grind glass so that the surfaces of the lenses are shaped like the surface you would be holding if you were to cut a small circle out of the surface of a large ball (Figure 24.3). Because of this, these lenses are called **spherical lenses**. In the rest of this chapter we will deal with lenses that have symmetrical spherical surfaces; asymmetrical lenses are also possible (Figure 24.4).

Figure 24.3
A spherical lens has a surface that is the shape of a portion of the surface of a sphere.

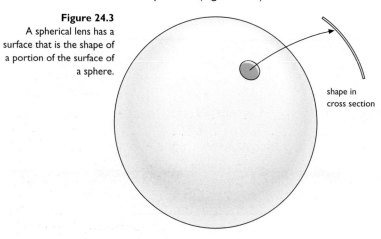

shape in cross section

Unfortunately a spherical converging lens does not have a geometrically simple effect on rays. Consider the simplest way in which we can shine light on to a lens – as a beam of light for which all rays are parallel to each other and also parallel to the **principal axis** of the lens. (The principal axis is a line passing through the centre of the lens perpendicularly to the plane of the lens.) Such a beam of light is converged by the lens, but sadly it is not converged to a neat single point on the principal axis, as you can see in Figure 24.2a. This effect is called **spherical aberration**.

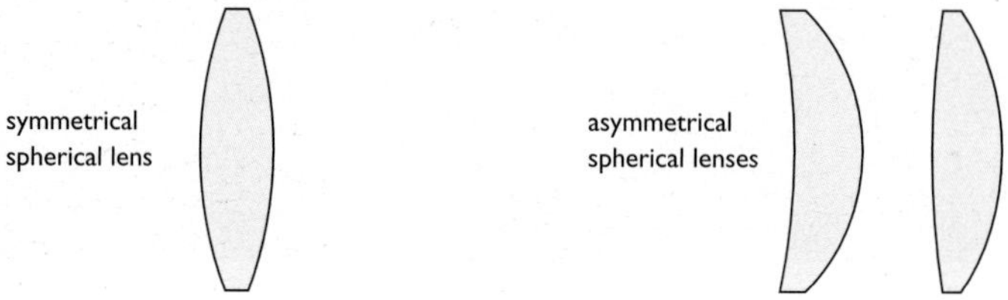

Figure 24.4
Spherical lenses may have symmetrical or asymmetrical surfaces.

symmetrical spherical lens

asymmetrical spherical lenses

Much more spherical aberration takes place with fat lenses than with thin lenses. For thin lenses it is more reasonable to ignore the effects of spherical aberration – that is, to 'idealise' for the sake of simplicity and to suppose that the lens focuses a parallel beam to a single point. If the incident beam is parallel to the principal axis then the single point lies on the principal axis and is called the **principal focus**. We need to express this as a formal definition – see Figure 24.5.

If the incident beam is not parallel to the principal axis of the lens then this single point lies at a certain distance from the lens in what is called the **focal plane** (Figure 24.6).

Figure 24.5
A formal definition of the principal focus of a thin converging lens.

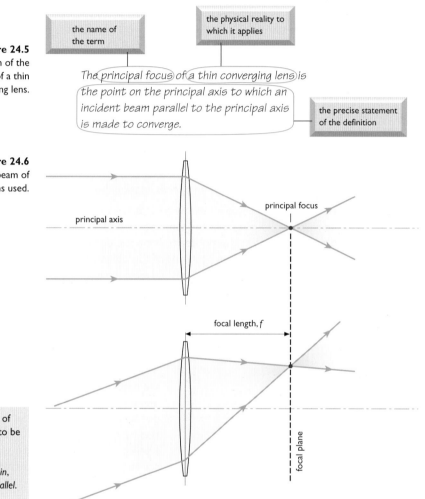

Figure 24.6
Focusing a parallel beam of light, and the terms used.

I Explain why each of these words has to be included in the definition of the principal focus: *thin, point, incident, parallel.*

Note that a tiny but bright light source placed at the principal focus of a lens will produce a parallel beam of light after refraction by the lens. This is consistent with the principle of reversibility (page 207). Note also that a lens has two principal foci, one each side, and that for symmetrical lenses the two foci are at equal distance from the centre of the lens. This distance is called the **focal length**, *f*, of the lens. The focal length depends on two things – the material of which the lens is made, quantified by its refractive index, and the shape of the lens. In general, as lenses of a particular material become fatter their focal lengths decrease.

Figure 24.7
The principal focus of a diverging lens.

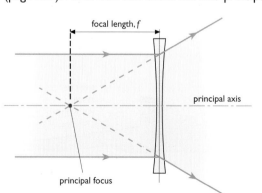

2 State the differences between *principal focus, focal length* and *focal plane.*

For a diverging lens we have to amend our definition of the principal focus. It is now the point on the principal axis from which a parallel incident beam diverges (Figure 24.7).

Ray diagrams for thin converging lenses

Most lenses are sufficiently thin for us to be able to ignore spherical aberration for many practical purposes. But there is a further complication, which has the potential to make it difficult to create ray diagrams that are simple enough to be useful. This is that refraction takes place at the two curved surfaces of the lens. We simplify this by visualising a central plane for the lens, which appears as a central line when we draw the lens in cross section. We then, in our drawings, assume that it is at this plane that all the refraction takes place. This has been done in Figures 24.6 and 24.7.

Thanks to our knowledge of the behaviour of a beam of light whose travel, before or after refraction, is parallel to the principal axis, we now have a considerable amount of information about the pathways of light through thin symmetrical converging lenses.

A If an incident ray of light is parallel to the principal axis then it is refracted so as to pass through the principal focus.

B If an incident ray is travelling away from a principal focus, then it is refracted so as to be parallel to the principal axis.

We can add to this:

C If an incident ray is travelling towards the centre of a thin lens then the curvature of the lens is too small to have a significant effect, and the ray travels in a straight line as if it were being refracted by a very thin plane sheet. This idea is an approximation, but it is a valid and very useful one.

A, B and C provide us with three known pathways that we can use to predict behaviour. Suppose that three rays spread out from the same point, and follow the pathways A, B and C (Figure 24.8). The detail of these pathways depends on the position of the object. After refraction, for most but not all possible positions of the object, the rays meet again at another point. It is the spread of rays from a single point, through a lens, and their convergence again to a single point, that results in the creation of a real image (see page 227).

Of course, rays spread out from a point on an object in other directions as well as along the routes A, B and C, and some of them pass through the lens. After refraction these also meet again at the same point, but these rays are harder to predict and it would be an unnecessary complication to include them in our diagrams. What we can do, however, is show a complete cone of light spreading from a point on the object, and a similar light cone converging to create a matching point on the image (Figure 24.9).

3 Make a sketch similar to Figure 24.8 to predict the position of the image of the cat's nose.

Figure 24.8
We can use known pathways to predict image creation.

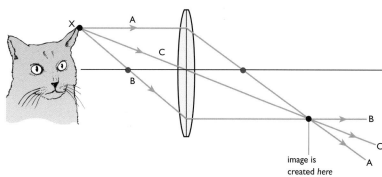

image is created *here*

Figure 24.9
A complete cone of light spreads from an object, passes through a lens, and is converged to a single point.

4 Make a sketch to show what happens to a cone of light that originates at the principal focus of a thin converging lens.

5 a Make a sketch of light cones spreading from a point on an object and being refracted by a diverging lens.
b Explain where the image lies.

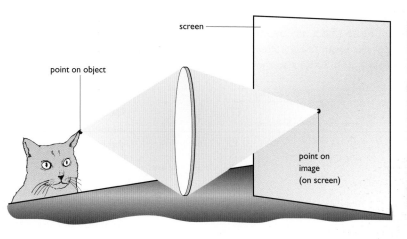

screen

point on object

point on image (on screen)

It is possible to place a screen such that the point of convergence of the cone of light lies on it. Light also spreads from other points on the object, and for each point on the object there is a corresponding single point on the screen. An image of the object appears on the screen. (The image may be impossible to see because of ambient light falling on the screen, unless the object is brightly illuminated.)

Real and virtual images

A **real image** (Figure 24.10) is one formed by actual intersection of rays; it can be projected on to a screen. A **virtual image** (Figure 24.11) is one formed only by apparent intersection of rays, and can only be seen by looking into the mirror or lens.

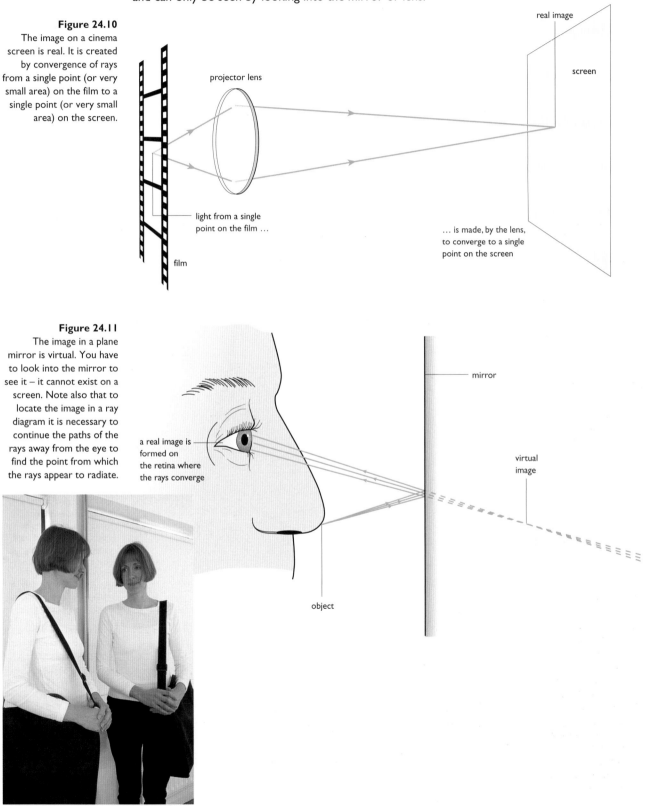

Figure 24.10
The image on a cinema screen is real. It is created by convergence of rays from a single point (or very small area) on the film to a single point (or very small area) on the screen.

real image

projector lens

screen

light from a single point on the film …

… is made, by the lens, to converge to a single point on the screen

film

Figure 24.11
The image in a plane mirror is virtual. You have to look into the mirror to see it – it cannot exist on a screen. Note also that to locate the image in a ray diagram it is necessary to continue the paths of the rays away from the eye to find the point from which the rays appear to radiate.

mirror

a real image is formed on the retina where the rays converge

virtual image

object

Images produced by converging lenses

The distance from object to lens can vary from zero to infinity. This distance, usually symbolised by u, makes a big difference to the image. It doesn't just affect the distance of the image from the lens, but also the nature of the image — whether it is real or virtual, magnified or diminished, upright or inverted. Image-to-lens distance is usually symbolised by v. This makes sense since u comes before v in the alphabet, and object comes before image.

Figure 24.12 uses the predictable pathways A, B and C (see page 226) to show different images produced by different object distances, u, relative to focal length, f.

Figure 24.12
Different kinds of image produced by converging lenses.

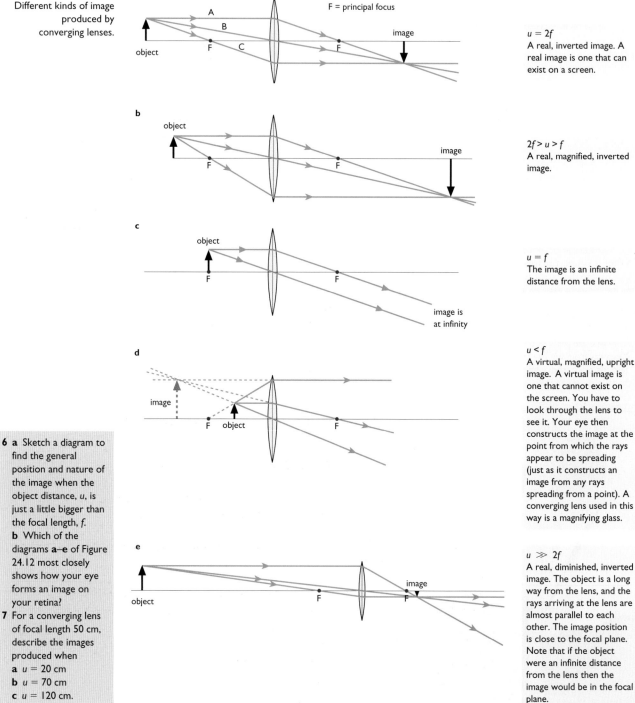

a
F = principal focus
image
object

$u = 2f$
A real, inverted image. A real image is one that can exist on a screen.

b
object
image

$2f > u > f$
A real, magnified, inverted image.

c
object
image is at infinity

$u = f$
The image is an infinite distance from the lens.

d
image
object

$u < f$
A virtual, magnified, upright image. A virtual image is one that cannot exist on the screen. You have to look through the lens to see it. Your eye then constructs the image at the point from which the rays appear to be spreading (just as it constructs an image from any rays spreading from a point). A converging lens used in this way is a magnifying glass.

e
image
object

$u \gg 2f$
A real, diminished, inverted image. The object is a long way from the lens, and the rays arriving at the lens are almost parallel to each other. The image position is close to the focal plane. Note that if the object were an infinite distance from the lens then the image would be in the focal plane.

6 a Sketch a diagram to find the general position and nature of the image when the object distance, u, is just a little bigger than the focal length, f.
b Which of the diagrams **a**–**e** of Figure 24.12 most closely shows how your eye forms an image on your retina?
7 For a converging lens of focal length 50 cm, describe the images produced when
a $u = 20$ cm
b $u = 70$ cm
c $u = 120$ cm.

Images produced by diverging lenses

All diverging lens images are virtual (Figure 24.13), no matter how far the object is from the lens.

Figure 24.13
Image produced by a
diverging lens.

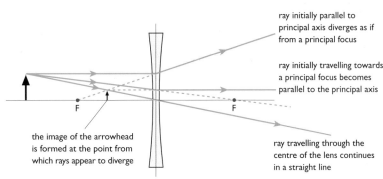

ray initially parallel to
principal axis diverges as if
from a principal focus

ray initially travelling towards
a principal focus becomes
parallel to the principal axis

the image of the arrowhead
is formed at the point from
which rays appear to diverge

ray travelling through the
centre of the lens continues
in a straight line

A mathematical model for lens action

We do not have to rely only on ray diagrams to predict relative object distances, u, and image distances, v. They are related together and to focal length by:

$$\frac{1}{f} = \frac{1}{u} + \frac{1}{v}$$

For a converging lens we can apply this to the situations shown in Figure 24.12a, b and d, using some sample figures with measurements in centimetres, and so work out image distances. In general, since $1/f = 1/u + 1/v$, then

$$\frac{1}{v} = \frac{1}{f} - \frac{1}{u}$$

Situation a	For $u = 2f$ ($u = 40$ cm, $f = 20$ cm): $1/v = 1/20 - 1/40 = 2/40 - 1/40 = 1/40$ $v = 40$ cm
Situation b	For $2f > u > f$ ($u = 30$ cm, $f = 20$ cm): $1/v = 1/20 - 1/30 = 3/60 - 2/60 = 1/60$ $v = 60$ cm The image is a comparatively long way from the lens.
Situation d	For $f > u$ ($u = 10$ cm, $f = 20$ cm): $1/v = 1/20 - 1/10 = 1/20 - 2/20 = -1/20$ $v = -20$ cm The negative sign here is no accident. It tells us that the image is on the same side of the lens as the object, and is virtual.

8 For a converging lens of focal length 50 cm, calculate the distances of the image from the lens when
a $u = 20$ cm
b $u = 70$ cm
c $u = 120$ cm.
Are your answers consistent with those of question 7?

9 Calculate the image distance for a diverging lens of $f = -20$ cm when the object distance is 15 cm.

The formula can be applied to diverging lenses as well as converging lenses; in this case the focal length must be given a negative value.

The general convention is to say that a negative sign indicates a virtual image or the focal length of a diverging lens. This is called the **'real is positive'** sign convention. Table 24.1 is a summary of the significance of the sign to the values of object distance, image distance and focal length.

Table 24.1

	Positive value means	Negative value means
u	real object	virtual object (we will not deal with this here)
v	real image	virtual image
f	converging lens	diverging lens

Linear and angular magnification

Lenses have the power to make images that are bigger or smaller than the object (Figure 24.14). Both can be useful. Larger images show more detail. Smaller images provide a larger field of view.

The simplest way to compare image with object is by the ratio of their sizes. This ratio is called the **linear magnification**:

$$\text{linear magnification, } m_L = \frac{\text{image size}}{\text{object size}}$$

$$= \frac{i}{o}$$

Figure 24.14
Magnified and diminished images.

You can see from Figure 24.15 that the ratio of image size to object size is the same as the ratio of image distance to object distance. So we can also say that

$$\text{linear magnification, } m_L = \frac{\text{image distance}}{\text{object distance}}$$

Using u for object distance and v for image distance,

$$m_L = \frac{v}{u}$$

Figure 24.15

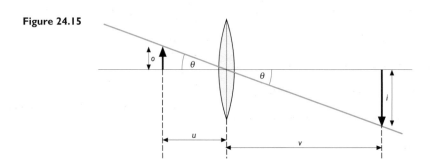

$$\tan \theta = \frac{o}{u} = \frac{i}{v}$$

where o is 'object size' and i is 'image size'

$\dfrac{i}{o} = \dfrac{v}{u} = $ linear magnification

Optical magnification effectively brings an object geometrically closer. A near object fills a larger proportion of the field of view, and its image falls on a larger area of the retina. The difference in size of the images formed on the retina by two identical objects, one near and one far away, can be expressed in terms of the angles that they make at the eye (Figure 24.16).

Figure 24.16
When nearer to the eye, the object makes a larger angle at the eye and so forms a larger image on the retina.

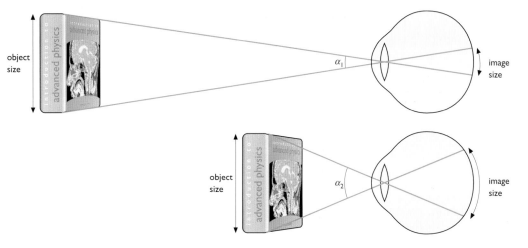

Where an object seen with the unaided eye makes an angle alpha, α, at the eye, but makes an angle beta, β, when seen with an optical aid, then the **angular magnification** is defined as the ratio of these angles:

$$\text{angular magnification, } M = \frac{\beta}{\alpha}$$

Linear magnification is a ratio of distances, and is dimensionless. Angular magnification, also known as the **magnifying power** of an optical instrument, is a ratio of angles and is also dimensionless.

The radian

Figure 24.17

Since angular magnification is a ratio of angles we will find the same ratio whatever units we measure the angles in (provided of course that we use the same units for both α and β). We could use degrees. One degree is 1/360 of the angle contained in a complete revolution. There is no fundamental reason for choosing the number 360 – it was a decision that was made a very long time ago. There is a better, and ultimately easier method of measuring angles. That is to take an arc of a circle – a portion of the circumference – that has the same length as the radius. The angle subtended by this arc at the centre of the circle is called a **radian** (Figure 24.17). Since there are 2π radii in the length of a complete circumference, there are 2π radians in a complete circle:

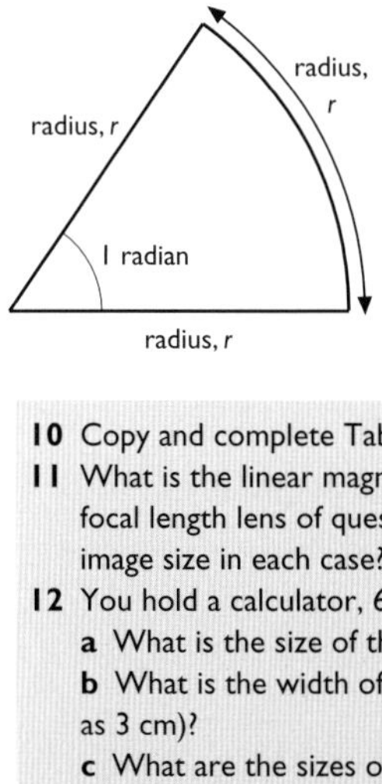

$$\text{angle contained in one complete revolution} = 360° = 2\pi \text{ radians}$$

The definition of the radian is related to the geometry of a circle in a fundamental way.

10 Copy and complete Table 24.2.
11 What is the linear magnification for each of the object distances for the $f = 50$ cm focal length lens of question 8? If the object is 5 cm across, what is the corresponding image size in each case?
12 You hold a calculator, 6 cm across, at arm's length, 75 cm from your eye.
  a What is the size of the angle that the calculator width makes at your eye?
  b What is the width of the image it makes on your retina (taking the image distance as 3 cm)?
  c What are the sizes of the angle at your eye and the image on your retina if you hold the calculator 10 cm from your eye?
  d What is the 'angular magnification' produced by moving the calculator from 75 cm away to 10 cm away?
  e Does the calculator *seem* that much bigger to you?

Table 24.2

Angle in revolutions	0.25	0.5	?	1	$0.16 = \dfrac{1}{2\pi}$
Angle in degrees	90	?	270	?	?
Angle in radians	$\dfrac{\pi}{2}$	π	$\dfrac{3\pi}{2}$	?	1

Remember that π is a number: 3.14 (to three significant figures)

Lens combination – the astronomical refracting telescope

Spectacles were in use as early as 1300, but it was almost another 300 years before methods of grinding glass were good enough to produce consistent images when one lens was placed in front of another. A simple **astronomical refracting telescope** has two converging lenses – the **objective** (nearer the object) and the **eyepiece** (nearer the eye).

Rays from distant objects are effectively parallel to each other – the rays do not spread significantly when considered over the short length of the telescope, relative to the long distance they have travelled. So the objective lens creates an image that is located at its focal plane. In **normal adjustment** of a telescope, the eyepiece is positioned so that its focal plane is in the same place (Figure 24.18). The image produced by the objective lens acts as the object for the eyepiece. Since this object is at the focal plane of the eyepiece, the rays emerging from the telescope are again parallel. But the angle between the rays and the axis of the object–telescope–eye system has been increased.

Figure 24.18
An astronomical refracting telescope in normal adjustment.

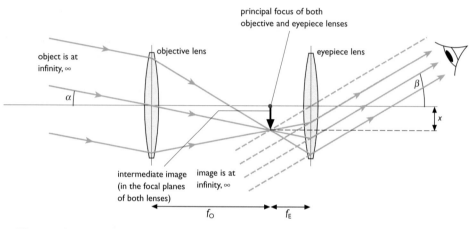

The angular magnification $M = \beta/\alpha$, where β and α are the angles shown in Figure 24.18.

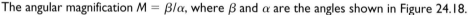

13 **a** Calculate the angular magnification of an astronomical refracting telescope in normal adjustment when the angle, α, subtended at the objective is 0.05° and the angle, β, subtended at the observer's eye is 0.90°.
b What are the values of tan β, tan α, and their ratio? How does this compare with the ratio of the angles measured in degrees? Is the approximation that these ratios are equal a reasonable one? Would we be justified in saying that $M = f_O/f_E$?
c If the focal length of the objective is 500 mm, give a figure for the focal length of the eyepiece.
14 Which lens of an astronomical telescope, the objective or the eyepiece, should
a have the longer focal length
b be fatter (that is, have greater thickness relative to diameter)?
15 The planet Jupiter makes an angle of x radians at the unaided eye.
a What angle does it make when viewed through a refracting telescope in normal adjustment made of lenses of focal lengths 100 cm and 20 cm?
b How long is the telescope?
16 Sketch a ray diagram to show rays passing through a telescope when it is used 'the wrong way round', with the eyepiece pointing towards the object.
17 Why can poor quality lenses make adequate spectacles but useless telescopes?
18 Suppose that an astronomical refracting telescope is used to view a ship that sails closer. What happens to the image produced by the objective lens? What, in consequence, happens to the image produced by the eyepiece?

Note that

$$\tan \beta = \frac{\text{opp}}{\text{adj}} = \frac{x}{f_E}$$

where x is the height of the image produced by the objective lens and f_E is the focal length of the eyepiece, and

$$\tan \alpha = \frac{\text{opp}}{\text{adj}} = \frac{x}{f_O}$$

where f_O is the focal length of the objective. So

$$\frac{\tan \beta}{\tan \alpha} = \frac{x/f_E}{x/f_O} = \frac{f_O}{f_E}$$

For small angles, $\beta/\alpha \approx \tan \beta/\tan \alpha$, so

$$\frac{\beta}{\alpha} \approx \frac{f_O}{f_E}$$

and therefore

$$M \approx \frac{f_O}{f_E}$$

So for large magnification the eyepiece should be a lens of much shorter focal length than that of the objective.

Resolving

Magnification is all very well, but magnification without a clear image is of little use. A test of an optical instrument is its ability to produce images of, or **resolve**, two closely spaced points on an object. If these images overlap then blurring occurs.

Figure 24.19
The appearance of the image of a point source, with diffraction effects.

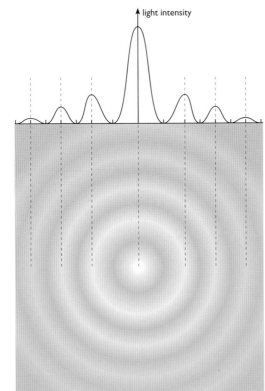

The resolving ability, or **resolution**, of an instrument will be affected by any mis-shapes or mis-adjustment in the lenses (or mirrors) of the optical system. But even with perfect optics, there is a limit to the resolution of the instrument. This is due to diffraction of light (see Chapter 2). Any instrument has a finite **aperture** – the opening through which the light must enter. This is usually circular. For an astronomical refracting telescope the objective lens is effectively the aperture. Diffraction takes place as it would at any gap through which waves pass. Each point on the object produces a small diffraction pattern at the image (Figure 24.19).

Neighbouring points on the object produce overlapping diffraction patterns. The **Rayleigh criterion** for deciding whether two points on an object are resolved is shown in Figure 24.20.

Figure 24.20
The Rayleigh criterion.

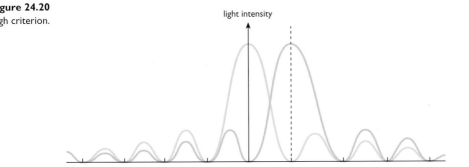

Two sources of light are said, by the Rayleigh criterion, to be just resolved when the central maximum of the diffraction pattern due to one of them coincides with the first minimum of the diffraction pattern due to the other. For a circular aperture this occurs when

$$\theta = \frac{1.22\lambda}{d}$$

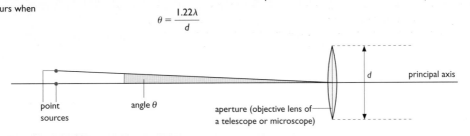

19 Why is it advantageous that the objective lens of a telescope should have a large diameter?

For a circular objective lens or aperture, if θ is the angular separation of two point sources of light of wavelength λ that are being viewed, and d is the diameter of the objective lens, then merging of images due to diffraction and interference at the aperture takes place if

$$\theta < \frac{1.22\lambda}{d}$$

If we write:

$$\theta = \frac{1.22\lambda}{d}$$

then θ is the minimum angle separating two points on an object for them to be resolved by the instrument. Note that a *small* value of θ is generally desirable.

Look at Figure 24.21. This shows a microscope image of part of an insect's leg. For objects much smaller than this, these processes of diffraction and interference make it impossible to produce any kind of image at all, no matter how good the microscope is.

Figure 24.21
This is a magnified image of an insect's leg. Two bristles very close together may have unresolved images.

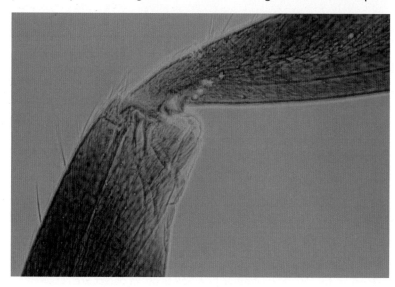

Comprehension and application

The eye as an optical system

The human eye includes a refracting system which we might call a lens system, except that in the case of the eye the word lens means the small lens-shaped disc that is just one component of the system. The absolute refractive index of the middle region of the lens is 1.40, and this decreases to 1.38 in the outer layers (Figure 24.22).

Figure 24.22
The refracting system of the eye.

The ray's path is simplified. The strongest refraction takes place as shown at the surface of the cornea, but refraction also takes place at other boundaries.

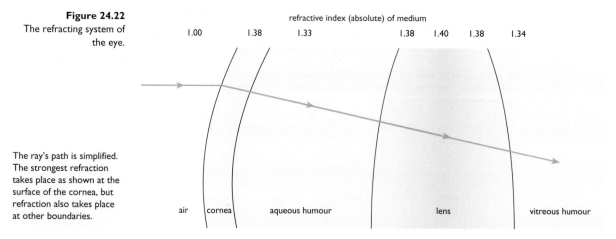

refractive index (absolute) of medium

| 1.00 | 1.38 | 1.33 | 1.38 | 1.40 | 1.38 | 1.34 |

air   cornea   aqueous humour   lens   vitreous humour

The human eye has a fixed image distance – from the refracting system to the retina. It is on the retina that the system produces a real image. The image must be a diminished image – smaller than the objects in the world outside; our eyes have to fit the whole world on to those small surfaces. It is also an inverted image – the image on the retina is upside-down compared with the object – a problem that our brains have to solve, turning the world the 'right way up' again.

The boundary between the air and the cornea has the largest change in absolute refractive index, and so most of the refraction achieved by the system happens at that surface. However, sufficient refraction takes place at the surfaces of the 'lens' for its shape to be important. The eye has to produce sharp images for a wide range of object distances, even though the image distance cannot change. By changing the shape of the lens, the eye varies its focal length. The ability to do this is called **accommodation**.

As optical systems the eye and the camera have a lot in common. They both produce images on screens – the retina and the film. The images are not only real but inverted and diminished. In both cases, the refracting system is acting as a converging lens. Accommodation, however, is achieved in two different ways. While the eye varies the focal length of the refracting system, the camera varies the image distance – the camera lens moves backwards and forwards. The formula relating focal length, f, object distance, u, and image distance, v, is:

$$\frac{1}{f} = \frac{1}{u} + \frac{1}{v}$$

The front-to-back diameter of the eyeball, which we can take as image distance, is about 3 cm.

The image on the retina is only sharp when light from a small area of an object falls on a single small area of the retina. Overlap of light from two nearby points on the object produces blurring. Such overlap occurs when the image is formed not at the retina but just in front of it – the image falls short of the retina. Light from a small area of the object then lands on a large area of the retina. This is **short sight**, and it happens because the refracting system is bending the light too much. It happens for distant objects and not for near ones. For near objects, a high degree of refraction is a good thing. A short-sighted person can see near objects, but cannot make clear images of far objects. The answer is to place a diverging lens in front of the eye (Figure 24.23).

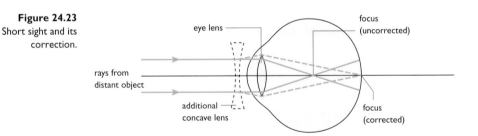

Figure 24.23
Short sight and its correction.

The eyes of a person with **long sight** cannot achieve enough refraction to create an image of a near object. The rays from the object tend to focus a long way from the refracting system, behind the retina. Again, light from a small area of object lands on a large area of retina, to produce blurring. The long-sighted eye needs help with its converging action. A converging lens in front of the eye provides this help (Figure 24.24).

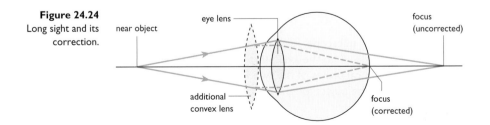

Figure 24.24
Long sight and its correction.

20 For light entering the eye, where does most of the refraction take place?

21 The human 'lens' has a graduated refractive index, with maximum absolute refractive index at the centre. Sketch a ray of light that travels from the aqueous humour, striking the 'lens' surface at a large angle of incidence, and out into the vitreous humour.

22 Laser treatment of the cornea can change its shape and correct defects of vision. What change in shape will compensate for:
a short sight
b long sight?

23 The eye has a fixed image distance of about 3 cm.
a What focal length does it need to produce an image of an object
i at infinity
ii at a distance of 20 cm?
b How does the eye vary its focal length?
c What are the differences in the accommodating methods of the eye and the camera?

24 Stan is 17 and his eyes each have a minimum focal length of 2.45 cm.
a What is the nearest object distance for which he can see clear images (assuming a constant image distance of 3 cm)?

b 40 years later his eyes have minimum focal length of 2.82 cm. How does Stan now read his newspaper? Provide detail of this.
c What components of Stan's eyes have deteriorated?

25 a If you have normal eyesight and gaze with one eye at a very distant object, what will happen if, without adjusting your eye, you place
i a converging lens just in front of the eye
ii a diverging lens just in front of the eye?
Draw ray diagrams to illustrate these situations.
b Will it be possible to adjust your eye to create a clear image in either case?

26 Human eyes are hardly affected by spherical aberration. This is achieved through the shape of the cornea. Draw a ray diagram to show how the shape of the cornea needs to differ from a spherical shape.

27 a What is the maximum possible resolving power, in terms of minimum angle subtended at your eye by two points of an object, when your pupil has a diameter of 3 mm and the wavelength of the light is 5×10^{-7} m?
b What happens to this resolving power when your environment becomes rather darker?
c Why might the actual resolving power of your eye be worse than that calculated in **a**?

Comprehension and application

Better telescopes and the compound microscope

Galileo has the credit for making an improvement to the astronomical refracting telescope, by replacing the converging eyepiece with a diverging one. This adaptation is called a **Galilean telescope** (Figure 24.25).

The advantage is that, for the same total telescope length, a rather larger magnification is achieved. It also produces an upright image, which is important for looking at terrestrial objects such as distant ships. So it was a useful development for commercial and military purposes, especially in Venice where Galileo lived, which was then a very powerful city state whose ships dominated much of the Mediterranean Sea.

Figure 24.25
The Galilean telescope.

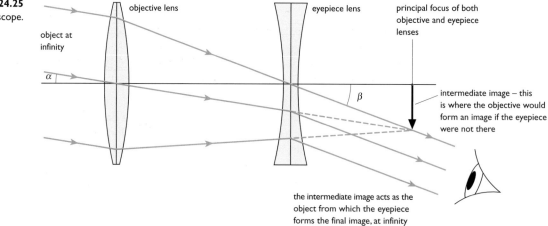

As in the astronomical telescope, rays entering and leaving the telescope are parallel, but they leave at an increased angle, β, to the axis. Angular magnification, as before, is β/α. And again, in normal adjustment the focal planes of the two lenses are in the same place and the image created by the objective acts as the object for the eyepiece.

$$M = \frac{\beta}{\alpha} \approx \frac{x/f_E}{x/f_O} = \frac{f_O}{f_E}$$ where x is the height of the intermediate image.

Since for the eyepiece the intermediate image acts as a virtual object, its object distance, like its focal length, is negative. Also, object distance and focal length are equal, and when the numbers are used in the formula $1/f_E = 1/u + 1/v$ we find that $1/v$ is zero, and image distance v is infinite. This is what we would expect, since the rays are parallel, as they would be for any body an infinite distance away.

Figure 24.26
A reflecting telescope.

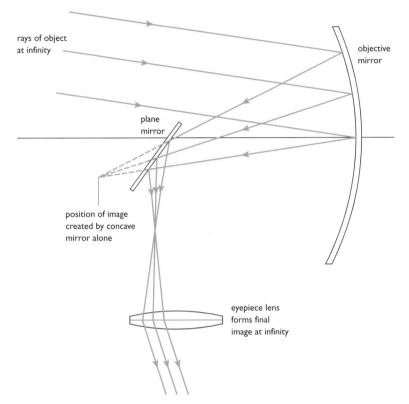

A **reflecting telescope** (Figure 24.26) replaces the objective lens with a concave mirror. The mirror can be made very large – much larger than is possible for a lens – and can therefore collect more light from faint sources. We say that the mirror has a large aperture, which also improves its resolving power. Reflecting telescopes can be made to be quite short, but another important advantage is that they are less susceptible to aberrations. Chromatic aberration does not take place at all at a surface that reflects rather than refracts. Spherical aberration can also be minimised by making the mirror parabolic rather than spherical. This is expensive, and not so useful for small amateur telescopes, but very useful for telescopes for astronomical research.

A simple converging lens is sometimes called a **simple microscope**. It can produce virtual, magnified images of near objects. The fact that the image is virtual means that the eye can be moved backwards and forwards and the image remains 'in focus'. A **compound microscope** has two lenses, and provides an image that is again virtual but with increased magnification. Like an astronomical refracting telescope, the compound microscope uses two converging lenses, but their relative focal lengths are reversed. The objective has a shorter focal length. Further differences are that the focal planes of the lenses are not in the same place, and the intermediate image does not lie in the focal plane of either lens (Figure 24.27).

Figure 24.27
A compound microscope.

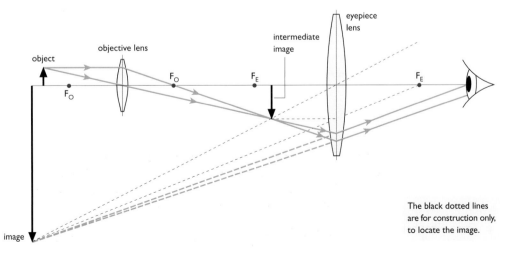

The black dotted lines are for construction only, to locate the image.

28 What is the main difference in the construction of an astronomical refracting telescope and a Galilean telescope?

29 Consider a pair of such telescopes which provide the same angular magnification of 8, and with identical objectives of focal length 100 cm. What must be the value of the focal length of the eyepiece in each case? How would you be able to say, even before picking up the telescopes, which was which?

30 A ship sails closer and closer to a Galilean telescope. What happens to the image?

31 **a** Explain why a reflecting telescope is better at resolving than an astronomical refracting telescope.
   b How do reflecting telescopes solve (at least partly) the problem of chromatic and spherical aberrations?

32 **a** Why is it possible to move a magnifying glass backwards and forwards and still see a clear image?
   b To increase magnification, should the magnifying glass be moved towards or away from the object? (Sketch a ray diagram to work this out.)
   c What is the maximum object distance relative to focal length?

33 To what extent is it reasonable to say that a compound microscope is an astronomical refracting telescope used the wrong way round?

34 If a compound microscope is made with an objective of focal length 0.50 cm and an eyepiece of focal length 6.25 cm, separated by a distance of 9.60 cm, what is the linear magnification for an object 0.56 cm from the objective?

35 **a** Have you benefited from use of any artificial lenses today? Where and how?
   b If artificial lenses did not exist, how might your day have been different? Think about how optical instruments have changed our general outlook on the world. Don't forget to consider the camera (for still and moving-picture making).

Examination questions

I An optical fibre carries coded information in the form of pulses of light. The optical fibre consists of a thread of glass whose core is surrounded by glass of a different refractive index called the cladding.

This question is about what happens as the light passes through a lens and enters the fibre.

a What is meant by the *focal length of a lens*? (1)

The light source is usually a diode emitting infrared radiation.

The diode, which acts as a point source, is placed 5 cm from a converging lens of focal length 5 cm.

b Draw a ray diagram to show what happens to the light rays as they pass through the lens and into the fibre. (2)

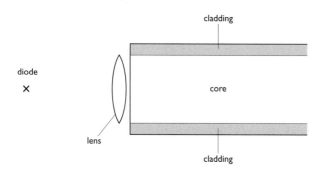

c State what happens to the speed, frequency and wavelength of the light as it enters the fibre. (3)

d As the fibre bends, the light will continue to travel in a straight line until it reaches the side of the core where it should be totally internally reflected at the boundary with the cladding. Explain what is meant by the term *critical angle* for a boundary between two media. (2)

e The core has a refractive index of 1.48. Suggest a suitable value for the refractive index of the cladding. (1)

f Explain why the light may not be totally internally reflected if a section of the fibre is bent sharply as shown below. (1)

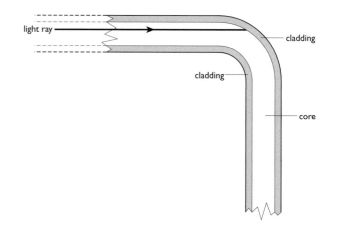

Edexcel (London), Physics, PSA1, June 1999

2 **a** A telescope is made from two converging lenses of focal lengths 2.50 m and 0.020 m.
i Show, with the aid of a labelled diagram, how the lenses would be placed for normal adjustment. Show, on the diagram, the principal focus of each lens.
ii The telescope is used to observe a planet which subtends an angle of 5.0×10^{-5} rad at the objective. Calculate the angle subtended at the eye by the final image. (4)
b State what is meant by chromatic aberration and explain the effect it would have on the image in an uncorrected refracting telescope. (3)

AQA (NEAB), Physics Advanced, Paper 1, June 1999 (part)

3 A photograph of a binary star system is viewed through a converging lens used as a magnifying glass.
a In the diagram, AB represents the photograph. Mark on the diagram the approximate positions of the principal foci and draw rays to show how a virtual magnified image is produced by the lens.

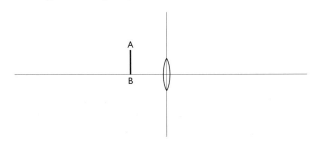

b If the focal length of the lens is 100 mm and the linear magnification is ×5, calculate the distance of the photograph from the lens. (4)

AQA (NEAB), Astronomy and Optics, June 1999

4 **a** A small object is placed on the principal axis of, and 150 mm away from, a diverging lens of focal length 100 mm.
i On the diagram below draw rays to show how an image is formed by the lens.

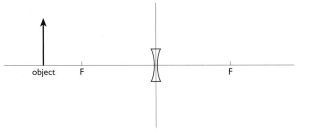

ii Calculate the distance of the image from the lens. (4)
b The diverging lens in part **a** is replaced by a converging lens, also of focal length 100 mm. The object remains in the same position and an image is formed by the converging lens. Compare *two* properties of this image with those of the image formed by the diverging lens in part **a**. (2)

AQA (NEAB), Astronomy and Optics, Mar 1999

5 A girl can only see objects clearly that are further away than 100 cm. Closer objects appear blurred.
a What is the name for her defect of vision? (1)
b Explain why, for her, close objects appear blurred. Illustrate your answer with a diagram of the eye. (3)
c The girl needs spectacle lenses so that she can view objects clearly that are 25 cm away, i.e. at the normal near point of vision.
i Does she need converging or diverging lenses? (1)
ii Draw a diagram to explain how an appropriate lens can rectify the girl's problem, by providing an image for her to view, at a distance of 100 cm from her eye. Your diagram should include the image formed 100 cm from her eye. (3)
iii Calculate the focal length of the lens that is needed to produce the image at 100 cm from her eye, when the object is 25 cm from her eye. (3)

International Baccalaureate, Physics, Higher Level, May 1998

The wave-like behaviour of light

● What behaviour of light can we interpret, predict and apply by use of the wave model?

amplitude   coherent source   constructive interference   destructive interference   diffraction grating   fringes   fringe separation   incoherent source   in phase   integer   monochromatic   optical activity   optical stress analysis   path difference   periodicity   phase angle   phase difference   plane polarisation   principle of superposition   sine curve   transverse plane   transverse waves   wavelets

There's more to light than meets the eye. Behaviour of light that depends on its wavelength – such as refraction, diffraction and polarisation – produces colour from white light, often with stunning effects as in the picture of vitamin crystals (Figure 25.1). Such pictures are not only good to look at but very informative. We use diffraction, for example, to spread light into its spectrum and so to look for emission and absorption lines. Study of starlight, and so our knowledge of stars, would be very limited without such techniques. Wavelength-dependent polarisation effects help us to analyse patterns of stress in materials, and to examine crystals like these.

Figure 25.1
These colourful patterns are a result of the wave behaviour of light. Vitamin crystals have rotated the plane of polarisation of the light waves, and the degree of rotation depends on the wavelength (colour) of the light.

Polarisation

A lens of a pair of Polaroid sunglasses clearly absorbs some radiation. When two such lenses are placed one behind the other, orientated as in the sunglasses, the total absorption is surprisingly small. But if one of the lenses is then rotated, absorption rapidly increases and eventually becomes complete. This is a polarisation effect. To analyse what is going on we can use James Clerk Maxwell's description of electromagnetic radiation.

When ripples travel across water the vibration of the surface is essentially vertical. (Water below the surface moves in a more complex way, but this is not of direct importance in this chapter.) The surface vibration is perpendicular to the direction of travel, and such waves are **transverse waves**. Maxwell suggested that electromagnetic radiation involves paired vibrations of electric and magnetic fields and that these paired vibrations are not only perpendicular to the direction of travel but perpendicular to each other. Each vertical vibration of the electric field is matched by a horizontal vibration of the magnetic field, for example (Figure 25.2).

Figure 25.2
A simplified representation of transverse vibrations that make up an electromagnetic wave.

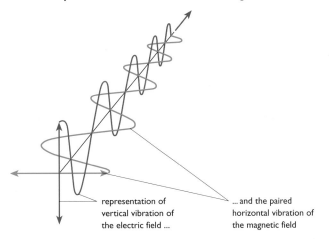

representation of vertical vibration of the electric field ...

... and the paired horizontal vibration of the magnetic field

But the complete picture is more complicated. Any particular sample of unpolarised light, like the light from the Sun, contains vibrations of the electric field that are not just in one direction, say up and down as is approximately the case for water ripples, but in any direction in a plane that is perpendicular to the direction of travel – the **transverse plane** (Figure 25.3). And for each vibration of the electric field, whether it is vertical vibration, horizontal vibration or any possibility in between, there is a paired vibration of the magnetic field.

Figure 25.3
Vibration of the electric field can take place in all directions in the transverse plane.

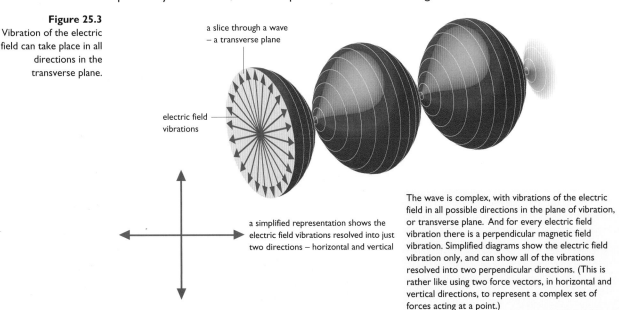

a slice through a wave – a transverse plane

electric field vibrations

a simplified representation shows the electric field vibrations resolved into just two directions – horizontal and vertical

The wave is complex, with vibrations of the electric field in all possible directions in the plane of vibration, or transverse plane. And for every electric field vibration there is a perpendicular magnetic field vibration. Simplified diagrams show the electric field vibration only, and can show all of the vibrations resolved into two perpendicular directions. (This is rather like using two force vectors, in horizontal and vertical directions, to represent a complex set of forces acting at a point.)

To be able to think about the waves we need to have a way of presenting the ideas as simply as possible. A first simplification is to leave the magnetic field vibrations out of the diagram. We know that they are there, always accompanying the electric field vibrations, so it is unnecessary to show them every time. Secondly, we can represent all of the electric field vibrations as two perpendicular components of vibration. As we would normally do when drawing components of force, we draw these components in horizontal and vertical directions (see Figure 25.3).

Now we are ready to see how this fits with the observations of the rotating Polaroid lenses. Maxwell's theory provides the explanation that Polaroid lenses do not absorb randomly, but absorb one component of the transverse vibration, and transmit the other. A second lens placed behind another in the same orientation will transmit the same component of the radiation. It can do little or no further absorption, because the radiation it can absorb has already been removed. But if this lens is turned then it starts to be able to absorb light from the other component of vibration. When it has been turned through 90° it can absorb all of this light.

The action of removing one component of vibration is called **plane polarisation** (Figure 25.4). In plane polarised light, vibration of the electric field effectively takes place in a single plane that runs along the direction of travel of the wave. (The wave has become more like a wave on a rope for which the vibrations are in a single plane.)

Figure 25.4
A Polaroid lens absorbs one component of the electric field vibration (and, at the same time, the corresponding component of the magnetic field vibration). The light that it transmits is plane polarised.

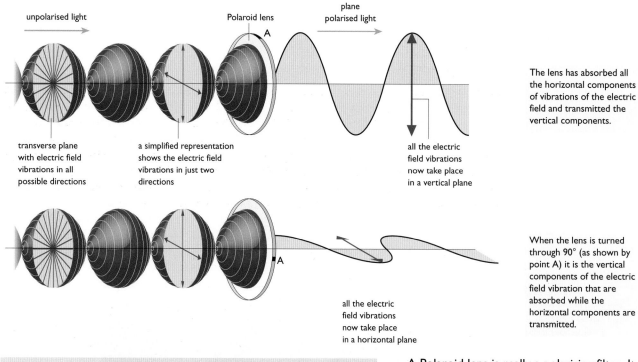

unpolarised light

Polaroid lens

plane polarised light

A

transverse plane with electric field vibrations in all possible directions

a simplified representation shows the electric field vibrations in just two directions

all the electric field vibrations now take place in a vertical plane

The lens has absorbed all the horizontal components of vibrations of the electric field and transmitted the vertical components.

A

all the electric field vibrations now take place in a horizontal plane

When the lens is turned through 90° (as shown by point A) it is the vertical components of the electric field vibration that are absorbed while the horizontal components are transmitted.

1 In what way is plane polarised light more like a water ripple than is unpolarised light?
2 A beam of light travelling in a horizontal direction is plane polarised such that its magnetic field vibration lies in a plane that is at 40° to the vertical. Describe the orientation of the plane of the electric field vibration. Draw a sketch to show this.
3 A single polarising filter reduces the intensity of the light falling on it by 50%. If a second polarising filter is placed in this weakened beam to give a further 50% reduction in intensity of transmitted light, how many possible relative orientations are there in which a third filter can be placed in order to give 100% reduction?
   (You will almost certainly need to sketch a diagram to help you to think about this. Drawing sketches that help you to think and explain is a good habit!)
4 When swimming under water, how might Polaroid lenses change what you see of the world?

A Polaroid lens is really a polarising filter. It is not the only means by which light can become polarised. Light reflected from surfaces is at least partially polarised. This means that a Polaroid filter in the right orientation can be very effective at absorbing reflected light. Polaroid filters greatly improve a view from above of an underwater world, which would otherwise be masked by light reflected from the water surface.

In partially polarised light the amplitude of one of the components of vibration of the electric field is reduced, but not to zero (Figure 25.5). Surfaces of glass and water generally cause partial polarisation.

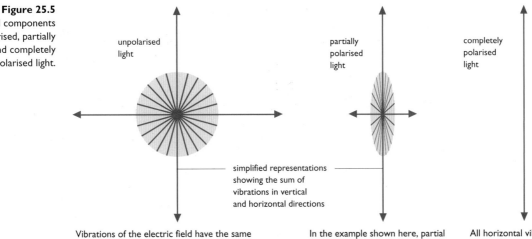

Figure 25.5
Electric field components for unpolarised, partially polarised and completely plane polarised light.

unpolarised light

partially polarised light

completely polarised light

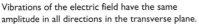

simplified representations showing the sum of vibrations in vertical and horizontal directions

Vibrations of the electric field have the same amplitude in all directions in the transverse plane.

In the example shown here, partial absorption of the horizontal components of electric field vibration has taken place. Vertical components have not changed.

All horizontal vibrations of the electric field have been absorbed, leaving only the sum of the vertical components.

Optical activity

Many materials, especially organic compounds, affect the direction of vibration of the travelling field of electromagnetic radiation. In terms of the electric field vibration, we can say that the material rotates the direction of vibration (Figure 25.6). This phenomenon is called **optical activity**. With unpolarised light this effect is not detectable – just as the rotation of a perfectly smooth cylinder, such as a metal rod, is not directly observable. But if the light is polarised, then a change in the direction of the polarisation is detectable. The further the light travels through the material, the further the vibration is rotated. Rotation may be wavelength dependent, and the emerging light may then be separated into colours (see Figure 25.1, page 240).

Figure 25.6
An optically active material rotates the plane of polarisation.

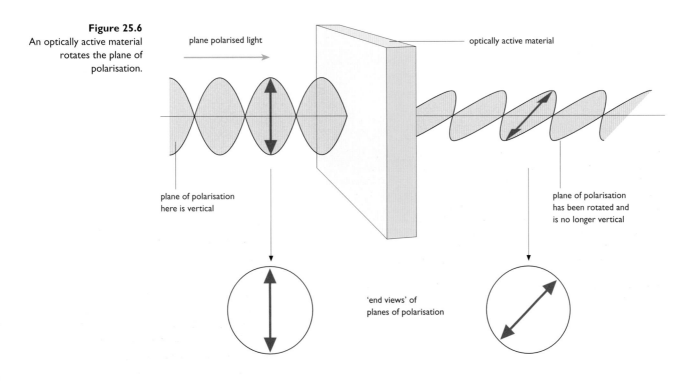

plane polarised light

optically active material

plane of polarisation here is vertical

plane of polarisation has been rotated and is no longer vertical

'end views' of planes of polarisation

The optical activity of plastic and glass is dependent on stress. Where the stress is greatest, the rotation of the vibration is greatest. Dark and light bands are produced when polarised light is shone through the material and viewed through a polarising filter (Polaroid). For this reason, engineers build plastic models of structures to help them to see areas of greater and less stress. The method is called **optical stress analysis** (Figure 25.7).

Figure 25.7
Optical stress analysis of an engineering structure model.

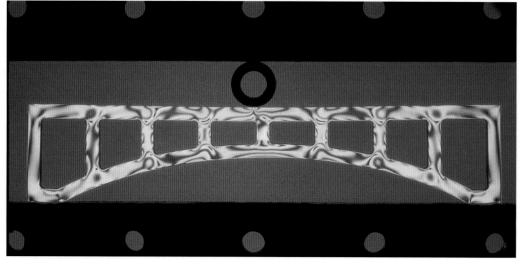

5 a How can you tell that stress changes the optical activity of a material such as perspex?
b Why are there coloured patterns in an optical stress analysis image?

Polarised waves and radio aerials

In a transmitting radio aerial, an alternating potential difference of a particular frequency causes electrons to oscillate with the same frequency. The energy that is supplied to the electrons is then transferred to the space around the aerial as an electromagnetic wave, which travels outwards at the speed of light. If the aerial is a vertical rod (or rods), then the oscillation of the electrons is vertical, and the radio wave is vertically polarised. That is, the oscillation of the electric (E) field is vertical, and the oscillation of the magnetic (B) field is horizontal (Figure 25.8).

Figure 25.8
In a vertical transmitting aerial, vertical electron oscillation results in a vertically plane polarised radio wave. Maximum electron oscillation in the receiving aerial is achieved when the aerial is aligned to be parallel to the wave's oscillating electric field.

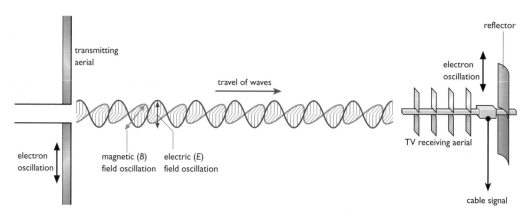

In the receiving aerial, the arriving radio waves induce electron oscillation. The intensity of the arriving radio waves might be small, and the purpose of the aerial is to produce the largest possible electron response. Electron oscillation is maximum if the alignment of the receiving aerial is parallel to the oscillation of the wave's electric field. So the receiving aerial should be fixed vertically. (Radio waves arrive at many frequencies, and so the electron oscillation is very complex. The design of the aerial can have some influence on its sensitivity to a particular range of wavelengths. However, a tuning circuit is needed to 'filter' out the required frequency band, which can be amplified as in a radio or TV.)

Graphic and mathematical snapshots of waves

Figure 25.9
The same wave at two times – time t and time $t + \delta t$.

A vertical slice through a gently disturbed water surface shows a familiar wave pattern. This can be plotted on a graph with the usual x- and y-axes. The x-axis then represents distance along the direction of travel of the waves across the water surface, and the y-axis represents the vertical displacement of the water surface. This graph is a 'snapshot' of the wave – it shows its shape at a particular instant in time, t. After a little more time, δt, the curve has moved to the right (Figure 25.9).

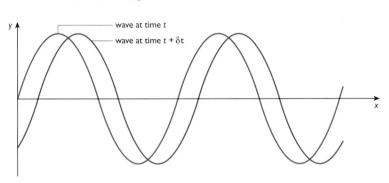

The curve is called a **sine curve**, after the mathematical relationship between x and y that generates such a shape. For a wave for which $y = 0$ and is increasing when $x = 0$, as in the blue curve in Figure 25.9, we can write this relationship as:

$$y = a \sin \frac{2\pi x}{\lambda} \text{ (with angle measured in radians)}$$

Only an angle can have a sine (or a cosine or a tangent). We use angles and the mathematics of circles when dealing with waves because, like circular motion, a wave is a phenomenon that repeats itself exactly over and over again. The mathematics of circles is the mathematics of identical repetition, or **periodicity**. Note that for a complete cycle of the wave x is the wavelength, λ, and the angle in the above formula is then 2π radians (or $360°$) – the angle turned through during one complete revolution, or one complete cycle of the wave motion. (You can read about circular motion and oscillations in *Further Advanced Physics*.)

Radians are much more useful (and easier, once you are familiar with them) than degrees for measuring angles when dealing with circular or other repetitive motion. Table 25.1 converts degrees to radians (see also page 231).

Table 25.1

Angle in degrees	0	57.3	90	180	270	360	450	540	630	720
Angle in radians	0	1	$\pi/2$	π	$3\pi/2$	2π	$5\pi/2$	3π	$7\pi/2$	4π
Number of revolutions	0	$1/(2\pi)$	1/4	2/4 = 1/2	3/4	4/4 = 1	5/4	6/4 = 3/2	7/4	8/4 = 2
	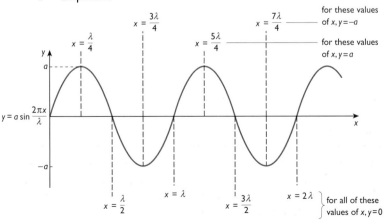									
Sine of the angle	0	0.84	1.00	0	−1.00	0	1.00	0	−1.00	0

The value of the sine of an angle varies between -1 and 1. This means that where $y = a \sin (2\pi x/\lambda)$, the value of y varies between $-a$ and a (Figure 25.10). For wave motion, this maximum value of displacement is the **amplitude** of the wave:

$a = $ amplitude

Figure 25.10
A snapshot of a wave. We can apply ideas of the radian and the sine function to this graph that shows the wave, and to the repeating patterns of displacement, y, at points, x, along the wave.

$y = a \sin \dfrac{2\pi x}{\lambda}$

for these values of x, $y = -a$

for these values of x, $y = a$

$x = \dfrac{3\lambda}{4}$ $x = \dfrac{7\lambda}{4}$ $x = \dfrac{\lambda}{4}$ $x = \dfrac{5\lambda}{4}$

$x = \dfrac{\lambda}{2}$ $x = \lambda$ $x = \dfrac{3\lambda}{2}$ $x = 2\lambda$ for all of these values of x, $y = 0$

6 Sketch displacement–distance, y versus x, graphs for a wave at time t and at time $(t - \delta t)$, assuming that the wave is travelling to the right.

7 For the function $y = \sin(2\pi x/\lambda)$, make a table of values of y when x is equal to:
$\lambda/4, \lambda/2, 3\lambda/4, \lambda, 5\lambda/4, 3\lambda/2, 2\lambda$.

8 Express 5 revolutions (5 complete circles) in
a radians
b degrees.

9 A piece of string the length of a diameter of a cylinder is wrapped around the circumference of the cylinder. How big is the matching angle at the centre of the circle?

10 Give the following in
a degrees
b numbers of revolutions.
$\pi/2$ radians, 2π radians, 4π radians, 5π radians, 6π radians, 1 radian, 2 radians

11 Give the following in
a radians
b numbers of revolutions.
60°, 180°, 270°, 720°

12 Two waves are described by:
$y_1 = a \sin(2\pi x/\lambda)$ and $y_2 = a \sin(2\pi z/\lambda)$,
where $z = 2x$.
Sketch snapshots of the two waves on the same axes.

Displacement–time graphs for wave motion

A speck of dust resting on a water surface bobs up and down repeatedly as waves pass by. Displacement is vertical, and we can represent it as y. It is also periodic, repeating itself at regular intervals and again we use sine functions to describe it:

$$y = a \sin \frac{2\pi t}{T}$$

We are not interested in what is happening elsewhere on the water surface, and so x does not appear here. We are interested in the variation of y with time, t, at one particular place. Not only has t replaced x, but T, the period or time for one complete cycle, has replaced λ, the wavelength or distance occupied by one complete wave.

Suppose that a second speck of dust rests on the water, quite nearby, also in the path of the waves. It also bobs up and down, but there is a time lag between the motions of the two specks. We can still write a formula for the displacement of this second speck at the same time, t, as the first:

$$y' = a \sin\left(\frac{2\pi t}{T} + \varphi\right)$$

φ is an extra angle added, called a **phase angle**, so that y and y' are identical except that, whatever the value of y, y' has the same value a short time earlier (Figure 25.11). We say that there is a **phase difference** in the motions of the specks of dust.

Where φ is zero then the two oscillations reach maximum, zero and minimum amplitude at the same time. They are **in phase**.

13 a How can you tell a 'snapshot' displacement–distance graph for a wave from a corresponding displacement–time graph?
b Write a concise description of the different information the two kinds of graph provide.

14 Sketch a displacement–time graph for:
$$y = a \sin\left(\frac{2\pi t}{T} + \frac{\pi}{2}\right)$$

Figure 25.11
The motions are the same in amplitude and period, but have a time difference which is quantified in terms of a phase angle, φ.

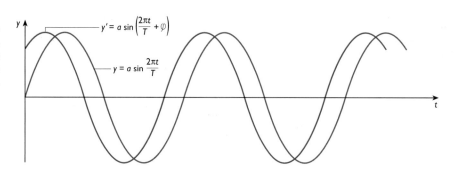

Superposition of waves

Two waves can exist in the same place at the same time. If they are travelling in the same direction, x, then we can superimpose snapshots of the two waves. The simplest examples involve

- two identical waves that are in phase (Figure 25.12) or
- two identical waves that are shifted by half a wavelength relative to one another, so that they are half a cycle or π radians out of phase (Figure 25.13).

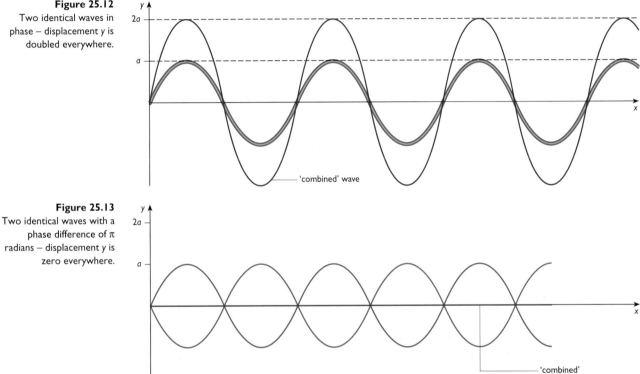

Figure 25.12
Two identical waves in phase – displacement y is doubled everywhere.

Figure 25.13
Two identical waves with a phase difference of π radians – displacement y is zero everywhere.

In both cases above, the total displacement is the sum of the individual displacements. In general (Figure 25.14) we can say that:

$$y_{\text{total}} = y_1 + y_2$$

This statement that the total displacement at any instant is the sum of the individual displacements is the **principle of superposition**. This principle continues to apply, no matter how complex the addition becomes due to differences of wavelength and frequency, amplitude or phase.

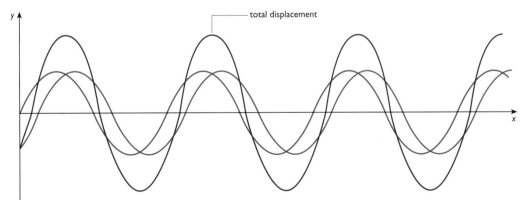

Figure 25.14
For all superimposed waves, total displacement y_{total} at any instant of time is the sum of the individual displacements.

The principle of superposition does not just apply to displacement–distance (y versus x) graphs, which are snapshots showing the variation of y along an x-direction at a single instant in time. It also applies to the motion of a single point in the path of the waves over a period of time. Again, we can represent the motion of a single point as a displacement–time graph, with y plotted against t rather than against x (Figure 25.15).

Figure 25.15
Total displacement y_{total} at any time is the sum of the individual displacements.

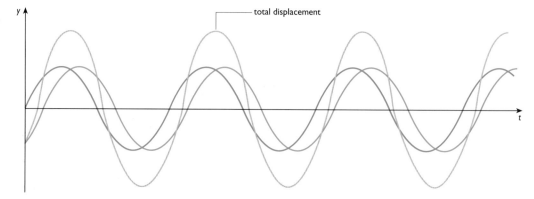

total displacement

15 a If $y_1 = a \sin (2\pi t/T)$ and $y_2 = a \sin \left(\dfrac{2\pi t}{T} + \pi\right)$, describe the result if the waves are superimposed.
b What would be the result if the amplitude of the first wave were increased to $2a$? Use sketch displacement–time graphs to help you answer.

For a point on a wave, whose motion is described by a displacement–time graph as here, when two waves are superimposed then

$$y_{total} = y_1 + y_2$$

That is, the principle of superposition applies to the displacement of a point at which waves are superimposed.

y_{total} can be found by calculation but the addition is complex. An indication of the value of y_{total} at any time t can be found by sketching graphs of y_1 and y_2 against time.

Double-source interference

Waves spread out from point sources, with spherical wavefronts. Unless their speed changes they have constant wavelength and the wavefronts are concentric spheres. In just two dimensions – either in a 2D representation of 3D waves, or in water surface waves – we can think of the waves spreading in circles rather than spheres. These circular waves are a simple form of geometry, and so it is not surprising that when two identical sets of concentric circles are superimposed a clear pattern is created. In some places, superposition results in reinforcement of waves to create increased amplitude of vibration. In other places superposition causes cancellation of vibration and zero amplitude. A pattern of increased and decreased amplitude, or reinforcement and cancellation, is an interference pattern.

Consider two point sources, vibrating in phase and generating identical sets of waves, which travel outwards and interfere (Figure 25.16).

Figure 25.16
Superposition at places of maximum and minimum disturbance in an interference pattern.

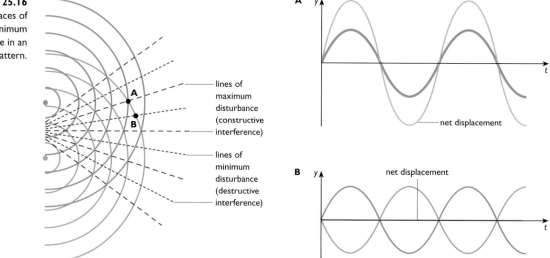

lines of maximum disturbance (constructive interference)

lines of minimum disturbance (destructive interference)

A

net displacement

B

net displacement

At point A in Figure 25.16, reinforcement or **constructive interference** is taking place, so that A is a location of maximum amplitude. This happens because crests of waves from both sources arrive at the same time, and troughs of waves also arrive at the same time. But the two crests arriving at A did not leave their sources at the same time and have not travelled the same distance. There is a **path difference**, l. This is related to the separation, d, of the two sources by:

$$\sin \theta = \frac{\text{opp}}{\text{hyp}} = \frac{l}{d}$$

or

$$l = d \sin \theta$$

where θ is as shown in Figure 25.17.

Figure 25.17
Calculating the path difference.

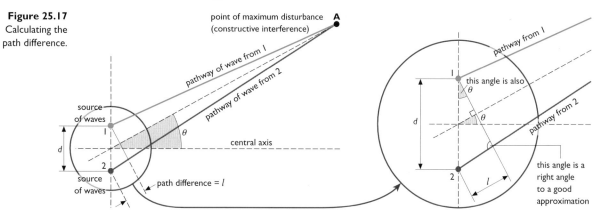

For an interference maximum, it is necessary that l is equal to a whole number of wavelengths, that is $l = n\lambda$ where n is a whole number, or **integer**. So, n can be 1, 2, 3, 4, 5 and so on. There is also an interference maximum when the path difference is zero – then the waves from the two sources have travelled equal distances and must be in phase if they started out in phase. When the path difference is zero, n is zero. In summary, for constructive interference:

$$n\lambda = d \sin \theta \quad \text{where } n = 0, 1, 2, 3, \text{ and so on}$$

At point B in Figure 25.16, a crest from one source arrives at the same time as a trough from the other. Now the path difference l must be half a wavelength, or $1\frac{1}{2}$, or $2\frac{1}{2}$ (and so on) wavelengths. Note that $1\frac{1}{2} = \frac{3}{2}$, $2\frac{1}{2} = \frac{5}{2}$, $3\frac{1}{2} = \frac{7}{2}$, etc. For completely **destructive interference**, that is, producing minimum amplitude, we need $l = (n + \frac{1}{2})\lambda$, so:

$$(n + \tfrac{1}{2})\lambda = d \sin \theta \quad \text{where } n = 0, 1, 2, 3, \text{ and so on}$$

The distance between adjacent maxima

Travelling across the interference pattern in a direction parallel to a line joining the two sources, we pass through alternating maxima and minima. We can calculate the distance between two adjacent maxima, for example. For an interference pattern of light the alternating maxima and minima are called **fringes**, and this distance is called the **fringe separation** or fringe width.

Figure 25.18
Calculating the fringe separation.

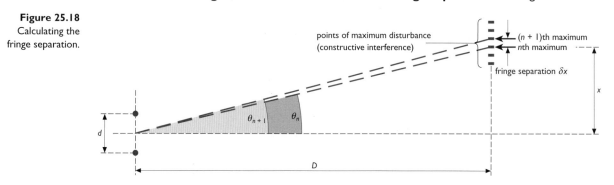

For the nth maximum, from Figure 25.18 (on the previous page),

$$n\lambda = d \sin \theta_n = \frac{dx}{D}$$

and for the $(n + 1)$th maximum,

$$(n + 1)\lambda = d \sin \theta_{n+1} = \frac{d(x + \delta x)}{D}$$

So

$$(n + 1)\lambda - n\lambda = \frac{d(x + \delta x)}{D} - \frac{dx}{D}$$

which simplifies to

$$\lambda = \frac{d\,\delta x}{D}$$

This is the basis for a useful method of measuring the wavelength of waves, as we shall see.

Interference resulting from diffraction at two gaps

A pair of narrow gaps in a barrier can cause a single source of waves to produce an interference pattern (Figure 25.19). Diffraction takes place and each gap acts as a source of circular waves. If the two sets of diffracted waves are in phase, which they will be if the wavefronts arriving at the barrier are parallel to it, then the interference pattern is the same as for two separate sources, and the same mathematics can be applied.

Figure 25.19
Two gaps can give rise to
an interference pattern
similar to that from
two sources.

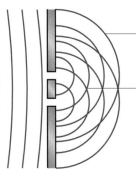

semicircular wavefronts
are created by diffraction

overlap of the two sets of waves produces the
dominant interference pattern – so that the two
gaps behave in a similar way to two sources

Note that interference also takes place between waves that
pass through a gap at different points – see the next section.

Diffraction and interference at a single gap

Figure 25.20
Diffraction at a single gap
can give rise to
interference.

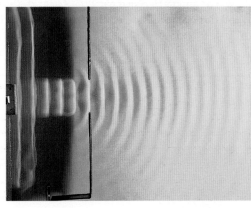

If one of the gaps in Figure 25.19 is blocked it is still possible for an interference pattern to develop (Figure 25.20). Each point on the wave passing through the single gap can be thought of as a new source of waves, and interference takes place between waves spreading from these new 'sources'. The notion of points on a wave acting as secondary sources of **wavelets** from which the new wave is constructed goes back to Christiaan Huygens.

17 Sketch, on the same
axes, displacement–
time (*y* versus *t*)
graphs for the wave
motion:
a at the central
maximum of a single-
gap interference
pattern
b at the first minimum
c at the first maximum.

The simplest way to begin to analyse this situation is to imagine the gap split into two halves. Then for each point in one half of the gap there is another point in the other half such that wavelets from the two points destructively interfere along a line that radiates from the gap. Likewise, the gap has four quarters, and at a particular angle destructive interference takes place between 'wavelets' from adjacent quarters. We can go on subdividing the gap in this way, discovering more and more angles at which destructive interference occurs.

Note also that, however we divide the gap, wavelets that travel along the central axis are always in phase. So we should expect constructive interference of particularly high amplitude along this central line. The diffraction/interference pattern has a strong central maximum (Figure 25.21).

Figure 25.21
The diffraction/interference
pattern shown by a
graph of intensity
against distance. θ, θ' and
θ'' are the angles
corresponding to the 1st,
2nd and 3rd minima,
respectively.

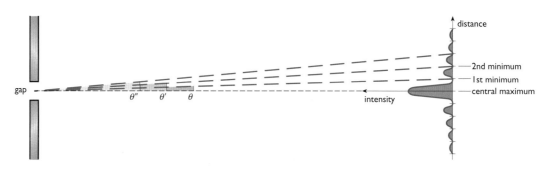

Diffraction and wavelength

Figure 25.22
Diffraction effects are
strongest when wavelength
and the obstacles or gaps
causing diffraction are
similar in size.

People who live in deep valleys often have problems with FM radio reception but not with medium wave reception, though both signals are carried by radio waves. The wavelengths are about 3 m and 300 m respectively. Both types of wave are diffracted by the contours of the ground, but for the 3 m diffracted waves the intensity falls much more rapidly as they spread (Figure 25.22).

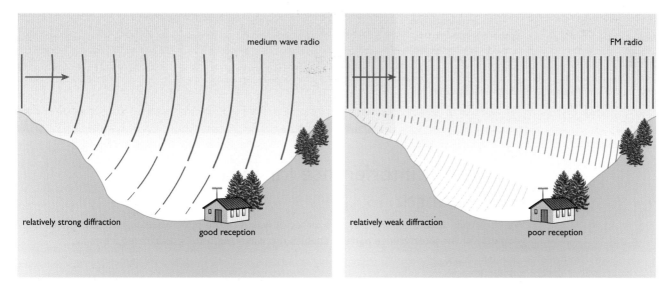

18 The frequency range
of speech is about
100 to 1000 Hz, and
the speed of sound in
air is 330 m s⁻¹.
Explain why speech
'travels around
corners' much better
than light.

Where obstacles are smaller than hills and valleys – such as buildings and trees – the shorter FM waves experience strong diffraction and so reception is good.

We do not see much of diffraction effects in everyday life. But if you look at a yellow sodium street light through condensation on a bus window, or through nylon mesh such as an umbrella (see Figure 6.9, page 51) or even through your eyelashes, you start to see diffracted light and the patterns of interference that are caused by superposition (cancellation and reinforcement) of diffracted waves. Nylon mesh and eyelashes provide quite small obstacles and gaps – bigger than the wavelength of the light but small enough for observable diffraction effects.

Interference of light waves – an elusive phenomenon

So far in discussing superposition we have dealt with waves in general. The easiest kind of wave to think about is the kind that we can see directly – waves on water. However, diffraction and interference gain their chief importance to us from what they tell us about light. The fact that light waves can experience diffraction and interference effects supports our use of wave models for thinking about light. But there are problems with observing these effects in light:

- The wavelength of visible light is small and so diffraction effects only take place where obstacles and gaps are also small. To observe diffraction and interference effects we must use small sources or small apertures. To maximise brightness we don't use small point sources or round apertures but long slits within absorbing barriers.
- White visible light is made up of a range of wavelengths, and diffraction and the resulting interference effects are dependent on wavelength. Maxima and minima of different wavelengths occur at different angles, and so an overall pattern is not clear unless we use a **monochromatic** light source – one emitting a single wavelength (or colour).
- If two adjacent sources of waves have a phase difference they will still produce a fixed interference pattern (slightly shifted 'off-centre' from the central axis, though that would have little real influence on what we could see). But ordinary sources of light emit their radiation with frequent and random changes in phase – they are **incoherent**. Whenever the waves from one or other source change in phase the interference pattern changes. The changes are far too fast to make the interference pattern observable. To observe double-source interference of light waves (Figure 25.23) it is necessary to use light from **coherent** sources for which the random phase changes are identical. A laser provides a pair of coherent sources when used with a double-slit arrangement. A sodium lamp can also do so if a single narrow slit is placed in front of the light source, and light travels from this to the double-slit arrangement.

Figure 25.23
A double-slit interference pattern produced with **monochromatic** and **coherent** light from two closely spaced slits.

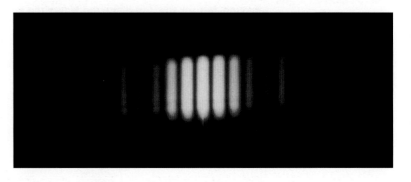

19 Explain what happens to the interference pattern produced by identical sources if the phase of the waves from one of them changes.

Double-slit interference for measurement of wavelength

We've already seen that for a pair of sources of waves of the same wavelength and in phase, the geometry of the interference pattern enables the wavelength to be calculated from:

$$\lambda = \frac{d\,\delta x}{D}$$

This applies to a light interference pattern as well as to any other. The light from the two sources must be monochromatic and coherent. Monochromatic light can be provided by a sodium lamp, which is essentially the same as a yellow street light. (The light is not perfectly monochromatic – it is made up of two distinct but very similar wavelengths of light that appear yellow to us, plus a few other wavelengths at much lower brightness. It is, however, close enough to a monochromatic source to produce a good interference pattern.) A single slit is placed in front of a single sodium lamp and the two pathways necessary for interference are provided by a double-slit arrangement, acting as a pair of coherent sources.

20 For slits whose centres are 0.2 mm apart, illuminated by a sodium lamp of wavelength 5.9×10^{-7} m, what is the fringe separation 0.1 m from the slits?

21 Explain why a 100 μm error in measuring the fringe separation, δx, has a much greater impact on the calculated value of wavelength than a 100 μm error in measuring the distance from the plane of the slits to the plane of the fringes.

The light and dark bands of illumination that can be seen either on a screen or through a microscope are called interference fringes. The distance between adjacent dark bands or adjacent light bands is δx. To measure δx we use a travelling microscope, which has a visible crosswire and can, with a turn of a screw, travel laterally across the interference pattern from fringe to fringe. The movement of the microscope also moves a Vernier scale along a steel rule on the stationary microscope mounting. Differences in initial and final readings of this scale provide measurements of the distance travelled. It is difficult to judge the centre of a bright band or a dark band, and this can produce a large error in the measurement of separation of just one pair of fringes. The answer is to measure the distance occupied by a larger number of fringes – as many as possible – and divide this distance by the number of fringes.

The separation of the centres of the two slits, d, can also be measured using the travelling microscope. Here it is not possible to make multiple measurements, and so there is the potential for significant error. Several repetitions of the measurement, and general care, can reduce error.

Measurement of D, the distance from the plane of the double slits to the plane of the observed pattern, which is the plane or screen on which the microscope is focused, is done by a simple rule. Since this is a larger distance, any likely error is a much smaller proportion of the total distance than is the case when measuring d or δx.

Finally, we can use the formula $\lambda = d\,\delta x/D$ to find the wavelength of the light.

Interference of light waves from a single aperture

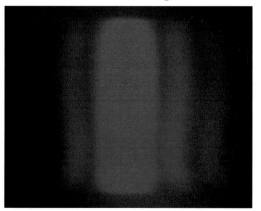

Figure 25.24
A diffraction/interference pattern produced by monochromatic light from a single slit.

The interference pattern (Figure 25.24) produced by diffraction at a single aperture (or illuminated opening in a barrier, which could be a long slit or circular hole) for light is similar in principle to that produced by any other kind of wave. There is a central maximum and a series of weakening fringes as predicted in Figure 25.21.

The resolution provided by an optical instrument, acting as a circular aperture, is determined by interference of diffracted waves which originate at closely spaced points on an object (see Chapter 24).

Interference of light waves from many slits – the diffraction grating

For a pair of slits, locations of maxima of interference obey the rule that

$$n\lambda = d \sin \theta_n = \frac{dx}{D} \qquad \text{(see page 250)}$$

For the first maximum in the pattern, $n = 1$ and

$$\lambda = d \sin \theta_1 = \frac{dx_1}{D}$$

θ_1 and x_1 can be made larger, for the same wavelength of light, by making the slit separation, d, smaller. Slit separation can be made a lot smaller only by making the width of each slit smaller. But then the intensity of light passing through the slits is reduced and the interference pattern is not so bright. The answer is to use a very large number of slits very close together, made by scratching a surface with fine parallel lines. It is possible to make thousands or even tens of thousands of such lines per centimetre on a glass sheet. This is called a **diffraction grating** (Figures 25.25 and 25.26, overleaf). With equally spaced slits, the above condition for maxima holds for all pairs of adjacent slits, and hence for the whole grating.

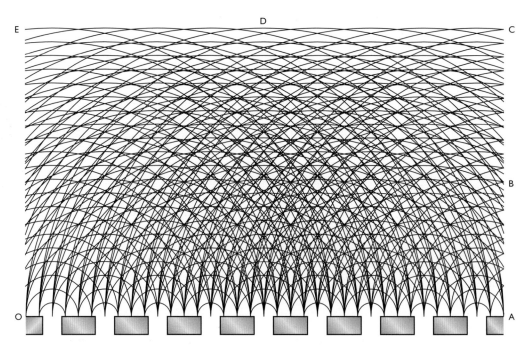

Figure 25.25
Obstacles and gaps comparable in size to the wavelength create diffraction. The diffracted waves overlap, resulting in interference. For an orderly array of obstacles or gaps, this produces high intensity in certain directions only, where the waves combine to produce straight 'wavefronts'. Here you can see such wavefronts by looking obliquely, from close to the page, along directions EC, OB, OC and OD.

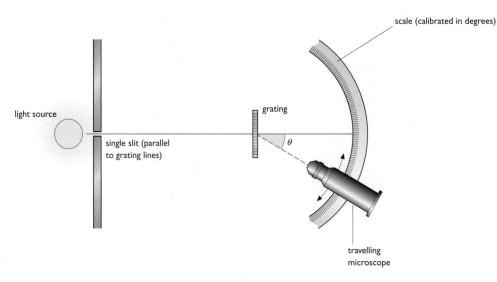

Figure 25.26
Viewing interference maxima from a diffraction grating.

The slit separation may be as small as, say, 10^{-6} m (corresponding to 10 000 lines per centimetre). This means that for the bright yellow light of a sodium lamp, with a wavelength of 5.890×10^{-7} m,

$$\sin \theta_1 = \frac{\lambda}{d} = \frac{5.890 \times 10^{-7}}{10^{-6}} = 0.5890$$

so

$$\theta_1 = 36.09°$$

The *first* maximum for this light thus makes a large angle with the central axis of the source–grating system. If the wavelength were 5.896×10^{-7} m, then θ would become 36.12°. So a tiny change in wavelength produces a change in angle that is big enough to see when viewing a diffraction pattern with a microscope. 5.890×10^{-7} m and 5.896×10^{-7} m are the two very similar wavelengths of the characteristic yellow light of a sodium lamp; the bright maxima for these can be resolved quite well with a diffraction grating (Figure 25.27).

Figure 25.27
A diffraction grating makes the characteristic spectrum of an element visible in the form of lines that are parallel to the rulings on the grating. This shows the line spectrum of sodium, compared with a continuous spectrum.

a Continuous spectrum from a white light source

b Line spectrum of sodium

22 a For a grating spacing of 10^{-6} m, what is the angle between the first maxima produced by the two similar wavelengths of yellow sodium light?
b What is the linear separation, in mm, of the two visible lines at a distance of 0.2 m from the grating?

23 If the two extremes of the visible spectrum have wavelengths 3.9×10^{-7} m and 7.0×10^{-7} m, what would be the total angle occupied by a first maximum of the spectrum of white light created by a grating with 5000 lines per cm?

24 Calculate the angle at which you would expect to find the second maximum of bright yellow sodium light, wavelength 5.890×10^{-7} m, using a grating with 4000 lines per centimetre.

25 Why are the characteristic wavelengths of an element, such as sodium, often called 'lines'?

The sodium in a lamp also emits other wavelengths, though at much lower intensity. For example, one of sodium's characteristic wavelengths is a blue/violet colour at 4.20×10^{-7} m. For this,

$$\sin \theta_1 = \frac{\lambda}{d} = \frac{4.20 \times 10^{-7}}{10^{-6}} = 0.420$$

so

$$\theta_1 = 24.8°$$

So a diffraction grating spreads the spectrum of light from a sodium lamp, or any other source, over a wide angle. It provides opportunities for analysis of spectra in great detail.

● **Comprehension and application**

Soap film interference

Figure 25.28
This everyday effect is explained by interference of light waves reflected from a soap film.

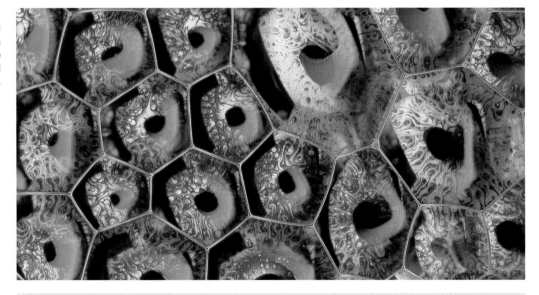

26 Use the extract on page 256 to help you create sketches to show
a the different effects of the two surfaces of a soap bubble on the phase of the reflected waves. (Show the 'half wavelength loss' that William Bragg mentions.)
b why there is destructive interference and not constructive interference when the thickness of the soap film is small compared with the wavelength of light
c why different wavelengths of light produce interference maxima at different angles of reflection.

27 From the text, what can you gather about the thickness of layers of oxide on heat-treated steel?

28 Does Bragg explain the 'principle of interference' as being the same, or not, as the principle of superposition?

29 Give examples of wavelength-dependent phenomena (such as rainbows, soap films, chromatic aberration, spectroscopy) under the headings *Dispersion effects* and *Superposition effects*.

When Thomas Young, lecturing at the Royal Institution in the first years of the nineteenth century, enunciated the principle of interference and applied it to explain certain remarkable optical effects his genius seized hold of an effect which any one may see at any time.

Let us take as an example of the effects of interference the colours of the soap film [Figure 25.28]. Young himself chose the same example, and gave its explanation in his book on Natural Philosophy.

The soap film is a thin layer of water held together by the mutual attractions of the molecules of which soap is composed. It is transparent to light. When a ray of light falls upon the film, a portion is reflected at the first surface it encounters, and another portion at the second. The two subsequently move away together and they interfere. There is a peculiar regularity in the soap film reflection. The set of waves reflected from the first surface is added to the second set which has traversed the film twice, and has therefore lagged behind a little. The two necessarily overlap: and where crest meets crest there is a double effect and so on.

Now the lag of one reflected system behind the other depends upon how much ground is lost in crossing the film twice: and this again depends upon the thickness of the film and the direction in which it is crossed. The lag may be reckoned in wavelengths of the particular quality of light. Suppose that it amounts to a whole number of wavelengths, one, two, three or more. Then the two sets of waves run absolutely together, crest and crest, trough and trough. They add together and make a wave of double the extent of movement up and down.

It must now be observed that it is necessary to add half a wavelength to the lag calculated geometrically. This is due to a certain physical effect. One reflection takes place in air at the surface of water, and the other in water at the surface of air. The two differ in character: the latter loses half a wavelength in the act of reflection. The effect is the same as that which has similar results in the case of organ pipes: where the reflections of the sound wave from open and closed ends show a like difference.

The half wavelength loss accounts perfectly for the absence of any reflection when the film is very thin, because it throws the two exactly out of step: the crests of one reflection fitting into the hollows of the other reflection, mutual interference and destruction being the result. It is a matter of ordinary observation that a very thin soap film reflects no colour: this part of the film is generally described as the black spot, though with care it can be made so large that the term 'spot' is quite inadequate.

If therefore all the causes of lag amount in all to a whole number of wavelengths there is a strong reflection. But the same lag will, in respect to some other wavelengths, amount to a whole number and a half. In that case the two reflected pencils [thin beams] destroy each other entirely, and there is no reflection of the light of that particular wavelength; all the energy involved in it passes on unchecked. Thus the film sorts out the various colours, reflecting some and transmitting others: it is coloured when looked at from either side. The colour reflected at a particular angle of reflection depends, as we have seen, upon the thickness of the film.

We do not see on a great scale the simple colour effects of the thin film, but in minor ways they occur often enough. They account for the colours of tempered steels which are coated with thin films of iron oxides. They make the bright colours which appear when petrol or other oils spread in thin sheets over water surfaces. We see them in cracks of glass or other transparent substances: and they are prettily shown in a form known as Newton's rings when a lens is laid upon a sheet of glass so that there is a thin film of air between the two bodies, and, as the thickness of the film increases from the centre outwards the colours appear in the form of concentric rings, the centre being at the point of contact.

The principle of interference appears again in a different part of our field when it explains the X-ray effects by which the structures of crystalline structures are determined. And yet again it is of importance to the electrical engineers who have to deal with the summation of alternating currents of electricity, which surge like waves along the conducting wires. The principle has a very strong application to acoustics accounting for beats and other phenomena of musical sounds.

(Adapted from *Universe of Light*, William Bragg, G. Bell and Sons (1933) pp. 138–145)

....

● **Extra skills task**   Information Technology

Develop a spreadsheet program that can predict the effect of superimposing two waves. You should be able to vary the amplitude of each, and their phase difference. Your program should draw the resulting wave on a graph.

....

Examination questions

1 a The figure represents the variation of the displacement y with distance x along a wave at a particular instant of time.

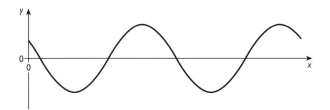

i By reference to the nature of motion of particles in transverse and in longitudinal waves, explain why the figure may be used to represent both types of wave.
ii State how two items of quantitative information about the wave may be obtained from the figure. (5)
b i State how an additional item of quantitative information may be obtained from a graph of the variation with time t of the displacement of one point in the wave shown in the figure.
ii State a feature which is the same in the graphs of **a** and **b i**. Explain how information from the graphs may be used to determine the speed of the wave. (5)
c There is an upper limit to the frequency of sound which can be heard and this limit is different for different people. Describe how, using a source of sound of variable but unknown frequency, a cathode-ray oscilloscope and a microphone, this upper limit is determined for an individual. (7)
d This figure shows the variation with time t of the displacements of two waves A and B.

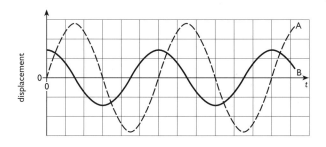

Use the figure to determine the phase difference between the waves. Suggest why, when two waves have different frequencies, the phase difference is not constant. (4)

OCR, Sciences, Physics, Foundation, June 1999

2 a When you look at water on a sunny day you often see just the bright surface because of the glare of reflected light. Wearing Polaroid sunglasses can remove this glare and allow you to see the fish swimming beneath the water.

i State the difference between polarised and non-polarised light. (1)
ii What do you see if you look at a light source through two parallel Polaroid filters which are placed so that their planes of polarisation are at 90° to each other (crossed Polaroids)? (1)
iii The sunlight is partially polarised on being reflected from the water surface. How can the Polaroid sunglasses remove the glare, allowing the fish to be seen? (1)
b As the angle of incidence changes, the proportion of the light which is polarised also changes. There is one particular angle, θ, at which the polarisation of the reflected ray is complete. It is also found that at this angle the reflected and the refracted ray are at right angles.
($_a\mu_w$, refractive index from air to water = 1.33)

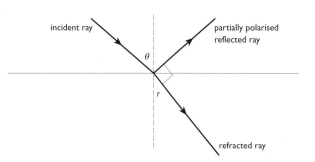

i Explain why the angle of refraction $r = 90° - \theta$.
ii Show that θ is about 53°. (3)

Edexcel (London), Module PSA2, June 1999

3 a Explain, with the aid of diagrams, the result of superposition of waves in the following cases.
i Two sound waves of the same frequency and amplitude but 180° out of phase with one another arrive at your ear together.
ii A low frequency sound wave of large amplitude and a high frequency sound wave of small amplitude arrive at your ear together. (5)
b Explain, with the aid of diagrams, how an interference pattern is the result of superposition when waves from two sources overlap. In your answer, make it clear how the conditions necessary in order to observe an interference pattern with water waves in a ripple tank and with light are met. (7)
c Explain the conditions necessary for the formation of a stationary wave using microwaves and describe an experiment to demonstrate such a wave. Suggest how this experiment may be used to determine the wavelength of the microwaves. (9)

OCR, Sciences, Basic 1, June 1999

4 An experiment using microwaves is set up as shown below.

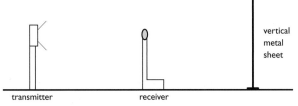

As the receiver is moved slowly towards the metal sheet a number of maxima and minima are detected by the receiver. Explain these observations. (4)

Edexcel (London), Physics, Module PH2, June 1999

5 a The light from a new type of laser appears to be yellow. This yellow light is, in fact, a mixture of red and green light, each of a single wavelength.
   i Describe how a diffraction grating may be used to make accurate determinations of the wavelengths of the red and green light. (8)
   ii Discuss quantitatively the relevant properties of the grating. (4)
   iii Explain why a double-slit method would not be suitable for making an accurate determination of these wavelengths. (4)
   b Visible light is an electromagnetic wave. Compare the properties of light waves with those of infrared and ultraviolet waves. (5)

OCR, Sciences, Further Physics, Mar 1999

6 a State *two* similarities between a radio telescope and an optical reflecting telescope. (2)
   b The dish of a radio telescope has holes of diameter 20 mm spaced close together in its reflecting surface in order to reduce the weight of the dish. Explain why the performance of this telescope will be far more satisfactory when receiving signals of frequency 7.5×10^8 Hz than when receiving signals of frequency 1.5×10^{10} Hz. (3)
   c Explain why the resolving power of a single dish radio telescope is normally much less than that of an optical telescope. (2)

AQA (NEAB), Astronomy and Optics, June 1999

7 This question is about the properties of different waves and their behaviour in different situations.
   a On the planet Venus, infrared radiation rays from a point on the surface S travel through the dense layer of atmosphere (where the index of refraction is n_1). They encounter a boundary with a much less dense layer of the atmosphere (where the index of refraction is n_2).
   The diagram (above right) shows the path travelled by a single ray R_1, and the directions of three other rays. The middle ray is the critical ray.

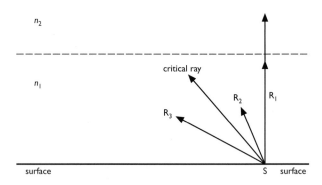

On the diagram,
   i Draw the paths followed by the three other rays. (3)
   ii If the relative refractive index between the two regions, n_1 and n_2, is 1.02, calculate the critical angle. (3)
   b When a patch of colourless oil floats on the surface of water, many colours can be observed at the surface.
   i With the aid of suitable diagrams, explain this phenomenon. (5)
   ii When a light ray meets the surface of a camera lens, part of it is reflected and part of it is transmitted. The reflected light is undesirable. In order to minimise the amount of light reflected off the camera lens, a thin transparent coating is often put on the front of the lens. This is called *blooming*.

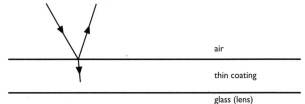

For a particular colour of light, how can *blooming* reduce the intensity of the light reflected from the surface of the lens? (2)
   c In a particular experimental set-up two *sound* waves, which are travelling in opposite directions, meet and a *standing wave* is produced.
   i With the aid of suitable diagrams, explain what is meant by a *standing wave*. (2)
   One of the original sound waves is described by the equation

$$displacement = A \sin(\omega t + kx)$$

where $A = 2.0$ mm, $\omega = 4100$ rad s^{-1}, $k = 13$ m^{-1}
   ii Calculate the frequency of this wave. (2)
   iii Calculate the speed of this wave. (2)
   iv State a possible equation for the other original wave. (1)
   v Derive the equation for the standing wave. (3)
   vi Calculate the distance between nodes on the standing wave. (2)

International Baccalaureate, Physics, Higher Level, Nov 1998 (part)

26 Wave–particle duality

THE BIG QUESTIONS
- From all the behaviour we can detect, what can we say about what electrons are really like?
- From all the behaviour we can detect, what can we say about what light is really like?

KEY VOCABULARY
black body   de Broglie relationship   discrete (values)   electron gun
massive particles   photoelectric emission   photon   Planck constant
quantised (frequency, energy)   quantum   quantum mechanics
stopping potential/voltage   threshold frequency   ultraviolet catastrophe
wave function   wave–particle duality   work function

BACKGROUND
At the start of the 1890s physics seemed a subject with a strong framework of well established knowledge about observable nature. This framework of understanding is called 'classical' physics. Newton's laws of mechanics were well accepted. Young and Fresnel had satisfied their fellow scientists that light travelled as a wave and not as a particle. Predictions of Maxwell's theories of electromagnetic radiation had been supported by the discovery of radio waves. It seemed that the framework of physics was well in place and that what now had to be done was to build on the detail. One area of detail was the need to create a theory that could provide explanations of atomic spectra. Another was to try to establish whether cathode rays – a phenomenon observed in vacuum tubes (see Chapter 7) – were waves or particles.

Then in the next 20 years came the discovery of radioactivity, the observation that cathode rays behaved as streams of particles, the exploration of structure within atoms including the suggestion of the existence of atomic nuclei, the theory of special relativity, and the realisation that particular phenomena – the ultraviolet catastrophe and photoelectric emission – defied the old theories. It was time to dismantle the framework of classical physics, to dig up the foundations and start again.

Finding support for the particle theory of cathode rays

When first discovered, it was not known whether the flow that came from a cathode – cathode rays – was some new kind of wave, perhaps a form of electromagnetic radiation, or a stream of particles. Debate about this raged among physicists near the end of the 19th century. The hypothesis that cathode rays were streams of particles predicted that they would obey the laws of mechanics. They would be subject to force and matching acceleration in the same way as any other body.

Scientists knew that they could apply magnetic force to the cathode rays and make them travel in circles (Figure 26.1a). Coils acting as electromagnets could provide the magnetic field.

Figure 26.1
Deflection of cathode rays in **a** magnetic and **b** electric fields.

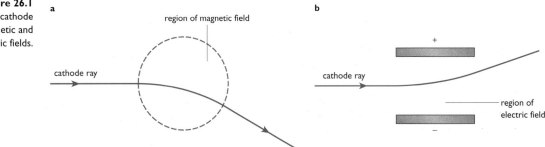

If cathode rays were streams of bodies of mass m and charge e, then since the centripetal force that produced circular motion was provided by a magnetic field,

magnetic force = centripetal force

$$Bev = \frac{mv^2}{r}$$

where B is the magnetic field strength or magnetic flux density, v is the speed of a 'particle' and r is the radius of the circular path. We don't have to worry about the detail of this equation here (the origins of the formulae for magnetic force and centripetal force are dealt with in *Further Advanced Physics*) – the important point is that it relates particle properties, mass and charge, to velocity. We can simplify the equation a little:

$$Be = \frac{mv}{r}$$

1 Why was it impossible to observe cathode rays before the development of vacuum tubes?
2 What effect do magnetic fields have on
a X-rays
b visible light?
What does this suggest about cathode rays?
3 Why does the deflection of cathode rays by magnetic fields give only modest, not convincing, support for the idea that the rays are beams of particles?

and transpose it to give:

$$\frac{e}{m} = \frac{v}{Br}$$

B and r could be measured. If the speed of the 'particles' were known it would be possible to work out a value for the ratio of their charge to mass. This would show that application of the laws of mechanics gave a real value relating to properties of individual units, or 'particles' of the cathode rays. But how do you measure the speed of particles, especially if you're not sure that they actually exist at all? The equation had too many 'unknowns' in it and could not be solved. So the existence of the circular motion under a magnetic force did not, on its own, provide enough evidence that the cathode rays were streams of particles.

Electrical force could also be applied to the cathode rays, using two metal plates with a potential difference between them (Figure 26.1b). It was possible to turn off the magnetic field and turn on an electric field. But this did not produce circular motion, and it wasn't possible to write down a useful equation that could be combined with $e/m = v/Br$ to get rid of the v term.

The problem solved

J. J. Thomson, working in Cambridge in 1897, had the idea of applying magnetic fields and electric fields to cathode rays at the same time, and adjusting the fields so that their effects cancelled each other out. The cathode rays then travelled in straight lines, with no deflection (Figure 26.2).

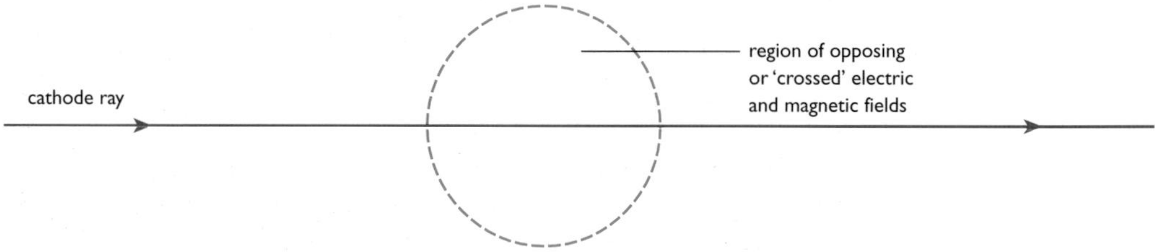

Figure 26.2
The electric and magnetic fields could be adjusted to cancel out the force on the cathode rays.

If the cathode rays were made of particles then they must be experiencing forces that are equal and in opposite directions. This would mean that

magnetic force = electric force

$$Bev = \frac{eV}{d}$$

where V is the potential difference between the plates and d is the distance between them. Both of these could be measured. The equation simplifies slightly:

$$Bv = \frac{V}{d}$$

and so

$$v = \frac{V}{Bd}$$

4 a What is the charge, in coulombs (C), of a kilogram of electrons?
b What is the charge, in coulombs (C), of one electron?
c What would happen to a beam of cathode rays with a wide range of velocities?
d How does the formula used for electric force, e*V/d*, relate to electric field strength? (See Chapter 18.)

($m = 9.1 \times 10^{-31}$ kg)

Thomson now had a way to find, or eliminate from the calculations, the speed of the particles. Substituting this into the equation that applies when only the magnetic field is switched on (see page 260):

$$\frac{e}{m} = \frac{V/Bd}{Br} = \frac{V}{B^2rd}$$

So, Thomson was able to find a ratio of charge to mass – providing support for the idea that cathode rays are particles (electrons) and not massless radiation.

The ratio of the charge to the mass of an electron is called the electron's specific charge. It is a very large number:

$$\frac{e}{m} = -1.76 \times 10^{11} \text{ C kg}^{-1}$$

The minus sign is present because electrons have negative charge. The large number suggests that electrons have a lot of charge packed into a very small mass. The mass of an electron is a very small quantity, just 9.11×10^{-31} kg.

Diffraction of electron beams

A crystal is a network or lattice of atoms, consisting of particles and inter-particle gaps. There is nothing surprising about diffraction of X-rays by crystals (see page 52), since these are electromagnetic radiation and we'd expect them to show wave-like behaviour.

J. J. Thomson's son, George P. Thomson, was one of the scientists who first observed that a crystal could cause a beam of electrons to show a kind of wave behaviour – diffraction (Figure 26.3). J. J. Thomson had identified electrons as particles and seemed to have defeated the alternative idea that cathode rays might be wave-like. Yet here was his son, discovering that though electrons have particle-like properties of mass and charge, they could also behave in a wave-like way.

This came in 1927, 30 years after J. J. Thomson's famous 'crossed-field' experiment which provided a charge/mass ratio for electrons. Now George Thomson, along with other scientists including the Americans Clinton Davisson and Lester Germer, had demonstrated **wave–particle duality** – the ability to behave in wave-like and in particle-like ways – of electrons.

Figure 26.3
Electron diffraction shows that an electron beam with a single velocity or energy can act as waves of one particular wavelength.

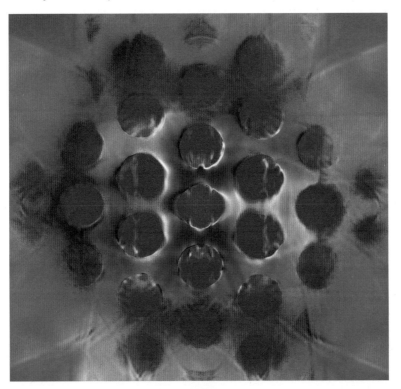

Diffraction and interference – one electron at a time

Detection of electron diffraction suggests the idea that they travel in a wave-like way, but arrive and interact with the screen in a particle-like way. If they arrive having passed through a double-slit arrangement, like the double slits that produce an interference pattern with light, then when only a few electrons have arrived and have been detected by their interaction with the screen there is little obvious pattern (Figure 26.4a). But as more and more electrons arrive, they are clearly bunched into interference fringes – lines of maximum and minimum electron intensity (Figure 26.4b).

Figure 26.4
Electrons produce points on a detecting screen, but they do not travel as 'points' with well defined locations. They travel in a more 'spread out' way, like a probability wave or wave function.

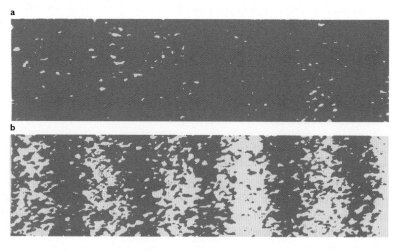

The same interference pattern results whether the electrons pass through the double-slit arrangement in large numbers together or one at a time. This remarkable fact led scientists to use the concept of 'probability' in describing an electron position. Though an electron arrives at the screen as a particle and makes a trace on the screen at a very particular place – a dot – its journey to the screen can only be described in terms of probability. Its exact position during the journey is un-knowable, and the best we can do is describe the probability of its arriving and interacting with our detecting apparatus at a particular position. It is this probability that behaves in a wave-like way. Though each electron can only make one dot on the screen, it seems to travel through the double-slit arrangement as a 'probability wave', and as such it cannot be said to pass through one slit or the other. The 'probability wave' passes through *both* slits, and electrons arrive at the screen and form a pattern as would be expected for a wave. The 'probability wave' can be described mathematically in terms of the **wave function**, Ψ, of the electron (Figure 26.5).

Figure 26.5
The wave function, Ψ, provides a snapshot of a wave/particle at a particular instant.

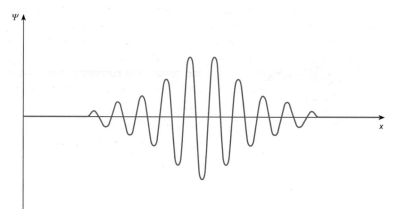

The probability of finding a particle at a particular single point is given by $|\Psi|^2$, which is the square of the absolute value of Ψ.

The representation of an electron in terms of a wave function is a feature of **quantum mechanics** – the physics of the very small that was developed largely during the early 20th century. Another feature concerns the nature of light, which we begin to consider next.

Photoelectric emission of electrons

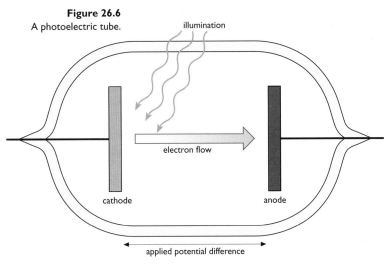

Figure 26.6
A photoelectric tube.

illumination

electron flow

cathode

anode

applied potential difference

A new phenomenon was discovered when energy was supplied to the cathode in a vacuum tube not by heating it but by shining light on to it. This new **photoelectric emission** was clearly worth studying carefully. Such study used a photoelectric tube – an evacuated tube with an anode and cathode. Light was shone on to the cathode to stimulate photoelectric emission of electrons (Figure 26.6).

The electrons were emitted with kinetic energy which could be measured by reversing the applied potential difference – making the illuminated cathode positive and the anode negative – so that escaping electrons were slowed down. With a large enough reverse potential difference, the electrons could be stopped completely and there was no current between the electrodes (Figure 26.7).

Figure 26.7
The photoelectric current could be stopped by applying a reverse potential difference.

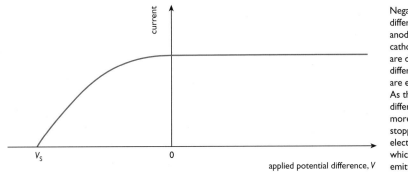

current

V_S

0

applied potential difference, V

Negative or reverse potential difference means that the anode is negative relative to the cathode, so that the electrons are decelerated. Electrons have different initial speeds so some are easier to stop than others. As the negative potential difference becomes larger, more and more electrons are stopped. When $V = V_S$ all electrons, including those which were fastest when emitted, are stopped.

The current did not decrease abruptly to zero when the voltage polarity was switched, but decreased gradually, suggesting that some electrons were more easily stopped than others. The electrons clearly did not all have the same energy – they had a range of energies up to a maximum value. The size of the potential difference that was big enough to stop *all* the electrons, the fast as well as the slow ones, and reduce the current to zero, was related to the maximum electron energy and was called the **stopping potential** or **stopping voltage**, V_S.

We know that any voltage is energy transferred per unit charge, so the voltage required to stop an electron with maximum energy is given by:

$$\text{stopping voltage} = \frac{\text{maximum electron kinetic energy}}{\text{electron charge}}$$

Using symbols:

$$V_S = \frac{E_k(\text{max})}{e}$$

or

$$eV_S = E_k(\text{max})$$

The relationship between stopping voltage and maximum electron kinetic energy is a simple one.

Photoelectric emission and frequency of light

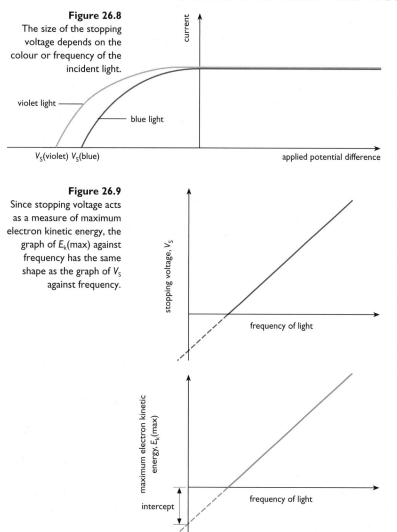

Figure 26.8
The size of the stopping voltage depends on the colour or frequency of the incident light.

violet light

blue light

V_S(violet) V_S(blue)

current

applied potential difference

Figure 26.9
Since stopping voltage acts as a measure of maximum electron kinetic energy, the graph of E_k(max) against frequency has the same shape as the graph of V_S against frequency.

stopping voltage, V_S

frequency of light

maximum electron kinetic energy, E_k(max)

intercept

frequency of light

One obvious investigation was to vary the frequency of the light shining on to the cathode and to observe the effect on stopping voltage, and therefore on maximum electron kinetic energy. Different colours, or frequencies, of light produced graphs of current against potential difference with different stopping voltages (Figure 26.8). Light of higher frequency required a larger stopping voltage, even if the light was not so bright.

Further systematic investigation of the relationship between stopping voltage and frequency of the light falling on the cathode revealed a clear relationship, as shown in Figure 26.9. We can say that increasing the frequency of the light falling on the cathode increases the stopping voltage and therefore also increases the maximum electron kinetic energy. The relationship is not a simple proportionality but is of the form $y = mx + c$ (with a negative value of the intercept, c). Also, below a certain frequency there is no photoelectric emission at all.

The relationship shown is of the form $y = mx + c$, with a negative intercept, c. The explanation of the simple relationship revolutionised our ideas about light.

Photoelectric emission and light intensity

Another investigation was of the relationship between stopping voltage and the intensity of the light shining. It was found that there was *no* relationship – light intensity has no influence on maximum electron energy (Figure 26.10).

Figure 26.10
An investigation with a negative result – the discovery of no relationship between maximum electron energy and light intensity. Negative results can be just as important as positive ones, and that is certainly true in this case.

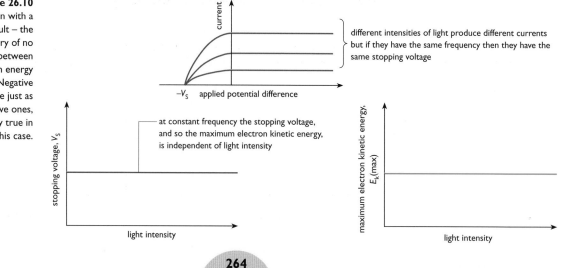

current

$-V_S$ applied potential difference

different intensities of light produce different currents but if they have the same frequency then they have the same stopping voltage

stopping voltage, V_S

at constant frequency the stopping voltage, and so the maximum electron kinetic energy, is independent of light intensity

light intensity

maximum electron kinetic energy, E_k(max)

light intensity

5 a Describe, in one sentence, the difference between thermionic emission (see Chapter 7) and photoelectric emission.
b Draw simple sketches to represent the two types of emission from a cathode.

6 In the photoelectric effect, the energy of light is transferred to electrons in a surface, which are then able to escape. In what domestic appliance is the energy of a beam of electrons transferred to a surface which then creates light?

7 a Sketch two graphs of photoelectric current against applied potential difference, on the same axes, showing what will happen if green light and violet light at the same intensity are used to illuminate a metal surface.
b On the same axes, sketch the line that would be obtained if the violet light were made much brighter.
c Sketch the shape of graphs of stopping voltage against
i frequency of light, and **ii** intensity of light.

8 Explain why the current in a photoelectric tube is not zero when the applied potential difference is zero.

It is always the surprising results that make the most impact, and this was surprising. Brighter or more intense light delivers energy to a metal surface more rapidly than does low intensity light. A simple wave would be expected to provide a continuous flow of energy on to the surface, and a faster rate of arrival of energy would be expected to produce both a faster rate of liberation of electrons *and* faster electrons. Investigations did show that brighter light produced more electrons, detectable as a bigger current between the electrodes. But brighter light did *not* produce faster electrons. Only increasing the *frequency* produced faster electrons. Wave theory could not explain this, nor why light of frequency below a minimum value could not induce photoelectric emission, however intense the light.

It took Einstein's highly original thinking to convert these findings into a more easily understood picture of what was going on. It was a very original picture, though it used an idea that had already been suggested by Max Planck. The story of this begins with a mismatch between theory and observation in another area of physics – in the study of electromagnetic radiation emitted by hot bodies, or thermal radiation. The story unfolds in the following pages.

Black body radiation and the ultraviolet catastrophe

Light shows wave-like behaviour when it travels, obeying rules of reflection, refraction, diffraction, interference and its own particular behaviour of polarisation. In the early years of the 20th century there was every reason to accept that light *was* a wave phenomenon and nothing more. But there was a problem.

All bodies emit electromagnetic radiation due to the random electrical interactions of their energetic particles. The particles are in motion, vibrating, in each other's electric fields. Increase in this vibration, as happens due to increase in temperature, in general increases the rate at which energy is transferred to the surroundings by electromagnetic radiation. An idealised emitter of radiation, with equal ability to emit radiation at all frequencies, is called a **black body**. The graph of the observed spectrum of a typical hot black body is shown in Figure 26.11a. A graph that is predicted by the assumption that matter emits radiation as a continuous wave-like flow looks like Figure 26.11b.

There is a huge disagreement between prediction and observation at short wavelengths. This mismatch was called the **ultraviolet catastrophe**, and it made it necessary to examine the existing theory very carefully. Two scientists solved the problem in two steps. They were Max Planck and Albert Einstein; their solution was revolutionary.

Figure 26.11
The observed spectrum of a hot black body has a peak of intensity not predicted by classical physics.

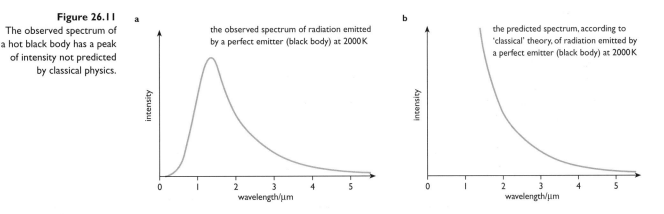

a   the observed spectrum of radiation emitted by a perfect emitter (black body) at 2000 K

b   the predicted spectrum, according to 'classical' theory, of radiation emitted by a perfect emitter (black body) at 2000 K

Planck's hypothesis

Max Planck made a hypothesis that particles within matter did not vibrate with continuously varying frequencies but only at certain fixed frequencies, like the fixed frequencies of vibrating strings. Frequencies, and other quantities, that change abruptly from one value to another but cannot vary continuously are said to have **discrete** values (Figure 26.12).

Figure 26.12
The meaning of discrete and continuously variable quantities.

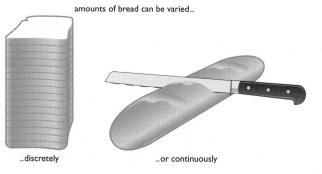

amounts of bread can be varied...

...discretely   ...or continuously

Planck found that, with these suppositions, he could show that there would be much less high-energy and high-frequency vibration than would otherwise be expected. This would result in much less high-frequency radiation than simple continuous wave theory predicted. The idea of particles vibrating only with certain discrete frequency values, like the strings of a violin or guitar, didn't seem to make much sense, but the mathematics worked. Planck was able to use this idea to produce an intensity graph that matched the observed spectrum of a black body.

The idea that the frequency and energy were not continuously variable but could have only certain values marked the beginning of quantum mechanics. Frequency and energy of the particles vibrating in a solid were said, by Planck, to be **quantised**.

Einstein's ideas on photoelectric emission

Planck had shown that the spectrum of radiation emitted by a hot black body could be explained if particles in the body may only vibrate at quantised frequencies and associated energies. Einstein took Planck's idea and applied it to the unsolved problem of the photoelectric effect.

He went further than Planck, and suggested that light leaves its source as 'discrete units' and also arrives and interacts with matter in such a quantised way. Einstein suggested that the energy of one unit, or **quantum**, is related to the frequency of the light by

$$E = h\nu$$

where ν is the frequency and h is a constant, now called the **Planck constant**, with the value 6.626×10^{-34} J s. He supposed that when one quantum, or **photon**, of light hits the metal surface it can induce one electron to escape. The photon is absorbed – it ceases to exist as a photon – but its energy is made available to the metal surface and the electron.

Figure 26.13
If you are trapped in the bottom of a hole then you need to gain potential energy in order to escape, and the amount you need depends on how far you are down the hole. If you have any energy left after this then you can run away.

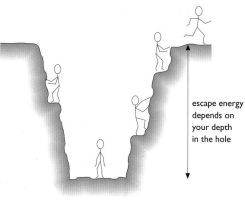

escape energy depends on your depth in the hole

Einstein was then able to explain the shape of the graph of maximum electron energy against light frequency (Figure 26.9) by supposing that the energy from the photon had two effects, which should be treated as mathematically distinct. First, the electron could be set free from the metal, effectively gaining potential energy (and the amount of energy an electron needs in order to escape depends on how strongly it is bound to the metal). Secondly, any remaining energy could become the kinetic energy of the electron, as in the analogy in Figure 26.13.

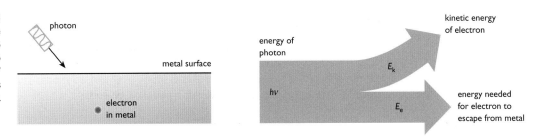

Figure 26.14
Some of the energy of the photon is transferred to the electron to allow it to escape, and any 'leftover' energy is then available as kinetic energy.

So we can say, in general (Figure 26.14):

photon energy = energy needed for electron to escape + kinetic energy of the electron

Representing the two quantities of electron energy as E_e and E_k,

$$h\nu = E_e + E_k$$

An electron that needs little energy to escape will have plenty of the energy that it has taken from the photon available for its kinetic energy. That is, when E_e is the least it can be, then E_k is the most it can be:

$$h\nu = E_e(\text{min}) + E_k(\text{max})$$

and so

$$E_k(\text{max}) = h\nu - E_e(\text{min})$$

which is an equation of the form $y = mx + c$ with a negative value of c. This is exactly the form we obtain when we plot $E_k(\text{max})$ against ν, as we've already seen (Figure 26.9). The gradient of the graph is therefore the Planck constant, h.

Graphs of maximum kinetic energy against incident light frequency for different metals all have the same gradient, h, as Einstein's equation predicts (Figure 26.15). But they have different intercepts. The minimum energy required for escape of an electron, $E_e(\text{min})$, is different for different materials. Einstein called this the **work function**, Φ, of the material. That is, $E_e(\text{min}) = \Phi$.

$$E_k(\text{max}) = h\nu - \Phi$$

Figure 26.15
Maximum electron kinetic energy against light frequency, showing the significance of the intercepts and the gradient in Einstein's photoelectric theory.

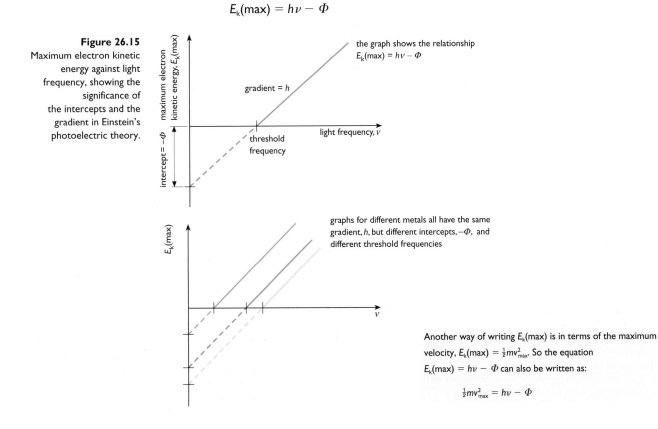

Another way of writing $E_k(\text{max})$ is in terms of the maximum velocity, $E_k(\text{max}) = \frac{1}{2}mv^2_{\text{max}}$. So the equation $E_k(\text{max}) = h\nu - \Phi$ can also be written as:

$$\tfrac{1}{2}mv^2_{\text{max}} = h\nu - \Phi$$

Threshold frequency

There is a minimum frequency of light which can induce emission from a particular surface. This is related to the work function. A photon with this minimum frequency provides the absolute minimum energy for escape – it provides $E_e(\text{min}) = \Phi$ and no more. The kinetic energy of the liberated electron is zero. The minimum frequency, given by

$$h\nu(\text{min}) = E_e(\text{min}) = \Phi$$

is called the **threshold frequency** ($\nu(\text{min})$). If the frequency of a photon is any smaller than this, escape will not take place at all (Figure 26.16).

Figure 26.16
If the photon doesn't supply enough energy then escape is not possible.

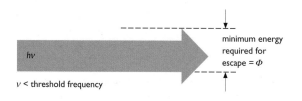

$h\nu$

minimum energy required for escape $= \Phi$

$\nu <$ threshold frequency

Threshold frequency is a property of a metal. Table 26.1 gives some values.

Table 26.1

Metal	Work function/eV	Threshold frequency/Hz
sodium	2.28	5.50×10^{14}
aluminium	4.08	9.84×10^{14}
iron	4.50	10.85×10^{14}
platinum	6.35	15.32×10^{14}

$1\ \text{eV} = 1.6 \times 10^{-19}\ \text{J}$

Models of light

This account of the photoelectric effect needed a very new and different way of thinking about light – that it did not arrive in a continuous flow as would be expected for a simple wave, but that it arrived in discrete units of energy. Einstein asked us to move on from thinking of light only as a wave. It is still fine to suppose that light travels as a wave, but it interacts with matter in quanta of energy, or photons. Light, it seems, is capable of particle-like behaviour. Like electrons, it shows wave–particle duality.

So the wave theory of light is falsified as a complete theory that we can use to explain all light behaviour. But thinking about light as a wave has proved very valuable. It has helped us to understand reflection, refraction, diffraction, interference and polarisation, and the existence of a wide spectrum of radiations. It is still an extremely useful model. We can no longer say that we accept a wave theory of light, since a good theory is consistent with *all* observations, but we can continue to use a wave *model* of light. The only requirement we have of a model is that it is useful at least some of the time.

9 Sketch energy transfer diagrams to show how the energy of a photon is shared when
a $h\nu < \Phi$ **b** $h\nu = \Phi$ **c** $h\nu > \Phi$.

10 The work function of potassium is 2.9×10^{-19} J.
a What is the minimum frequency of light required to induce photoelectric emission from its surface? (Take $h = 6.6 \times 10^{-34}$ J s.)
b Calculate the work function in eV and compare it with those of the metals in Table 26.1.

11 a Sketch graphs of the form $y = mx + c$ for:
i positive m and $c = 0$
ii positive m and positive c
iii negative m and positive c
iv positive m and negative c
v negative m and negative c
b Explain the relevance of this to photoelectric emission and in particular to:
i the Planck constant
ii work function
iii threshold frequency.

12 Make a sketch of the electromagnetic spectrum showing typical frequency values of radio, infrared, visible, ultraviolet and X-rays. Add matching photon energy values.

13 DISCUSS each of these statements:
a Einstein's interpretation of the photoelectric effect falsified wave theory.
b Einstein's interpretation of the photoelectric effect resulted in the demotion of the wave theory of light to a wave model of light.
c Light behaves as a wave phenomenon when it travels but as a particle phenomenon when it interacts with matter.

The de Broglie relationship

14 Electrons are liberated by thermionic emission, accelerated by an electron gun, pass through a double slit arrangement, and hit a screen which detects them individually by producing scintillations or dots of light. At which stages does it make more sense to explain their behaviour in terms of waves and at which stages is it better to use particle ideas?

In 1923, a few years before the experimental work of George Thomson and of Clinton Davisson and Lester Germer (see page 261), a young French postgraduate student called Louis de Broglie (who also had the title of Prince de Broglie, pronounced de Broy) wrote up his work for presentation for his qualification as a doctor of philosophy, PhD. It included a simple but new idea – that if, as Einstein had shown, light could behave in wave-like and in particle-like ways then perhaps particles like electrons could sometimes behave in wave-like ways. In other words, perhaps wave–particle duality applied to electrons as well as to light.

The idea was so simple it seemed silly at the time, and de Broglie's examiners thought so. De Broglie sent his work to Einstein, however, who provided a very supportive response. De Broglie got his doctorate and when his ideas were confirmed four years later with the observation of electron diffraction (see page 261), he won a place in history. Later the wave–particle duality of electrons would help to explain the energy levels of electrons in atoms (see Chapter 27).

The **de Broglie relationship** provides a simple match of electron wavelength to electron momentum, p. The wavelength of an electron was found to be inversely proportional to its momentum, the constant of proportionality being the Planck constant, h:

$$\lambda = \frac{h}{p}$$

So electrons can behave in particle-like ways and in wave-like ways. That is, we can use particles as models, in our minds, of electrons at some times and at other times it is more appropriate to use waves as models. Electrons cannot be said to be either waves or particles.

Exactly the same can be said of photons. We can't picture photons, but we can picture waves and particles so we find it useful to use waves and particles as models. Photons are neither waves nor particles, but sometimes they are wave-like and sometimes they are particle-like.

Just as the de Broglie relationship applies to electrons, it applies to photons. This is supported by interactions between photons and **massive particles**. (Massive here means 'having mass', however small that mass might be. Electrons are massive, photons are not.) Momentum is exchanged between the massive particles and the photons (Figure 26.17). That is, photons can gain momentum from or lose momentum to electrons, and electrons can gain momentum from or lose momentum to photons. This tells us that photons are capable of having momentum, even though they have no mass. Their momentum is related to their wavelength by the de Broglie relationship:

$$p = \frac{h}{\lambda}$$

Figure 26.17
In a photon–electron collision, the photon gives energy and momentum to the electron.

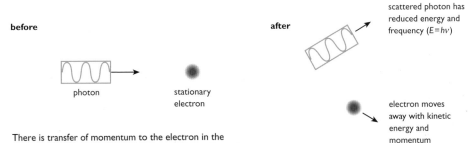

before

photon

stationary electron

after

scattered photon has reduced energy and frequency ($E=h\nu$)

electron moves away with kinetic energy and momentum

There is transfer of momentum to the electron in the collision. The need for momentum conservation tells us that the photon must have momentum.

Neutron beams and neutron diffraction

Exposure to radiation can kill cancer cells. But some of these cells can survive X-ray bombardment and continue to divide to produce ever more cells and so a growing tumour. A beam of neutrons, however, can shunt into nuclei within the DNA of the cells, knocking out protons and alpha particles. These are energetic, and cause ionisation in the region of their origin, causing local damage to the DNA so that the cells can no longer divide. So, neutron beams can be useful for cancer treatment.

An interesting feature of a beam of neutrons is that it has wave properties as well as particle properties. A neutron, just like an electron, can be described by a wave function. A neutron beam can experience diffraction. Just as for electrons, the wavelength and the momentum of a neutron are related by the de Broglie relationship.

15 A neutron is almost 2000 times more massive than an electron.

a Compare the wavelength of a neutron with that of an electron at the same speed.

b Thus compare the resolving capability of a fast neutron beam with that of a fast electron beam as used in an electron microscope.

c What practical obstacles exist to using neutron beams for useful imaging?

● **Comprehension and application**

The electron microscope

Like a TV, an electron microscope contains an **electron gun** consisting of a heated cathode, from which electrons are set free by thermionic emission, and a cathode–anode potential difference to provide acceleration. It also contains coils which act as magnetic lenses to provide a very fine beam of electrons, and more coils to produce magnification of the image of the specimen that is placed in the pathway of the beam.

The specimen must be placed inside the vacuum in which the electrons travel. In a transmission electron microscope (TEM) the electrons pass through the specimen and on to a screen (Figure 26.18). 'Shadows' are created where the specimen absorbs or deflects the electrons. To transmit the beam at all the specimen must be very thin – a maximum of 100 nm.

Figure 26.18 (left) Electron pathways through a transmission electron microscope (TEM).

Figure 26.19 (right) Electron pathways in a scanning electron microscope (SEM).

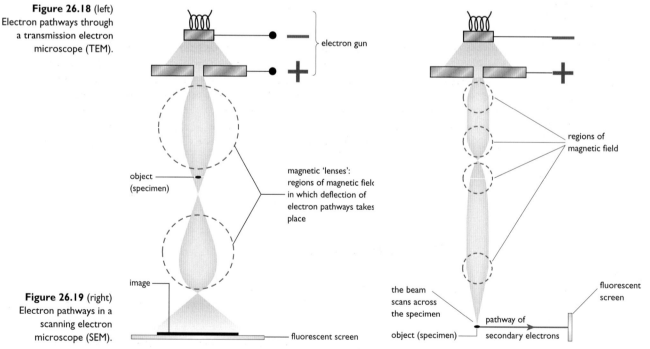

electron gun

object (specimen)

magnetic 'lenses': regions of magnetic field in which deflection of electron pathways takes place

image

fluorescent screen

regions of magnetic field

the beam scans across the specimen

object (specimen)

pathway of secondary electrons

fluorescent screen

In a scanning electron microscope (SEM) it is 'secondary' electrons that are detected, and this makes it possible to create '3D' images. The energetic electrons of the main beam of the instrument bombard the specimen, knocking secondary electrons out of the surface atoms (Figure 26.19). The result is an image of the object's surface that looks like that produced by reflected light, but hugely magnified (Figure 26.20). A TEM image shows internal structure, while an SEM image shows surface detail.

Figure 26.20
A scanning electron microscope image of part of a printing plate (×200).

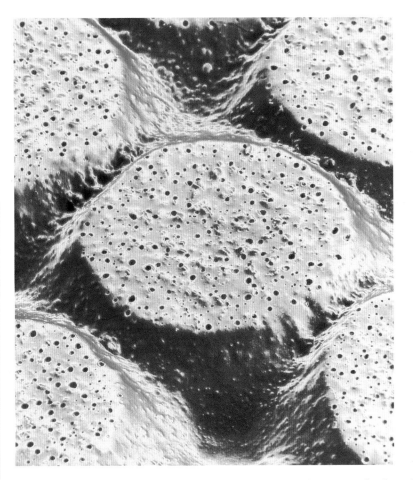

16 **a** What is the momentum of
i an electron with a wavelength of 5×10^{-12} m
ii a photon with a wavelength of 5×10^{-7} m?
b What is the energy of the photon
i in joules
ii in electronvolts?
(Take $h = 6.6 \times 10^{-34}$ J s.)

17 Why does an electron microscope that uses fast electrons produce sharper images than one that uses a beam of slow electrons?

18 At what electron momentum would the resolution of an electron microscope be as poor as that of a light microscope? Estimate the electron speed.

The unaided human eye can resolve points that are about a tenth of a millimetre apart; a light microscope with a magnification of ×1000 can resolve points separated by a thousandth of a millimetre, or about 10^{-6} m. There is no point in trying to build better optical instruments, because diffraction around objects much smaller than 10^{-6} m apart will blur the images (see page 233). Resolution cannot be improved further. But electron microscopes use electron waves with a wavelength of, typically, 5×10^{-12} m. At such a small wavelength diffraction does not blur images until objects are similar in size to molecules; a TEM can provide clear magnification up to ×500 000.

Examination questions

1 **a** In the photoelectric effect, photons cause the prompt emission of free electrons from metal surfaces.
i What is a photon? (1)
ii What is a *free* electron? (1)
b Explain why the wave model of light is unable to account for the prompt emission outlined in **a**. (2)
c A light photon of wavelength 4.00×10^{-7} m transfers all its energy to a stationary free electron in the surface of a metal.
The Planck constant is 6.63×10^{-34} J s
The speed of light in free space is 3.00×10^{8} m s^{-1}

Calculate:
i the momentum of the photon (2)
ii the energy gained by the free electron. (2)
OCR, Materials and Waves, June 1999

2 **a** Calculate the speed of electrons which have a de Broglie wavelength of 1.5×10^{-10} m. (2)
b Would you expect the electrons in part **a** to be diffracted by crystals in which the atom spacing is 0.10 nm? Explain your answer. (2)
AQA (NEAB), Particles and Waves, June 1999

3 a The figure shows schematically an arrangement for producing interference fringes using a double slit.

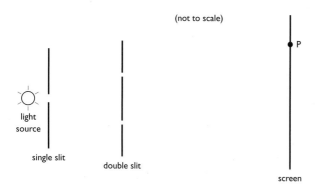

(not to scale)

light source

single slit

double slit

screen

P

A dark fringe (minimum intensity) is observed at the point labelled **P**.

i Show clearly on the diagram the distance that is equal to the *path difference* between the light rays from the two slits to the point **P**. (1)

ii Explain how the path difference determines that the light intensity at point **P** is a minimum. (3)

iii Explain briefly the role of diffraction in producing the interference patterns. (You may draw a sketch to support your explanation if you wish.) (2)

b In one experiment the separation of the slits is 4.0×10^{-4} m. The distance from the slits to the screen is 0.60 m.

Calculate the distance between the centres of two adjacent dark fringes when light of wavelength 5.5×10^{-7} m is used. (2)

c A student has learned that electrons behave like waves and decides to try to demonstrate this using the arrangement in the figure. The lamp is replaced by a source of electrons and the system is evacuated.

The student accelerates the electrons to a velocity of 1.4×10^6 m s^{-1}. The beam of electrons is then incident on the double slits. The electrons produce light when incident on the screen.

mass of an electron = 9.1×10^{-31} kg
Planck constant = 6.6×10^{-34} J s

i Calculate the de Broglie wavelength associated with the electrons. (3)

ii Explain briefly, with an appropriate calculation, why the student would be unsuccessful in demonstrating observable interference using the slit separation of 4.0×10^{-4} m. (2)

AQA (AEB), Physics Paper 1, June 1999

4 a Describe the principles of the photoelectric effect. Your answer should include

i what is meant by a photon and how to calculate its energy

ii the significance of threshold frequency

iii the meaning of work function energy

iv the evidence for the particulate nature of electromagnetic radiation. (14)

b The figure shows the variation with frequency f of the maximum kinetic energy E_{max} of photoelectrons emitted from sodium and zinc.

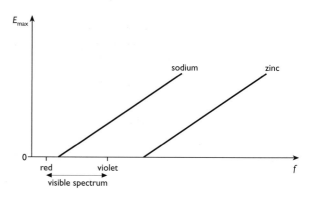

E_{max}

sodium

zinc

red   violet

visible spectrum

f

i Compare the photoelectric behaviour of sodium and zinc.

ii Suggest how the Planck constant may be obtained from the figure. (7)

OCR, Sciences, Basic 1, March 1999

5 This question is about the dual nature of light.

In the history of physics, the nature of light has been a subject of much debate and investigation. In Young's experiment, light was incident on two closely spaced slits and a pattern of light and dark fringes was observed on a screen beyond.

a Explain why Young's experiment provided evidence for the *wave* nature of light. (2)

b What pattern, if any, would be observed on the screen if light behaved like a *stream of tiny particles having no wave nature*? Explain briefly. (2)

c If the intensity of the incident light is *extremely low*, it is observed that individual dots occur on the screen, one at a time, at different places. A fringe pattern is gradually built up by the accumulation of such dots, if recorded on photographic film. Explain how this low-intensity experiment provides evidence for *both* the particle-like *and* wave-like behaviour of light. (4)

International Baccalaureate, Physics, Higher Level, Nov 1998

27 Spectra and atoms

THE BIG QUESTIONS
- How does light interact with atoms?
- What does this interaction tell us about atoms?

KEY VOCABULARY
absorption lines   absorption spectrum   Bohr atom   discrete energy levels
Doppler effect   electronvolt   emission line spectrum   excited state
Fraunhöfer lines   ground state   ionisation energy   laser
Pauli exclusion principle   periodic table   periodicity (of chemical behaviour)
population inversion   principal quantum number   red shift   stimulated emission
thermal radiation   transition

BACKGROUND   The time line below shows ideas about the atomic nature of matter.

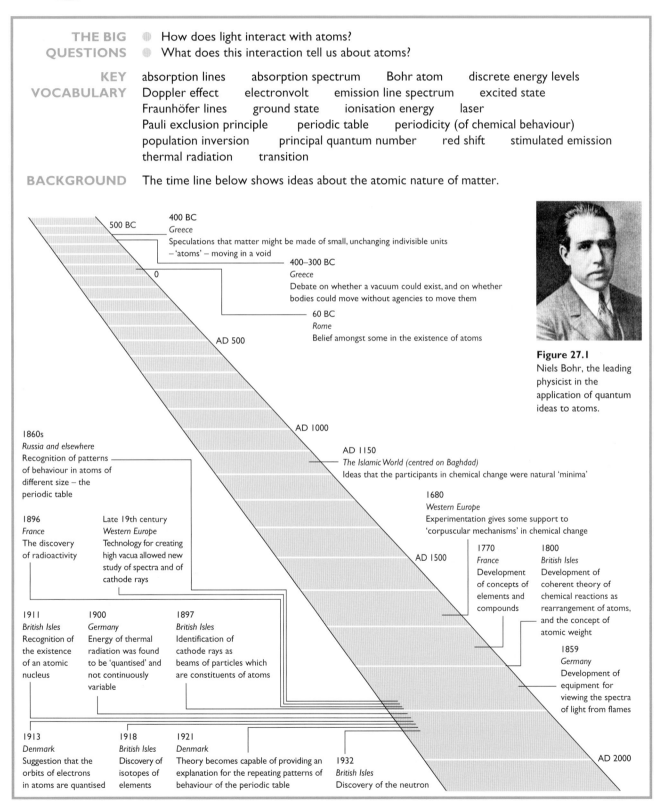

500 BC

400 BC
Greece
Speculations that matter might be made of small, unchanging indivisible units – 'atoms' – moving in a void

0

400–300 BC
Greece
Debate on whether a vacuum could exist, and on whether bodies could move without agencies to move them

60 BC
Rome
Belief amongst some in the existence of atoms

AD 500

AD 1000

AD 1150
The Islamic World (centred on Baghdad)
Ideas that the participants in chemical change were natural 'minima'

1860s
Russia and elsewhere
Recognition of patterns of behaviour in atoms of different size – the periodic table

1680
Western Europe
Experimentation gives some support to 'corpuscular mechanisms' in chemical change

AD 1500

1770
France
Development of concepts of elements and compounds

1800
British Isles
Development of coherent theory of chemical reactions as rearrangement of atoms, and the concept of atomic weight

1896
France
The discovery of radioactivity

Late 19th century
Western Europe
Technology for creating high vacua allowed new study of spectra and of cathode rays

1859
Germany
Development of equipment for viewing the spectra of light from flames

1911
British Isles
Recognition of the existence of an atomic nucleus

1900
Germany
Energy of thermal radiation was found to be 'quantised' and not continuously variable

1897
British Isles
Identification of cathode rays as beams of particles which are constituents of atoms

1913
Denmark
Suggestion that the orbits of electrons in atoms are quantised

1918
British Isles
Discovery of isotopes of elements

1921
Denmark
Theory becomes capable of providing an explanation for the repeating patterns of behaviour of the periodic table

1932
British Isles
Discovery of the neutron

AD 2000

Figure 27.1
Niels Bohr, the leading physicist in the application of quantum ideas to atoms.

Emission of electromagnetic radiation by matter

Electromagnetic radiation emerges from matter, carrying energy away from it. You and all the materials around you emit, and absorb, radiation with a range of wavelengths (Figure 27.2). The mean wavelength of the radiation that you emit is smaller than that of the radiation the walls are emitting. If your temperature were to increase then this mean wavelength would get still smaller, and the total intensity of the radiation emitted would increase. Eventually, if drastic chemical and physical changes did not take place first, you would begin to glow red.

Figure 27.2
Radiation emitted by hot bodies at different temperatures.

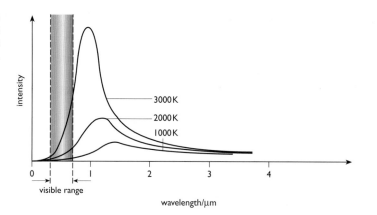

This process of **thermal radiation** is a result of random acceleration and deceleration of charged particles in each other's electric fields. These accelerations and decelerations are the effects of forces between the randomly moving particles. The randomness explains the wide spread of wavelengths of thermal radiation that a body emits.

There are other processes. For example, detectable radio waves are emitted when large numbers of charged particles in a material vibrate in unison (Figure 27.3).

Figure 27.3
Production of radio waves.

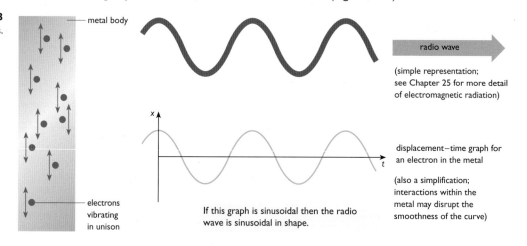

Electromagnetic radiation is also emitted by atoms, not by their random influences on each other nor by vibrating in unison, but in a third way (Figure 27.4). This happens, for example, in yellow sodium street lights and inside fluorescent tubes. The radiation cannot be explained by random interactions between particles and their neighbours – it is emitted by atoms behaving independently of each other, as in a gas. It comes from within individual atoms.

To emit this radiation an atom must be energetic. Energy can be given to atoms by heating but then they also emit a lot of thermal radiation due to interactions, unless they are so far apart that there can be little interaction. In a near-vacuum a small amount of material can be given energy, and the radiation that is observed is then the radiation emitted by individual atoms. The study of this type of emission of radiation only became possible with the improvement in vacuum technology that came in the late 19th century.

Figure 27.4
Some different types of electromagnetic emissions.

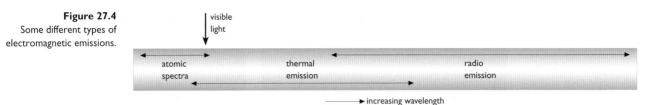

visible light

| atomic spectra | thermal emission | radio emission |

→ increasing wavelength

1 How is it possible to generate radio waves at a particular frequency while thermal radiation is emitted with a wide range of frequencies?
2 The speed of electromagnetic radiation in a vacuum and in air at all wavelengths is $3.0 \times 10^8 \, \text{m s}^{-1}$.
a What is the frequency of the radio waves transmitted by a radio station if their wavelength is 300 m?
b Use the formula $E = h\nu$ to calculate the energy of a photon that interacts (along with very many others) with your radio aerial to make it possible for you to listen to this station.
c By what factor is this smaller than the minimum energy of a photon that could induce photoelectric emission from the aerial? (For the metal of the aerial, $\Phi = 6.4 \times 10^{-19} \, \text{J}$.)
(Planck's constant $h = 6.6 \times 10^{-34} \, \text{J s}$)

Electromagnetic radiation emitted by thermal processes (random emission) overlaps in wavelength with radiation from radio sources and with atomic emission spectra (which are non-random).

Emission line spectra

Atoms behaving independently, in a gas at very low pressure so that they are a long way apart, emit patterns of light which can include infrared, visible and ultraviolet frequencies. But first they must be 'excited' by being given extra energy, which can be done by heating or by the effects of high potential difference.

The light that the excited atoms emit is not a continuous mix of frequencies. Instead it is made up of a number of very definite or discrete colours. Each chemical element has its own individual pattern of discrete colours. The patterns can best be detected by use of a diffraction grating (see Chapter 25), which creates maxima at angular positions that are highly wavelength-dependent. The maxima lie parallel to the lines etched on to the grating and so they appear as lines (Figure 27.5). The pattern of lines produced by a particular element is called its **emission line spectrum**. An element's characteristic emission line spectrum acts rather like a fingerprint or a DNA profile (Figure 27.6, overleaf).

3 List as many differences as you can between the emission of radio waves and the emission of line spectra by matter.

Figure 27.5
This schematic diagram shows the production of an emission line spectrum.

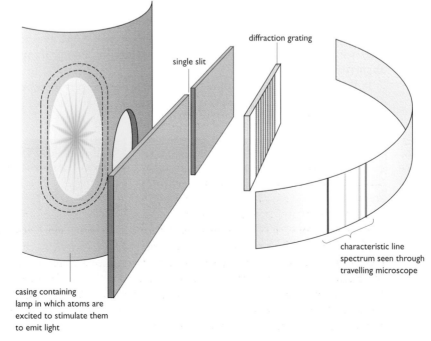

diffraction grating

single slit

characteristic line spectrum seen through travelling microscope

casing containing lamp in which atoms are excited to stimulate them to emit light

Figure 27.6 a
Each element can be identified by its characteristic emission line spectrum. The spectra for **a** hydrogen and **b** sodium are shown here.

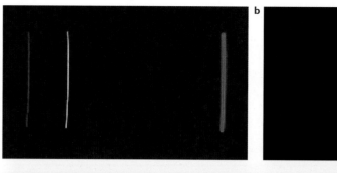

Figure 27.7
Early vacuum tubes in an illustration from 1903.

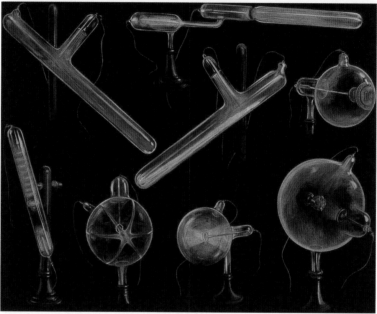

Investigation of the spectra of elements became possible once the technology existed to create a high vacuum inside glass tubes. Elements could exist at low density inside the vacuum tubes, and they could be given energy, or excited, by strong electric fields applied using metal plates.

Absorption spectra

Individual atoms absorb radiation as well as emit it. When white light shines through a sufficiently large sample of an element in its gaseous state it emerges with the element's characteristic wavelengths missing. There are dark lines or bands in the continuous spectrum. The positions of these dark bands is the same as the positions of the lines in the emission spectrum (Figure 27.8). They are **absorption lines**. An element's characteristic lines or frequencies make up its **absorption spectrum**.

In the creation of an absorption spectrum, the wavelengths absorbed are re-emitted, but in all directions, so that the original beam of light has a very much reduced (but not zero) intensity of light at these wavelengths.

Figure 27.8 a
a The emission spectrum of helium and
b the absorption spectrum of the Sun.

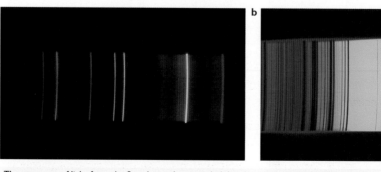

4 If you examined white light after it had passed through sodium vapour, what would you expect to observe?

The spectrum of light from the Sun shows the same dark bands as the spectrum of light that has passed through helium in a laboratory here on Earth. We conclude that helium must exist in the outer layers of the Sun.

Discrete energy carried by light

Albert Einstein was the first to suggest the idea that the energy of electromagnetic radiation does not interact with matter as a continuous flow but in discrete units, or quanta. This was based on Max Planck's ideas that energy of thermal vibration within material varies in jumps, or discretely, rather than continuously. A quantity that varies discretely is said to be 'quantised' (see page 266). A quantum of energy is related to the frequency, ν, of a photon of light by the simple formula:

$$E = h\nu$$

where h is the Planck constant.

Photons are interacting with the retinas in your eyes as you read this. There are a lot of photons, but the energy of each one is small. The Planck constant is a very small number, about 6.6×10^{-34} J s. For measuring such small quantities of energy the **electronvolt** (eV) is more useful than the joule. (1 eV = 1.6×10^{-19} J. It is the energy transferred to an electron when it is subject to, and accelerated by, a potential difference of 1 volt.)

5 a Calculate the energy, in J, of a photon with a frequency of 10^{16} Hz.
b Calculate the frequency of a photon that carries 10^{-20} J of energy.
c Quote the energies of **a** and **b** in eV.

Atoms and energy levels

In 1913 the Danish scientist Niels Bohr suggested that the electrons in atoms had quantised energies; that is, they could orbit the nucleus only with particular values of energy and so particular radii. This answered the biggest question in the physics of the day – why didn't electrons in atoms spiral towards the nucleus and emit energy continuously as they did so? They didn't, said Bohr, because their energy could not vary continuously, but only in jumps. This was another extension of Max Planck's ideas from a dozen years earlier, that particles could only vibrate and emit thermal radiation with certain energy and certain frequencies. Bohr took this idea of quantisation and applied it specifically to the electrons in orbit in atoms, and in doing so explained the existence of characteristic spectra of elements.

Bohr introduced the idea of **discrete energy levels** for electrons in atoms (Figure 27.9), and the idea that the movement of an electron from a higher energy level to a lower one, losing energy, results in the emission of one photon of light. The energy carried away by the photon is equal to the size of the energy difference between the levels.

Figure 27.9
A simplified representation of energy levels in an atom.

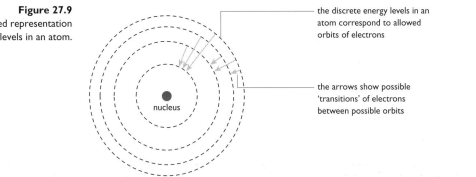

nucleus

the discrete energy levels in an atom correspond to allowed orbits of electrons

the arrows show possible 'transitions' of electrons between possible orbits

This model is simplified, for example, in that it suggests that electron orbits are simple circles.

The electron, and so the atom, has gone through an energy **transition**. If the energy of the electron at the higher energy level is E_2, and at the lower energy level is E_1, then the energy lost by the electron is $E_2 - E_1$. If the energy of the photon that is created is $h\nu$ then we can write an equation:

electron energy change = energy of the new photon
$$E_2 - E_1 = h\nu$$

Atoms of a particular element have their own particular arrangement of energy levels and this is why they emit their own particular range of frequencies of light, producing an emission spectrum that is characteristic of the element.

Energy transitions and absorption of light

Essentially the same formula applies when a photon loses its energy to an electron in an atom. The electron gains energy and jumps away from the nucleus. We can write:

electron energy change = energy of the absorbed photon

The electron has gained energy and an existing photon has lost energy and ceased to exist. The electron starts with energy E_1 and finishes with energy E_2 (Figure 27.10):

$$E_2 - E_1 = h\nu$$

For absorption to happen the photon must have just the right energy. So when light with a range of frequencies, such as white light, passes through a substance only photons with 'just the right energy' are absorbed. The beam of light emerges from the material with those particular photon energies, and so particular frequencies, missing (or of significantly reduced intensity).

Figure 27.10
Exchanges of electron energy with photon energy. Atoms of different elements have different patterns of energy levels, and hence different spectra.

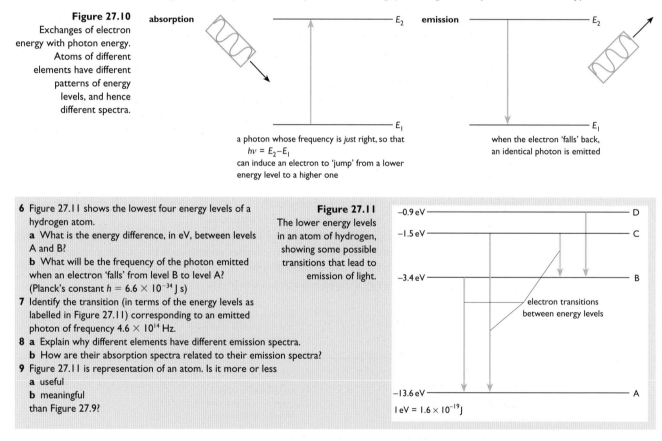

absorption ——————— E_2 emission ——————— E_2

——————— E_1 ——————— E_1

a photon whose frequency is *just* right, so that
$h\nu = E_2 - E_1$
can induce an electron to 'jump' from a lower energy level to a higher one

when the electron 'falls' back, an identical photon is emitted

6 Figure 27.11 shows the lowest four energy levels of a hydrogen atom.
  a What is the energy difference, in eV, between levels A and B?
  b What will be the frequency of the photon emitted when an electron 'falls' from level B to level A? (Planck's constant $h = 6.6 \times 10^{-34}$ J s)
7 Identify the transition (in terms of the energy levels as labelled in Figure 27.11) corresponding to an emitted photon of frequency 4.6×10^{14} Hz.
8 a Explain why different elements have different emission spectra.
  b How are their absorption spectra related to their emission spectra?
9 Figure 27.11 is representation of an atom. Is it more or less
  a useful
  b meaningful
than Figure 27.9?

Figure 27.11
The lower energy levels in an atom of hydrogen, showing some possible transitions that lead to emission of light.

−0.9 eV ——————— D
−1.5 eV ——————— C
−3.4 eV ——————— B

electron transitions between energy levels

−13.6 eV ——————— A
$1 \text{ eV} = 1.6 \times 10^{-19}$ J

The Bohr atom

Bohr's idea of saying that electrons in atoms only have certain 'allowed' energies seems at first to be an inadequate way of explaining why the electrons don't spiral in to the nucleus. It is because the idea fits the observations so very well that it is convincing. Most of all, it provides an explanation for the line spectra produced when atoms emit or absorb electromagnetic radiation. The **Bohr atom** with discrete electron orbits each of a particular energy level, as depicted in Figure 27.9, is a good model. It shows that the frequencies seen in atomic spectra are the frequencies of photons that have just the right energy to match the energy jumps the electrons make between orbits.

    The wave behaviour of electrons goes some way to providing an explanation of why electrons in atoms are only 'allowed' certain orbits. If, as de Broglie suggested and electron diffraction observations supported, electrons have wave behaviour and so a wavelength, then only particular patterns of the waves can fit into orbits around the atomic nucleus (Figure 27.12).

Figure 27.12
An 'allowed' orbit of an electron 'wave'.

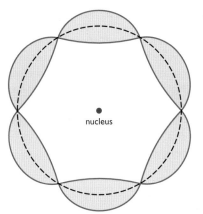

nucleus

Niels Bohr's ideas have had to be refined since he first suggested them. For example, many spectral lines come in pairs. To explain this, physicists introduced the idea of electron spin, which can be thought of as rather like the Earth's daily spin on its own axis. Electrons can spin one way or the other, with slightly different energies. Also, electron orbits can have different shapes and orientations, and this also affects energy level possibilities. Perhaps most importantly, we know that it is impossible to find out the exact position of an electron. Electrons seem to be fuzzy characters, and quantum ideas of the wave function (see page 262) need to be invoked for a full explanation. The simple Bohr atom, however, provides the foundation for all these later ideas.

Ground state and excited atoms

Just as a ball falls towards the Earth and stops when it has the least possible amount of gravitational potential energy, so an electron 'falls' towards the nucleus, provided that there is somewhere to fall to. When all the electrons in an atom have the lowest possible energy then the atom is in its **ground state**. Almost all the atoms in the materials around you are in their ground state.

But some atoms have more than ground state energy. They are in an **excited state**. They do not usually stay in an excited state for very long. Electrons 'fall' like the ball, and when they do the atom emits one photon for each electron transition (Figure 27.13).

Atoms can be excited by an electric field, as in a spark or lightning flash. Photons of the right frequency can also excite atoms, as can bombardment by other high-energy particles. They can also be thermally excited – the yellow colour of a flame is due to the strong light emitted by sodium atoms as they return from excited state to ground state. (Sodium is moderately plentiful in the environment. Other elements are also present in the flame, but the characteristic yellow colour of sodium is exceptionally bright compared with other colours.)

Figure 27.13
Energy levels in an atom of lithium.

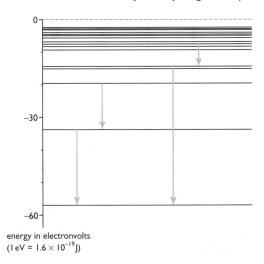

energy in electronvolts
$(1\,eV = 1.6 \times 10^{-19}\,J)$

An atom can be excited so that one (or more) electrons can exist in higher energy levels.

During an energy transition, represented by the arrows, an electron falls back to a lower energy level and a photon is emitted. The frequency of the photon depends on the size of the energy transition (shown by the length of the arrows). Since the atom has a 'fixed set' of energy levels, it can only emit a 'fixed set' of photon frequencies.

Ionisation

Ionisation is similar to excitation in that it involves an increase in electron energy, and can be caused by light of high enough frequency, by bombardment by a beam of energetic particles, by an electric field or by a thermal supply of energy. For ionisation an electron gains enough energy to escape completely from the atom (Figure 27.14). If it escapes *and* has energy left over then it can escape at high speed – the 'spare' energy provides the free electron with kinetic energy. To take an atom from its normal ground state to an ionised state takes a minimum amount of energy, which is the same for all atoms of an element but different for different elements. It's called the element's **ionisation energy**.

Figure 27.14
Energy levels and ionisation energy for a hydrogen atom.

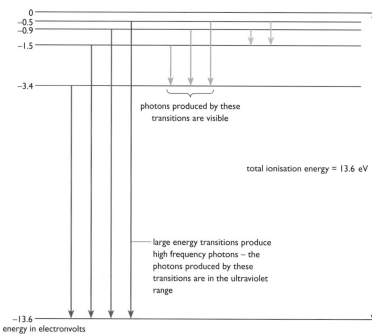

photons produced by these transitions are visible

total ionisation energy = 13.6 eV

large energy transitions produce high frequency photons – the photons produced by these transitions are in the ultraviolet range

−13.6
energy in electronvolts
$(1\,eV = 1.6 \times 10^{-19}\,J)$

If a hydrogen atom in its ground state gains 13.6 eV or more, its electron can escape.

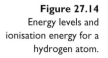

10 A hydrogen atom is in its ground state and gains energy. What happens if the energy is:
a less than the ionisation energy
b equal to the ionisation energy
c more than the ionisation energy?

Quantum numbers

The energy levels in the Bohr atom can be numbered according to their radii. An electron at the least distance from the nucleus and lowest energy level has a **principal quantum number**, n, of 1. An electron at the next available radius has a principal quantum number of 2 (Figure 27.15).

But real atoms are more complicated than this. Orbits can have different shapes, electrons can have spin in one direction or in the other. It takes not one but a set of four quantum numbers to describe completely the status of an electron in an atom. No two electrons in any atom can have the same set of four numbers. This is called the **Pauli exclusion principle**.

Figure 27.15
The four lowest energy levels of a hydrogen atom, with principal quantum numbers.

| −0.9 eV ———————————————————— n = 4 |
| −1.5 eV ———————————————————— n = 3 |
| −3.4 eV ———————————————————— n = 2 |
| −13.6 eV ———————————————————— n = 1 |

11 What is the frequency of a photon emitted when an electron makes a transition between energy levels of principal quantum numbers 3 and 1 in a hydrogen atom?
$(1\,eV = 1.6 \times 10^{-19}\,J;\ h = 6.6 \times 10^{-34}\,J\,s)$

The Bohr atom and the periodic table

Before the discovery of the electron, scientists had constructed a table of the elements according to their chemical behaviour. Each vertical column of the table shows a family or group of elements with some similarity of chemical properties. This repetition of behaviour is called **periodicity**, and arranging the elements according to these families forms the **periodic table** (Figure 27.16). The Bohr atom, combined with the idea that no two electrons in the same atom can have the same set of four quantum numbers, was able to provide at least partial explanation of these patterns.

Figure 27.16
Part of the periodic table of the elements, showing the number of electrons with each principal quantum number, n, for the ground state atom. Notice the patterns in these numbers.

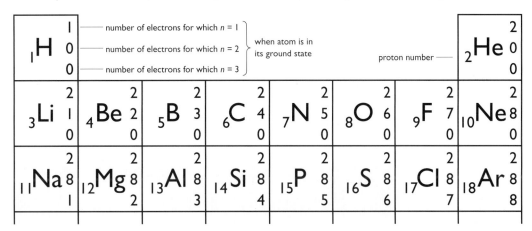

X-ray spectra

The source of X-rays used by medical radiographers for routine X-rays in hospitals contains a high voltage electron gun which accelerates electrons towards a tungsten target (Figure 27.17). Tungsten has a high proton number, Z, so that large forces exist between tungsten nuclei and the energetic electrons. It also has a high melting point and can withstand the high temperatures resulting from the bombardment.

The very rapid deceleration of electrons in the electric fields of the tungsten nuclei results in emission of X-rays. Interactions between individual electrons and individual nuclei vary, depending, for example, on the pathway of each electron relative to the nucleus. So X-rays are emitted with a continuous range of photon energies.

Some electrons, however, not only interact with the nuclei but also cause excitation of the tungsten atoms. The atoms' electrons are excited to higher energy levels, and they then fall back to the ground state. This results in emission of a photon with an energy matching the difference in energy levels. So superimposed on the continuous spectrum of radiation emerging from the target are specific photon energies that are characteristic of the target element.

Figure 27.17
Principle of an X-ray tube and typical spectrum of X-ray photon energies.

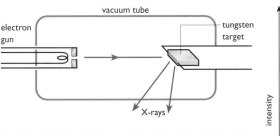

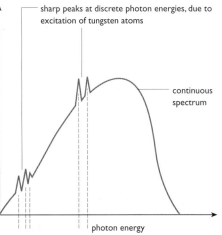

12 Why do very few electrons follow pathways that head directly towards a nucleus?

13 Why does an electron gun in an X-ray tube require a higher voltage than one in a television?

14 **a** Explain why the spectrum of photon energies provided by an X-ray tube has both continuous and discrete components.
b If a tungsten target were replaced by one of a material of similar nuclear size and melting point, what differences would you expect to see in the spectrum?

● **Comprehension
and application**

Starlight

We can analyse the absorption spectrum of an element in a laboratory. The principle is simple enough. We shine light with a full range of wavelengths through the element, in a gaseous state, and then shine the light through a diffraction grating to see which wavelengths are greatly reduced in intensity, appearing as dark lines in the spectrum. We can also point a diffraction grating at a star (the actual experimental arrangement is a little more complex, but the principle is simple) and see what wavelengths are missing from the starlight.

We can then compare missing wavelengths in the starlight with the absorption spectra of elements. Where there is a match in the patterns it suggests to us that the light from the star has passed through the matching elements.

Some absorption, of course, takes place in the Earth's atmosphere, but we are familiar with the absorption spectra of materials in the atmosphere. Where there are other lines in the starlight absorption spectrum we conclude that the outer layers of the star, through which emerging lights must pass, contains the matching elements. The characteristic absorption lines in the spectrum of light from the Sun (the solar spectrum, Figure 27.18) are called **Fraunhöfer lines** after their discoverer, Joseph von Fraunhöfer, who made optical instruments in the early 19th century.

Figure 27.18
The solar absorption spectrum. Dark lines also exist in the spectra of light from other stars.

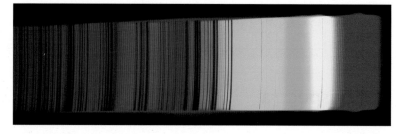

Helium was then unknown on Earth, but an unexplained set of lines in the solar spectrum caused chemists to seek out an element that possessed matching lines. A gaseous element was found, and named after Helios, the Greek for Sun.

The development of the study of the absorption spectra of starlight made it possible to make new studies of the stars – to classify them and to build theories of the processes that take place within them. Absorption spectra lie at the heart of such astrophysics.

There is one important feature about the spectra from distant stars, and the more distant they are the stronger this feature is. The lines in the spectra are not quite where they are in spectra of our near-neighbour stars, or in spectra seen in Earth laboratories. All of the lines are shifted towards the red end of the spectrum. This is the famous **red shift** (Figure 27.19), and is the main evidence that we are part of an expanding Universe.

Figure 27.19
The red shift in the light from distant galaxies tells us that the Universe is expanding.

These galaxies are a long, long way away. Their light is redder than it would be if they were not in relative motion with us, the observers. This can be seen from the shift in lines in the absorption spectrum of stars, relative to the frequencies of the absorption lines of elements observed in Earth laboratories. The light is red-shifted due to the **Doppler effect**, which is a result of relative motion of a source of waves and the observer. The further a galaxy is from us, the bigger the red shift usually is. This tells us not only that objects in space are moving away from us, but that the further away they are, the faster the relative speed. Galaxies seem to obey a simple proportionality here – the speed at which they are moving away seems to be proportional to their distance from us. This is expressed as $v = Hd$, where v is the apparent relative speed and d is the distance away. The constant of proportionality, H, is called the Hubble constant.

The Doppler effect

In a uniform medium waves spread out from a source with spherical wavefronts – or circular wavefronts for waves on a two-dimensional surface. For a moving source the waves do not all spread from the same point, so the spheres (or circles) do not have the same centre. The result is that the wavelength in the direction of motion of the source is less than for a stationary source, and the wavelength of the waves 'behind' the moving source is greater (Figure 27.20). These changes in wavelength are matched by changes in frequency (Figure 27.21).

Figure 27.20
The Doppler effect for a moving source.

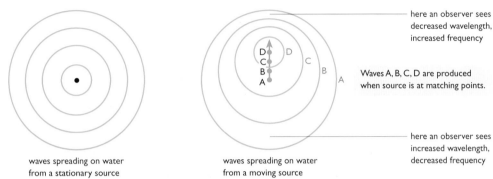

here an observer sees decreased wavelength, increased frequency

Waves A, B, C, D are produced when source is at matching points.

here an observer sees increased wavelength, decreased frequency

waves spreading on water from a stationary source

waves spreading on water from a moving source

Figure 27.21
The drop in frequency of sound we hear as a fast vehicle goes by is an example of the Doppler effect.

If an observer moves towards a stationary source of waves they meet the wavefronts more frequently than they would if they stayed still. If they move away from the source they pass through the wavefronts less frequently (Figure 27.22).

Figure 27.22
The Doppler effect for a moving observer.

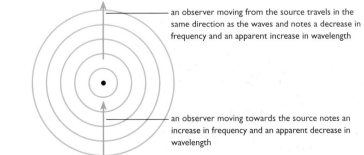

an observer moving from the source travels in the same direction as the waves and notes a decrease in frequency and an apparent increase in wavelength

an observer moving towards the source notes an increase in frequency and an apparent decrease in wavelength

waves spreading on water from a stationary source

For the moving observer, their own motion changes the frequency of the waves. The *speed* of the waves depends on the nature of the waves and the medium, not on what the observer is doing. The speed stays the same while the frequency changes. It follows that since $c = f\lambda$, where c is the speed of the waves, f is frequency and λ is wavelength, any change in frequency must result in a change in wavelength. To the observer, the wavelength seems shorter when moving towards the source, and longer when moving away.

For *all* kinds of waves, and whether it is the source or the observer that is moving, the shift in measured frequency of waves, Δf, is given by:

$$\frac{\Delta f}{f} = \frac{v}{c}$$

where f is the frequency of emission of the source, v is the speed of *relative* source–observer motion, and c is the speed of the waves in the relevant medium. (Read more about the Doppler effect on page 453.)

15 The absorption spectrum of a particular star is seen to show evidence of the presence of iron.
  a What form will this evidence take?
  b How can we feel satisfied that this evidence does not arise due to the action of the Earth's atmosphere?
16 Explain how it was possible to suppose that a previously undiscovered element exists by looking at the solar spectrum.
17 DISCUSS
  A star appears as a scintillating (twinkling) point of light. How is it possible to establish ideas about the nature of a star when there is apparently so little data from the star itself?
18 a Calculate the wavelengths emitted (and absorbed) by atoms of strontium, relating to the transitions to (or from) the lowest energy level as shown in Figure 27.23.
  b What will be the effect of red shift on the wavelengths absorbed by this element in a distant star and detected on Earth?
19 Explain why frequency of detected sound *decreases* as a vehicle goes by.
20 Look at question 6 on page 278, relating to the transition between levels A and B in Figure 27.11. What Doppler shift of this frequency will be observed in light from a distant star, which contains a high density of hydrogen in its outer layers, if it is travelling
  a towards us at a speed of 0.1% the speed of light
  b away from us at a speed of 10% the speed of light?

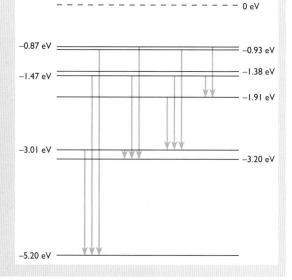

Figure 27.23 Some of the energy levels in an atom of strontium, ^{38}Sr, with some possible electron transitions.

- - - - - - - - - - - - - - - - - 0 eV
−0.87 eV ———————————— −0.93 eV
−1.47 eV ——————— −1.38 eV
—————— −1.91 eV
−3.01 eV ——————— −3.20 eV
−5.20 eV ———————

· ·

● Comprehension and application

Lasers

Lasers are everywhere: in CD players, in supermarkets, in telephone systems. They provide narrow beams of monochromatic light. Monochromatic means single colour, and single colour means single frequency.

The single frequency is achieved by electron transitions between the same energy levels in atoms all of the same element. Identical transitions produce identical photons.

Of course, to emit energy, the electrons have to 'fall' from a higher to a lower energy level. There's no problem with that – when electrons are given extra energy and the atoms become 'excited' then the electrons usually fall back very quickly, and photons are emitted. But in most materials very few of the atoms are in excited states.

Some elements have excited states that last for longer than usual. Neon, for example, has an excited state that is relatively stable. Large numbers of neon atoms in a tube can be excited to this state by supplying them with energy. If this produces more excited than ground-state atoms then that is a reversal of the normal situation. The neon is then in a state of **population inversion**. The 'population' refers to the neon atoms; the 'inversion' to the fact that there are more excited than not-excited (ground state) atoms.

21 In the context of lasers, explain the meaning of the following:
  a light **d** amplification
  b stimulated **e** radiation.
  c emission
22 a From the energy transition shown in Figure 27.24, calculate the frequency of light from a neon laser.
  (1 eV $= 1.6 \times 10^{-19}$ J; $h = 6.6 \times 10^{-34}$ J s)
  b What is the colour of this light?
  (The range of visible wavelengths is approximately 400–700 nm.)
23 a What does 'monochromatic' mean?
  b How is it possible to produce laser beams with different colours?
24 What support does laser technology give to the Bohr theory of the atom?

Though the excited state of these neon atoms is relatively stable, the probability of an electron falling to a lower level is greatly increased in the presence of photons of just the right frequency which are not absorbed. 'Just right' here means exactly the same frequency as that emitted when the electron makes its fall (Figure 27.24). So radiation can stimulate these excited-state neon atoms to emit radiation of the same frequency. This is called **stimulated emission**.

In a helium–neon laser the gases are held in a tube. Photons of the right frequency move backwards and forwards between mirrors at each end of the tube, multiplying in number as they stimulate electrons in more atoms to fall back to the lower energy state. The light amplifies itself. The mirror at one end is semi-transparent, allowing the laser beam to escape. The name 'laser' makes sense – it's an acronym for 'light amplification by stimulated emission of radiation'.

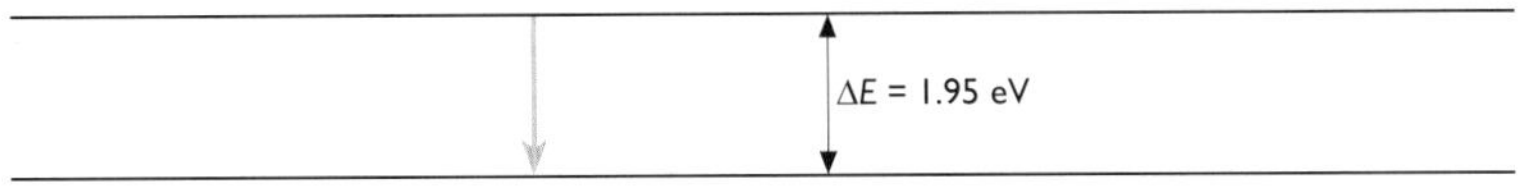

Figure 27.24
The transition associated with laser action in neon.

$\Delta E = 1.95$ eV

. . . .

● **Extra skills task** Communication

With your tutor group, take part in a discussion about wave–particle duality and quantum theory. Everybody should contribute to the discussion by carrying out independent research and each providing 'expert evidence' on one or more of the following topics:

- the ultraviolet catastrophe
- the work of Max Planck
- photoelectric emission as evidence for quanta of light
- Einstein's role in the development of quantum theory
- the Bohr atom
- Schrödinger and the atom
- the importance of atomic quantum numbers
- the Heisenberg Uncertainty Principle
- electron diffraction
- the electron microscope
- de Broglie wavelength
- variation of diffraction patterns with beam intensity
- the wave function.

Sources to use in your research may include some of the many 'popular science' books on the atom and quantum theory. Through your discussion, try to establish:
  a in what circumstances it is more appropriate to use wave models
  b in what circumstances it is more appropriate to use particle models
  c whether a model can ever be a complete description of the aspect of the physical world that it represents
  d why many people feel uncomfortable with the co-existence of very different models to describe the same feature of the physical world
  e whether they are right to feel uncomfortable.

Examination questions

1 The diagram represents some of the energy levels of an isolated atom. An electron with a kinetic energy of 2.0×10^{-18} J makes an inelastic collision with an atom in the ground state.

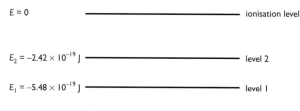

$E = 0$ ──────────────────── ionisation level

$E_2 = -2.42 \times 10^{-19}$ J ──────────── level 2

$E_1 = -5.48 \times 10^{-19}$ J ──────────── level 1

$E_0 = -2.18 \times 10^{-18}$ J ──────────── ground state

a Calculate the speed of the electron just before the collision. (2)

b i Show that the electron can excite the atom to level 2.
ii Calculate the wavelength of the radiation that will result when an atom in level 2 falls to level 1 and state the region of the spectrum to which this radiation belongs. (6)

c Calculate the minimum potential difference through which an electron must be accelerated from rest in order to be able to ionise an atom in its ground state with the above energy level structure. (2)

AQA (NEAB), Particles and Waves, June 1999

2 The figure shows four energy levels for electrons in a hydrogen atom. It shows one transition, which results in the emission of light of wavelength 486 nm.

increasing energy

a On the figure, draw arrows to show
i another transition which results in the emission of light of shorter wavelength (label this transition L)
ii a transition which results in the emission of infra-red radiation (label this transition R)
iii a transition which results from absorption (label this transition A). (4)

b Calculate the energy change which an electron has to undergo in order to produce light of wavelength 486 nm. (4)

OCR, Physics, Paper 2, June 1999

3 A muon is a particle which has the *same charge* as an electron but its *mass* is 207 times the mass of an electron.

An unusual atom similar to hydrogen has been created, consisting of a muon orbiting a single proton. An energy level diagram for this atom is shown.

0 eV ──────────────────────

−312 eV ──────────────────────

−703 eV ──────────────────────

−2810 eV ─────────────────── ground state

a i State the ionisation energy of this atom.
ii Calculate the maximum possible wavelength of a photon which, when absorbed, would be able to ionise this atom.
iii To which part of the electromagnetic spectrum does this photon belong? (5)

b Calculate the de Broglie wavelength of a muon travelling at 11% of the speed of light. (3)

Edexcel (London), Physics PH2, June 1999

4 In a model of a hydrogen atom, the single electron normally orbits around the nucleus in a low energy state (ground state). In an excited hydrogen atom, the electron has a higher energy and orbits at a greater distance from the nucleus. The model is illustrated below.

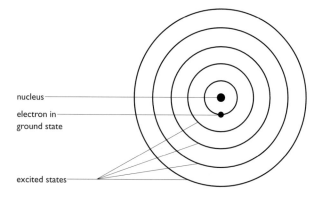

nucleus

electron in ground state

excited states

Show how the model is used to provide an explanation for
a the line spectrum of atomic hydrogen (3)
b the difference between emission and absorption spectra. (2)

OCR, Sciences, Basic 1, June 1999

5 a i State two postulates of the Bohr model of the hydrogen atom. (4)

   ii Describe two limitations of the Bohr model. (2)

b The three lowest energy levels of a fictitious atom are shown.

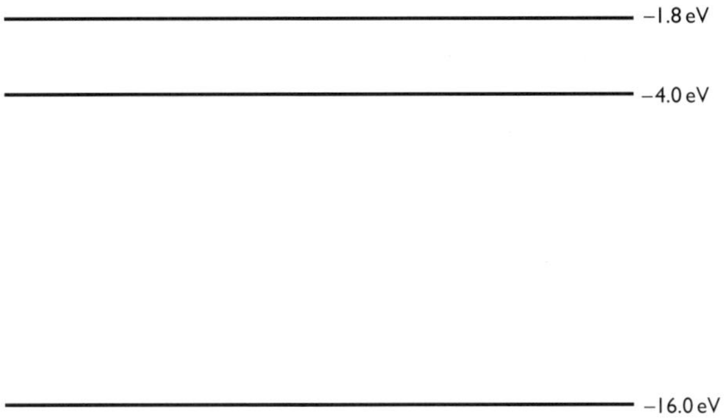

—————————————————— −1.8 eV

—————————————————— −4.0 eV

—————————————————— −16.0 eV

i Determine the minimum energy required in joules to eject an electron in the lowest state from the atom. (3)

ii Assuming that energy level n has energy k/n^2 determine the energy of level $n = 4$ in electronvolts. (3)

iii Determine the wavelength of the radiation associated with a transition from level $n = 2$ to level $n = 3$. (3)

iv Name the region of the electromagnetic spectrum in which this radiation is found. (1)

v A tube containing this fictitious element in gaseous form is placed between a beam of white light and a prism. The spectrum produced is crossed by dark lines. Explain the existence of these lines. (4)

International Baccalaureate, Physics, Subsidiary Level, Nov 1997

28 Nuclear stability and decay

THE BIG QUESTIONS

- What holds nuclei together?
- Why do some nuclei change?
- What kinds of change take place?
- What are the effects of nuclear change on the material in which it takes place?

KEY VOCABULARY

alpha particle   antimatter   antineutrino   antiproton   beta⁻ particle
beta⁺ particle   charge number   conservation rules   decay curve
differential equation   electron capture   excited state (of nucleus)
exponential decay   fission   fusion   ground state (of nucleus)   half-life
mutual annihilation   neutrino   neutron-rich   nucleon emission   nuclide
positron   proton-rich   *Q*-value   random process   strong nuclear force
weak nuclear interaction

BACKGROUND

There is nothing new about radioactive materials. They exist in the natural environment, and even our own bodies contain radioactive materials with huge numbers of decays taking place every second. The interior of the Earth would have cooled long ago without the energy released by radioactive emissions in the material from which the planet is made. What *is* relatively new is our ability to detect the ionising radiations, and to make use of radioactive materials for their ionising and their penetrating abilities.

Figure 28.1
We live on a radioactive planet. It is energy made available by radioactive processes that keeps the interior of the Earth at high temperature, resulting in volcanic activity.

The strong nuclear force

In a world where only the forces of gravity and electricity existed, there would be no nuclei. The force of gravity is far too weak to balance the huge electrical repulsion between protons and so it couldn't hold the nucleons close together. This alone is enough to suggest that there is another kind of force responsible for the attraction that nucleons exert on each other. It must be a strong force to overcome the electrical repulsion. It has the simple and sensible name of the **strong nuclear force**.

This name suggests that the force can be stronger than the electric force, which it can, but it is a force that acts over a very small range of distances – in the region of 10^{-15} m or 1 fm (femtometre). At distances bigger than this the force rapidly decreases and so there are no nuclei with diameters bigger than a few femtometres. We do not experience the strong nuclear force directly in everyday life, and even at the scale of the atom and its orbiting electrons the force is effectively absent.

Models of the effects of the strong nuclear force

I What are the differences, apart from the obvious scale, between the behaviour of a cluster of Velcro-coated balls and the behaviour of a nucleus?

Velcro provides a fairly strong force between two surfaces. The force only acts over short distances, and so it provides a simple model for the force that acts between two nucleons (Figure 28.2). It is, however, a model of only limited usefulness because a nucleus does not behave like a collection of Velcro-covered balls. Nucleons in a nucleus have energy – as if they are in constant motion, like the molecules in a tiny drop of liquid. The liquid drop is a more useful model for the nucleus. It helps us to think about nuclear stability, nuclear emissions and radioactive decay, as well as nuclear fission and fusion, as we shall see in the following sections.

Figure 28.2
Models of the nucleus: the 'Velcro' model is a useful way of drawing nuclei to show their neutron and proton structures, but nuclei behave more like drops of liquid.

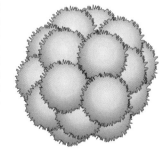
'Velcro' balls

liquid molecules – particles in continuous random motion

Nuclides

The term 'isotope' is used in connection with a named element (see page 72). We can talk about the isotopes of carbon or the isotopes of iron, for example. **Nuclide** is a useful term that describes *any* possible nuclear structure, and not just that of a particular element (Figure 28.3).

Figure 28.3
A selection of nuclides.

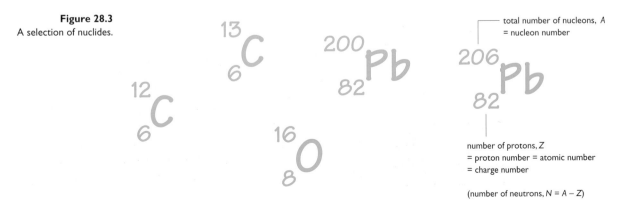

$^{13}_{6}C$ $^{200}_{82}Pb$

$^{12}_{6}C$

$^{16}_{8}O$

total number of nucleons, A
= nucleon number

$^{206}_{82}Pb$

number of protons, Z
= proton number = atomic number
= charge number

(number of neutrons, $N = A - Z$)

Alpha emission and nuclear change

An arrangement of two protons and two neutrons is particularly stable. A helium nucleus is such a cluster. In a larger nucleus in which nucleons are moving like molecules of a liquid drop, two protons and two neutrons can exist as a cluster at least for a short time within the nucleus. Almost always, such a group of nucleons will not have nearly enough energy to escape from the influence of the strong nuclear force. But sometimes the cluster does get away (Figure 28.4). It speeds into the world outside the nucleus, where, in the unlikely event that it encounters human detection equipment, we can recognise it and call it an **alpha particle**. The parent nucleus has simply lost two protons and two neutrons, and, like a gun firing a bullet, it recoils.

Figure 28.4
A representation of alpha decay.

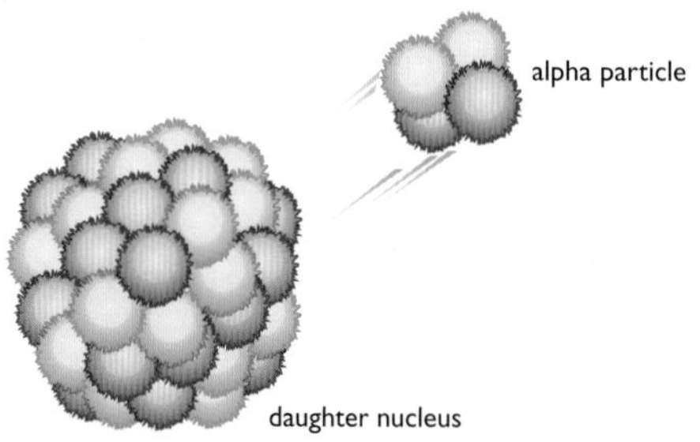

An example of alpha decay is:

$$^{220}_{86}\text{Rn} \longrightarrow \ ^{216}_{84}\text{Po} + \ ^{4}_{2}\alpha + Q$$

parent   daughter
nuclide   nuclide

2 Is there a comparable process to alpha emission in the behaviour of liquid drops?

3 $^{238}_{92}\text{U}$, a uranium isotope, decays by alpha emission. Identify the daughter nuclide of this process.

Note that emission of an alpha particle by a nucleus reduces its nucleon number by four and its proton number by two.

Q is a quantity of energy, called the **Q-value** of the decay. It is the energy that is carried away as the kinetic energy of the alpha particle and the recoiling nucleus. Some alpha decays also involve emission of gamma rays (see page 294), and then the gamma ray photons carry away some of the Q-value energy. This energy is related to their frequency, ν, by $E = h\nu$.

The energy spectrum of alpha decays

The Q-value of the decay of a particular alpha-emitting nuclide is the same for all of its decays. A simple alpha decay involves just two particles – the nucleus and the alpha particle. Both gain kinetic energy from the decay. The alpha, being the smaller body, has the higher speed, and we say that the nucleus recoils. Since there are only two particles, which are the same size in all decays and always have the same amount of energy to share, they always share the energy in the same way. So all of the alpha particles emitted in such a process have the same energy.

An energy spectrum shows the number of alpha particles plotted against their energy. For this simple two-particle process the graph is a simple one (Figure 28.5).

Figure 28.5
Energy spectrum for the alpha decay of radon-220.

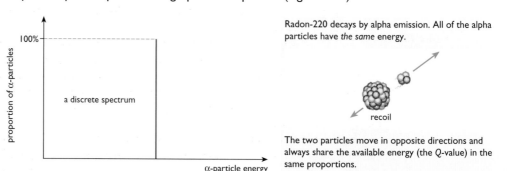

a discrete spectrum

proportion of α-particles

100%

α-particle energy

Radon-220 decays by alpha emission. All of the alpha particles have the *same* energy.

recoil

The two particles move in opposite directions and always share the available energy (the Q-value) in the same proportions.

(When gamma rays are emitted at the same time as an alpha particle, the situation is a little more complicated.)

The energy spectrum of beta decays

Decays of beta-emitting nuclides also have Q-values which are characteristic of each nuclide and its decay. That is, in the decay of a particular nuclide there is a fixed amount of energy that is made available to the decay products. But the energy spectrum for beta decay is very different from that for alpha decay. The beta particles do not all carry away the same amount of kinetic energy, but have a continuous range of kinetic energies up to the maximum possible, which is the total Q-value of the decay (Figure 28.6).

Figure 28.6
Typical energy spectrum for beta decay.

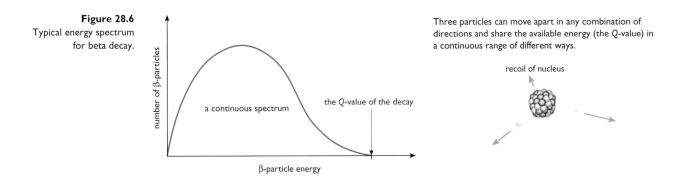

This spectrum was detected in the early 1930s, and a solution was suggested by an Austrian physicist called Wolfgang Pauli. This suggestion was that there was a third particle involved, which nobody had found because it was difficult to detect. Since the three particles can travel away from the site of decay in all possible combinations of directions, they share the momentum (see Chapter 14) and the energy in all possible ways. Sometimes it is the beta particle that travels away with most of the kinetic energy, and sometimes it is the nucleus or the third particle which carries most of the energy away, leaving little for the beta particle. The third particle has come to be called the **antineutrino**, for which the standard symbol is $\bar{\nu}$.

Beta⁻ emission and nuclear change

Beta emission seems to be not quite so simple as alpha emission in another way. The emission takes place from the nucleus but whereas an alpha particle is made of nucleons, and is itself identical to a small nucleus, the beta particle involved here is an electron. We can call these electrons or negatively charged beta particles **beta⁻ particles**, to distinguish them from the less common positively charged beta⁺ particles (see the next section). The structure of nuclei does not include electrons, so we are forced to recognise that beta⁻ particles must be created inside the nucleus at the time of decay.

There are few free neutrons in the Universe. That is because neutrons are unstable. The average lifetime of a free neutron is a matter of minutes. They decay, creating a proton, an electron and an antineutrino. Only inside nuclei are neutrons relatively stable. In the nuclei of some materials they are as stable as it is possible to be, while in others they are not. These latter materials emit beta⁻ radiation. And along with the electron or beta⁻ particle, a nucleus of such a material also emits an antineutrino.

Inside the unstable nucleus, it is an individual neutron that decays. The newly created proton remains in place. The antineutrino is not detectable directly (see Figure 28.7 overleaf). What we can detect, by its ionising effect in the human-sized world, is the high-energy electron, or beta⁻ particle:

$$\mathstrut_{0}^{1}\mathrm{n} \rightarrow \mathstrut_{1}^{1}\mathrm{p} + \mathstrut_{-1}^{0}\mathrm{e} + \bar{\nu} + Q$$

Figure 28.7
A neutrino detector.

Neutrinos and antineutrinos were impossible to detect in the 1930s when Pauli suggested their existence, and detection is not much easier now.

At Los Alamos in southwest USA, this detector is now full of 200 tonnes of mineral oil. Neutrino events within the oil create blue flashes of 'Cerenkov' light, which can be detected by the 1220 photomultipliers around the walls of the chamber.

The participation of a third particle in beta decay, the antineutrino, results in a varying three-way split of the available energy, or Q-value, leading to a continuous spectrum of beta particle energies.

4 What is the charge number of a neutron?

5 When is a neutron more stable – when it is free or when it is part of a nucleus of carbon-12? (Carbon-12 is not a radioactive material.)

6 Carbon-14, $^{14}_{6}$C, decays by beta⁻ emission. Write down the change to identify the daughter material.

Note that the product particles of neutron decay are energetic. The change has a positive Q-value. The process of neutron decay cannot be explained by gravitational, electrical or the strong nuclear force. Again, nature forces us to invent a new way in which matter interacts with matter. It's called the **weak nuclear interaction**.

An example of beta⁻ decay is:

$$^{40}_{19}\text{K} \rightarrow ^{40}_{20}\text{Ca} + ^{0}_{-1}\text{e} + \bar{\nu} + Q$$

When a nucleus emits a beta⁻ particle then inside the nucleus a neutron effectively turns into a proton; the proton number increases by one, while the total number of nucleons doesn't change.

The beta⁻ particle seems to have a proton number of −1. This by itself does not make a lot of sense. Some people prefer to call this the **charge number** rather than the proton number. Since the electron has a negative charge which is of the same size as that of a proton, this value of −1 makes good sense.

Some exotic modes of decay

Some nuclides decay by the emission of a particle similar to an electron but with positive charge. This is called a **beta⁺ particle**, or a **positron**. It is an example of an **antimatter** particle – the electron is matter, the positron is antimatter (Figure 28.8).

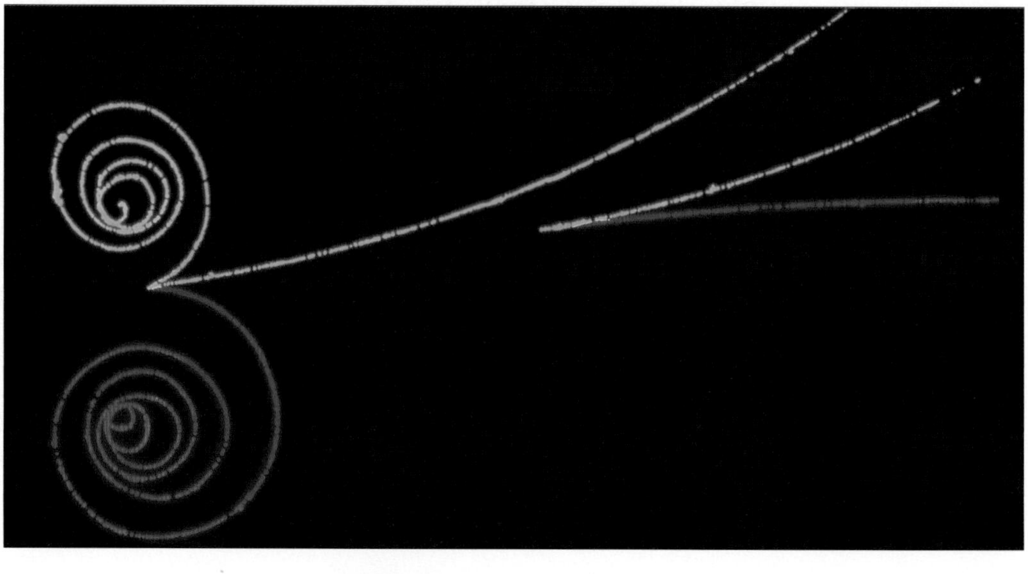

Figure 28.8
An electron and a positron, created together from the energy of a photon. Their paths in a bubble chamber have similar general shapes but curve in opposite directions in the magnetic field, as we would expect for two particles of the same mass but opposite charge.

We live in a Universe which has the possibility of two kinds of material. One of these is everyday material, or 'matter'. Every particle of matter has a 'mirror image' particle which is its antimatter equivalent. Matter–antimatter pairs of particles include the electron and the positron, the proton and the **antiproton**, and the neutrino and the antineutrino. There is far more matter than antimatter in the Universe that we can detect. There seem to be far more electrons than positrons, far more protons than antiprotons. Scientists are still trying to understand why this is, but the Universe would be very different from the one we are part of, if quantities of matter and antimatter were the same.

An example of beta$^+$ or positron decay is:

$$^{15}_{8}O \rightarrow {}^{15}_{7}N + {}^{0}_{1}e + v + Q$$

The proton number of the nuclide has decreased, though the nucleon number has not changed. A proton has changed into a neutron by emission of the positron:

$$^{1}_{1}p \rightarrow {}^{1}_{0}n + {}^{0}_{1}e + v + Q$$

Note that a positron has a charge number of 1, the same as that of a proton. Also note that the creation of a positron is accompanied by the creation of a **neutrino**, v, rather than an antineutrino, for which we use $\bar{v}$.

Nuclei, of course, are normally surrounded by electrons. Some nuclei can absorb one of their orbiting electrons in a process called **electron capture**. For example:

$$^{55}_{26}Fe + {}^{0}_{-1}e \rightarrow {}^{55}_{25}Mn + v + Q$$

Another example of electron capture is

$$^{81}_{36}Kr + {}^{0}_{-1}e \rightarrow {}^{81}_{35}Br + v + \gamma + Q$$

In this case, note that gamma emission also takes place.

In both situations the nucleon number has stayed the same, but the proton number has decreased. The electron has entered the nucleus where it is has ceased to exist as such, but has joined with a proton to create a neutral neutron:

$$^{1}_{1}p + {}^{0}_{-1}e \rightarrow {}^{1}_{0}n + v + Q$$

There are even a few nuclei that can emit single nucleons. An example of **nucleon emission** is:

$$^{16}_{6}C \rightarrow {}^{15}_{6}C + {}^{1}_{0}n + Q$$

Note there is an important pattern in *all* decays. The nucleon number and the charge number have the same *totals* before and after the change. We say that nucleon number and charge number obey **conservation rules**.

Some more examples of decay processes, with their Q-values

The decay of carbon-11 by β^+ emission:

$$^{11}_{6}C \rightarrow {}^{11}_{5}B + {}^{0}_{1}e + v + Q \, (= 0.97 \text{ MeV})$$

The decay of oxygen-13 by β^+ and proton emission:

$$^{13}_{8}O \rightarrow {}^{12}_{6}C + {}^{0}_{1}e + {}^{1}_{1}p + v + Q \, (= 6.97 \text{ MeV})$$

The decay of thallium-207 by β^- and gamma emission:

$$^{207}_{81}Tl \rightarrow {}^{207}_{82}Pb + {}^{0}_{-1}e + \bar{v} + \gamma + Q \, (= 2.32 \text{ MeV})$$

The decay of radon-222 by α emission:

$$^{222}_{86}Rn \rightarrow {}^{218}_{84}Po + {}^{4}_{2}\alpha + Q \, (= 5.49 \text{ MeV})$$

Gamma emissions and energy

Some nuclei are excited. Almost all of the nuclei around us are not. But evidence for the existence of excited states comes from gamma ray emission. Gamma emission does not change nuclear structure in terms of numbers of protons and neutrons, but it carries away energy. The energy carried away is the energy of the photon, given by $E = hf$ (or $E = h\nu$, but we are avoiding the use of ν here since we've used it to represent the neutrino).

Having emitted such a gamma ray photon, a nucleus does not repeat exactly the same process. If it has lost all of its excess energy it now stays inactive. The process of the photon emission seems to have brought the nucleus down from its **excited state** to a more settled state in which it has no excess energy to lose – its **ground state**.

The terms used here are similar to the terms used when talking about the energy levels of complete atoms and the energy changes that take place when electrons make transitions between levels. In both cases 'ground state' means having no energy to lose, and 'excited state' means having energy to lose. But take care – do not confuse atomic processes with nuclear processes! A nucleus is tiny compared with the size of the atom with its electrons in orbit, but the sizes of the energy changes that are involved in nuclear processes are generally very much larger than those involved in electron transitions. The consequence of this is that photons emitted by nucleons carry more energy, and have higher frequencies, than photons emitted as a result of electron transitions.

Some nuclei have a number of possible excited states. It is unusual for a nucleus to go directly from its ground state to an excited state – it would need to accept a lot of energy for this to happen. But emission of alpha and beta radiations can leave the daughter nucleus in an excited state. So the alpha or beta emission is then followed by gamma ray emission (Figure 28.9).

7 Why are gamma rays in a different part of the spectrum to (almost all) electromagnetic radiation that is emitted by atoms due to electron transitions?

Figure 28.9
Decay of a nucleus of carbon-15 by beta⁻ emission can leave the daughter nucleus, nitrogen-15, in an excited state.

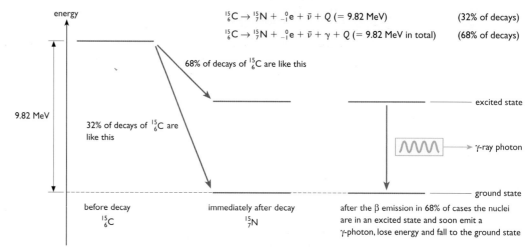

$$^{15}_{6}\text{C} \rightarrow {}^{15}_{7}\text{N} + {}^{0}_{-1}\text{e} + \bar{\nu} + Q \ (= 9.82 \text{ MeV}) \qquad \text{(32\% of decays)}$$
$$^{15}_{6}\text{C} \rightarrow {}^{15}_{7}\text{N} + {}^{0}_{-1}\text{e} + \bar{\nu} + \gamma + Q \ (= 9.82 \text{ MeV in total}) \qquad \text{(68\% of decays)}$$

energy

9.82 MeV

68% of decays of $^{15}_{6}\text{C}$ are like this

32% of decays of $^{15}_{6}\text{C}$ are like this

excited state

γ-ray photon

ground state

before decay
$^{15}_{6}\text{C}$

immediately after decay
$^{15}_{7}\text{N}$

after the β emission in 68% of cases the nuclei are in an excited state and soon emit a γ-photon, lose energy and fall to the ground state

Fusion and nuclear change

Nuclear **fusion** is a process of merging of pairs of nuclei. It is a difficult process to start, because two nuclei repel each other very strongly by electrical force. However, if they have enough energy then they can overcome this repulsion and get close enough together for the strong nuclear force to take control. When small nuclei fuse then energy is transferred outwards. This is the process that takes place in stars, and seems to be the process by which all the large nuclei (bigger than hydrogen and helium) in the Universe were created. Examples of fusion reactions in stars (ignoring energy changes) are the fusion of nuclei of two isotopes of hydrogen to create helium, and the fusion of hydrogen and carbon to create nitrogen:

$$^{1}_{1}\text{H} + {}^{2}_{1}\text{H} \rightarrow {}^{3}_{2}\text{He}$$

$$^{1}_{1}\text{H} + {}^{13}_{6}\text{C} \rightarrow {}^{14}_{7}\text{N}$$

Fission and nuclear change

Very large nuclei – those with more than about 200 nucleons – are all unstable. Some emit ionising radiation. Some can wobble and then fall apart, forming two smaller product or daughter nuclei and usually setting a few neutrons free in the process. This is nuclear **fission**. Like the fusing of small nuclei, the fissioning of large nuclei transfers energy into the surroundings.

Fission is the process by which nuclear power stations operate. A problem that continues to blight the value of nuclear power is that the daughter nuclei are almost always unstable and therefore radioactive. They are a very troublesome component of nuclear waste. A typical fission reaction in a nuclear reactor is:

$$^{235}_{92}U + ^{1}_{0}n \rightarrow ^{236}_{92}U \rightarrow ^{141}_{56}Ba + ^{92}_{36}Kr + 3^{1}_{0}n$$

The nuclear fuel here is uranium-235. Note that this does not fission directly, but is induced to fission by absorbing a neutron to become uranium-236. Most such nuclei of uranium-236 survive very short times before fissioning. The barium, Ba, and krypton, Kr, are the daughter nuclides here. Fission happens very quickly, and the nucleons do not have time to be sorted into any particular arrangements, so many different daughter nuclides are created in nuclear reactors and bombs. The tendency is for them to be **neutron-rich** – that is, to have a high proportion of number of neutrons relative to number of protons in comparison with stable nuclides of the same size. Neutron-rich nuclides are usually beta⁻ emitters.

Patterns of stability and instability

We can write a list of all stable nuclides – that is, all nuclides that do not experience any measurable radioactive decay. It's not a very long list. Only about 80 elements have any stable isotopes. Most of these elements have one stable isotope; some have two or three. Elements such as uranium with large nuclei have no stable isotopes.

We can go further than a simple list and start to look for patterns of structure that relate to stability. We can do this by plotting number of neutrons N against proton number Z. Each nuclide is a point on the chart, and the points form a remarkably thin line (Figure 28.10).

8 Make a list of similarities and differences between fusion and fission.

9 Apply conservation rules to complete the following, and identify the daughter nuclides:

$$^{1}_{1}H + ^{14}_{7}N \rightarrow ^{15}_{?}?$$

$$^{236}_{92}U \rightarrow ^{120}_{?}? + ^{?}_{50}Sn + 3^{1}_{0}n$$

10 We never see raindrops bigger than peas. What would happen to a bigger drop? How does this compare with the behaviour of nuclei?

Figure 28.10
A chart showing the stable nuclides. Only the nuclides shown by red dots are stable. The fact that the chart forms such a thin line suggests that the balance of proton number to neutron number is important for the stability of a nucleus.

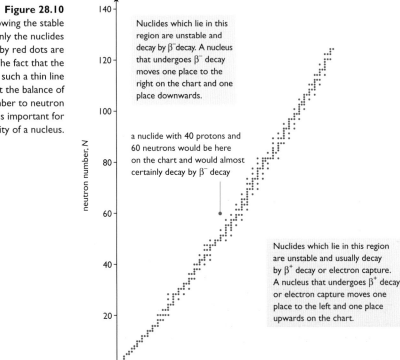

Nuclides which lie in this region are unstable and decay by β⁻decay. A nucleus that undergoes β⁻ decay moves one place to the right on the chart and one place downwards.

a nuclide with 40 protons and 60 neutrons would be here on the chart and would almost certainly decay by β⁻ decay

Nuclides which lie in this region are unstable and usually decay by β⁺ decay or electron capture. A nucleus that undergoes β⁺ decay or electron capture moves one place to the left and one place upwards on the chart.

neutron number, N

proton number, Z

We can begin to match the shape of the line with what we already know. Smaller stable nuclei have more equally balanced numbers of protons and neutrons than do large stable nuclei. Large nuclei with roughly equal numbers of protons and neutrons are, the line tells us, not stable. For larger nuclei the strong nuclear force, which acts over very short distances only, has more of a problem overcoming the electrical repulsion of the protons. The mix of nucleons seems to have to be strongly diluted by neutrons in order to create stable nuclei.

We could go further with our chart of nuclides and add the unstable ones. This would give us far more dots on the chart. In fact we could draw a dot at any point on the chart to represent a potential nuclide – but if the dot is far from the stability curve then it is unlikely that such a nucleus could exist for more than an extremely short time. As a general trend, but not an absolute rule, the further the dot is from the line of stability then the more unstable the nuclide.

Nuclides that fall apart before we can ever hope to detect them are not very interesting in themselves. But there are many unstable nuclides that last long enough to be detected. These are the radioactive nuclides (or radionuclides) that are listed in tables in data books. We could add these to our chart, but there are far more of them than there are stable nuclides so it would be a time-consuming exercise. However, it is worthwhile to look at the patterns that emerge when we examine just two sections of the chart and add the unstable nuclides (Figure 28.11).

There are clear patterns. Nuclides above and to the left of the line of stability are neutron-rich, and tend to decay by processes that reduce the number of neutrons. This is usually beta⁻ decay. Nuclides below and to the right of the line are **proton-rich**, and tend to decay by processes that increase the ratio of neutrons to protons – such as beta⁺ decay, electron capture and alpha decay.

11 Use the chart of nuclides (Figure 28.10) to explain why the daughter nuclides of fission tend to be neutron-rich.

12 Find out the structures of stable isotopes of barium and krypton.
  a Does the information suggest that the daughter nuclides of the fission of uranium-236 as shown on page 295 will be only slightly or highly unstable?
  b Sketch a section of a chart of nuclides showing the positions of the six stable isotopes of krypton.

13 Explain why spent nuclear fuel is a strong source of beta⁻ radiation.

14 Suggest likely modes of decay for the following nuclides:

  $^{216}_{84}$Po $^{13}_{8}$O $^{131}_{53}$I (iodine-131)

15 Suggest why larger nuclides have a higher neutron:proton ratio than smaller ones.

16 Say whether the following increase or decrease the neutron:proton ratio of a nuclide:
  a beta⁻ emission
  b beta⁺ emission
  c alpha emission
  d electron capture.

Figure 28.11
Patterns of instability for **a** large and **b** small nuclei.

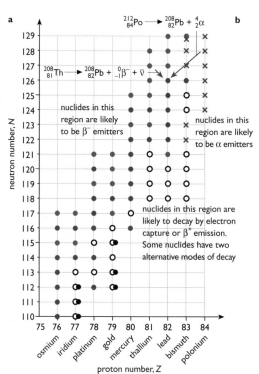

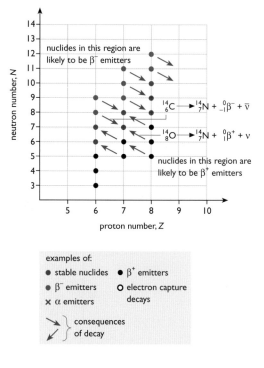

examples of:
● stable nuclides   ● β⁺ emitters
● β⁻ emitters   ○ electron capture decays
× α emitters

↘ ↙ consequences of decay

Levels of instability and randomness of decay

The material around you is made mostly (though not entirely) of atoms with stable nuclei. Unstable nuclei were created by the same processes of fusion in stars as were stable nuclei, but most of the unstable nuclei that were present in the newly-formed Solar System have decayed and turned into stable nuclei a long time ago. The Earth is still radioactive, but not nearly as radioactive as it once was.

Consider uranium-238, $^{238}_{92}U$. Its nuclei are mildly unstable. They change, or decay, by the following process:

$$^{238}_{92}U \rightarrow {}^{234}_{90}Th + {}^{4}_{2}\alpha + Q$$

But the decay of a particular nucleus is completely unpredictable. If you could observe a single nucleus it might decay within an instant, or it might remain unchanged for billions of years. We say that its decay is a **random process**.

Consider polonium-212, which decays as follows:

$$^{212}_{84}Po \rightarrow {}^{208}_{82}Pb + {}^{4}_{2}\alpha + Q$$

Again, if you could watch a single nucleus the decay could be immediate or it might not happen for a very long time. But the polonium-212 nuclei are much more unstable than the uranium-238 nuclei, so the probability is that you will not have to wait so long for this.

If, rather more realistically, you observe a small sample, big enough to see, of the polonium, then enough nuclei will decay in each second for the sample to be highly radioactive. A sample of the same number of uranium-238 atoms would be less radioactive.

To compare the instability or the level of radioactivity of different nuclides we need to consider such samples rather than individual nuclei, whose behaviour is both unobservable and unpredictable. So consider a sample of a particular nuclide that contains a population, or number, of N identical nuclei that have not yet decayed. In a material that is radioactive this population changes with time. The population changes due to decay of nuclei – they change into new types of nuclei (the daughter nuclei) and though these stay in the sample we do not count them as part of the original population. The rate of change of the original population is dN/dt and can be measured in becquerel, Bq. This has a negative value because N is decreasing and not increasing with time, t. The activity of the sample, a positive quantity, is $(-dN/dt)$ (see page 73). Activity is also measured in becquerel.

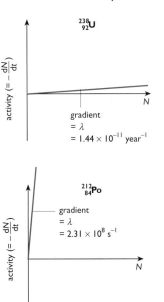

Decay events happen randomly – it is impossible to predict exactly when a particular nucleus will decay. But each radioactive nucleus has a fixed *probability* of decaying within a certain time, and in a large population there *are* predictable patterns of decay. The larger the population, the more nuclei will decay during a given interval of time, so the activity $-dN/dt$ is bigger when N is bigger. In fact the mathematical relationship between these quantities is quite simple – it is a proportionality (Figure 28.12):

$$-\frac{dN}{dt} \propto N$$

or

$$\frac{dN}{dt} = -\lambda N$$

Note that the two graphs in Figure 28.12 have different scales – the gradients of the two lines are very different. If they were drawn to the same scale it would not be possible to distinguish each line from the nearby axis.

17 Ink is a suspension of coloured grains of material in liquid. You take a sample of ink and dilute it by, say, 50%. You take the diluted sample and repeat the dilution process, and you do this several times, always by 50%.

a What will happen to the appearance of the ink?

b Will the appearance of the ink change more during the first dilution process or during the tenth?

c Does the population of coloured grains per millilitre, N, change most during the first dilution or the tenth?

d If time, t, is measured as number of dilutions, would you expect the process to obey the relationship $dN/dt \propto -N$?

18 DISCUSS

During the 20th century the world population rose from about 1 billion people to about 6 billion.

a Was the rate of change of world population bigger at the beginning or the end of the century?

b Would you expect the rate of change of the human population to be always, sometimes or never proportional to the size of the population?

For any radioactive sample, the rate of change of population is proportional to the population itself. The constant of proportionality is $-\lambda$. λ is the decay constant (see page 74), and is a characteristic of the nuclide.

Note that N is a simple number and so is dimensionless, and dN/dt has unit s^{-1}. So, for the equation to be homogeneous, λ also has SI units s^{-1}. On one of the graphs on page 297 we used year $^{-1}$, not an SI unit, but one that is nevertheless easily understood.

Exponential decay

In radioactive decay the rate of change of a population is always proportional to the population. Proportionality is a simple relationship, but when we write this relationship down as:

$$\frac{dN}{dt} = -\lambda N$$

we see that it is a **differential equation** – one which involves calculus. Fortunately, this is a relatively simple differential equation, and equations of this form have a standard non-differential companion equation. (This companion equation is called the 'solution' to the differential equation.) Suppose that we have an initial population of N_0 at time $t = 0$, which becomes N after time t. We can say that, if $dN/dt = -\lambda N$, then

$$N = N_0 e^{-\lambda t}$$

e is called the 'base of natural logarithms' and has a value of 2.718. The relationship is the **exponential decay** relationship. We can represent it as a graph of population against time. This shows the decline in the population, and is called a **decay curve** (Figure 28.13). The gradient of the decay curve at any time t is the rate of change of population, dN/dt, at that time.

Figure 28.13
Exponential decay of the population of samples of uranium-238 and polonium-212. Exponential change is the inevitable result of any proportional relationship linking rate of change of a quantity with its own value.

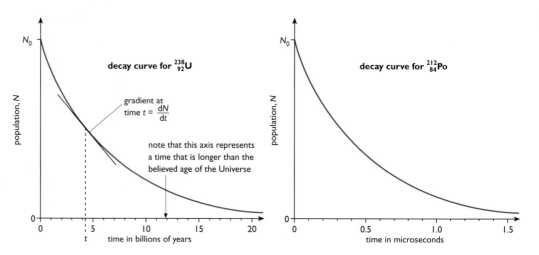

Exponential curves are the result of a simple initial proportionality linking a quantity to its own rate of change. Because of this, exponential curves are common – you'll find them in chemistry, biology, geography, economics and so on, as well as in physics.

19 Imagine a hall full of 500 students, each with a die which they throw. Everybody throws their dice at the same time. Those with a six leave the room. Those with other numbers stay and throw again, and again the students throwing a six can leave. They repeat this until there is nobody left.

a Sketch a graph of number of students still in the room (the population of the room) on the y-axis against number of throws on the x-axis.

b Use the graph to estimate how many throws will be needed before half the students have left.

c Is it possible to predict exactly how many throws will be needed to empty the hall?

Note that since activity $(-dN/dt)$ is proportional to population, a graph of activity against time (Figure 28.14) has the same shape as a graph of population, N, against time.

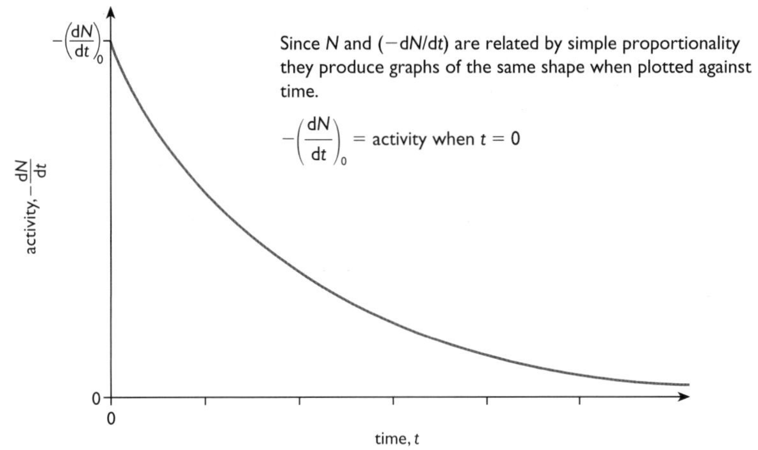

Since N and $(-dN/dt)$ are related by simple proportionality they produce graphs of the same shape when plotted against time.

$$-\left(\frac{dN}{dt}\right)_0 = \text{activity when } t = 0$$

Figure 28.14
The activity of a sample falls off exponentially.

Half-life

If time is plotted on the x-axis and population of undecayed nuclei on the y-axis then the time taken for the population to change by any particular proportion is always the same. This is a feature of an exponential decay curve. The time taken for an initial population of a radioactive nuclide to halve is called the **half-life** (Figure 28.15) and it is a characteristic of the nuclide. We can abbreviate half-life as $T_{1/2}$.

Half-lives of some unstable nuclides are shown in Table 28.1.

Figure 28.15
For a particular nuclide the time taken for a population to halve is always the same – it is its half-life, $T_{1/2}$.

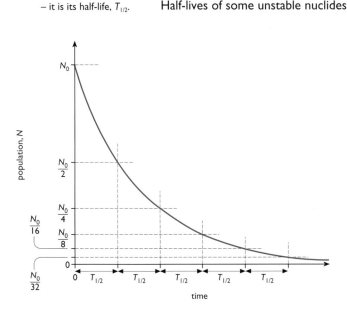

Table 28.1
Modes of decay and half-lives of some nuclides.

Nuclide	Mode of decay	Half-life, $T_{1/2}$
$^{14}_{6}C$	beta$^-$	5730 years
$^{15}_{6}C$	beta$^-$	2.4 s
$^{13}_{8}O$	beta$^+$	8.7 ms
$^{15}_{8}O$	beta$^+$	2.06 min
$^{19}_{8}O$	beta$^-$	27 s
$^{40}_{19}K$	various	1.3×10^9 years
$^{55}_{26}Fe$	electron capture	2.6 years
$^{90}_{37}Rb$	beta$^-$	2.6 min
$^{212}_{84}Po$	alpha	3×10^{-7} s
$^{216}_{84}Po$	alpha	0.15 s
$^{228}_{90}Th$	alpha	1.9 years
$^{235}_{92}U$	alpha	1.6×10^5 years
$^{238}_{92}U$	alpha	4.5×10^9 years

Half-life and decay constant – an inverse relationship

The decay constant is the constant of proportionality λ in the relationship:

$$-\frac{dN}{dt} = \lambda N$$

A nuclide with a larger decay constant is more unstable. Such a nuclide also has a shorter half-life. There is an inverse relationship between decay constant and half-life. We need to look at this more closely to see the precise nature of the relationship.

The half-life is the time that elapses for the undecayed (unchanged) population to halve. If the initial population is N_0 then after one half-life this has become $N_0/2$. That is, when $t = T_{1/2}$, $N = N_0/2$. Substituting these into the general formula (page 298):

$$N = N_0\,e^{-\lambda t}$$

we get:

$$\frac{N_0}{2} = N_0\,e^{-\lambda T_{1/2}}$$

so:

$$\frac{1}{2} = e^{-\lambda T_{1/2}}$$

and, taking the inverse:

$$2 = e^{\lambda T_{1/2}}$$

From this, the mathematics of logarithms then tells us that:

$$\log_e 2 = \lambda T_{1/2}$$

The logarithm of a number to the base e is just another number; $\log_e 2$ is 0.693, for example. Your calculator may allow you to calculate more such logarithms. Note that $\log_e x$ can also be written $\ln x$, and so if your calculator has a button marked $\ln x$ that is the one you should use here. For your physics, you do not need to understand this last step, in which

$$2 = e^{\lambda T_{1/2}} \quad \text{became} \quad \log_e 2 = \lambda T_{1/2}$$

unless you wish to study the maths further.[1] You *do* need to know and to be able to use the relationship between decay constant and half-life:

$$\lambda = \frac{\log_e 2}{T_{1/2}}$$

$\log_e 2$ is a number, 0.693, and so:

$$\lambda = \frac{0.693}{T_{1/2}}$$

Note that λ has the unit s^{-1} and half-life is measured in seconds, s. The equation is homogeneous.

---

20 Calculate the decay constants of each of the nuclides in Table 28.1.

21 Stable isotopes of oxygen have mass numbers of 16, 17 and 18. Why is it unsurprising that oxygen-13 has a much shorter half-life than oxygen-15?

22 **a** The half-life of uranium-238 is very similar to the believed age of the Earth. What proportion of the uranium-238 that was present when the Earth formed is still here now?

b About 99.3% of the Earth's uranium is uranium-238 and the rest is uranium-235. Would this figure of 99.3% have been larger or smaller when the Earth was formed?

23 **a** Sketch a graph of decay constant against half-life for ranges of half-life from 0 to 30 s. Mark the positions of the nuclides carbon-15 and oxygen-19 on the line (see Table 28.1).

b What will be the gradient of a graph of λ against $1/T_{1/2}$?

---

[1]The logarithm of a number is the power to which the base, here e (which is equal to 2.718), must be raised to give the number (see *Further Advanced Physics*).

Figure 28.16

Granite contains small but significant quantities of rubidium compounds. Measurements of rubidium–strontium content show how old the rock is.

Radioactive dating of rocks

Rubidium–strontium

There are about 90 elements in and on the Earth. Many have two or more stable isotopes, while others have just one. Any sample of earthly material – rock, air or human tissue – is made almost entirely of stable nuclides. But unstable nuclides are present. These are either ones with very long half-lives (hundred of millions of years or longer) or they are ones which are produced continuously on Earth, for example by the radioactive decay of other materials.

Rubidium is a soft and very reactive metal, found in compounds in the rocks of the Earth's crust (Figure 28.16). Most rubidium on the Earth is rubidium-85, the rest being rubidium-87.

Since the two isotopes are chemically identical they always exist together. Rubidium-87 is radioactive, but only just. It has a very long half-life of 47 billion years, and so its activity is very low. It decays to strontium-87 by beta$^-$ decay:

$$^{87}_{37}\text{Rb} \rightarrow {}^{87}_{38}\text{Sr} + {}_{-1}^{0}\text{e} + \bar{\nu} + Q$$

Rubidium decays to strontium in exactly the same way in molten rock as in a solid. A liquid that contains rubidium will also contain some strontium. But, on crystallisation, the rubidium and strontium compounds separate, so that newly crystallised rubidium-bearing rock contains very little strontium. Decay of rubidium continues, and new strontium nuclei are created within the rock. The ratio of strontium-87 to rubidium-87 in a rock sample shows how long the strontium has been accumulating (Figure 28.17). It shows how long the rock has been solid.

Figure 28.17
Decay curve for rubidium-87.

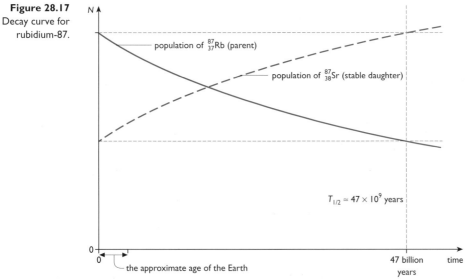

population of $^{87}_{37}$Rb (parent)

population of $^{87}_{38}$Sr (stable daughter)

$T_{1/2} \approx 47 \times 10^9$ years

the approximate age of the Earth

47 billion years

time

Rubidium-87 is only slightly radioactive and has a very long half-life, which can be measured from the activity of a sample containing a known number of nuclei:

$$\text{activity} = \lambda N = \frac{0.693N}{T_{1/2}}$$

$$T_{1/2} = \frac{0.693N}{\text{activity}}$$

Nevertheless, in solid rock the population of rubidium-87 nuclei decreases and that of strontium-87 increases. 'Older' rock has a higher proportion of strontium nuclei than rock that became solid more recently.

Uranium–lead

Silicon and oxygen are by far the commonest elements in the crust of the Earth. Their isotopes, the ones that we can find now, are stable. Time does not seem to affect them. Their abundances have changed very little since the Earth was formed, and so they tell us nothing about the age of the Earth.

The proportion of two isotopes of uranium, uranium-235 and uranium-238, however, has certainly changed over time. Both are radioactive, but uranium-238 is more stable, having a longer half-life. It is not surprising, then, that there is now much more uranium-238 (99.27% of all uranium) than uranium-235 (0.72%) in the Earth's crust. Unfortunately, without knowing the age of the Earth there is no way of knowing how much of each of these isotopes was present at the beginning. So, present-day proportions, which can be found using a mass spectrometer (Figure 28.18), say nothing reliable about the Earth's age or the age of particular rocks.

For geological dating that leaves the possibility of measuring the proportion of a nuclide such as uranium-238 to its decay product trapped in a rock, as is done for rubidium and strontium. But in this case it is made more complicated by the fact that uranium-238 does not decay to a stable nuclide. In fact a uranium nucleus must go through no less than 14 decays before it becomes lead-206, a stable nucleus. This series of decays is called a decay chain (Figure 28.19).

When thinking about an individual nucleus, the half-life can be thought of as the time in which the nucleus has a 50% probability of decay. Most of the nuclides in the decay chain that runs from uranium-238 to lead-206 have half-lives of a few seconds, a few minutes, or a few days. The most stable of them has a half-life which is about 10 000 times shorter than the half-life of uranium-238. So all of these nuclides are only present in tiny amounts in uranium-bearing rock. The overall process is a slow decrease in the proportion of uranium-238 relative to lead-206. It can be summarised as:

$$^{238}_{92}U \rightarrow {}^{206}_{82}Pb + 8\,{}^{4}_{2}\alpha + 6\,{}^{0}_{-1}e + 6\bar{\nu} + Q$$

Thus radioactive dating with uranium and lead is essentially similar to dating with rubidium and strontium. Moon rocks and many meteorites have been dated by the uranium–lead method. The same result keeps turning up each time: the rock seems to have been solid for 4.6 billion years. If the Moon, the meteorites and the Earth all formed from a swirling disc at about the same time, then the Earth is also 4.6 billion years old.

---

24 What is the approximate mass of 10^{18} atoms of rubidium-87:
   a in atomic mass units
   b in kg?
   (The atomic mass unit, u, is useful for comparing masses of nuclides. It is *roughly* (see Chapter 34) the same as the mass of one nucleon:
   $1\,u = 1.66 \times 10^{-27}\,kg$.)

25 a What is the atomic number of strontium-90?
   b What is the mass number (nucleon number) of the most common isotope of rubidium?
   c What is the neutron number of rubidium-87?
   d What is the fundamental difference between rubidium-87 and strontium-87?

26 Explain why the following decays are unsuitable for measuring the age of solid rock:
   a rubidium-90 to strontium-90
   b thorium-228 to radium-224.
   (Refer to Table 28.1, page 299.)

27 Krypton-90 decays by β^- decay into rubidium-90, with a half-life of 33 s.
   a Write out the nuclear reaction using standard symbols.
   b If it were possible to obtain a pure sample of krypton-90, what would be the ratio of the number of atoms of krypton-90 to the number of atoms of rubidium-90 after
   i 11 minutes
   ii 1 second?
   c Why is there no natural krypton-90 on the Earth?

28 One reason that there is more uranium-238 than uranium-235 on the Earth is that uranium-238 is the more stable nuclide. Can you suggest another possible reason?

29 a Sketch a graph to show the change in abundance of uranium-238 in a rock with time.
   b On the same axes, sketch a graph to show the change in abundance of lead-206 with time, assuming that no lead-206 was present when the rock first crystallised.

30 'Unstable nuclides with half-lives of less than 100 million years are not found on Earth unless they are continuously being produced.'
   a What processes on Earth might continuously produce short half-life nuclides?
   b Does the statement provide any evidence relevant to the debate on the age of the Earth?
   c Do you believe that the Earth is a few thousand years old, 4.6 billion years old, or something in between?

Figure 28.18
This mass spectrometer printout shows the relative abundances of isotopes of lead in one sample of rock. Rock can be similarly analysed to determine relative abundances of uranium isotopes.

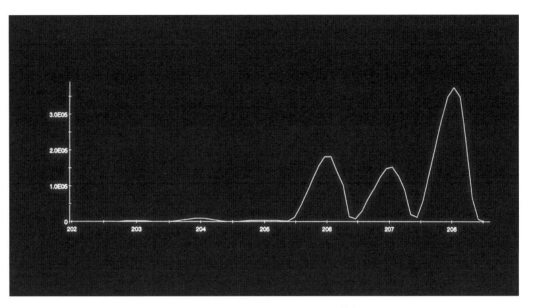

Figure 28.19
Since the Earth formed, about half of its nuclei of uranium-238 have set off on the 'journey' along the decay chain. Once on the decay chain, there is a high probability that a nucleus will reach the other end, and become lead-206, relatively quickly.

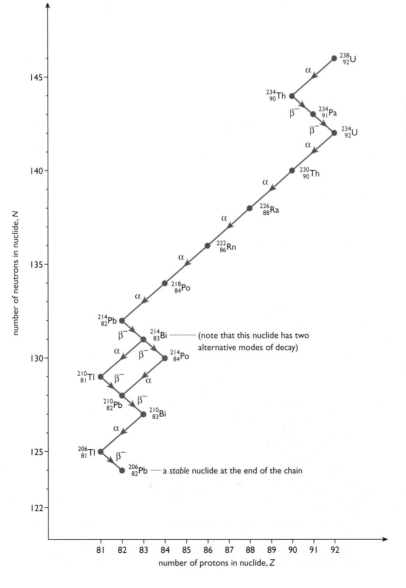

Most of the decay processes shown here have short half-lives, measured in seconds, minutes or days.

(note that this nuclide has two alternative modes of decay)

a *stable* nuclide at the end of the chain

● **Comprehension and application**

Medical radioisotopes

During a positron emission tomography (PET) scan (see page 312), neuroscientists interested in the principles of how our brains work can investigate which parts of the brain are most active at different times – and so they can find out which parts of our brains are involved in different activities, from simple 'motor' activity such as pressing a button to more sophisticated processes such as listening to jokes.

It is gamma radiation that emerges from the head during a PET scan, and detectors arrayed around the head can respond to the emerging photons. But the neuroscientists don't use a simple gamma-emitting radioisotope, as there are no gamma emitters that have suitable properties. A key requirement for PET scans is that the radioisotope should naturally accumulate at places of high brain activity. Another key property for all radioisotopes is half-life. Nuclides with half-lives of a few seconds are of little medical use – by the time a sample of material has been introduced into a patient and has spread through their body the activity has fallen drastically. It would be possible to compensate for this simply by using a bigger sample of material, with bigger total activity. But that would expose the patient to a high dose of the radiation, and it is essential that exposure is kept low. If, on the other hand, a nuclide with a long half-life is used, its activity is low and so large amounts of material have to be used in order to provide a detectable level of emission. If this material stays in the body for a long time then the body is exposed to a high total level of radiation.

The radioisotope for detecting mental activity travels to the places where the brain is working hardest and needs the most energy, so that from these places there is more detectable emission. Oxygen is an element normally found in glucose, and a high concentration of glucose is found in the parts that are most active. It is possible to 'label' glucose molecules with radioactive oxygen, and it so happens that there is an isotope of oxygen that is ideal: oxygen-15.

Oxygen-15 is a beta$^+$ emitter. That is, it emits positrons. It has a half-life of just over two minutes – long enough for glucose to be introduced to the body and distribute itself, including to the energy-hungry brain. The positrons emerge from the oxygen nuclei, and interact with electrons. Antimatter meets matter and **mutual annihilation** takes place, resulting in the creation of a pair of gamma ray photons which travel in opposite directions, through the brain, out through the skull and into the scientists' detectors – showing up the places in the brain where glucose is more concentrated.

Radioisotopes can in principle be introduced into tumours to kill cancerous cells, although healthy cells are also killed, and the patient is exposed to a high total dose of radiation. But by far the greatest use of radioactive materials is as tracers. The oxygen-15 in PET scanning is acting as a tracer, though it is unusual in that it is not the emission directly from the oxygen nuclei that is detected, but the secondary gamma emission resulting from mutual annihilation.

Technetium-99m is an important medical tracer; it is a direct emitter of gamma radiation. The 'm' stands for metastable. It is produced from the decay of molybdenum-99, by beta$^-$ emission. The daughter nucleus in such a decay, the technetium, is in a metastable state – it has an excess of energy relative to the nuclear ground state, but it does not lose this energy immediately after the decay of the parent molybdenum. The technetium 'falls' to its ground state, after some time, with an effective half-life of six hours. So it's ideal for medical purposes because it is a gamma emitter with a half-life that allows it to be introduced into a patient's body in modest amounts, and it can be produced quite simply in a hospital by using a supply of molybdenum.

Figure 28.20
Gamma scanning can be used to monitor the action of the heart. The sensitive camera is detecting emissions from the heart and transmitting images to the screen. Heart muscle can be seen as a purple/orange ring or horseshoe, depending on the angle of the scan, and the chambers as black or dark blue at the centre of these rings.

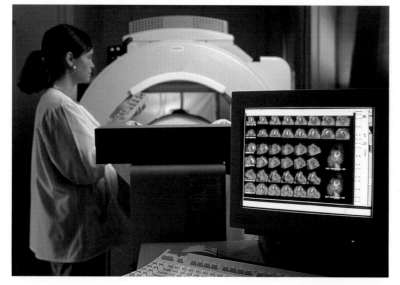

Some other medical radioisotopes are shown in Table 28.2.

Table 28.2
Medical radioisotopes.

Radioisotope	Mode of decay	Half-life
oxygen-15	β^+	2.06 minutes
technetium-99m	γ	6 hours
hydrogen-3 (tritium)	β^-	12.3 years
sodium-24	β^- and γ	15 hours
phosphorus-32	β^-	14.3 days
iodine-131	β^- and γ	8 days

31 a Why does a long half-life make a material unsuitable for use as a medical tracer?
b Why does a very short half-life make a material unsuitable as a medical tracer?
c Why does the half-life of technetium-99m make it very suitable as a medical radioisotope?
32 Suggest reasons why an isotope of hydrogen might be a useful medical radioisotope despite its relatively long half-life.
33 In use as a medical radioisotope, why does a nuclide that emits gamma radiation only (such as technetium-99m) give a lower dose of radiation to the patient's body than does a beta emitter of the same activity?
34 Why are alpha emitters not suitable as medical tracers?
35 A sample of sodium-24 is produced and isolated in a nuclear power station, with a total initial activity of 10^{14} Bq, and transported to a hospital, arriving 10 hours later.
a i What is the decay constant of sodium-24?
ii What was the initial mass of the sample? (The mass of 1 atom of sodium-24 is 24.0 u, and $1\,u = 1.66 \times 10^{-27}$ kg.)
iii Sketch a straight-line graph relating activity to sample population.
iv Sketch a graph relating activity to time.
b Estimate the activity of the sample on arrival. Explain your method.
c If an amount of sodium-24 with a total activity of 10^8 Bq is to be given to a patient in a drink, what mass of the material should be used?

....

Extra skills task Information Technology and Application of Number

Develop a computer program that can produce a decay curve for any decaying sample. It should allow you to produce curves for samples of different size (populations) and decay constant.

• •

Examination questions

1 Nuclei of $^{218}_{84}$Po decay by the emission of an α particle to form a stable isotope of an element X. You may assume that no γ emission accompanies the decay.
a i State the proton number and the nucleon number of X.
ii Identify the element X. (2)
b Each decaying nucleus of Po releases 8.6×10^{-13} J of energy.
i State the form in which this energy *initially* appears.
ii Using *only* the information provided in the question, calculate the difference in mass between the original $^{218}_{84}$Po atom and the combined mass of an atom of X and an α particle. (speed of light in vacuum $= 3.0 \times 10^8$ m s^{-1}) (3)

AQA (NEAB), Particles and Waves, Mar 1999

2 a Define the following terms:
i 'atomic number' (1)
ii 'mass number' (1)
iii 'isotope' (2)
iv 'radioactive half-life'. (2)

b A radioactive isotope has a half-life of 6 hours. A sample of the isotope has an initial activity of 1000 disintegrations per second.
i Draw a graph to show how the activity varies with time over four half-lives. (3)
ii Does the probability of decay of a single atom decrease with time in this way? Explain. (2)
c The isotope ^{215}Po can decay by emitting either an alpha or a beta particle.
i Write down the equation for each of these possible decay sequences. (5)
ii It is known that an alpha particle from this decay can ionise about a million atoms. Estimate the energy of the alpha particle. (2)
d Describe briefly one use of radioactive isotopes. (2)

International Baccalaureate, Physics, Subsidiary Level, Nov 1997

3 The decay of radioactive materials is a random process. On average, nuclides which decay rapidly exist for a shorter time than nuclides which decay slowly. It is common practice when making calculations on decay to make use of the half-life of a nuclide. One difficulty that arises with these calculations is when the radioactive material is a mixture of two or more nuclides. This question considers the case when a mixture of two radioactive nuclides is present. In decommissioning a nuclear power station, this difficulty is compounded by the presence of about a hundred different radioactive nuclides in significant quantities.

a Explain what it means to say that radioactive decay is a *random process*. (2)

b State two physical quantities which cause change of phase of matter but which do not cause a change in the rate of decay of a radioactive material. (2)

c The table gives the variation with time of the total activity A_{mix} of a mixture of cobalt and nickel together with the separate activities A_C and A_N due to the cobalt and nickel.

Time/year	A_C/Bq	A_N/Bq	A_{mix}/Bq	$\ln(A_{mix}/Bq)$
0	6900	250	7150	8.87
5	3540	241	3781	8.24
10	1820	232	2052	7.63
20	479	215	694	6.54
30	126	199	325	5.78
40	33.3	185	218	5.38
50	8.79	172	181	5.20
60	2.32	159	161	5.08
70	0.611	147	148	5.00
80	0.161	137	137	4.92
90	0.0425	127	127	4.84
100	0.0112	118	118	4.77

A graph showing how $\ln(A_{mix}/Bq)$ varies with time is plotted below.

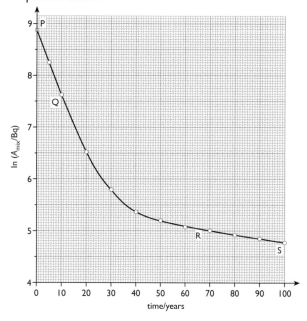

i Explain the following.

1 PQ on the graph corresponds mainly to the decay of cobalt.

2 RS on the graph corresponds mainly to the decay of nickel.

3 The shape of QR is a curve.

ii Determine the following gradients.

1 the gradient of PQ

2 the gradient of RS

iii Given that the general decay law is of the form $x = x_0\exp(-\lambda t)$, use the gradients found in **ii** to *estimate* values of the decay constants for the cobalt and the nickel nuclides.

iv Use your answer to **iii** to calculate the half-life of the cobalt. (10)

d Suggest whether these two nuclides, with these activities, would pose any hazard if found when de-commissioning a nuclear reactor. (2)

e In an actual reactor, activities of radioactive materials can often be 10^{12} times larger than those given in the table. Explain when and why each of these two nuclides would pose the greater hazard. (4)

OCR, Physics, Paper 2, June 1999

4 a i State *two* differences between a proton and a positron.

ii A narrow beam of protons and positrons travelling at the same speed enters a uniform magnetic field. The path of the positrons through the field is shown in the figure.

Sketch the path you would expect the protons to take.

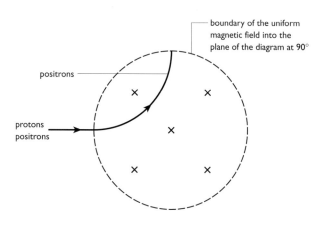

iii Explain why protons take a different path to that of the positrons. (5)

b The next figure shows five isotopes of carbon plotted on a grid in which the vertical axis represents the neutron number *N* and the horizontal axis represents the proton number *Z*. Two of the isotopes are stable, one is a beta minus emitter and two are positron emitters.

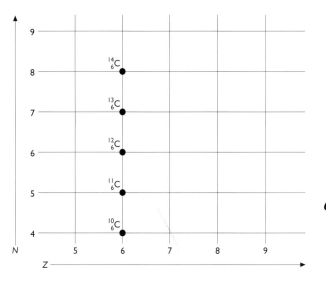

i Which isotope is a beta minus emitter?
ii Which of the two positron emitters has the shorter half-life? Give a reason for your choice. (3)

c A positron with kinetic energy 2.2 MeV and an electron at rest annihilate each other. Calculate the average energy of each of the two gamma photons produced as a result of this annihilation. (2)

AQA (NEAB), Particle Physics, Mar 1999

5 a The nucleus of an atom of gold may be represented as

$${}^{197}_{79}\text{Au}$$

Describe fully a model of a neutral atom of gold, making reference to the numbers of protons, neutrons and electrons, and their relative masses and charges. (9)

b The α-particle scattering experiment provided evidence for the model of the atom in **a**. Describe this experiment, making particular reference to
i the apparatus and observations made
ii the interpretation of the observations. (8)

c A naturally occurring sample of neon gas is analysed. Make suggestions why the proton (atomic) number (10) is found to be a whole number whereas the average of the nucleon (mass) numbers of the nuclei (20.2) is not a whole number. (4)

OCR, Sciences, Physics Foundation, Mar 1999

6 One isotope of potassium, ${}^{40}_{19}\text{K}$, has a half-life of 1.4×10^9 years and decays to form argon, ${}^{40}_{18}\text{Ar}$, which is stable. A sample of rock taken from the Sea of Tranquillity on the Moon contains both potassium and argon in the ratio

$$\frac{\text{number of potassium-40 atoms}}{\text{number of argon-40 atoms}} = \frac{1}{7}$$

a Define *half-life*. (2)

b The decaying potassium nucleus emits a particle X.
i Write down the nuclear equation representing this decay.
ii Suggest the identity of X. (2)

c Assume that when the rock was formed, there was no argon-40 present in the sample and that none has escaped subsequently.
i Estimate the age of the rock.
ii State, with a reason, whether your answer in **i** is an overestimate or an underestimate of the age of the rock if some escape of argon has occurred. (5)

OCR, Physics, Paper 2, Nov 1999

29 Particle physics

THE BIG QUESTIONS	● What is everything made of?
	● How can we tell?
	● What does the question 'what is everything made of?' mean?

KEY VOCABULARY   antilepton   antiquark   baryon   baryon number   colour charge   dark matter   eightfold way   exchange (or gauge) boson   Feynman diagram   fundamental particle   gluons   Grand Unified Theory, GUT   graviton   hadron   Higgs boson   lepton   lepton number   meson   pair production   paradigm shift   quantum chromodynamics, QCD   quantum electrodynamics, QED   quark   standard model   strangeness   supersymmetry   unification of forces

BACKGROUND   It's been said that all the best stories have a beginning, a middle and an end, but not necessarily in that order. So for the story of particle physics we'll first look at the middle; that is, at the present state of the subject. Then we can go back to see how we got here, and we can also look right back to **the** beginning – the big bang. Since the story is still developing we can't make any conclusions about how it will finish, but we can look forward to see what problems particle physicists hope to solve in the future, and why. Where we are now is called the **standard model**.

● Particles and their interactions – the standard model

The standard model describes what we know about energy and the interactions of matter at the scale of the very, very small. In fact, the standard model is based on three types of particle that are so small that they can be called **fundamental**, meaning that they do not have an internal structure. These three types of fundamental particle are **leptons**, **quarks** and **exchange bosons**.

If asked 'what are fundamental particles of which matter is made?' we might miss out the exchange bosons. They do not combine together as permanent building blocks of bigger particles. What they seem to do is to travel between other fundamental particles – the leptons and the quarks – allowing these to interact with one another. Leptons and quarks seem to throw bosons backwards and forwards, which is why the word 'exchange' is used. Many bosons have short lifetimes, and are created and destroyed during exchange processes, as we shall see. Without the bosons there would be no interactions – no forces between particles. (This includes the forces of attraction between the molecules and atoms, and within the nuclei, of your body. You are stuck together by bosons! Prepare to see the world in some very new ways.)

We know that there are different types of force in nature. We know that the behaviour of the gravitational force and the electric force, for example, differ. They act at different scales, and one acts between masses while the other acts between electric charges. We also know that nuclei are held together by a strong nuclear force – strong enough to overcome the electrical repulsion of protons. And within nuclei, particles interact in another way which cannot be explained by gravity, electricity or strong nuclear force, and this is called the weak nuclear force, or weak interaction.

1 Does
a a molecule,
b an atom,
have internal structure? Are they fundamental particles?

The four forces and the standard model

For each type of force there is a particular kind of exchange boson. The photon is the most familiar of these. At the level of fundamental particles, it is by exchange of photons that particles exert electric force on each other. Only particles that have the property that we call electric charge can do this. The study of photon exchange and the electrical interactions between particles is called **quantum electrodynamics**, or QED. We can draw simple diagrams, called **Feynman diagrams**, to represent such interactions (Figures 29.1 and 29.2).

Figure 29.1
A Feynman diagram showing the electrical interaction between two charged particles.

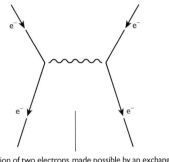

repulsion of two electrons, made possible by an exchange photon which exists for only a very short time

Electric (and magnetic) force is due to exchange of bosons. The bosons that 'carry' the electric force are photons. Without this exchange, there would be no electric force. The atoms of your body are held together by these photons.

Figure 29.2
Examples of **a** weak and **b** strong interactions between particles, shown by Feynman diagrams. Interactions and participants in them, such as bosons and quarks, are explained further on the following pages.

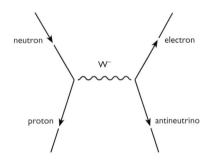

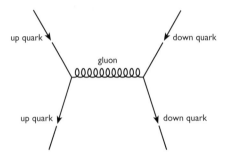

a The decay of a neutron – known as β^- decay when it happens inside a nucleus. There are three 'out-going' particles. The exchange boson, W^-, exists only for a short time.

b Quarks interact by exchanging bosons called gluons. The gluon exists only during the interaction.

In the 1970s, theoretical physicists – those who explore the world with pencil and paper, with ideas and with the results of the experimental physicists – developed an idea that the electric force and the weak force were two manifestations of the same thing. They said that for particles with very high energies we should expect these two forces to have identical behaviour, even though at 'normal' energies they seem very different. This **unification** of two forces is part of a tradition in physics. Isaac Newton showed that the gravity we feel on Earth is one manifestation of the same type of force as that which keeps the Moon in orbit and the planets in orbit around the Sun. In the middle of the 19th century, James Clerk Maxwell showed that electricity and magnetism were a single phenomenon.

The physicists who suggested the unification of electric and weak interactions – now sometimes called electroweak theory – suggested that the exchange bosons for the weak interaction were not photons but a 'new', unheard-of type of short-lived particle. They called them W bosons and Z bosons, and they predicted exactly how much mass these should have. Experimental physicists blasted protons together to try to find such particles. They found them, and so the electroweak theory won considerable support from the physics community in general.

The strong nuclear force is a short-range force – it acts only over short distances – and it acts only between quarks. Only quarks have got what it takes to feel the strong force. 'What it takes' to feel the force is called **colour charge**. (For electric force, 'what it takes' is called electric charge, and for gravitational force it is mass that is needed.) The tiny scale over which the strong interaction works makes it difficult to study, but nevertheless a coherent theory has been developed. This is called **quantum chromodynamics**, or QCD, and according to this theory the force is again carried by exchange particles. The effect of these exchange particles is to stick quarks together into larger, non-fundamental particles. The exchange particles are called **gluons**.

That leaves gravity. There are some very profound questions that left Newton less than satisfied with his own ideas. How does gravitational 'action at a distance' work? How does the Moon 'know' that the Earth exists? What is going on in the space between? Einstein attempted to answer these questions in his theory of general relativity, which claims that the presence of mass changes the shape of spacetime, so that the orbit of the Moon is just 'straight line' motion in a region of spacetime that is warped by the Earth's gravity. This challenges some assumptions of Newtonian physics, and for those of us brought up on Newton it takes some effort to begin to understand. There is another problem. Einstein's general relativity and the quantum theory of the very small do not mesh together very well at all. There is something missing, and it is bound to be something very profound. Perhaps some new young genius will soon see through problems that older brains can't solve. (This could be your big opportunity – Galileo, Newton, Maxwell, Einstein, *your name could go here*.) In the meantime, for consistency, particle physicists have suggested that gravity, like the other interactions, is carried by an exchange boson. They've called them, in advance of finding direct evidence for their existence, **gravitons**.

2 Sketch a table to summarise similarities and differences between gravitational, electric and strong forces.

The fundamental particles of matter

So much for exchange particles. As for the fundamental particles of matter itself – the quarks and the leptons – these come in a range of sizes (masses). There seem to be six types of quark paired up into three 'generations', and we will look at some of the evidence for this in the following pages. These need names, and the chosen names are *up* and *down* (first generation), *charmed* and *strange* (second generation), *top* and *bottom* (third generation). Everyday matter seems to contain only the up quark and the down quark (the first generation quarks). These are the lightest of the quarks. There are also six types of lepton, and these also seem to come in pairs. The first pair is the electron and its matching neutrino (which is the one involved in beta$^+$ decay), termed the electron-neutrino, ν_e. And as if to emphasise that the Universe is an interesting place, there are six different **antiquarks** and six different **antileptons**. We will go on to see how this picture is arrived at, but in the meantime Table 29.1 summarises the standard model.

Table 29.1
Fundamental particles and their interactions according to the standard model.

Type of force	Exchange boson
strong	gluon
electric	photon
weak	W and Z bosons
gravity	gravitons

Fundamental matter particles			
	quarks		
up, u	charmed, c	top, t	These are subject to the strong force.
down, d	strange, s	bottom, b	
	leptons		
electron, e	muon, μ	tau, τ	These are not subject to the strong force.
neutrino, ν_e	neutrino, ν_μ	neutrino, ν_τ	

Quarks, it seems, interact by means of the strong force, but leptons do not. Perhaps now is the time to look back to see how the standard model was developed.

Particle physics up to the early 20th century

We can trace ideas about particles back to the civilisations of the Eastern Mediterranean, about 2500 years ago. Ideas about the deep nature of matter, beyond what we can see directly, were discussed then. But the technology did not exist to explore the hidden world of the very small, and the process of testing hypotheses by experiment was not part of the culture.

So it wasn't until the 'modern' era, starting about 400 years ago, that particle explanations for behaviour such as the compressibility of gas re-emerged. And it's only a little more than 100 years since the entire scientific community accepted the particle theory of matter.

In the 1890s came the discovery of radioactivity, and then J. J. Thomson's experiment which showed that cathode rays possessed mass and charge. The 'rays' were seen as particles (electrons) emerging from atoms. From this it was necessary to conclude that atoms were not fundamental themselves but had internal structure. Atoms were little things that had even littler things inside them.

3 Why are the crew of aircraft exposed to higher levels of ionising radiation than most other people experience?

In 1910, just 15 years after the discovery of radioactivity, Father Theodor Wulf, a priest, went up the Eiffel Tower to test for the presence of ionising radiation (as we now call it). He found more than expected, and hypothesised that some radiation came from space. His suggestion for testing this hypothesis was to measure the intensity of radiation using high balloons, and others did this in the years that followed. Wulf's idea was supported by their findings. Radiation intensity is up to five times greater at 5000 m than it is at sea level.

Just a little later, Rutherford and his colleagues discovered the nucleus by watching what happened to alpha particles fired at a thin target. Hydrogen nuclei were found to be the simplest and seemed to be fundamental particles, which were called protons. For many years – until the development of quark theory and quantum chromodynamics – protons were thought of as fundamental.

All of this, along with Bohr's first use of quantum theory to describe atoms and then the Pauli exclusion principle (see Chapter 27), provided a perfect match to the whole of chemical behaviour. Atoms came in different sizes (masses) and their electron structures built up according to relatively simple rules. Chemical reactions of material were interactions involving their electron structures.

Figure 29.3
Carl Anderson, a Californian scientist, was the first person to detect antimatter. These tracks of electrons and positrons, with pathways curving in opposite directions due to a magnetic field around the apparatus, were produced by the action of cosmic radiation in Carl Anderson's cloud chamber on the top of Pike's Peak, Colorado, in 1932.

After Rutherford's success with the bombardment of material by alpha particles, many other scientists tried similar activities. One of these was the bombardment of a beryllium target by alpha radiation. This had the curious effect of producing a flow of protons, not from the beryllium but from nearby paraffin wax. It was eventually accepted that the protons were being shunted out of the paraffin – where they existed as nuclei of the paraffin's hydrogen – by some other particles that were being emitted from the beryllium. These other particles were not detected directly, but clearly they were big enough and/or fast enough to give significant momentum to protons. They had no charge. They were recognised as components of the nuclei of atoms, contributing much of the mass of nuclei larger than hydrogen. They were, of course, neutrons.

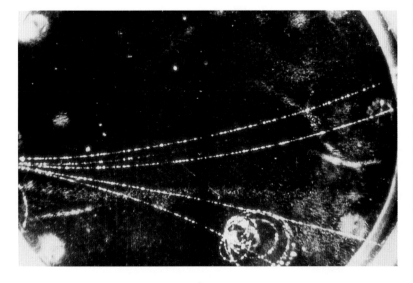

In the 1930s, electron tracks were observed in cloud chambers. By putting a cloud chamber into a strong magnetic field the tracks could be seen to curve. Some of these tracks were seen curving 'the wrong way' (Figure 29.3). That is, they seem to be being made by particles like electrons, but with positive charge. A little earlier, theoretical physicist Paul Dirac had predicted that for every kind of fundamental particle of matter there exists a particle with the same mass but opposite charge. These 'opposite' particles were described as particles of antimatter. The 'new electrons' were then recognised as antimatter particles – anti-electrons, or positrons.

Later observations showed that new particles and antiparticles can appear when there is a big enough supply of energy. The energy might be the energy of a photon, which ceases to exist and instead we see an electron–positron pair, as in Figure 28.8, page 292. The available energy and the amount of mass that can be created from it are related by the equation:

$$E = mc^2$$

This process of creation of an electron and a positron is called **pair production**. The opposite of pair production is mutual annihilation, in which a particle and its antiparticle cease to exist as such (Figure 29.4). Then their mass may disappear but the energy does not cease to exist – photons are created. The total energy of the produced photons is related to the mass of the annihilated particles by $E = mc^2$.

Figure 29.4
Mutual annihilation in the brain – making use of particle physics to study how our brains work.

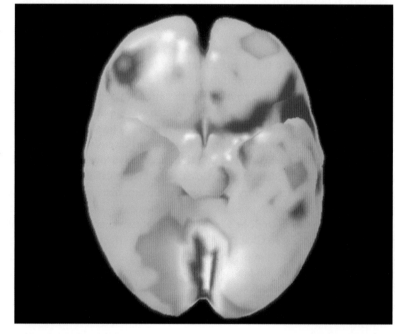

This is a PET scan of a brain. PET stands for 'positron emission tomography'. A radioactive source, a tracer, which emits β^+ particles, or positrons, is provided to the brain via a drink. Positrons and electrons mutually annihilate, emitting pairs of photons which can be detected from outside the head. Distribution of the tracer in the brain can then be seen.

At about the same time as the first observation of antimatter, the continuous energy spectrum of beta particles was suggesting that the process did not involve two particles but three. But the third particle in beta decay was not easy to detect. The name 'neutrino' was given to it, to suggest both that it was electrically neutral and that it was very small, with little or no mass.

It might not be too surprising that a 'particle' with no charge and little or no mass should be hard to detect. Indeed, neutrinos from the Sun are passing through your body as you read, and at high intensity. You do not need to be in sunshine for this to happen. They pass through your body as you lie asleep at night, with the same intensity, having travelled through the Earth and out the other side. Neutrinos do not interact much with other particles. Yet a photon also has no charge or mass but you carry a pair of very effective photon-detectors with you. Photons are exchange bosons, and good at interacting.

One of the problems that scientists are investigating now is the mismatch that exists between theoretical predictions of the rate of emission of neutrinos by the Sun and the observed rate of emission. Mismatches between theory and observation are always interesting because they suggest that either the theory needs to be tweaked a little or it needs to be abandoned and a new theory found. To abandon one theory and develop a new one is always a difficult challenge. In physics, or any other field of study, a major change in fundamental approach that is required for a new theory can be called a **paradigm shift**. Only those with open minds are capable of experiencing paradigm shifts. It is likely that further investigation will show that current theory can be developed enough to incorporate an explanation of neutrino emission by the Sun, but nobody can tell what level of new ideas could emerge.

4 DISCUSS
What major changes – or personal 'paradigm shifts' – have you experienced through studying post-16 physics?

5 According to the summary of the standard model, what kind of particles are neutrinos?

Accelerators

Most people have a particle accelerator in the corner of a room at home. Not that they're much interested in the particles. It's the patterns that the huge numbers of fast particles make on a specially coated screen as they collide with it that people seem to find so fascinating. The patterns are patterns of photons. People have been known to watch these patterns for hours on end.

The accelerated particles are electrons, emitted from a hot cathode by thermionic emission and subjected to electric force due to a potential difference, of about 1 kV, between the cathode and an anode. Fluctuations in the electric force vary the intensity of the electron beams, while magnetic forces make the beams scan across the screen. An electron microscope is just a little more sophisticated with a rather higher potential difference – up to about 25 kV.

Charged particles are accelerated when in regions of changing potential; that is, where there is a potential difference between one point and another. Potential difference is defined in terms of the work done per unit charge, as follows:

potential difference between two points = work done (or energy transferred) per unit of charge in transferring charge between them

$$\Delta V = \frac{\Delta W}{Q}$$

So
$$\Delta W = Q\,\Delta V$$

The electronvolt is the energy, ΔW, gained by an electron (or other particle of charge $\pm 1.6 \times 10^{-19}$ coulomb) when accelerated by a potential difference of 1 volt. The electronvolt is therefore the product of 1 V and 1.6×10^{-19} C, which is 1.6×10^{-19} J.

The energies of the electrons in televisions and electron microscopes are therefore about 1 keV and 25 keV respectively. Other types of accelerator are shown in Figure 29.5.

Figure 29.5
Types of accelerator.

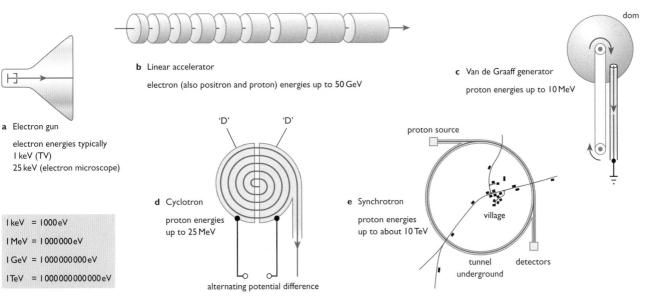

a Electron gun

electron energies typically
1 keV (TV)
25 keV (electron microscope)

b Linear accelerator

electron (also positron and proton) energies up to 50 GeV

c Van de Graaff generator

proton energies up to 10 MeV

d Cyclotron

proton energies
up to 25 MeV

alternating potential difference

e Synchrotron

proton energies
up to about 10 TeV

proton source

village

tunnel   detectors
underground

dom

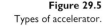

1 keV = 1000 eV

1 MeV = 1 000 000 eV

1 GeV = 1 000 000 000 eV

1 TeV = 1 000 000 000 000 eV

6 What is
  a 10 MeV
  b 1 GeV
in joules?

7 a What is the total energy carried by 5×10^{17} electrons that have been accelerated by a potential difference of 1 kV,
  i in eV   **ii** in J?
  b What is the rate, in watts, at which energy is transferred to a TV screen if 5×10^{17} electrons hit it every second?
  c How does this compare with the power of a domestic light bulb?

A linear accelerator can give an energy of as much as 50 GeV (= 50×10^9 eV) to each electron that travels along it.

The energy of each proton in an accelerated beam in a circular synchrotron becomes very large – up to 1 TeV (10^{12} eV) – and so experimental particle physics is sometimes called 'high energy physics'.

Why accelerate particles?

In a TV electrons are accelerated so that they can induce the creation of photons by collisions in a fluorescent screen. In an electron microscope electrons are given high energy and high momentum so that their wavelength is small, as described by the de Broglie relationship, and diffraction blurring of images is reduced (see Chapter 26).

Both a Van de Graaff generator and a cyclotron can accelerate protons, and other small ions, enough to induce changes in nuclei with which they collide. This creates nuclei of nuclides, or radioisotopes, that have uses, such as medical tracers. The first cyclotron was developed in 1930 by E. O. Lawrence in California. His first motivation was to study proton interactions, but later his cyclotron was also used to create isotopes for medical purposes in nearby hospitals. It was in these hospitals that the uses of radioactive materials for medical purposes was pioneered.

After the discovery of pions and muons in secondary cosmic radiation, scientists recognised that they could study other particle interactions more directly by supplying the energy artificially. And so the process of building bigger and bigger accelerators began. In order to study the interactions a good particle detector was needed, and this came in 1952 in the form of the bubble chamber. This is an enclosed body of liquid close to its boiling point, so that in sudden and brief expansions bubbles begin to form, this process starting most easily in the vicinity of charged particles. Bubble chambers can provide snapshots of particle activity (Figures 29.6 and 29.7) (also see Figures 9.5, 17.2 and 28.8).

We've seen that the energy of a photon can be 'repackaged' as mass of an electron and a positron. It is not only the energy of a photon that can be reorganised. Particles can change into different kinds of particles. The faster a particle is going, the more energy it has for creation of new particles. Energy converts to mass, according to $E = mc^2$. So it's possible that the new particles can have more mass than the initial particle. A single particle can 'turn into' several particles that each have as much mass as it had. Or the end of one particle can result in the creation of another one with a much greater mass. It is in such processes that particle physicists see new phenomena and test their theories.

Figure 29.6 (left) This Big European Bubble Chamber was installed at CERN in 1971. It took photographs of tracks produced by particles accelerated to high energy.

Figure 29.7 (right) A typical bubble chamber photograph of the tracks of sub-atomic particles.

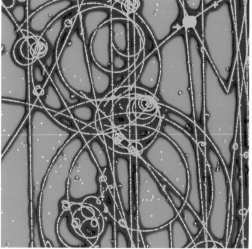

For more than half a century, scientists have accelerated protons and electrons in order to collide them with other particles and observe the results. It was through observing collisions of energetic protons that it was discovered that the constituents of atoms – protons, neutrons and electrons – together with a few other known particles, like antineutrinos and pions, were just a small proportion of the total number of 'sub-atomic' particles that exist.

Looking for patterns – mass

The focus of research and thought in particle physics came to be the search for patterns amongst so many different kinds of particles. The first step was to sort observed particles according to their masses. They fell into three groups – the **baryons**, the **mesons** and the leptons. The habit of naming particles with 'on' at the end was started by the Irish physicist Johnstone Stoney, when he suggested the name for negative carriers of electric charge – 'electrons' – in the 1890s. It's a habit that was used throughout the 20th century, though not for every new particle. By the 1960s it was possible to sort particles as shown in Table 29.2. There were also, of course, the antiparticles.

Table 29.2
The three groups of sub-atomic particles as they could be classified by mass in the 1960s.

Classification of particle	Examples	Mass/MeV	
baryon	xi, Ξ^-	1321	
	xi, Ξ^0	1315	
	sigma, Σ^-	1197.3	
	sigma, Σ^0	1192.5	
	sigma, Σ^+	1189.4	These all feel the strong force.
	lambda, Λ	1115.6	
	neutron, n	939.6	
	proton, p	938.3	
meson	kaon, K^0	497.7	
	kaon, K^+	493.7	
	pion, π^+	139.6	
	pion, π^0	135.0	
lepton	muon, μ	105.7	These do not feel the strong force.
	electron, e	0.511	
	neutrino, ν	0	

Note that the electric charge of particles with the same name is indicated by a superscript $+$, $-$ or 0.

Looking for patterns – conservation rules

A next step in the search for underlying patterns was to examine the conservation rules that particle interactions follow. In Chapter 28 we saw that in nuclear changes, charge number and nucleon number are conserved. They are the same before and after every nuclear change.

While nuclear changes involve protons and neutrons (and also electrons and neutrinos and their antiparticles, in the case of beta decay), we are now dealing with a bigger number of different particles than just the proton and neutron. The idea of conservation of nucleon number had to be amended to conservation of all baryons including the lambda, the sigma and the xi. All baryons have a **baryon number** of 1, and their antiparticles have a baryon number of −1 (see Table 29.3, page 316). Baryon number obeys, like charge number, a conservation rule.

An interaction that does not conserve charge number or baryon number, the physicists found, just did not happen. The following is an example of an interaction that can happen, provided that one or both of the original particles is moving fast enough:

$$p + n \rightarrow p + p + n + \bar{p}$$

charge numbers, C	1	0	1	1	0	−1
total charge number	1	=	1			
baryon numbers, B	1	1	1	1	1	−1
total baryon number	2	=	2			

The following interaction, on the other hand, is never seen. Nature doesn't seem to 'allow' it.

$$p + n \rightarrow p + p + n + n$$

charge numbers, C	$\begin{matrix} 1 & 0 \\ \end{matrix}$	$\begin{matrix} 1 & 1 & 0 & 0 \end{matrix}$

Charge number is not conserved.

total charge number $1 \neq 2$

baryon numbers, B $\begin{matrix} 1 & 1 \end{matrix}$ $\begin{matrix} 1 & 1 & 1 & 1 \end{matrix}$

Baryon number is not conserved.

total baryon number $2 \neq 4$

In looking at electrons and neutrinos, it became clear that a similar conservation rule applies – conservation of **lepton number**, L (see Table 29.3). All particle interactions conserve lepton number. If an interaction does not conserve lepton number then it can't happen. β^- decay is an example of an allowed interaction:

$$n \rightarrow p + e^- + \bar{\nu}_e$$

C $\begin{matrix} 0 & 1 & -1 & 0 \end{matrix}$

total charge number $0 = 0$

B $\begin{matrix} 1 & 1 & 0 & 0 \end{matrix}$

total baryon number $1 = 1$

L $\begin{matrix} 0 & 0 & 1 & -1 \end{matrix}$

total lepton number $0 = 0$

Table 29.3
Some baryon and lepton numbers.

Particle	Antiparticle	Baryon number, B	Lepton number, L	
proton, p		1	0	
	antiproton, $\bar{p}$	−1	0	
neutron, n		1	0	
	antineutron, $\bar{n}$	−1	0	baryons
lambda, Λ^0		1	0	
xi$^-$, Ξ^-		1	0	
	xi$^+$, Ξ^+	−1	0	
pion, π^+		0	0	mesons
	pion, π^-	0	0	
electron, e$^-$		0	1	
	positron, e$^+$	0	−1	
neutrino, ν_e		0	1	
	antineutrino, $\bar{\nu}_e$	0	−1	leptons
muon, μ^-		0	1	
	muon, μ^+	0	−1	
neutrino, ν_μ		0	1	
	antineutrino, $\bar{\nu}_\mu$	0	−1	

8 Explain why the interaction

$$p + n \rightarrow p + p + n + \bar{p}$$

can only happen if one or both of the original particles is/are moving fast.

9 Count charge numbers and baryon numbers to decide whether particle physicists will ever find these interactions:
a $p + p \rightarrow p + n$
b $p + p \rightarrow p + p + \bar{n} + p$

10 Refer back to the account of β^+ decay in Chapter 28 and show that such a decay obeys conservation of charge, baryon number and lepton number.

11 Does the process of pair production

$$\gamma \rightarrow e^- + e^+$$

obey rules of conservation of charge, baryon number and lepton number?

Strange events

Even the set of conservation rules, invented it seems by nature and not by the physicists, could not explain why some interactions that physicists expected to see didn't seem to happen. The following interaction was allowed, according to the conservation rules for charge number, baryon number and lepton number:

$$\pi^- + p \rightarrow K^0 + n$$

	π^-	p	K^0	n
C	-1	1	0	0
B	0	1	0	1
L	0	0	0	0

but it was never seen. This seemed strange. To try to explain why events like this never happened, physicists tried assigning a new property to particles, and supposing that this new property followed a new conservation rule. The word they chose for this new property was **strangeness**. You shouldn't worry about the use of such words. Particle physicists are exploring new regions of reality, so they're bound to need to use words in new ways. Words like 'charge' were new once but have now become familiar.

The particle physicists guessed that some interactions or events didn't happen because they didn't conserve strangeness. They assigned a number to each particle – its strangeness, S. Many particles are not strange, and have a strangeness of 0. Of those that are, their interactions show that the value of their strangeness is not always 1. Table 29.4 lists the strangeness numbers of the sub-atomic particles.

Note that none of the leptons have strangeness. Some mesons and some baryons are strange, while others are not.

The following interactions do take place:

$$\pi^- + p \rightarrow K^0 + \Lambda^0$$

	π^-	p	K^0	Λ^0
C	-1	1	0	0
B	0	1	0	1
L	0	0	0	0
S	0	0	1	-1

$$\pi^0 + n \rightarrow K^+ + \Sigma^-$$

	π^0	n	K^+	Σ^-
C	0	0	1	-1
B	0	1	0	1
L	0	0	0	0
S	0	0	1	-1

Table 29.4 Strangeness numbers.

Particles	Strangeness, S	
photon	0	
electron, e	0	
neutrino, ν	0	leptons
muon, μ	0	
pion, π^+	0	
pion, π^0	0	
pion, π^-	0	mesons
kaon, K^+	1	
kaon, K^0	1	
kaon, K^-	-1	
proton, p	0	
neutron, n	0	
lambda, Λ^0	-1	
sigma, Σ^+	-1	
sigma, Σ^0	-1	baryons
sigma, Σ^-	-1	
xi, Ξ^0	-2	
xi, Ξ^-	-2	
omega, Ω^-	-3	

Nature makes the rules. Nature seems to be telling us that when particles change, then at least in some kinds of change their total strangeness cannot and must not change (but also see pages 319–20). The reasons for this are deep, and so far unknown. We know that strangeness – like charge, baryon number and lepton number – is conserved in some interactions, but we don't know why. There's still plenty of physics left to do.

12 Are these allowed?
 a $\Lambda^0 \rightarrow p + \pi^-$
 b $\Xi^0 \rightarrow p + \pi^+$
 c $\pi^+ + p \rightarrow \Sigma^+ + K^+$
13 Which conservation rules show that each of these is not allowed?
 a $\Lambda^0 \rightarrow \pi^+ + \pi^-$
 b $\gamma + p \rightarrow n + \pi^0$
 c $p + p \rightarrow \pi^+ + \pi^+$
14 a Write down the events shown in Figure 29.8 as 'equations' and say whether they happen.
 b What does the wavy line represent in each case?

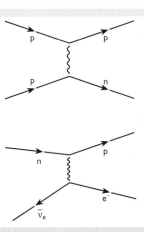

Figure 29.8

Searching for particles – new fundamentals

Does history repeat itself? In the 19th century Mendeleev arranged the chemical elements into neat patterns and went so far as to assume that there were elements nobody had discovered yet. He left gaps in the neat patterns. In fact, it was only by leaving gaps that he could get the patterns to make sense. But they did make sense and he was even able to predict some of the properties of the missing elements. When chemists found such new elements it was hard not to accept Mendeleev's patterns – now called the periodic table. Something similar happened in particle physics in the 20th century.

In the 1940s and 1950s, more and more new particles were found in observations of cosmic rays and in accelerator beams. There seemed at first to be little order, except that it was possible to make some classifications of particles by mass and by the conserved quantities of charge, baryon number, lepton number and strangeness. However, physicists found that if they plotted strangeness against charge, for baryons and mesons, then some neat patterns appeared (Figure 29.9). They called the patterns the **eightfold way**, a term that comes from Buddhism. And just as in Mendeleev's first attempt to draw up a periodic table of elements, there were gaps. From the general pattern it was possible to predict the properties of the missing particles. When scientists found particles with just these properties, it seemed that the eightfold way was a meaningful way of organising the particles.

Figure 29.9
Patterns of the eightfold way.

Note that these representations are graphs of strangeness against charge, with the strangeness axis drawn at a non-standard angle so that the patterns of particles are aesthetically pleasing.

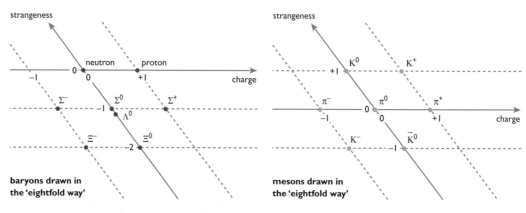

baryons drawn in the 'eightfold way'

mesons drawn in the 'eightfold way'

The existence of a pattern suggests an underlying reason. If particles could be arranged in neat patterns, perhaps there were patterns in the particles themselves – just as there are patterns within atoms (in particular, the number of protons) that determine their behaviour. The challenge was to work out *why* the baryons and mesons fell into patterns.

Physicist Murray Gell-Mann was one of many who were looking for the answer to the riddle of the eightfold way. He made a very imaginative and bold suggestion. This bold suggestion was that baryons and mesons are not 'fundamental' at all but are made of smaller particles that come in twos, for mesons, or in threes, for baryons. What was really bold, because it went against all previous experience, was the suggestion that these new particles do not have whole-number charge numbers, but charges that come in units of $\pm 1/3$ or $\pm 2/3$ of the proton charge. Also, said Gell-Mann, they are incapable of existing, and being observed, on their own. They *only* come in twos and threes. It all sounds very unlikely – except that it all fits with the observations. A few years after Gell-Mann's suggestion, the observed deflection of a bombarding stream of electrons and of neutrons within nuclei provided evidence that supported the idea that mesons and baryons had internal structure. With an imaginative flair for the curious, Murray Gell-Mann gave the new particles a curious name – he called them quarks.

By supposing that there were three types of quark, Gell-Mann was able to draw the eightfold way for baryons and for mesons, showing how each one is made up of different quark combinations (Figure 29.10). The three types of quark were called *up*, *down* and *strange*, each with an antiquark of opposite properties. We now know that there are three more quarks, but these are more massive and are created only where very high energies are available. We don't see these heavy quarks in ordinary matter. Table 29.5 shows the properties of the up (u), down (d) and strange (s) quarks and their antiquarks.

Figure 29.10
The eightfold way and the quark structure of hadrons.

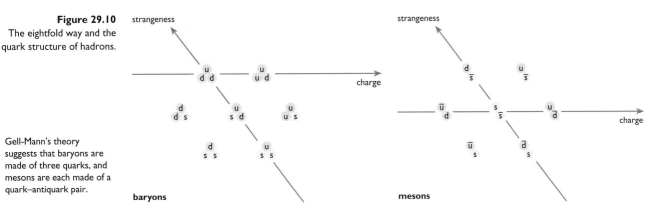

Gell-Mann's theory suggests that baryons are made of three quarks, and mesons are each made of a quark–antiquark pair.

Table 29.5
Quark properties.

Quark	Antiquark	Baryon number, B	Charge number, C	Strangeness, S
u		$\frac{1}{3}$	$\frac{2}{3}$	0
	ū	$-\frac{1}{3}$	$-\frac{2}{3}$	0
d		$\frac{1}{3}$	$-\frac{1}{3}$	0
	d̄	$-\frac{1}{3}$	$\frac{1}{3}$	0
s		$\frac{1}{3}$	$-\frac{1}{3}$	-1
	s̄	$-\frac{1}{3}$	$\frac{1}{3}$	1

Lepton number is not included. All quarks have lepton number of zero.

So baryons and mesons are not fundamental themselves but are made up of combinations of quite a small number of different quarks, which we believe *are* fundamental. Particles made of quarks – all the baryons and all the mesons – are known collectively as **hadrons**. All hadrons feel the strong force.

The weak interaction – changing quarks and breaking a rule

Charge, baryon number and lepton number are conserved in *all* particle changes, or interactions. There are, however, interactions in which strangeness is not conserved. These are all weak interactions – that is, they involve W and Z bosons as exchange particles. (Interactions that involve gluons as exchange particles are strong interactions and they all conserve strangeness.)

We need to look again at the weak interaction. An example is β^- decay, which can be summarised as:

$$n \rightarrow p + \beta^- + \bar{\nu}_e$$

But we have seen that neutrons and protons are hadrons. That is, they have quark structures. The neutron has three quarks, an up and two downs – udd. The proton has a structure of two ups and a down – uud. The electron and the antineutrino are not hadrons but leptons, and they do not have quark structures. So in the decay of a neutron into a proton, a down quark turns into an up quark (Figure 29.11). The number of quarks stays the same, but one quark has changed. At the heart of the process is the brief life of a W boson, the signature of a weak interaction. β^- decay involves no strange particles, so there is no breaking of strangeness conservation here.

Figure 29.11
Feynman diagram of β^- decay, showing the quark structure of the neutron and proton.

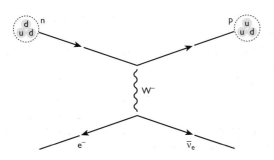

Figure 29.12
Decay of a π^-.

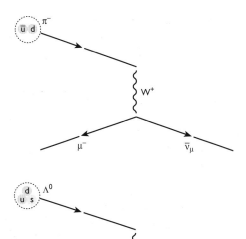

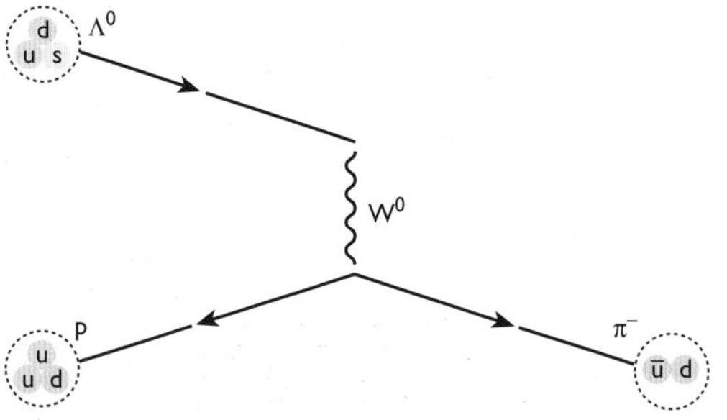

A pion is a hadron with a structure of a quark and an antiquark. The π^-, for example, has a structure of an anti-up quark and a down quark – $\bar{u}d$. It decays into two leptons, a muon and an antineutrino, which have no quarks:

$$\pi^- \rightarrow \mu^- + \bar{\nu}_\mu$$

Since the pion has a quark and an antiquark, its total effective quark number is zero, and there are zero quarks after the decay. The total number of quarks can be said to stay the same, but there is no denying that changes have taken place to the nature of the initial quark and antiquark. And the interaction involves the creation, for a short time, of a W boson (Figure 29.12). So the decay of a pion is another example of a weak interaction. Again, as for β^- decay, strangeness is not involved. But there are other weak interactions that do involve strange particles.

The decay of a lambda particle into a proton and a pion is another weak interaction:

$$\Lambda^0 \rightarrow p + \pi^-$$

Note that while the lambda particle includes a strange quark in its structure, the proton and pion do not (Figure 29.13). Strangeness has not been conserved in the lambda decay. In weak interactions, strangeness conservation may be disobeyed.

Figure 29.14 shows the decay of the strange particles Λ^0 and K^0 via weak interactions.

Figure 29.13
Decay of a Λ^0, a weak interaction that disobeys strangeness conservation.

Figure 29.14
This is a bubble chamber picture with added colour to identify the various particles. A proton in the bubble chamber is hit by a π^- (green, bottom of picture), creating a Λ^0 and a K^0. These neutral particles leave no tracks but the Λ^0 decays into a proton (red) and a π^- (green), while the K^0 decays into a π^+ (yellow) and a π^- (green).

Jets and quarks

In an area such as particle physics, theoreticians often develop new ideas about the fundamental nature of things, and practical physicists then devise the experiments that will test these hypotheses. In some cases, however, it is the data that come first, before the ideas. That was true in the overall development of the standard model of particle physics. An entire zoo of new particles was discovered, and the theoretical physicists then developed testable hypotheses that might lead to a coherent theory of the relationships between them all. Patterns were recognised in the behaviour of hadrons. The 'eightfold way' provided a hypothesis that the hadrons were not themselves fundamental particles but had internal structure and this structure was based on quarks. So practical physicists began the search for internal structure and for quarks themselves.

15 a What kind of interactions break the rule of conservation of strangeness?
b Does the interaction between the proton and the π^- shown in Figure 29.14 obey strangeness conservation?
c From the quark structures of the particles involved in the decays of Λ^0 and K^0 in Figure 29.14, show that the decays must be weak interactions.

16 DISCUSS
What similarities and differences are there in the searches for underlying patterns in the chemical behaviour of elements and in the behaviour of fundamental particles?

Within a few years, scattering experiments, involving bombardment of protons by very energetic electrons, confirmed that protons were not simple hard spheres but had active internal structure. But if free quarks can exist, it seems that they can do so only for incredibly short times.

Electron–positron collisions result in mutual annihilation, creating gamma ray photons. The higher the energy of the electrons and the positrons when they collide, the higher the gamma ray energies. These very high energy photons tend not to last very long themselves, but their energy becomes mass again, in the form of pairs of particles which each then produce a 'jet' of still more particles. The higher the energy of the photon, the more mass and the more particles that can be created (Figure 29.15).

Figure 29.15
This is a representation on a computer screen of the result of a very high energy collision between an electron and a positron, which took place in the ALEPH detector at CERN. The dotted yellow and blue lines show jets of particles and the yellow bars represent their energies.

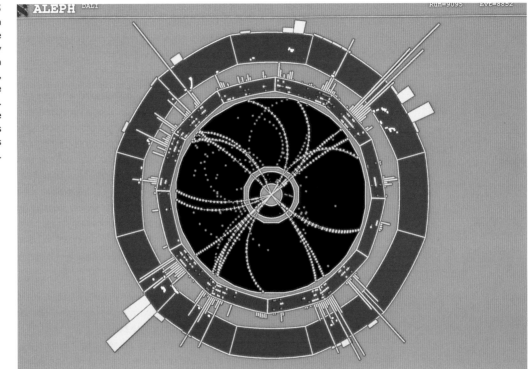

The creation of such jets of particles can only be explained by the existence of an intermediate stage of short-lived particles. These last only for about 10^{-23} s, between their creation from the energy of the gamma photon and their decay into jets. These processes can only be explained if the short-lived particles are quarks and gluons.

17 DISCUSS
Why is it inevitable that an area of study such as particle physics will need to invent its own new vocabulary? Does the vocabulary surprise you? In what other areas of present-day human activity is new vocabulary being created rapidly?

18 a Show that mutual annihilation of an electron–positron pair obeys:
  i charge conservation
  ii lepton conservation.
  b How can it obey energy conservation?

19 Why can particles such as electrons and quarks be called fundamental but particles such as protons and pions cannot?

20 Explain how electron–positron collisions provide evidence for the existence of quarks.

21 In the photograph of the jet event in Figure 29.15, explain
  a why the particle pathways are curved
  b how so many particles can be generated by a single event.

Future questions for high energy research

Particle physicists deal with deep questions. They are exploring the nature of matter, energy, light, and asking one of the biggest questions of all – 'why is the world the way it is?' Some of the specific riddles that they are working on are discussed below.

Where does mass come from?

Mass of a body is related to its gravitational interactions and to its inertia; in Einstein's general relativity these are the same thing. Attempts are being made to solve the biggest problem of all in present-day theoretical physics – how Einstein's ideas can be fitted together with what we know about very small particles. The answer may lie in the study of mass, and what causes it. According to the standard model, particles have mass because they interact with a particle called the **Higgs boson**. There is not yet any direct evidence to support the hypothesis that this particular particle exists. One of the reasons why physicists want to accelerate particles to higher and higher energy is so that they can look for this evidence.

What is supersymmetry? Does it exist?

In the 1980s, Abdus Salam and Stephen Weinberg developed a theory which indicated that the weak force that acts inside nuclei and the electric force are fundamentally the same, even though they don't seem very similar at first. The discovery of particles such as the W boson, predicted by the theory, provided strong supporting evidence. Particle physicists now talk about the 'electroweak force'. A next step is to try to see if the strong nuclear force and the electroweak force can be traced back to the same fundamental origins. This search is called a search for a **Grand Unified Theory**, or GUT. A favourite but very provisional theory is called **supersymmetry**. This links the fundamental matter particles, quarks and leptons, with the force particles, or exchange bosons. The theory predicts that heavy superparticles exist, but it hasn't been possible to try to create these so far because it would need acceleration of particles to higher energies than can be achieved by existing accelerators.

What is dark matter?

Most – as much as 90% – of the mass of galaxies is invisible. That is, it does not exist in the form of stars and so it emits no light. Yet we know from measuring the rotation of galaxies that the mass must be there. This matter is called **dark matter**.

It's rather embarrassing after so much study of science to have to admit that we don't know what 90% of the Universe is made of, and in fact we can't even see it or detect it directly in any way. Various theories exist for the nature of this dark matter. One of these is that it exists as new kinds of particles, perhaps heavy superparticles, held to galaxies by gravity.

Why are there three generations of quarks and leptons?

You and everything around you are made from quarks of one generation only, and one kind of lepton. But scientists with their accelerators and detectors have discovered that there are three kinds of quark and three kinds of lepton (Figure 29.16). The question is, why *three* generations?

Figure 29.16
The second and third generation quarks and leptons are heavier than their everyday equivalents. It takes a lot of energy to create such particles.

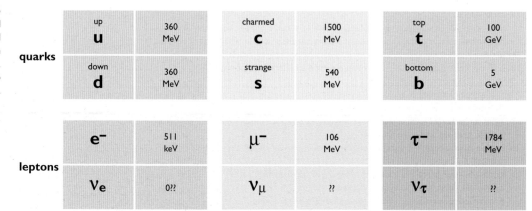

22 What is the obstacle that scientists have in studying particles with a lot of mass, like the Higgs boson, and the top quark? What would they like to do about this?

Why is the Universe lop-sided?

We know that matter exists. We know that antimatter exists. What we don't know is why the world is made of matter and not antimatter, or an exciting mixture of equal amounts of both, continuously mutually annihilating. In fact one theory claims that that is how the Universe once was, but there was just a little bit more matter than antimatter. When all the mutual annihilation was over it was the excess of matter that was left behind. Perhaps it has got something to do with the three generations of quarks and leptons ...

Seeking answers in high energy research

Figure 29.17
Physicist Alison Wright works at CERN, Europe's centre for particle physics in Geneva, Switzerland.

CERN is home to the Large Electron–Positron Collider, or LEP, a massive particle accelerator running 27 km in circumference more than 100 m underground. Inside LEP, beams of electrons and their antiparticles, positrons, are accelerated to energies of 100 GeV – which is why Alison Wright (Figure 29.17) calls herself a high energy physicist. Electrons and positrons collide at four places around the ring. At each collision point a huge detector waits to record what happens when matter and antimatter meet.

Alison belongs to an international team of almost 500 physicists who run the ALEPH experiment at LEP, sharing responsibility for looking after the 750 m³ apparatus and analysing data from many millions of collisions.

The ALEPH detector collects very precise information about the particles created in the aftermath of electron–positron collisions (see Figure 29.15) – information such as which types of particle, their trajectories and their energies. By piecing all of this information together, Alison and her colleagues are building up a more detailed understanding of fundamental particles such as the W and Z bosons, or the families of quarks, and how they interact with each other.

There are still many unanswered questions, and the physicists at CERN are already planning the next generation of experiments – with the 14 000 GeV Large Hadron Collider (LHC) shown in Figure 29.18. This new accelerator needs new detectors. The design of a special system to detect and measure muons is keeping Alison busy.

Figure 29.18
Inside the 27 km tunnel at CERN.

23 Suggest how Alison Wright might be able to identify muons amongst particles created during collisions.
24 a Particle physicists also call themselves 'high energy' physicists. What is 100 GeV in joules? Why is the term 'high energy' used?
b Why do particle physicists give high energies to particles such as electrons and positrons?

● **Extra skills task** Communication

Bohr's theory of the atom fitted some observations, but provided an incomplete picture; Einstein's theories of special and general relativity were controversial when published. Particle physics is a relatively new branch of physics. Many new theories and models exist, such as the Grand Unified Theory, quantum electrodynamics, quantum chromodynamics, supersymmetry, etc. For each of these theories, scientists have produced new and controversial ideas.

Research one theory of your choice from the physics of the past 100 years or so. Examine evidence from contemporary or later writings, papers and secondary sources. Write a brief report which identifies examples of scientific opinion, describes the debates that took place and noting any bias from either the researcher or the reviewer.

● Examination questions

1 Predict and explain how, if at all, the quark composition of a nucleus changes during
  a β^+ emission (2)
  b γ emission. (2)
      OCR, Nuclear Physics, June 1999

2 A carbon-14 nucleus undergoes β^- decay, forming a new nucleus, releasing a β^- particle and one other particle which is difficult to detect.
  a Write down the proton number and the nucleon number of the new nucleus.
  b Name the particle which is difficult to detect.
  c Name the baryons and leptons involved in the decay.
  d **i** Give the quark structure for the neutron and the proton.
    ii Hence state the quark transformation that occurs during β^- decay. (7)
    AQA (NEAB), Physics Advanced, Paper 1, June 1999 (part)

3 The two decays shown below may both conserve energy and momentum.

$$\mu^+ \rightarrow e^+ + \nu_e$$
$$\mu^+ \rightarrow e^+ + \nu_e + \overline{\nu}_\mu$$

The first cannot occur but the second can.
  a State a further conservation law which *both* decays obey.
  b Which conservation law forbids the first decay?
  c Show how this conservation law allows the second decay. (4)
    Edexcel (London), Physics PH4, June 1999 (part)

4 **a** Explain why a bubble chamber is more efficient at detecting high-energy particles than a cloud chamber. (2)
  b Two pions π^+ and π^0 travelling at the same speed enter a bubble chamber. State and explain which pion is more likely to be detected, giving *two* reasons for your choice. (3)

    lifetime of π^+ is 2.6×10^{-8} s
    lifetime of π^0 is 0.8×10^{-16} s

c Explain why the tracks produced in a bubble chamber by electrons are spiral in form. (2)
    AQA (NEAB), Particle Physics (PH06), June 1999 (part)

5 The diagram shows the main features of a bubble chamber photograph in which a pion has collided with a stationary proton (reaction A), followed by two subsequent decays (reactions B and C).

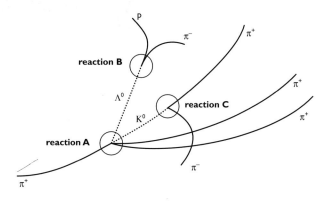

The following information may be useful:

particle	baryon number	strangeness
π	0	0
Λ^0	1	-1
p	1	0
K^0	0	1

a Write an equation for reaction A. (1)
b Show that charge, baryon number and strangeness are all conserved in reaction A. (3)
c How is it possible for there to be more pions at the end than at the beginning of the reactions? Where did they come from? (2)
d In addition to the quantities mentioned above, what else must be conserved in all three reactions? (1)
    International Baccalaureate, Physics Higher Level, May 1998

VIII
ADVANCED MOTION AND FORCE

30 Forces in equilibrium

THE BIG QUESTION

● In a world where so many forces act, how is it ever possible for bodies to be in equilibrium?

KEY VOCABULARY

adhesion   centre of gravity   centre of mass   co-planar forces   dynamic friction
lubrication   moment   normal reaction   polygon of forces
principle of moments   rotational acceleration   rotational equilibrium
simple couple   static friction   statics   translational equilibrium

BACKGROUND

Every static body is subject to forces that are balanced. This includes everything from you as you sit or stand, to a pylon standing in a gale, to a pyramid or a mountain. Consideration of such balanced forces is called **statics**, and this is at the heart of civil engineering (Figure 30.1).

Figure 30.1
The forces at each point of this structure are balanced.

Normal reaction

Thanks to gravity the mass of an object, including a person, has weight. The weight is a 'downwards' force (that is, it acts towards the centre of the Earth). A force equal to the weight acts on any surface on which the body rests. The surface experiences force. Any body that experiences force exerts an equal and opposite force. This is in agreement with Newton's Third Law (see Chapter 11).

Figure 30.2
The weightlifter does not fall through the floor, before or after lifting the extra masses.

In the case of the weightlifter and the floor (Figure 30.2), the weight of the weightlifter acts down on the floor, and the floor exerts an equal and opposite force. This force is called a **normal reaction**. It acts at 90° to the surface.

When the weightlifter picks up the masses the force acting *on* the floor increases. The force exerted *by* the floor increases to match this. It 'reacts' to the change in weight.

The origins of normal reaction

In our everyday experience we feel gravitational and electric forces. Occasionally we are aware of magnetic force (which is a force with electrical origins).

Gravity can 'act at a distance' – the Earth and the Moon pull each other without the need for close contact. Likewise with magnets – we can feel the forces between them while they are still some distance apart. Electric forces of attraction and repulsion also act between bodies that are separated, and the forces increase as the bodies get closer together.

Atoms and molecules are not simple particles but collections of particles. The net forces between them can be attractive when they are some distance apart, but when their charged particles are very close together the net effect is repulsive force. The more they are pushed together, the stronger the repulsive force becomes. That is what happens when the weightlifter picks up the masses. The particles of the soles of his shoes are pushed closer to the particles of the surface of the floor by the increased weight. The repulsive force between them increases accordingly – increasing the normal reaction. Total weight and normal reaction remain in balance.

1 a What are the origins of the forces acting when you touch a table surface with your fingertip?
b Sketch this situation, using vector arrows to represent these equal and opposite forces.
c Why do the forces increase when you press harder?

● Friction

Figure 30.3
Friction prevents the
weightlifter sliding.

2 Write a sentence to
describe the relative
directions of the
normal reaction and
the frictional force
exerted by a surface.

3 Explain the role of
friction in:
a keeping you in your
seat (it might help to
try to imagine a totally
frictionless seat)
b leaning against a wall
(imagine a frictionless
wall and floor).

4 a Why does a heavy
object jolt into motion
when pushed across a
floor?
b Once moving at a
steady speed, are the
forces on the object
balanced or
unbalanced? Explain.

5 a Sketch a bicycle
brake block in contact
with a wheel rim
during braking. Add
arrows to show the
forces of normal
reaction and friction
that are acting.
b In terms of the
nature of the surfaces,
explain why increasing
the force of normal
reaction increases the
force of friction.

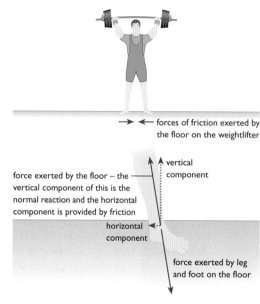

→ ← forces of friction exerted by
the floor on the weightlifter

force exerted by the floor – the
vertical component of this is the
normal reaction and the horizontal
component is provided by friction

vertical
component

horizontal
component

force exerted by leg
and foot on the floor

The floor can exert frictional force, as well as a
normal reaction, on the weightlifter's feet.
Frictional forces are important in keeping him
standing. They act in directions such as to
oppose sliding, always parallel to the floor
(Figure 30.3).

Static friction

Frictional forces acting on a static body –
static friction forces – usually balance the
forces that tend to make it slide. However, a
point can be reached where the force that is
tending to cause sliding is greater than the
maximum possible frictional force. Then the
body starts to slip. This is clearly dangerous
for a weightlifter, so the surface of a
weightlifting area needs a suitably rough
texture.

Dynamic friction

It can be hard to make a body start moving. There are two reasons for this. One is the body's
inertia – it requires an unbalanced force to accelerate the body from its condition of rest.
Secondly, friction resists the motion, but becomes smaller once the body is moving. Static friction
acts on the stationary body but a smaller **dynamic friction** acts on the moving body. This
provides an explanation of the sudden start in movement of one surface across another.
Frictional resistance to motion is greater when still than once the surfaces have started to slide.

The origins of frictional force

Friction between surfaces can be thought of as being due to interlocking of the irregularities on
the surfaces and to forces of attraction, or **adhesion**, between the surfaces. This is consistent
with the observations that

- friction between rough surfaces (with large irregularities) tends to be greater than between
smooth surfaces (with small irregularities), and
- liquids can be used to fill the depressions in surfaces and push the surfaces a little apart, thus
reducing the level of interlocking – that is, liquids can **lubricate** (Figure 30.4).

Friction involves interaction between materials at a particle level. It is therefore a surprisingly
complex subject, and not completely understood. It is still a subject of research.

Figure 30.4
A lubricant separates the
surfaces and so reduces
friction between them.

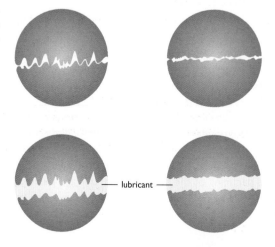

— lubricant —

Friction and the action of lubricants are complex areas of
study, but are commercially important because of the role of
friction in all relative motions of surfaces. These diagrams
provide a simple understanding.

Normal reaction and frictionless collisions

In a collision between an object such as an ice-hockey puck and a wall, the wall exerts forces on the puck; the puck exerts equal and opposite forces on the wall. There are two kinds of force involved – frictional force, which can only act in a direction parallel to the surfaces in contact, and normal reaction, which can only act normally (at right angles) to the surfaces (Figure 30.5).

For a puck, the frictional force is usually a lot smaller than the normal force. So it is worthwhile to think about what happens in an 'idealised' situation, when friction can be ignored (Figure 30.6). Now the only force acting is perpendicular to the wall. This means that the acceleration of the puck is also perpendicular to the wall. (Since acceleration is firmly linked to force by $F = ma$, acceleration must always be in the same direction as net force.) There is no component of acceleration parallel to the wall, because there is no force acting parallel to the wall. The component of the velocity parallel to the wall therefore does not change. The component perpendicular to the wall does change.

Figure 30.5
The forces exerted by a side wall on an ice-hockey puck.

Figure 30.6
An 'idealised' frictionless collision.

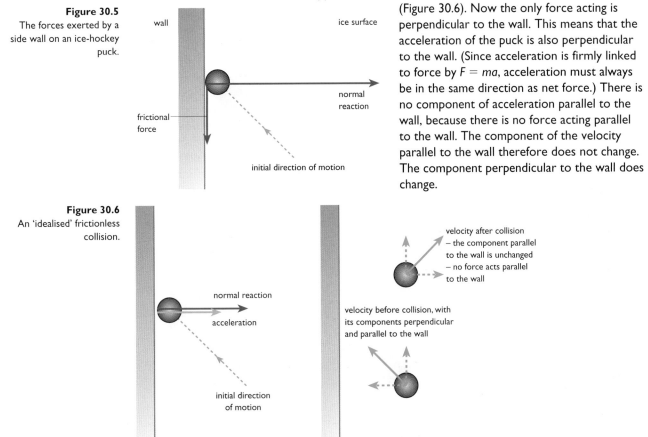

Translational equilibrium

A body can move from place to place experiencing *translational* motion. Or it can go round and round, experiencing *rotational* motion.

A body can be subject to several forces at once. This might apply to a bolt holding components of a bridge together, for example. A body such as the bolt is said to be in **translational equilibrium** if it has no linear acceleration. 'Linear' here means in a straight line, and the word tells us that we are not dealing with rotational effects.

To be in translational equilibrium – to have no linear acceleration – a body must be subject to a total force of zero. If there are several forces acting their *net* effect on translational motion must be zero. We can write this briefly as the following statement, which is called the condition for translational equilibrium:

for translational equilibrium of a body it is necessary that $\Sigma F = 0$

There are two ways to find out if a number of forces acting on a body result in equilibrium – by scale drawing and by calculation.

If the forces, drawn to scale and in the appropriate directions, can be drawn as a complete polygon then the body will be in equilibrium (Figure 30.7a). This is called a **polygon of forces**. If, however, having attempted this it turns out that the 'ends' of the polygon don't meet, then the resultant force is given by a vector joining these ends (Figure 30.7b).

Figure 30.7
The polygon of forces – only a closed polygon results in zero net force, or equilibrium.

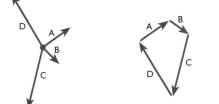

a A body with a set of forces acting on it is in translational equilibrium if the forces can be drawn as a closed polygon.

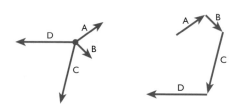

b These forces do not result in equilibrium of the body.

To calculate the net force of a number of forces we first choose two mutually perpendicular directions. It is not critical which two directions we use, so it makes sense to use the x direction (horizontal) and the y direction (vertical). We then calculate the components of all the forces along these two directions (as described in Chapter 12), and add the sets of components in each direction. The body will be equilibrium if the sum of the components of all the forces along each of the two directions is zero (Figure 30.8).

In some situations it may be more helpful to choose a different pair of directions. For example, for a car on a slope we might be interested in components of the forces parallel and perpendicular to the road surface.

Figure 30.8
Using calculations to look for equilibrium – if the total forces acting in any two mutually perpendicular directions are zero, then the body is in equilibrium.

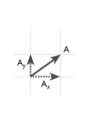

A set of forces acting at a point ...

... and their individual components in the x and y directions

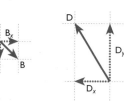

If the sum of the components in the x direction is zero,

$$A_x + B_x + C_x + D_x = 0$$

and if the sum of the components in the y direction is zero,

$$A_y + B_y + C_y + D_y = 0$$

then the forces are in translational equilibrium

Moments

The **moment** of a force about a fixed point is a measure of its turning effect about that point. The size of a moment is the product of the force and the shortest distance from the fixed point to the line of action of the force.

It is also equal to the product of the distance of the point of action of the force from the fixed point, multiplied by the component of the force that is perpendicular to a line joining these two points.

For either of these ways of calculating moment, the language has a rather technical complexity. It's easier to see what it means with the help of diagrams (Figure 30.9).

A moment is a product of a force and a distance. That means that its SI unit is the newton metre, N m. Note that a newton metre is equivalent to a joule, the unit of energy. Moment and energy have the same dimensions. It would not be incorrect to use the joule as a unit of moment, but conventionally the newton metre is used.

6 Why wouldn't it be complete enough to say simply that 'the moment of a force about a pivot is the force times the distance from the pivot'?

Figure 30.9
Calculating the moment of a force F about a point P. Note that both methods provide the same answer.

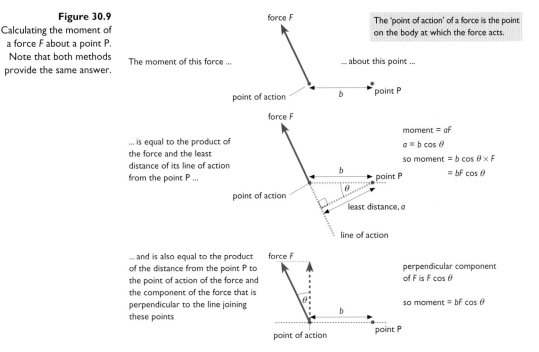

The moment of this force ...

... about this point ...

The 'point of action' of a force is the point on the body at which the force acts.

force F

point of action

b

point P

... is equal to the product of the force and the least distance of its line of action from the point P ...

force F

point of action

b

point P

θ

least distance, a

line of action

moment = aF
$a = b \cos \theta$
so moment = $b \cos \theta \times F$
$= bF \cos \theta$

... and is also equal to the product of the distance from the point P to the point of action of the force and the component of the force that is perpendicular to the line joining these points

force F

θ

b

point of action

point P

perpendicular component of F is $F \cos \theta$

so moment = $bF \cos \theta$

It is possible to calculate the moment of a force about any point that you choose. If you choose a point close to the line of action of the force then its turning effect about that point will be relatively small. It will have a larger moment about a point that is further away (Figure 30.10).

Figure 30.10
The same force has a greater turning effect about a point when applied further away from the point.

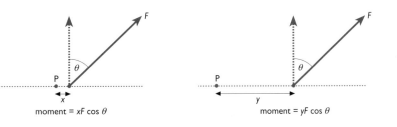

moment = $xF \cos \theta$

moment = $yF \cos \theta$

Rotational equilibrium

Turning effect can be clockwise or anticlockwise. Clockwise moments and anticlockwise moments have opposite effects. If they are equal to each other in size, the total moment will be zero. There will then be no **rotational acceleration** – rotation will not become faster and faster. The moments are in balance, and the body on which they act is in **rotational equilibrium**.

Two quantities that are equal in size can only add up to equal zero if one is positive and the other is negative. Clockwise and anticlockwise moments must have opposite signs. It doesn't matter, in fundamental terms, which we call positive and which we call negative. A reasonable convention is to say that anticlockwise moments are positive.

Rotational equilibrium (that is, zero rotational acceleration) occurs when

total clockwise moments = $-$(total anticlockwise moments)

This is the condition for rotational equilibrium (Figure 30.11), which is also called the **principle of moments**.

7 A 40 N force acts perpendicularly downwards on a uniform beam at a distance of 2 m from a central pivot. At what distance from the pivot must a single downward 44 N force act for the beam to be in rotational equilibrium?

Figure 30.11
The condition for rotational equilibrium.

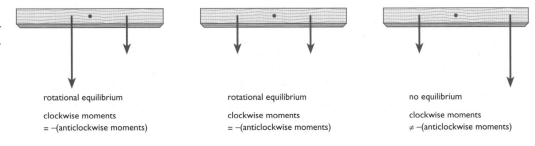

rotational equilibrium

clockwise moments = $-$(anticlockwise moments)

rotational equilibrium

clockwise moments = $-$(anticlockwise moments)

no equilibrium

clockwise moments $\neq$ $-$(anticlockwise moments)

Centre of mass

Usually, one force acting on a body will be its weight. Gravitational force acts on all parts of a body, but for simplicity we can draw a single arrow acting at a particular point. This is the point at which the whole weight of the body effectively acts. It is called the body's **centre of gravity**.

In the Earth's gravitational field, the weight of a body, from a bacterium to an oil tanker, is proportional to its mass. The body has a **centre of mass** – a point at which all of its mass can be considered, as a useful simplification, to be. The centre of mass can usually be taken to be in the same place as the centre of gravity.

Figure 30.12
In a non-uniform gravitational field the centre of gravity and centre of mass of a body do not coincide.

However, centre of gravity and centre of mass do not always have a shared location. In a gravitational field in which the field strength changes from one place to another over short distances relative to the size of the object, then the distribution of the object's weight is not identical to the distribution of its mass. The centre of gravity is shifted away from the centre of mass (Figure 30.12).

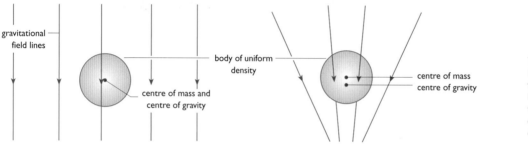

gravitational field lines

body of uniform density

centre of mass and centre of gravity

centre of mass
centre of gravity

The 'lower' half of this body experiences a greater force of gravity – that is, has greater weight – than the 'upper' half. The centre of gravity is not in the centre of the body.

Figure 30.13 (below)
Only if weight and upwards force are co-linear will there be no rotational acceleration.

A body suspended in the Earth's gravity by its centre of mass, or by any point directly above or below, will show no tendency to rotate. The upwards suspending force is co-linear with – in the same straight line as – the weight (Figure 30.13). This provides a simple way of locating the centre of mass of an object.

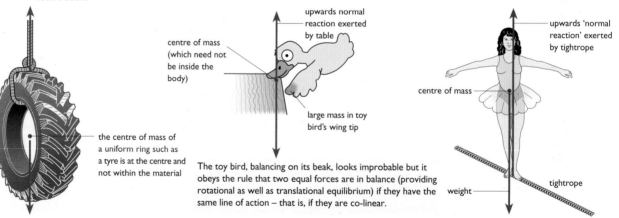

the centre of mass of a uniform ring such as a tyre is at the centre and not within the material

upwards normal reaction exerted by table

centre of mass (which need not be inside the body)

large mass in toy bird's wing tip

The toy bird, balancing on its beak, looks improbable but it obeys the rule that two equal forces are in balance (providing rotational as well as translational equilibrium) if they have the same line of action – that is, if they are co-linear.

upwards 'normal reaction' exerted by tightrope

centre of mass

weight

tightrope

8 Explain, in terms of moments and centre of mass, why it takes less arm strength to do press-ups when your knees are on the floor than when, instead, your toes are on the floor.

9 Explain why one arrangement shown in Figure 30.14 is essentially unstable while the other is essentially stable.

10 A novice tightrope walker carries a pole which has a bucket of water hanging by a long rope from each end.
   a Explain how this arrangement can increase the stability of the performer.
   b What advice would you give to the performer about the best way to use the arrangement (that is, the best values for length of pole, length of rope, mass of water in each bucket, and how and where the pole should be held)?

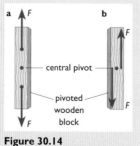

a F

b F

central pivot

pivoted wooden block

F

F

F

Figure 30.14

However, if the body is suspended at a point on any other vertical line then it will rotate. The suspending force and the weight are not co-linear. The weight has an unbalanced moment about the point of suspension.

Note that the centre of mass of a body is not necessarily within the material body. The centre of mass of a tyre, for example, is in the centre of the circle. It is possible to suspend a tyre from a single rope attached at any point so that the tyre hangs in a vertical plane, but it is not possible to hang it by the one rope alone in a horizontal plane.

Simple couples

Figure 30.15
A couple does not cause
linear acceleration but
rotational acceleration.

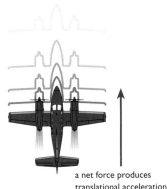

a couple produces
rotational acceleration

a net force produces
translational acceleration

We have seen that equal and opposite forces acting on a body are 'balanced' – the net force is zero, no acceleration takes place, and the body is in equilibrium. However, it *is* possible for forces to be equal in size and opposite in direction but for them to have an effect on motion. This happens when the forces are not co-linear. Such a pair of equal and opposite forces is called a **simple couple**.

A simple couple cannot produce acceleration of a body along any straight line – it cannot produce linear acceleration. It has no influence on translational motion. But it can produce rotational acceleration (Figure 30.15), making the body go round faster and faster.

A simple couple does not act along a single straight line but does lie in a single plane. The two forces are **co-planar**.

11 State
   a a similarity
   b a difference
   between linear acceleration and rotational acceleration.
12 Forces forming a simple couple are co-planar but not co-linear. Is it possible for two forces to be co-linear but not co-planar? (Try this with a couple of pencils.)

The moment of a simple couple

13 Two 25 N forces act in opposite directions at points 0.8 m apart. What is the moment of the couple?
14 Two forces, both equal to F, act at two points on an object a distance x apart.
   a What is the moment of one of the forces about the point of action of the other?
   b What is the sum of the moments of the two forces about a point $\frac{3}{4}x$ from one of them?
15 Explain how a high-diver can use a couple to make her body spin while falling. Once falling, is it possible to stop spinning? How could the diver avoid spin?

Clearly a couple has a combined turning effect, and a combined moment. We can calculate the moment of each force separately, if we wish. Or we can calculate the sum of their moments. When we do this we must calculate their moments about the *same* point and then add them together. As long as we use the same point for both forces we can choose any point we wish. One option is to consider moments about a point on the object half-way between their lines of action. If we label the forces F_1 and F_2, and they are each distance $x/2$ from this half-way point, then:

$$\text{moment of } F_1 \text{ about the half-way point} = F_1 \frac{x}{2}$$

$$\text{moment of } F_2 \text{ about the half-way point} = F_2 \frac{x}{2}$$

$$\text{total moment of the couple} = F_1 \frac{x}{2} + F_2 \frac{x}{2}$$

F_1 and F_2 are equal, and can be called simply F, so

$$\text{total moment of the couple} = F \frac{x}{2} + F \frac{x}{2}$$
$$= Fx$$

where F is the value of each force, and x is the distance between their points of action (Figure 30.16). Note that we find the moment of a couple is Fx whatever point we choose to calculate the moment about.

moment of couple = Fx

Figure 30.16
The moment of a
simple couple.

More complex couples

The two forces in Figure 30.17 also make a couple. To calculate their moment we have two options, which are illustrated in a and b.

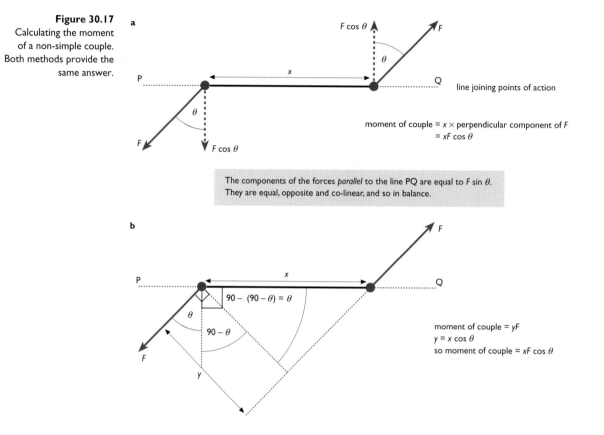

Figure 30.17
Calculating the moment of a non-simple couple. Both methods provide the same answer.

a

moment of couple = x × perpendicular component of F
                  = $xF \cos \theta$

The components of the forces *parallel* to the line PQ are equal to $F \sin \theta$. They are equal, opposite and co-linear, and so in balance.

b

moment of couple = yF
$y = x \cos \theta$
so moment of couple = $xF \cos \theta$

90 − (90 − θ) = θ

· ·

● **Comprehension and application**

Ladder forces

The purpose of a ladder is to support weight in places where such support doesn't otherwise exist. The ladder has to exert an upward force to balance weight.

When you are high above the ground, it's reassuring to think that the ladder is going to stay in translational and rotational equilibrium. For such absence of acceleration, all forces must be in perfect balance. To see how this is possible we need to consider the forces acting. We can start with the forces acting vertically (Figure 30.18a).

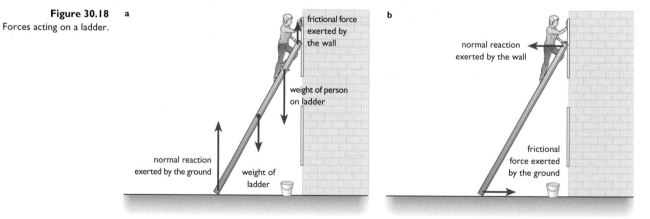

Figure 30.18
Forces acting on a ladder.

a

frictional force exerted by the wall

weight of person on ladder

normal reaction exerted by the ground

weight of ladder

The vertical forces acting on a ladder.

b

normal reaction exerted by the wall

frictional force exerted by the ground

The horizontal forces acting on a ladder.

Downward forces acting on the ladder are the weight of the person and the weight of the ladder. If the ladder is to avoid vertical acceleration (up or down) these forces must be in balance with forces acting upwards. These are provided by the normal reaction exerted by the ground on the foot of the ladder, and by an upwards frictional force acting parallel to the surface of the supporting wall.

It is possible that the combined turning effect of these forces will just happen to be zero. But this is unlikely also. In the situation shown in Figure 30.18 (where frictional force is small compared with total weight), there is a net clockwise turning effect, whichever point is used for calculation of total moments. (Note that to calculate moments, we consider the components of force that act perpendicularly to the length of the ladder. Components that act parallel to the ladder can have no turning effect because their line of action is at zero distance from any possible pivot.)

Fortunately there are horizontal forces that can compensate for this clockwise moment. There are two forces, one acting at each end of the ladder, that act as an anticlockwise couple. These forces are the friction acting parallel to the surface of the ground at the foot of the ladder, and the normal reaction acting on the top (Figure 30.18b). Since there are only two horizontal forces they must be equal and opposite in direction, or the ladder would accelerate horizontally.

The equilibrium of the ladder depends on a combination of normal reactions and friction, and the weights of the ladder and its load.

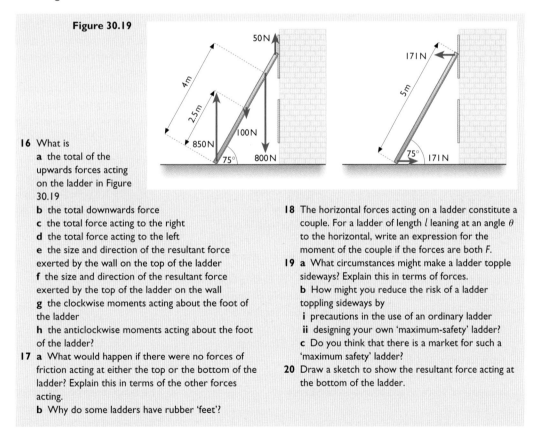

Figure 30.19

16 What is
  a the total of the upwards forces acting on the ladder in Figure 30.19
  b the total downwards force
  c the total force acting to the right
  d the total force acting to the left
  e the size and direction of the resultant force exerted by the wall on the top of the ladder
  f the size and direction of the resultant force exerted by the top of the ladder on the wall
  g the clockwise moments acting about the foot of the ladder
  h the anticlockwise moments acting about the foot of the ladder?

17 a What would happen if there were no forces of friction acting at either the top or the bottom of the ladder? Explain this in terms of the other forces acting.
  b Why do some ladders have rubber 'feet'?

18 The horizontal forces acting on a ladder constitute a couple. For a ladder of length l leaning at an angle θ to the horizontal, write an expression for the moment of the couple if the forces are both F.

19 a What circumstances might make a ladder topple sideways? Explain this in terms of forces.
  b How might you reduce the risk of a ladder toppling sideways by
    i precautions in the use of an ordinary ladder
    ii designing your own 'maximum-safety' ladder?
  c Do you think that there is a market for such a 'maximum safety' ladder?

20 Draw a sketch to show the resultant force acting at the bottom of the ladder.

● Examination questions

Questions 1 and 2 relate to Chapter 5.

1 a A materials scientist needs to be able to determine several different properties of a new material. Some of these properties are
  i density
  ii the Young modulus
  iii the ultimate tensile stress.
    Explain the meaning of each of these terms and outline how the Young modulus and the ultimate tensile stress may be determined. (12)

b The figure shows the force–extension graphs for identically-shaped pieces of steel and a new material.

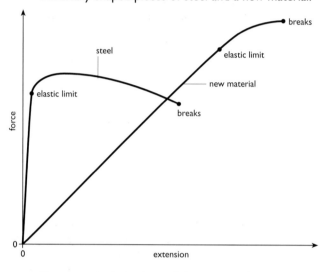

Compare the behaviour of the two specimens.
    Suggest, with two reasons, which of the two materials would be the more suitable for use in a car bumper. (9)

OCR, Sciences, Basic 1, Mar 1999

2 As part of a quality check, a manufacturer of fishing line subjects a sample to a tensile test. The sample of line is 2.0 m long and is of constant circular cross-section of diameter 0.50 mm. Hooke's law is obeyed up to the point when the line has been extended by 52 mm at a tensile stress of 1.8×10^8 Pa.
    The maximum load the line can support before breaking is 45 N at an extension of 88 mm.

a Calculate
  i the value of the Young modulus
  ii the breaking stress (assuming the cross-sectional area remains constant)
  iii the breaking strain. (5)

b Sketch a graph to show how you expect the tensile stress to vary with strain. Mark the value of stress and corresponding strain at
  i the limit of Hooke's law
  ii the breaking point. (4)

AQA (NEAB), Mechanics and Electricity, Mar 1999

3 An athlete is analysing his shot putting technique so as to improve his performance. He finds that the optimum performance is achieved when the angle which his leg makes with the ground is 57° immediately before releasing the shot. The maximum force he can exert on the ground is 650 N at an angle of 57° to the ground.

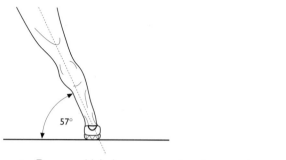

a Draw and label arrows on the diagram above to represent
  i *T*, the force the foot exerts on the ground
  ii *N*, the normal reaction of the ground on the foot
  iii *F*, the frictional force of the ground on the foot. (3)

b Calculate the magnitude of
  i the frictional force *F*
  ii the normal reaction of the ground *N*. (2)

AQA (NEAB), Mechanics and Electricity, June 1999

4 a Explain what is meant by
  i the moment of a force
  ii the torque of a couple. (4)

b A desk lamp is illustrated below.

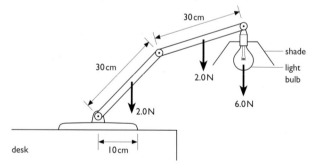

The lamp must be constructed so that it does not topple over when fully extended as shown below. The base of the lamp is circular and has a radius of 10 cm. Other dimensions are shown on the figure. The total weight of the light bulb and shade is 6.0 N and each of the two uniform arms has weight 2.0 N.

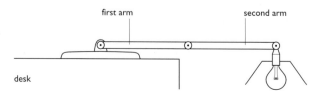

i On the second figure, draw an arrow to represent the weight of the base.

ii The lamp will rotate about a point if the base is not heavy enough. Mark this point and label it P.

iii Calculate the following moments about P.

   1 moment of first arm

   2 moment of second arm

   3 moment of light bulb and shade

iv Use the principle of moments to calculate the minimum weight of base required to prevent toppling. (7)

OCR, Physics, Paper 2, June 1999

5 A uniform ladder AC of weight 400 N and length 3.5 m is held horizontally by two people. One person supports the ladder at end A and the other supports it at point B, a distance of 1.0 m from C, as illustrated below.

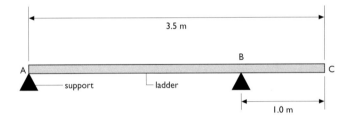

a On the figure, draw the three forces acting on the ladder. (2)

b Calculate the supporting force provided by the person

   i at end A

   ii at B. (5)

c The person at A releases the ladder. Calculate the moment which the person at B must provide in order to keep the ladder horizontal. (3)

OCR, Sciences, Physics Foundation, Mar 1999

6 The diagram shows a free-body force diagram for an aircraft flying along a straight path and climbing at constant velocity.

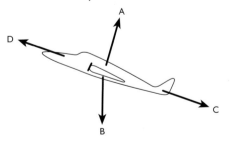

a Name each of the four forces shown and identify in each case what exerts the force. (4)

b State whether or not the resultant force is zero. Explain your answer. (2)

Edexcel (London), Physics PH3, June 1999

7 a Explain how the rotors of a helicopter produce a lift force. (3)

b i On the figure below, mark and name the vertical forces acting on the helicopter when it is hovering at rest.

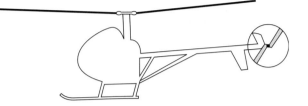

ii State how the forces are related. (2)

c To make the helicopter move forward, the pilot tilts the aircraft as illustrated.

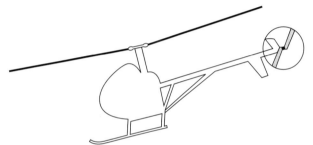

i Explain why the helicopter moves forward.

ii The pilot wishes to maintain level flight. Suggest why the rotor speed must be increased. (4)

OCR, Sciences, Physics of Transport, June 1999

31 Straight line motion

THE BIG QUESTION	● What help does mathematics, in particular graphical and algebraic representation, give us in understanding straight line journeys?
KEY VOCABULARY	equations of uniformly accelerated motion inertial frame of reference
BACKGROUND	They don't make them like this any more. Will the train stop before it reaches poor Madeleine, tied to the tracks by her merciless Uncle? Read on to find out.

Figure 31.1

Displacement and velocity – the story so far

Displacement is the change in position of a body from a specified initial position measured in a specified direction. It is a vector with the same dimensions and the same units as distance. The displacement of a body and the distance travelled are numerically the same for straight line (or linear) motion, though displacement can be positive or negative depending on its direction. For non-linear motion, displacement and distance are *not* numerically the same (Figure 31.2).

Figure 31.2
When the motion is not in a straight line, the displacement is not equal to the distance travelled.

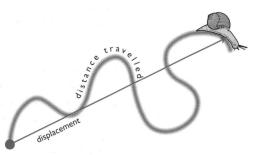

I How can you tell from the diagram in Figure 31.3 that acceleration is taking place?

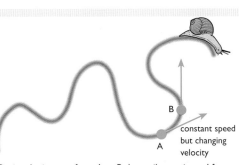

Figure 31.3
On a curved path the direction of the velocity is continually changing.

constant speed but changing velocity

During the journey from A to B the snail experienced force and accelerated, and the effect was to change the direction of motion.

We use the letter s for displacement. Velocity, v, corresponds to displacement as speed corresponds to distance. Velocity is rate of change of displacement and is a vector quantity. Speed is rate of change of distance and is a scalar quantity. A body moving in a curved path can have constant speed but it cannot have constant velocity. It is subject to force that is changing its motion – it is accelerating.

For linear motion, speed and velocity are numerically the same, but again with the direction of velocity being indicated by a positive or negative sign, while speed is always positive.

Relative velocity

Figure 31.4
The velocity of an object depends on the inertial frame of reference of the observer.

Two people do not always tell the same story of the same event. Imagine that a person on a train that is travelling with constant velocity drops a ball, and describes its motion as a vertical fall to the ground. Another person, standing by the track sees the ball fall. Their descriptions of the ball's journey will be very different (Figure 31.4).

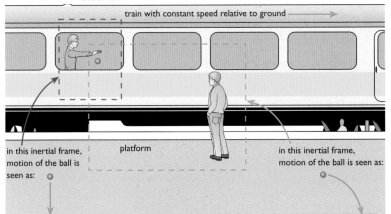

train with constant speed relative to ground

in this inertial frame, motion of the ball is seen as:

platform

in this inertial frame, motion of the ball is seen as:

We say that the passenger on the train and the bystander by the track are in two different **inertial** (which means non-accelerating) **frames of reference**. One has velocity relative to the other.

Common sense might tell us that the frame of reference that includes more space or more matter is the superior one. The bystander's frame of reference includes the whole landscape around, but the passenger's frame of reference includes only the train. Common sense seems to tell us that bigger is better. Common sense can be wrong. (See *Further Advanced Physics*.)

Graphical representation of straight line motion

Displacement and velocity are vector representations of motion. A positive sign suggests displacement or velocity in one direction, and a negative sign suggests motion in the opposite direction. That is, where motion is, say, backwards and forwards or up and down then our maths needs to be stable to distinguish one direction from the other, which it can do by use of signs. We will use the standard Cartesian convention, as used when plotting graphs – vectors with rightward and upward directions have positive signs, and vectors with leftward or downward directions have negative signs (Figure 31.5).

Figure 31.5
The convention used for the direction of motion.

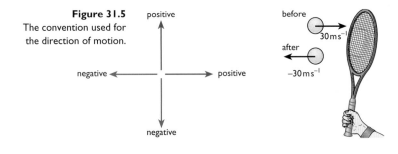

positive

negative

positive

negative

before

$30\,\mathrm{m\,s^{-1}}$

after

$-30\,\mathrm{m\,s^{-1}}$

A particular straight line motion of a particular body can be represented on a graph of displacement against time. The shape of the displacement–time graph provides a story of a journey. There are many possible journeys, and many possible shapes of displacement–time graphs. We usually say that the start of the journey is at time zero (though it is also possible to produce graphs showing only parts of journeys, and the graph does not then have to start at time $t = 0$).

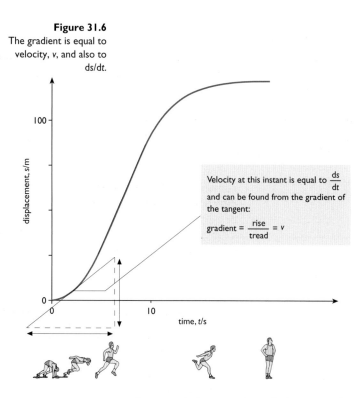

Figure 31.6
The gradient is equal to velocity, *v*, and also to ds/dt.

Velocity at this instant is equal to $\dfrac{ds}{dt}$ and can be found from the gradient of the tangent:

$$\text{gradient} = \frac{\text{rise}}{\text{tread}} = v$$

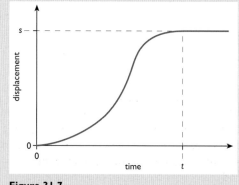

Since velocity is rate of change of displacement we can interpret the gradient of a displacement–time graph as velocity. This is easiest for a simple straight line graph – one for which displacement is proportional to time, which means that the body is on a constant velocity journey. However, we can also measure the gradient at a chosen point on a curving graph. In this case the gradient, and the velocity, change in the course of the journey.

The gradient is then given by $\dfrac{ds}{dt}$, which can be found by drawing a tangent to the curve and using this straight line to determine suitable matching values of the change in *s* and the change in *t* (Figure 31.6).

In general, for a portion of a journey of time Δt for which the corresponding change in displacement is Δs,

$$\text{average } v = \frac{\Delta s}{\Delta t}$$

Where Δs and Δt are very small, we can choose to write them as δs and δt. Then, over the short period of time δt,

$$\text{average } v = \frac{\delta s}{\delta t}$$

This is the average velocity during a short period of time.

In calculus notation, where δs and δt are so small that they approach zero, we write:

$$v = \frac{ds}{dt}$$

This does not provide us with a value of average velocity but the instantaneous velocity.

In the simple case of a straight line graph that starts at the origin – showing a constant speed journey – if we measure Δs as displacement from the journey's beginning then we can write it as *s*, and if Δt is the time since the start of the journey then it is equal to the actual time passed, simply *t*. Then:

$$v = \frac{s}{t}$$

where *v* is not only the average velocity but the velocity at every instant in the particular case of such very simple motion.

(We should remember that a perfectly constant velocity journey is an idealisation for the sake of simplicity. Velocity cannot rise instantaneously at the start of a journey, nor fall in an instant at the end.)

2 Two cannonballs have velocities *v* and −*v*. What is the difference between their
  a speeds
  b velocities?

3 Figure 31.7 is a displacement–time graph showing a displacement *s* at time *t*.

Figure 31.7

  a What is the velocity at time *t*?
  b Is it equal to the quantity *s*/*t*?
  c Does the quantity *s*/*t* provide any useful information about the journey?

Alternative motion pictures

Chapter 10 dealt with displacement–time graphs, which provide a very useful way of representing motion in pictorial form. The gradient of a displacement–time graph is equal to the velocity of the moving body – so that just by looking at the graph we can see, for example, whether velocity is constant, increasing or decreasing.

If we can plot graphs of displacement against time then there is no reason why we shouldn't plot graphs of velocity against time. We can plot the two different kinds of graph for the same motion (Figure 31.8). We shouldn't be surprised that though the journey of the body is the same, the two graphs look rather different.

Figure 31.8
Two representations of the same event.

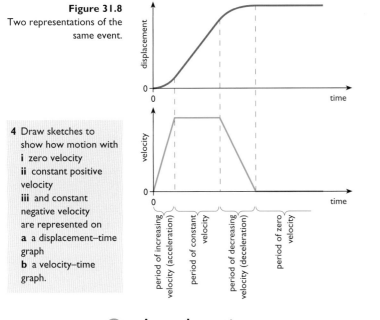

When we show this information in these alternative presentations together, it is worth remembering that at every instant of time, the value of the velocity is equal to the value of the gradient of the displacement–time graph.

Two ways of showing the same thing – a displacement–time and a velocity–time graph for a 100 m race.
At any particular time, *t*, the gradient of the first graph provides the *value* of the velocity shown on the second graph.

Note, for example, that during the period of constant velocity, the displacement–time graph has a constant gradient.

The transitions from changing to constant velocity are shown in simple form as 'corners' on the velocity–time graph, on the assumption that they take place instantaneously. In reality the 'corners' will be a little more rounded than this.

4 Draw sketches to show how motion with
i zero velocity
ii constant positive velocity
iii and constant negative velocity
are represented on
a a displacement–time graph
b a velocity–time graph.

Acceleration

Acceleration is rate of change of velocity. For a portion of a journey of duration Δt, this statement can be written in mathematical language as:

$$\text{average } a = \frac{\Delta v}{\Delta t}$$

where Δv and Δt are corresponding values of change in velocity and change in time.

Note also that:

$$\text{since average } v = \frac{\Delta s}{\Delta t}, \text{ then } \Delta v = \Delta\left(\frac{\Delta s}{\Delta t}\right)$$

$$\text{average } a = \frac{\Delta(\Delta s/\Delta t)}{\Delta t}, \text{ which we can write as } \frac{\Delta^2 s}{\Delta t^2}$$

Figure 31.9
The gradient of a velocity–time graph is equal to acceleration.

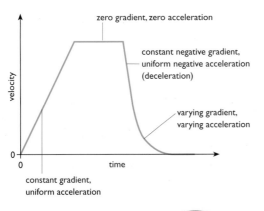

zero gradient, zero acceleration

constant negative gradient, uniform negative acceleration (deceleration)

varying gradient, varying acceleration

constant gradient, uniform acceleration

The unit of acceleration is the m/s², normally written as m s⁻². In calculus notation, instantaneous $a = dv/dt = d^2s/dt^2$. Acceleration at any instant is equal to the gradient of the velocity–time graph at that instant (Figure 31.9).

Acceleration is a variable. It can change. In this chapter we will deal for the most part with acceleration that stays the same, at least for one stage of a journey. Constant acceleration is also called uniform acceleration.

Graphical representation of uniform and changing acceleration

Examples of uniform acceleration and changing acceleration are shown graphically in Figure 31.10.

Figure 31.10
a A velocity–time graph for a body that gets steadily faster – that is, one that has uniform acceleration. Since acceleration = $\Delta v / \Delta t$, it is equal to the gradient of the straight line.

b A velocity–time graph for a body with changing acceleration. The acceleration is still equal to the gradient at any time, but now the gradient is the tangent to the curve.

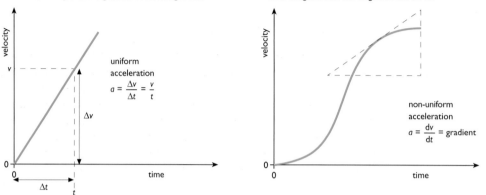

Average and instantaneous acceleration

Note that, just as for velocity, we can speak about average and instantaneous values of acceleration. It helps to put the ideas into a visual format, as velocity–time graphs.

Figure 31.11
Velocity–time graphs for two bodies with the same initial velocity u and the same final velocity v after time t.

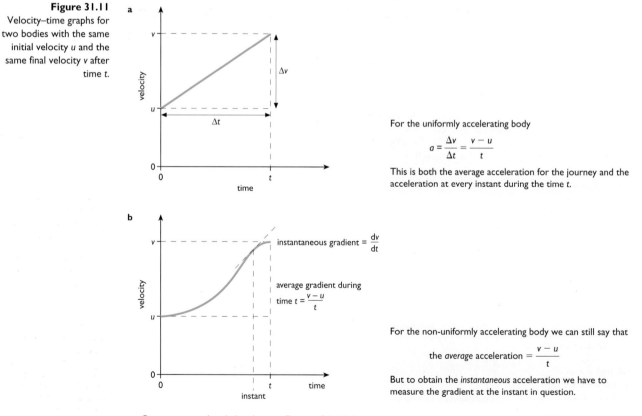

For the uniformly accelerating body

$$a = \frac{\Delta v}{\Delta t} = \frac{v - u}{t}$$

This is both the average acceleration for the journey and the acceleration at every instant during the time t.

For the non-uniformly accelerating body we can still say that

$$\text{the } \textit{average} \text{ acceleration} = \frac{v - u}{t}$$

But to obtain the *instantaneous* acceleration we have to measure the gradient at the instant in question.

Over time t, both bodies in Figure 31.11 have the same average acceleration. This is equal to:

$$\frac{v - u}{t}$$

For the uniformly accelerated body its acceleration is always equal to its average acceleration, so we can write

$$\text{uniform } a = \frac{v - u}{t}$$

For the non-uniformly accelerated body, instantaneous acceleration must be found from the gradient of a point on the graph. For very small values of Δv and Δt (see Figure 31.10b), we write $\Delta v = \delta v$ and $\Delta t = \delta t$. Then

$$\text{average acceleration over time } \delta t = \frac{\delta v}{\delta t}$$

To deal with acceleration further we have to deal with calculus. Newton was aware that the old maths of $+$, $-$, $\times$ and $\div$ wasn't enough to deal with instantaneous values of changing quantities. So he invented his own new kind of maths, based on infinitesimal concepts, and he called this maths 'fluxions'. A few years later Gottfried Wilhelm von Leibnitz also developed new maths for dealing with changing quantities. The fundamental principles of the two versions were the same, but Leibnitz's version was more similar in presentation to the calculus we use now. (Though they were working in similar areas, Newton and Leibnitz did not agree about very much, and especially about who deserved the credit for the new maths.)

For an infinitesimal time, where δt and δv are vanishingly small, we use calculus notation:

$$a = \frac{dv}{dt}$$

This is instantaneous acceleration, and the formula applies to both uniform and non-uniform acceleration. It is thus a more universally useful formula than the formula above for uniform acceleration.

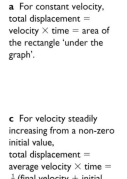

Use $a = (v - u)/t$ for the following.

5 What is the average acceleration of a cyclist whose velocity changes from zero to 5 m s^{-1} in 3 s?

6 What is the average acceleration of a car whose velocity changes from zero to 24 m s^{-1} in 18 s?

Average velocity and total displacement

Average velocity for a complete journey is defined as total displacement divided by total time:

$$\text{average velocity} = \frac{\text{total displacement}}{\text{total time}}$$

This means that:

$$\text{total displacement} = \text{average velocity} \times \text{total time}$$

In all cases the area under a velocity–time graph is equal to total displacement, as Figure 31.12 illustrates.

Figure 31.12

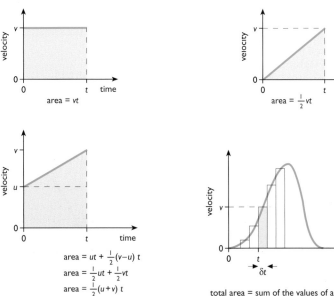

a For constant velocity, total displacement = velocity × time = area of the rectangle 'under the graph'.

area = vt

b For velocity steadily increasing from zero, total displacement = average velocity × time = $\frac{1}{2}$ (final velocity + initial velocity) × time = $\frac{1}{2}$ (final velocity × time) = area of the triangle 'under the graph'.

area = $\frac{1}{2}vt$

c For velocity steadily increasing from a non-zero initial value, total displacement = average velocity × time = $\frac{1}{2}$ (final velocity + initial velocity) × time = area of the trapezium 'under the graph'.

area = $ut + \frac{1}{2}(v-u)\,t$
area = $\frac{1}{2}ut + \frac{1}{2}vt$
area = $\frac{1}{2}(u+v)\,t$

d For a varying velocity, displacement during short time δt = (velocity at time t) × δt = area of the small shaded rectangle. Total displacement at time t is approximately equal to the sum of the areas of all such rectangles up to time t. As δt tends to zero the approximation becomes more and more true.

total area = sum of the values of all ($v\,\delta t$)

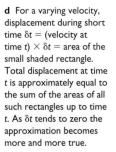

343

More graphical representations of uniform acceleration

Figure 31.13
From Chaplin to Keaton –
a journey in three stages.

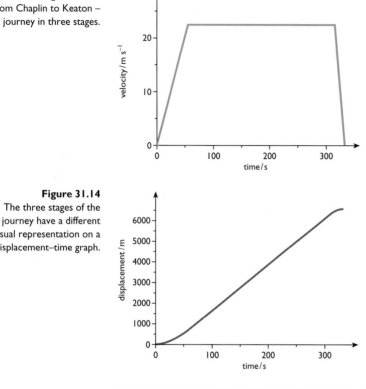

A train travels from town A, Chaplin, to town B, Keaton. It accelerates uniformly away from Chaplin station, at 0.4 m s^{-2} for 56 s. It then travels for 260 s at the constant velocity it has reached, and then decelerates uniformly, at acceleration -0.7 m s^{-2} for 16 s into Keaton station. All of this information can be shown on a velocity–time graph (Figure 31.13).

Figure 31.14
The three stages of the journey have a different visual representation on a displacement–time graph.

Exactly the same information can also be shown on a graph of displacement against time (Figure 31.14). The graph looks different, but it is the same journey.

7 Sketch graphs to show how motion with zero acceleration (constant velocity) and positive uniform acceleration are represented on
   a velocity–time graphs
   b displacement–time graphs.
8 Figure 31.15 shows a velocity–time graph for a city bicycle journey from one set of traffic lights to another. Sketch the general shape of the corresponding displacement–time graph.
9 Figure 31.16 shows a displacement–time graph for a country bicycle journey up a hill. Sketch the general shape of the corresponding velocity–time graph.
10 An arrow is fired from a bow, experiencing a force that decreases as the tension in the string decreases.
   a Is acceleration uniform?
   b Sketch the general shapes of acceleration–time and velocity–time graphs for the period of the acceleration.
11 a What is the final velocity of a car which has an initial velocity of 10 m s^{-1} and which accelerates at 2 m s^{-2} for 6 s?
   b Sketch a velocity–time graph for the car's period of acceleration.
   c Sketch a corresponding displacement–time graph for the same period (assuming that initial displacement is zero).
12 With the same acceleration and initial velocity as in question 11, how long would it take the car to reach a velocity of 32 m s^{-1}?

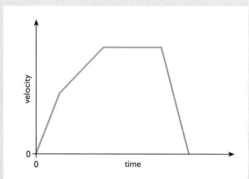

Figure 31.15

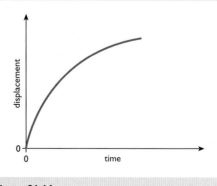

Figure 31.16

Motion where the acceleration is due to gravity

Ignoring the effects of the air, the free fall of a dropped ball is a journey with uniform acceleration and in a straight line. Displacement–time and velocity–time graphs for such a journey are shown in Figure 31.17.

Figure 31.17
The displacement–time and velocity–time graphs for a dropped ball. Note that the graphs follow the Cartesian convention – vectors acting downwards have negative values.

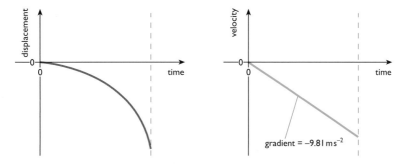

We can go further and consider the graphs for a ball that is thrown vertically upwards (ignoring, for simplicity, the periods when the ball is accelerated and decelerated by the hands that throw and catch it).

Figure 31.18
If the ball is thrown vertically upwards and is caught again by the thrower then the final displacement is zero. Acceleration is the same for the whole journey.

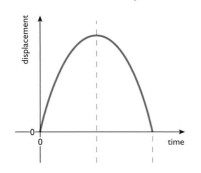 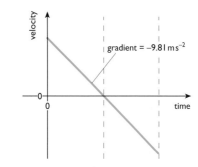

13 A stone is thrown vertically upwards at 4 m s^{-1} and hits the ground after 1.1 s. Assume its acceleration to be -10 m s^{-2}.
 a Why is the acceleration negative?
 b With what velocity does it hit the ground?
 c Sketch velocity–time and displacement–time graphs for the motion of the stone, assuming that initial displacement is zero.

Between them, the graphs in Figure 31.18 show a change in direction, velocity that decreases to zero and then increases again, but constant downward acceleration – the velocity–time graph has a constant gradient of -9.81 m s^{-2}, the acceleration due to gravity. So for a ball thrown upwards:

- displacement increases and decreases again
- velocity decreases to zero and then increases in the negative direction
- acceleration is constant at -9.81 m s^{-2}.

Equations of uniformly accelerated motion

It is useful to be able to deal with situations in which bodies have some initial velocity at time zero, and accelerate uniformly to a new velocity at time t. We will use u to denote the initial velocity and v to denote the final velocity. (To remember which is which, note that u comes before v in the alphabet – initial velocity must exist before final velocity.) Assuming that the acceleration is positive, the velocity–time graph for such motion is that shown in Figure 31.11a, page 342. We can see from the graph that for this motion:

14 Explain why the equation $a = \dfrac{v - u}{t}$ is of no use for working out instantaneous acceleration, except when acceleration is uniform.

$$a = \frac{v - u}{t}$$

which we can rearrange to

$$at = v - u$$

and

$$v = u + at$$

This last equation is called an **equation of uniformly accelerated motion**, and is a useful tool for the analysis of such journeys. It contains four variables, but these do not include displacement. Note that three of the four variables are vectors that can have positive or negative values, depending on their direction.

We can develop equations of motion that include displacement, s, by thinking about average velocity. The average velocity of uniformly accelerated motion can be calculated from the initial and final velocities (Figure 31.19):

$$\text{average velocity of uniformly accelerated motion} = \frac{u + v}{2}$$

Figure 31.19
Where velocity changes uniformly we need only use the first and last values for calculation of average velocity.

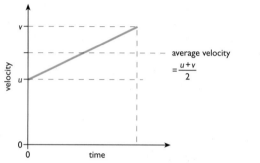

We also know that the average velocity of any journey is the total displacement divided by the total time:

$$\text{average velocity for any journey} = \frac{s}{t}$$

So

$$\frac{s}{t} = \frac{u + v}{2}$$

We can rearrange this to make v the subject, or to make t the subject:

$$\frac{2s}{t} = u + v$$

$$\frac{2s}{t} - u = v$$

$$v = \frac{2s}{t} - u$$

$$\frac{2s}{t} = u + v$$

$$2s = t(u + v)$$

$$t = \frac{2s}{(u + v)}$$

and we can substitute these into the first equation of motion, $v = u + at$:

$$\frac{2s}{t} - u = u + at$$

$$v = u + a\frac{2s}{(u + v)}$$

These need some tidying:

$$\frac{2s}{t} = 2u + at$$

$$2s = 2ut + at^2$$

$$s = ut + \tfrac{1}{2}at^2$$

$$v(u + v) = u(u + v) + 2as$$

$$vu + v^2 = u^2 + uv + 2as$$

$$v^2 = u^2 + 2as$$

These provide us with two more equations of motion – and with two more tools for analysis of motion. Note that the three equations of motion deal with five variables, but each of the equations contains only four variables. It would be possible to derive two more equations of motion – one that excludes initial velocity, u, and one that excludes acceleration, a. However, three equations, alone or in combination, can deal with all situations in which there are up to two unknowns. The equation $s = ut + \tfrac{1}{2}at^2$, for example, is useful when v is unknown and we wish to calculate s.

15 A lorry accelerates from rest ($u = 0$) at traffic lights, with acceleration 2.2 m s^{-2}. How far does it travel in 2 s?

16 A cyclist starts off from the traffic lights at the same time, and travels the same distance in only 1.5 s, with uniform acceleration.
 a What is the cyclist's acceleration?
 b Use $F = ma$ to help you to explain how it is unsurprising that the cyclist has a bigger acceleration.

17 A train travels at 20 m s^{-1}. It brakes with a uniform deceleration of 1.6 m s^{-2} over a distance of 80 m.
 a If initial velocity is positive, is the train's acceleration positive or negative?
 b Calculate the train's final velocity.

18 For what length of time was the train in question 17 decelerating?

19 A careless bunjee jumper has a pocket full of coins which fall out at different stages of the jump. How long does a coin take to hit the surface 15 m below when the jumper is:
 a travelling downwards at 4 m s^{-1} at a height of 15 m above the surface?
 b travelling upwards at 4 m s^{-1} at the same height?

20 a The person who threw the stone in question 13 released it when it was level with the top of her head. How tall is she?
 b A person 1.85 m tall throws a stone in the same way. How long does it take to hit the ground?

Comprehension and application

On the track of some dastardly deeds

A train leaves Arbuckle station at 09.10 and travels to Keystone, calling in at Keaton and Chaplin.

Jim Henchman, a rogueish employee of Algernon De Ville of Bounder Hall, attempts to take control of the train from the driver. Following a particularly dramatic fight the driver knocks Henchman unconscious, and kicks Henchman's revolver horizontally backwards from the cab, from a height of 3 m above ground level.

Figure 31.20
The train's scheduled journey.

	Departure	Arrival
Arbuckle	09.10	
Keaton		09.25
Chaplin		09.35
Keystone		09.50

Henchman's actions are part of a plot to rob sweet Madeleine of her rightful inheritance to Bounder Hall. The poor girl has been tied to the tracks, midway between Keaton and Chaplin, by her evil Uncle Algernon. The train is 80 m away, travelling at 23 m s^{-1} when the driver notices the poor girl. It takes the driver (who is, of course, called Roger and is the dashing young hero on his way to confront Algernon at Bounder Hall) 0.7 s to respond, and he then applies the brakes to produce a constant deceleration of 4.5 m s^{-1}.

Meanwhile, Aunt Agatha, sitting regally in a first-class compartment, drops a boiled egg, newly taken from her hamper. From her window, Aunt Agatha had previously noted a metal object, perhaps a revolver, hit the ground alongside her.

Figure 31.21

21 Plot a displacement–time graph for the complete journey shown by the timetable (Figure 31.20). (You will have to estimate the curvature of the graph that is due to acceleration and deceleration on approaching and leaving each station.) Assume that the train spends 1 minute at each station.

22 Figure 31.22 shows a velocity–time graph for the start of the journey which presents an idealised version of events, showing instantaneous start and finish to accelerations and decelerations.

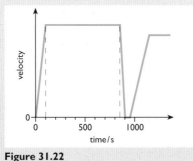

Figure 31.22

a What is happening between 900 s and 960 s?
b Explain how a more completely realistic graph would differ from this.

23 Draw diagrams of the trajectory of the egg relative to Aunt Agatha:
a if dropped when the train has constant velocity
b if dropped just as Roger applies the brakes.

24 a How long does the revolver take to hit the ground? (Assume $g = 10$ m s^{-2}.)
b Neglecting air resistance, calculate the horizontal velocity relative to the train with which the revolver was kicked, if Aunt Agatha was sitting 5 m behind the cab.

25 Does Roger untie the terrified Madeleine and hold her safely in his manly arms, or does he squash her with several tonnes of locomotive?

● **Comprehension and application**

Parachuting on the Moon

Parachuting on the Moon is a sport that seems unlikely to catch on. The Moon's lack of atmosphere makes the use of a parachute rather pointless. But we can sit comfortably on Earth and imagine what falling to the Moon's surface *might* be like, and to think how different it might be from a parachute jump on Earth.

The acceleration due to gravity on Earth, neglecting air resistance, is 9.8 m s^{-2}. The acceleration due to gravity on the Moon is 1.6 m s^{-2}. The graphs in Figure 31.23 show the relationship between acceleration and time for parachutists on the Moon and the Earth.

Figure 31.23
Acceleration–time graphs for parachutists **a** on the Moon and **b** on the Earth.

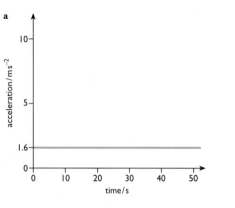

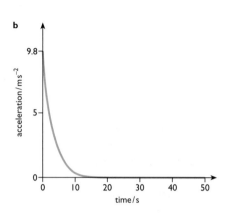

26 Explain the very different forms of the graphs in Figure 31.23.
27 Sketch corresponding velocity–time graphs. For the Earth, mark your graph with the terminal velocity.
28 Sketch corresponding displacement–time graphs. Show carefully how your graphs distinguish the different types of velocity as the parachutists approach the two surfaces.
29 A free-fall jumper on Earth reaches a terminal velocity of 55 m s^{-1} after falling for about 12 s.
　a At what time after starting their falls do a free-fall jumper on the Earth and a free-fall jumper on the Moon have the same speed?
　b How far have each of them fallen?
30 In which case will you hit the ground at higher speed – when jumping (without a parachute) from a 30-storey skyscraper on the Earth, or from a 30-storey skyscraper (yet to be built) on the Moon? (Assume 1 storey ≈ 3 m.)

● **Extra skills task**

Information Technology and Application of Number

Develop a spreadsheet program that can produce acceleration–time graphs for any gravitational field that you put in. The program should also produce corresponding velocity–time graphs and displacement–time graphs. Use your program to generate sets of graphs for a selection of planets in the solar system. State any necessary assumptions. Refine your graphs to show the effects of the planetary atmosphere, as appropriate, for each planet.

Examination questions

1 a Collisions between objects are said to be either *elastic* or *inelastic*. Complete the table by placing a tick (✔) in the relevant places to indicate which quantities are conserved in these collisions. (3)

Collision	Momentum	Kinetic energy	Total energy
elastic			
inelastic			

b i A fast-moving neutron of mass m collides head-on with a stationary atom of hydrogen, also of mass m, as illustrated in the figure.

The neutron is captured by the atom to form a 'heavy' isotope of hydrogen of mass $2m$ which moves off with a speed of $3.0 \times 10^7 \, \mathrm{m \, s^{-1}}$.

1 State whether the collision process whereby the neutron is captured is elastic or inelastic.
2 Calculate the speed of the neutron before capture.

ii A similar neutron to that in **i** now collides head-on with a stationary nitrogen atom of mass $14m$ to form a 'heavy' isotope of nitrogen. Calculate the speed of this 'heavy' nitrogen atom. (5)

OCR, Physics, Paper 2, Nov 1999

2 A tennis machine serves tennis balls over the net for a player to practice return shots.

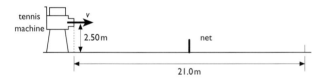

The ball emerges horizontally 2.50 m above the ground with velocity v and hits the ground 21.0 m away from the machine as shown in the diagram above.

a Ignoring air resistance show that the ball takes 0.714 s to reach the ground. (2)
b Calculate the velocity v with which the ball left the machine. (2)
c Calculate the vertical component of the velocity of the ball when it reaches the ground. (1)
d State the horizontal component of the velocity of the ball when it reaches the ground. (1)
e Find the angle at which the ball strikes the ground. (3)
f What would be the effect on this angle of taking air resistance into account? Explain your answer. (1)

Edexcel (London), PSA1, June 1999

3 a Water flows from a nozzle with an initial velocity of $5.8 \, \mathrm{m \, s^{-1}}$ at an angle of 45° to the horizontal, as shown.

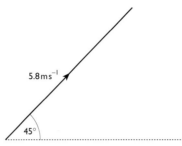

For this velocity, use a vector triangle or calculation to
i show that the horizontal component is $4.1 \, \mathrm{m \, s^{-1}}$
ii explain why the vertical component has the same magnitude as the horizontal component. (3)

b The nozzle is part of a sprinkler used to water a lawn. The sprinkler is at ground level and when the water leaves the nozzle at an angle of 45° to the horizontal, the path of the jet of water is as shown below.

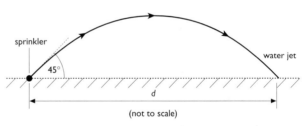

(not to scale)

The initial speed of the water is $5.8 \, \mathrm{m \, s^{-1}}$ and air resistance may be assumed to be negligible. Using the information in **a**, calculate, for one drop of water in the jet,
i the time taken to reach its maximum height
ii the total time between leaving the nozzle and hitting the ground
iii the horizontal distance d from the nozzle to the point of impact with the ground. (5)

c On the figure above, sketch paths of the jet of water when the water emerges from the nozzle at a velocity of $5.8 \, \mathrm{m \, s^{-1}}$ and at an angle to the horizontal of
i 60° (label this path H)
ii 30° (label this path L). (3)

d Suggest the angle to the horizontal at which the nozzle on the sprinkler will give the maximum value of d. (1)

OCR, Sciences, Physics Foundation, Mar 1999

4 a Define *acceleration*. (2)

b A body has an initial velocity *u* and an acceleration *a*. After a time *t*, the body has moved a distance *s* and has a final velocity *v*. The motion is summarised by the equations

$v = u + at$

$s = \frac{1}{2}(u + v)t$

i State the assumption made about the acceleration *a* in these equations.

ii Use the equations to derive an expression for *v* in terms of *u*, *a* and *s*. (3)

c A photographer wishes to check the time for which the shutter on a camera stays open when a photograph is being taken. To do this, a metal ball is photographed as it falls from rest. It is found that before the shutter opens, the ball falls 2.50 m from rest and, during the time that the shutter remains open, the ball falls a further 0.12 m, as illustrated.

Assuming that air resistance is negligible, calculate

i the speed of the ball after falling 2.50 m

ii the time to fall the further 0.12 m.

[You may wish to use an equation of the form

$$x = \frac{-b \pm \sqrt{b^2 - 4ac}}{2a} .]$$

iii The time for which the shutter stays open is marked on the camera as 1/60 s. Comment on whether the test confirms this time. (6)

OCR, Physics, Paper 2, Nov 1999

5 A racing team is preparing its car for a circuit whose layout is shown below.

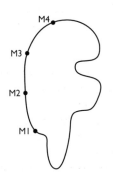

Two different sets of gears are being considered, set A and set B. Trial runs are conducted to determine the best set to use over the 'fast' part of the circuit, from marker M1 to M4. The markers are equally spaced, 0.400 km apart.

Race car speeds measured at the markers are given in the table below.

Marker	Speed with gearset A (km h^{-1})	Speed with gearset B (km h^{-1})
M1	64	64
M2	144	162
M3	160	184
M4	192	188

a Assuming that the accelerations are constant between markers, determine the average speed between markers and thus complete the table below. (2)

Marker interval	Average speed with gearset A (km h^{-1})	Average speed with gearset B (km h^{-1})
M1 → M2		
M2 → M3		
M3 → M4		

b Determine the time intervals between markers and thus complete the table below. (2)

Marker interval	Time with gearset A (s)	Time with gearset B (s)
M1 → M2		
M2 → M3		
M3 → M4		

c Which gearset should be used to obtain the shortest time from M1 to M4? (1)

d Draw a graph to show how the speed, as presented in the original data table, varies with time. (4)

e Which gearset provides the greatest acceleration

i initially (i.e. between M1 and M2)?

ii finally (i.e. between M3 and M4)? (2)

International Baccalaureate, Subsidiary/Standard Level, May 1998

32 Force and motion on the road

THE BIG QUESTION	● How can we apply knowledge of force and motion to the particular example of the car?
KEY VOCABULARY	braking force driving force hazard motive power risk rolling friction
BACKGROUND	Whether we use them ourselves or not, motor-driven vehicles are a big feature of our lives. They offer personal freedom, and provide work for people in many parts of the world. But they also produce noise, pollution and safety hazards, both to vehicle-users and to other people. This chapter looks at the car as a context for applying ideas about force and motion, and in particular it looks at reducing the risks.

● Getting around – driving force of cars

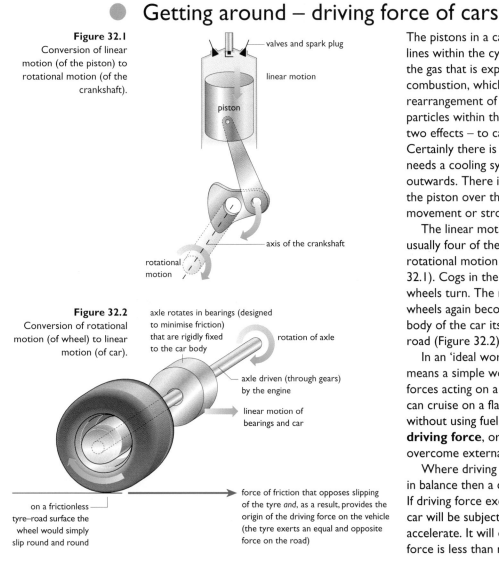

Figure 32.1
Conversion of linear motion (of the piston) to rotational motion (of the crankshaft).

valves and spark plug

linear motion

piston

axis of the crankshaft

rotational motion

Figure 32.2
Conversion of rotational motion (of wheel) to linear motion (of car).

axle rotates in bearings (designed to minimise friction) that are rigidly fixed to the car body

rotation of axle

axle driven (through gears) by the engine

linear motion of bearings and car

on a frictionless tyre–road surface the wheel would simply slip round and round

force of friction that opposes slipping of the tyre *and*, as a result, provides the origin of the driving force on the vehicle (the tyre exerts an equal and opposite force on the road)

The pistons in a car engine move in straight lines within the cylinders, taking energy from the gas that is expanding. Energy is supplied by combustion, which is essentially a rearrangement of the potential energies of particles within the materials. This can have two effects – to cause heating and to do work. Certainly there is heating, and the engine needs a cooling system to transfer this energy outwards. There is also work – force acts on the piston over the downwards distance of its movement or stroke.

The linear motion of the car's pistons – usually four of them – is converted to rotational motion of the crankshaft (Figure 32.1). Cogs in the gears turn, and so the drive wheels turn. The rotational motion of the wheels again becomes linear motion – of the body of the car itself as it travels along the road (Figure 32.2).

In an 'ideal world' – and to a physicist that means a simple world – there are no resistive forces acting on a car. In such a world a car can cruise on a flat road at a constant speed without using fuel. But in the real world a **driving force**, or motive force, must act to overcome external resistive force.

Where driving force and resistive force are in balance then a car will keep a steady speed. If driving force exceeds resistive force then the car will be subject to a net force and will accelerate. It will decelerate whenever driving force is less than resistive force.

Figure 32.3
Forces on a car.

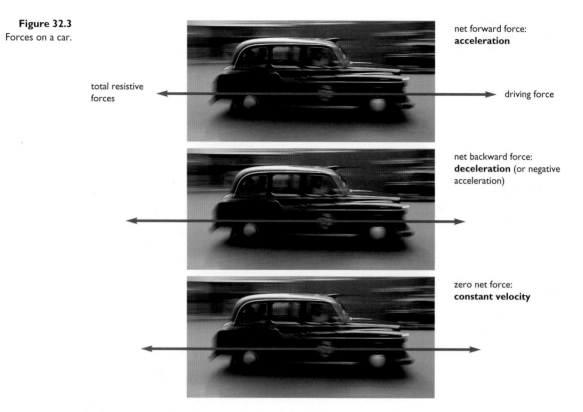

net forward force:
acceleration

total resistive
forces

driving force

net backward force:
deceleration (or negative
acceleration)

zero net force:
constant velocity

We can summarise the situation in Figure 32.3 as:

$$\text{net force} = \text{driving force} - \text{resistive force}$$
$$= F_D - F_R$$

where F_D and F_R are the simple magnitudes of the forces.

Any net force results in acceleration, and it is related to this by the fundamental and familiar equation $F = ma$. So,

$$ma = F_D - F_R$$

The external resistive force has two causes. One is the drag force of the air acting on the moving car body. The other is the **rolling friction** or rolling resistance between the tyres and the ground, due to distortion of the tyre as it rolls, different parts of it coming into contact with the ground. At low velocity, drag force is quite small, and rolling friction makes the bigger contribution to total resistive force F_R. However, drag force is proportional to the square of the velocity, and so drag forces dominate at higher velocities.

Coping with hills

An additional force influences a car that is moving uphill or downhill – the force of gravity (Figure 32.4). For a car going uphill there is an additional force opposing motion, and this is related to the mass of the car and to the slope of the hill.

Figure 32.4
The effect of gravity on a
car on a slope.

component of weight
parallel to slope which
may *or* may not be
balanced by driving
force or frictional force

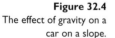

weight of car

component of weight
perpendicular to slope
– this is balanced by the
normal reaction exerted
by the road on the car

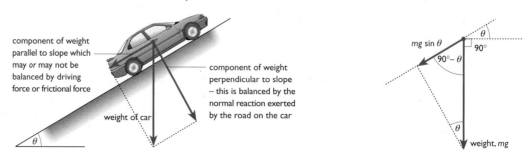

$mg \sin \theta$

θ

$90°$

$90° - \theta$

θ

weight, mg

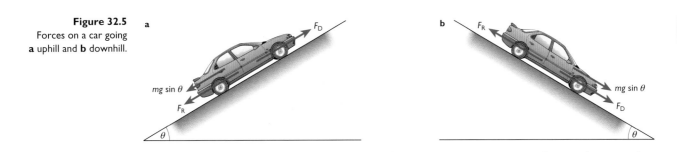

Figure 32.5
Forces on a car going
a uphill and **b** downhill.

I a Calculate the acceleration along a flat road of a car of mass 1800 kg for which the driving force is 1.6 kN and the resistive force is 2.0 kN.
b What does this acceleration become if driving force and resistive force stay the same as in part a and the car is going down a hill which makes an angle of 3° to the horizontal?
c Rewrite $ma = F_D - mg \sin \theta - F_R$ and $ma = F_D + mg \sin \theta - F_R$ for $a = 0$ and so show that the equations agree with the 'common sense' idea that more driving force is needed to go uphill at constant speed than downhill.
d Imagine that the car moves slowly up a hill so that resistive force is very small compared with driving force and with the component of its weight down the slope. If driving force has a maximum value of 10 kN, what is the steepest hill the car can climb at a constant velocity (assuming that the tyres don't slip on the surface)?

If the mass of the car is m then its weight is given by mg where g is the gravitational field strength. The component of this force that is perpendicular to the road surface is balanced by the normal reaction exerted by the road. The car does not accelerate either into or away from the road surface. The component that is parallel to the road surface acts down the hill, and tends to produce acceleration in this direction.

For a car going uphill (Figure 32.5a) this component of the car's weight acts in the opposite direction to the driving force so that the net force, ma, is given by:

$$ma = F_D - mg \sin \theta - F_R$$

where F_R, as before, is the combination of resistive forces of drag and rolling friction, θ is the angle between the slope and the horizontal, and g is the gravitational field strength of 9.8 N kg^{-1}.

When the car is going downhill (Figure 32.5b) the same component of the car's weight now acts in the direction of the driving force, making a positive contribution to the force in the direction of travel:

$$ma = F_D + mg \sin \theta - F_R$$

Motive power

The useful energy output of a car is equal to the work it does on moving forwards, both in overcoming resistive force and in accelerating. It is equal to the average net forward force multiplied by the distance travelled. This quantity, however, applies to a particular section of a particular journey, and we must specify the length of this section. It is more useful in general to think about what is going on moment by moment, and we can do this in terms of *power*.

We know that, in general, by definition, power is the rate of transfer of energy:

$$P = \frac{dW}{dt} \quad \text{or, if power is constant over time } \Delta t, \quad P = \frac{\Delta W}{\Delta t}$$

For a car travelling along a road in a straight line under the action of a constant driving force, F_D:

$$\Delta W = \text{driving force } (F_D) \times \text{distance } (\Delta x)$$

We are assuming that driving force and distance are parallel. For straight line motion this is true. So, since power is constant,

$$P = \frac{F_D \times \Delta x}{\Delta t}$$

And since $\Delta x / \Delta t$ is the constant velocity of the car, v,

$$P = F_D v$$

2 a Sketch a graph of motive power against velocity for a car with a constant driving force.
b Sketch a graph of drag force against velocity (see page 352).
c Resistive force increases with velocity. To maintain a particular velocity, a car's driving force must increase to match this. So the graph in part **a** represents an over-simplified situation. Sketch a graph that shows the general relationship between motive power and car velocity, taking account of the non-constant driving force.

Note that this power relates to driving force, and is the rate at which the car does work in order to move. It is called the **motive power**. It is not the same thing as the power output of the engine, which relates to the rate at which the *engine* does work. Some energy is dissipated in the car's transmission system, so not all of the energy made available as work by the engine is usefully transferred.

Trailers

A trailer copies the motion of the towing vehicle, if the tow bar is of fixed length as we shall assume. The pulling force acting on the trailer is exerted on it by the towing vehicle, through the tow bar. The trailer exerts an equal and opposite force on the towing vehicle (Figure 32.6).

Figure 32.6
Forces between a car and trailer.

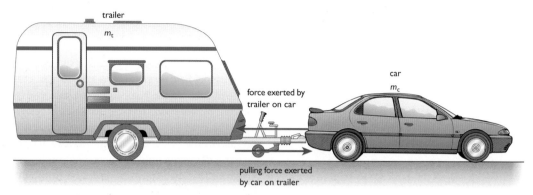

---

3 A car of mass 2000 kg tows a caravan of mass 800 kg along a level road, and accelerates at 0.5 m s^{-2}.
a What is the net force acting on the caravan?
b If the caravan experiences resistive forces of 5.6 kN, what is the pulling force that the car must exert on it through the tow bar?
c What is the force that the caravan exerts on the car?
d Use knowledge of the dependence of resistive force on velocity to explain why a car's maximum velocity is significantly reduced when it tows a caravan.

$$m_{total} = m_t + m_c$$

Car and trailer have the same acceleration, a.
On a level road, net force on the car and trailer (considered as a single body) is given by

$$\text{net force} = \text{driving force} - \text{total resistive force} = F_D - F_R$$

and also

$$\text{net force} = m_{total}a = m_t a + m_c a$$

So

$$F_D - F_R = m_t a + m_c a$$

and

$$F_D = F_R + (m_t + m_c)a$$

When pulling a trailer, a car must supply increased driving force both to compensate for increased resistive force and to accelerate a larger mass.

Application of mathematics to analysis of car motion

4 a For the car in question 1a, what is the final velocity if the initial velocity is 15 m s^{-1} and the acceleration continues for 8 s?
b What is the car's change in momentum? How is this consistent with the principle of conservation of momentum?
c What is the change in the car's kinetic energy?

5 Which of the listed equations tells you in mathematical form that velocity is constant when net force is zero?

6 Explain why the velocity of a moving car is not constant when driving force is zero.

Motion can be analysed by a number of means:

- in terms of momentum:

   momentum, $p = mv$

- in terms of energy, usually kinetic energy:

   kinetic energy $= \frac{1}{2}mv^2$

- in terms of the relationship between force and acceleration:

   $F = ma$

- in terms of the equations of uniformly accelerated motion (Chapter 31):

   $v = u + at$

   $v^2 = u^2 + 2as$

   $s = ut + \frac{1}{2}at^2$

or by a combination of the above! To solve many problems, the skill needed is the ability to choose the best approach.

Stopping

Figure 32.7
An opposing frictional force provides the braking force, which adds to the resistive force F_R.

direction of travel

frictional force acting on tyre rim when axle is driven by engine

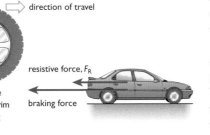

direction of travel

resistive force, F_R

frictional force acting on tyre rim during braking

braking force

It is possible to stop a car simply by removing the driving force and allowing the resistive force to act alone. The car continues to do work against these resistive forces, and energy transfers away from it. Its kinetic energy decreases.

But putting trust in resistive forces to produce the required deceleration is not often enough. What is almost always needed is an active **braking force**. This acts in the opposite direction to the motion, and is achieved through tyre–road friction (Figure 32.7).

To achieve a frictional force in the required direction, acting on the rim of the tyre, it is necessary to oppose the rotation of the wheel. This is what the brake pads do, by applying frictional force to a drum or disc which rotates with the wheel (Figure 32.8a and b).

Figure 32.8
Drum brakes and disc brakes.

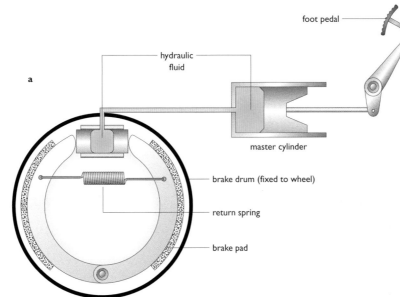

a

hydraulic fluid

foot pedal

master cylinder

brake drum (fixed to wheel)

return spring

brake pad

7 Imagine that the tow bar in question 3 becomes detached when the velocity is 22 m s^{-1}. The resistive force acting on the caravan then maintains an average value of 2.4 kN.
a What is the average value of the caravan's deceleration?
b How long does it take for the caravan to stop?
c How far does it travel before stopping?

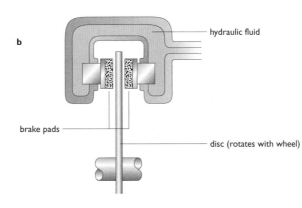

b

hydraulic fluid

brake pads

disc (rotates with wheel)

Energy transfer for a stopping car

A stopping car must lose its kinetic energy. To do this it must do work on something. It could do work on the air around it until it comes to a halt. It could do work against the forces of resistance that act on its own moving parts. It could do work on an object with which it collides, or it could do work on the road by applying its brakes. By one means or another, the energy of the car is transferred to the surroundings – it is dissipated (Figure 32.9). But also, the car could do work against the force of gravity by climbing a hill until it stops. In this case it could be said that some energy is available for later use and the energy is stored in a useful way rather than being dissipated.

Figure 32.9
Energy transfer for a car stopping on a flat road.

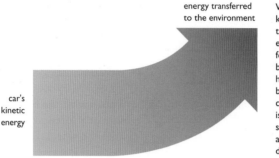

energy transferred to the environment

When a car stops, its kinetic energy is transferred to its environment. On braking, for example, friction in the braking system results in heating. First the discs and brake blocks get hot. In cooling down, their energy is transferred to the surroundings, mostly the air. The energy is dissipated.

car's kinetic energy

When a car's brakes are applied, friction between the road and the tyres provides a braking force, in the opposite direction to the car's motion. The braking force acts on the car over a certain distance, x_b, which is the car's braking distance. For simplicity we can suppose that the force, F, is constant. Then

change in kinetic energy = work done

$$\Delta E_k = -Fx_b$$

$$\tfrac{1}{2}m\,\Delta(v^2) = -Fx_b$$

Note that the minus sign is required because the change in kinetic energy is negative.

If the car's velocity before braking is v and its final velocity is zero, then

$$\Delta(v^2) = (\text{final value of velocity})^2 - (\text{initial value of velocity})^2$$

$$= 0^2 - v^2$$

$$= -v^2$$

Figure 32.10
The relationship between braking distance and a car's initial travelling speed is not a linear one.

and so

$$\tfrac{1}{2}m(-v^2) = -Fx_b \quad \text{or} \quad \tfrac{1}{2}mv^2 = Fx_b$$

Knowledge of this relationship allows us to plot a graph of braking distance against a car's travelling velocity, Figure 32.10. Braking distance is the output variable (dependent variable) and so it helps to make this the subject of the equation:

$$x_b = \frac{mv^2}{2F}$$

For a car of constant mass experiencing a constant braking force we can simplify this equation by writing

$$\frac{m}{2F} = \text{constant, } c$$

Then

$$x_b = cv^2$$

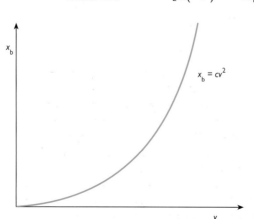

x_b

$x_b = cv^2$

v

The graph does not tell the whole story for a stopping car. In an emergency stop, the car will travel some distance while the driver reacts and applies pressure to the brake pedal. During this period of time the car's velocity will hardly change, and the formula

$$v = \frac{x_t}{t}$$

applies, where x_t is the distance travelled during this 'reaction time' and t is the reaction time itself. So,

$$x_t = vt$$

This is a simpler relationship than the one that applies during actual braking. The distance travelled during the reaction time is proportional to the time, and a graph of 'reaction distance' against travelling velocity of the car when the driver first sees the danger is a simple straight line. The relationship between reaction distance and travelling velocity is a linear one. The gradient of the graph is the reaction time, which may be different for different drivers, and different for the same driver under different conditions of tiredness.

Driving manuals show information about braking distance, reaction distance and total stopping distance in the form of diagrams like that in Figure 32.11. Note that total stopping distance is reaction distance plus braking distance, $x_t + x_b$. Presentation of the information in such a visual format is more powerful, especially to readers who have little knowledge of mathematical relationships, than stating the relationship in words or algebraic symbols.

8 How much kinetic energy must a car of mass 1800 kg lose to stop when its initial travelling velocity is
  a 20 km h^{-1}
  b 40 km h^{-1}?
9 A runaway truck rolls down a road on one side of a valley and starts to move up the hill on the other side.
  a If its mass is 5 tonnes, how much potential energy does it lose in falling through a height of 20 m to the bottom of the valley? (1 tonne = 1000 kg)
  b If all of this energy became kinetic energy, what would be its velocity when at the bottom of the valley?
  c Why will it not achieve this velocity?
  d If, by the time the truck reaches its highest point having passed through the valley bottom and rolled up the hill, 60% of its initial potential energy has been dissipated, what height will it reach above the valley bottom?
10 Explain why a graph of initial travelling velocity against reaction distance is a linear one, but a graph of initial travelling velocity against braking distance is not.
11 Sketch graphs of reaction distance against a car's initial travelling velocity for a driver who is fresh and alert and for a driver who has drunk a pint of beer. (Alcohol, consumed even in modest quantities, increases reaction time.)

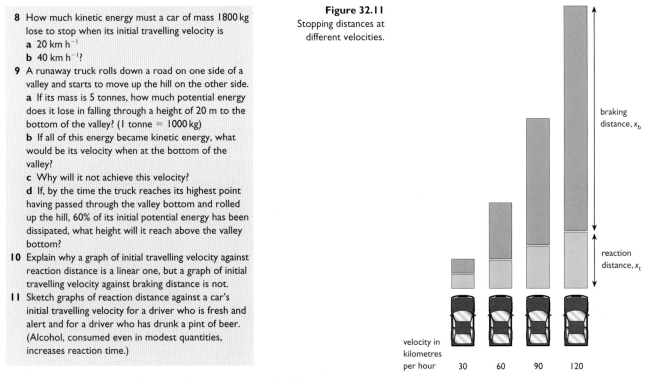

Figure 32.11
Stopping distances at different velocities.

braking distance, x_b

reaction distance, x_t

velocity in kilometres per hour   30   60   90   120

Stopping too quickly

In the early days of the motor car a person had to walk in front, waving a red flag. This was because the human body is not designed to withstand impacts by fast-moving heavy objects; nor is it designed to be fast-moving itself and to withstand impact with something that is large and stationary. The law-makers thought that people could get killed. They were right. We all know that cars present **hazards**, or the potential for accidents. What we can do is recognise the hazards and reduce the **risk**, or probability, of actual harm. We can then decide for ourselves whether we want to accept the level of risk. We also have to remember that our personal presence on the road adds to the hazards that other people, including pedestrians, face. The risk of harm to others must also be reduced. This requires road safety measures, such as speed limits, together with cars that are engineered to take account of hazards and minimise risk.

It is force that does damage. Safety features in cars are designed to reduce the force that acts on the body of a passenger during an accident. One means of reducing the maximum force that you might experience during a collision is the use of seat belts. Your body, like any other mass, resists acceleration or deceleration. This resistance is called inertia, and is measured in terms of mass, where $m = F/a$. In an accident the mass of your body will keep going until decelerated by a suitable force. A seat belt stops you with a smaller force than you would experience if your body slammed into the windscreen or the back of the seat in front of you.

Seat belts are carefully designed to achieve this reduction in force. Though a seat belt can be worn many times, it can be used in an accident only once. The webbing is designed to unravel when the force (strictly, the tension that is stretching the belt) reaches a certain level. It is the passenger's body exerting this force, while the seat belt exerts an equal and opposite force on the passenger. The belt must not unravel at levels of force that are experienced in normal conditions, but must unravel before the force becomes so big that injury becomes likely. As the passenger's body causes the webbing to unravel it does work on it, and loses kinetic energy:

$$\text{work that you must do on your surroundings in order to stop} = \text{kinetic energy that you must lose}$$

Or:

$$\text{mean force you exert on your surroundings} \times \text{distance over which you stop} = \text{kinetic energy you must lose}$$

$$\text{mean force exerted } by \text{ your surroundings } on \text{ you} \times \text{distance over } which \text{ you stop} = \text{kinetic energy you must lose}$$

$$Fx = \tfrac{1}{2}mv^2$$

$$F = \frac{mv^2}{2x}$$

Minimising the force that acts requires minimising the right-hand side of this last equation. Driving at a low speed, v, has the most dramatic effect. It also helps to have a smaller mass. But another way to minimise the force is to maximise the distance x over which you stop. Seat belts, crumple zones and air bags are all designed to do this (Figure 32.12).

Figure 32.12
Safety features that reduce the force acting on a passenger in an accident, by increasing the distance over which the force acts.

seat belt and body exert equal and opposite forces

airbag and head exert equal and opposite forces

wall and car exert equal and opposite forces

A seat belt that unravels increases the distance over which the body is decelerated.

An air bag increases the distance over which the head is decelerated.

The car's crumple zone increases the distance over which the car is decelerated.

12 a Calculate the ratio

$$\frac{\text{(work that a car must do on its surroundings to stop at an initial velocity of } 30 \text{ m s}^{-1})}{\text{(work that the same car must do to stop at an initial velocity of } 15 \text{ m s}^{-1})}$$

Use

work that must be done = kinetic energy lost

b How does this result provide strong support for the idea that lower velocity limits mean reduced risk for all road users?

13 Explain why the fitting of cosmetic 'bull bars' to the fronts of vehicles increases the risk of harm to pedestrians and cyclists with whom the vehicle collides.

14 Imagine that you and a fly crash into a wall at the same velocity.
a Use the equation

$$F = \frac{mv^2}{2x}$$

to explain why you are more injured than the fly.
b Show that the equation is consistent with the more familiar $F = ma$.

Motor efficiency

Efficiency in general (see Chapter 15) is the ratio of useful power output to power input, normally multiplied by 100 and then expressed as a percentage:

$$\text{efficiency} = \frac{\text{useful power output}}{\text{power input}} \times 100\%$$

For a car, the motive power can be thought of as the useful power output (page 353), and the power input can be measured in terms of fuel used:

$$\text{useful power output} = F_{\mathrm{D}}v$$

$$\text{power input} = \begin{array}{c} \text{mass of fuel used} \\ \text{per second, } \mathrm{d}m/\mathrm{d}t \end{array} \times \begin{array}{c} \text{'energy content' of the fuel} \\ \text{in joules per kilogram, } C \end{array}$$

$$= \frac{\mathrm{d}m}{\mathrm{d}t} \times C$$

The ratio of these, multiplied by 100, gives a measure of a car's efficiency.

Of course, the useful end result of car use is distance travelled, and the input is the quantity of fuel used. So one measure of car 'value for money' is in terms of the ratio of energy input from fuel to distance travelled, measured in MJ km^{-1} (Figure 32.13).

Figure 32.13
In general, energy input per km increases with speed, i.e. 'value for money' decreases with speed.

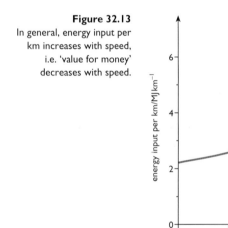

. . . .

Comprehension and application

ABS braking

Cars kill people. Around the world, hundreds of thousands of drivers, passengers and innocent bystanders die every year. Yet most of us feel safe in cars, until the moment of panic when something goes wrong. Usually the problem is driver error – failure to see something, failure to respond quickly enough, failure to anticipate what could go wrong. Such error becomes both more likely and potentially more damaging as speed increases. Less often, but still too often, the problem is mechanical failure, which is obviously more likely when maintenance standards are low.

We have to make judgements about risk. Most of us decide that the risk is acceptable relative to the benefits of using cars. Cars can be built to reduce risk of harm, by incorporating design features such as ABS braking that can help to prevent impact, and others that mitigate the consequences of impact, such as crumple zones, seat belts and air bags.

ABS braking reduces the risk of skidding. Skidding can be caused by wheels 'locking'. That is, wheels stop turning while the car is still moving and the consequence is a skid. It's most likely during sudden braking in an emergency situation, when the driver's instinct may be to press as hard as possible on the footbrake. An ABS system has sensors on the wheels which sense the wheels' lack of rotation and send a signal to a control unit. This then reduces the pressure on the brake to the affected wheels. The hydraulic pressure can be applied in pulses to prevent a skid. ABS allows the driver to maintain control of the steering.

A car of mass 1.6 t travels at 72 km h^{-1} on a straight road and the driver suddenly sees a stationary vehicle 20 m ahead. (1 t = 1 tonne = 1000 kg)

15 Calculate the momentum and the kinetic energy of the car in SI units.

16 a What force is needed to stop the car in a horizontal distance of 20 m?
b What force would be needed on a slope that rises 1 m for every 10 m of horizontal distance?
c What force would be needed on a slope that falls 1.5 m for every 10 m of horizontal distance?
d What force acts on a person of mass 70 kg as the car stops in a horizontal distance of 20 m?
e If the road is horizontal but wet so that the maximum possible braking force that can exist between the road and the car tyres is 12 kN, what happens?
f For such a maximum force, what is the minimum upward slope (in terms of increased height per 10 m of horizontal distance) that will allow the car to stop within a distance of 20 m?

17 a Suppose that a collision takes place. Will it be elastic or inelastic? Explain this.
b All collisions obey momentum conservation, but both cars are at rest soon after the accident. Explain this, and so explain why collisions between cars on roads are very different to collisions between spacecraft.

18 a ABS cannot improve road conditions or driving standards. How might it help to prevent a collision in this scenario?
b Draw a plan showing likely outcomes with and without ABS.

19 Sketch a block diagram to illustrate the principles of ABS. Blocks should represent wheel, brake, sensor and control unit, and you should show the flow of information between them.

20 DISCUSS
a To what extent might safety features such as seat belts, which reduce the risk of damage to a car's occupants, reduce the safety of others?
b To what extent might safety features such as ABS, which reduce the risk of serious accident at a given speed, actually undermine road safety?

● **Extra skills task** Communication

The law imposes limits on the speed of motorists. Some methods are crude: catching and fining offenders. Other methods are designed to prevent motorists from speeding in the first place. Many shock slogans, posters and TV adverts have been devised to persuade drivers that a small increase in speed can have a devastating effect on others. Find out about government advertising programmes if you can.

Identify and explain the physical principles that cause the seriousness of accidents to rise very rapidly as speed rises.

Present a report to other students as a verbal presentation, with visual aids, designed to make them too become safer drivers. Make sure that your presentation wins and retains the attention of your audience.

● # Examination questions

1 a Describe how the driving wheels of a vehicle may generate a forward force at the ground. (3)
b A cycle and cyclist of total mass 70 kg move along a straight horizontal road. At a particular time, the speed is 8.8 m s^{-1} and the acceleration is 0.12 m s^{-2}. At this time, the cyclist experiences a pressure drag force of 45 N and a resistive friction force of 15 N at the ground. Calculate
i the resultant force required to give the cyclist this acceleration
ii the power required by the cyclist at this time, taking into account the resistive forces. (6)

c The cyclist now moves up a hill inclined at an angle of 6.0° to the horizontal as shown in the figure, and slows to a steady speed of 2.2 m s^{-1}. Assume that the road surface is unchanged.

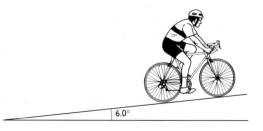

i State how the pressure drag force and resistive friction force depend on speed. (2)

ii Calculate

  1 the new drag force, assuming $F = A\rho v^2$ (3)

  2 the new power required by the cyclist to move at the constant speed of 2.2 m s^{-1}. (4)

OCR, Sciences, Physics of Transport, Nov 1999

2 This question is about the drag force exerted on a car.

The drag force on a car is caused by the air flow around the car and is related to the speed, *v*, of the car by the following equation:

$$F_D = KA\rho v^2$$

where F_D is the drag force, *A* is the front cross-sectional area of the car, ρ is the density of air and *K* is a constant. The equation assumes that there is no wind blowing.

The graph below shows how the drag force varies with the square of the speed for a certain car whose cross-sectional area *A* is 2.0 m^2. The density of air = 1.2 kg m^{-3}.

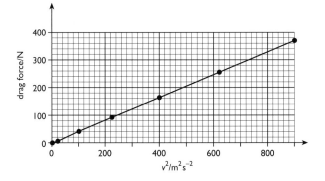

a From the graph determine

  i the drag force on the car when its speed is 20 m s^{-1}. (1)

  ii the value of *K*. (3)

b Calculate the engine power that is required to maintain a steady speed of 20 m s^{-1}. (2)

c By what factor would the engine power increase if the car were travelling at 20 m s^{-1} into a head wind of speed 20 m s^{-1}? (2)

International Baccalaureate, Subsidiary/Standard Level, Nov 1998

3 a Cars must be fitted with rear warning lights (stop-lights) to indicate when a driver applies the brakes. Conventional stop-lights take 0.50 s to switch on to full brightness. A car manufacturer claims that a new type of stop-light 'switches on 25 times faster' and that this is an added safety feature.

  i Calculate

    1 the time taken for the new type of stop-light to switch to full brightness

    2 the difference in the distances moved by a car travelling at 30 m s^{-1} during the times that the two types of stop-light reach full brightness.

  ii Suggest why the new type of stop-light may lead to greater road safety. (4)

b The legal speed limit on motorways is approximately 30 m s^{-1}. In an incident on a motorway, a car of mass 900 kg leaves a skid mark 75 m long when stopping. The maximum deceleration of the car when skidding is approximately 10 m s^{-2}.

  i Show that, before the incident, the car must have been travelling above the legal speed limit. (3)

  ii Calculate, for this skid, the maximum average braking force between each of the four tyres and the road. (2)

  iii When the motorway is wet, the braking force provided by each wheel is reduced to 50% of that calculated in **ii**. Discuss quantitatively the effect of this reduced braking force on stopping distance. In your discussion, assume that the speed of the car, before braking, is the same on the dry road and on the wet road. (3)

OCR, Sciences, Physics Foundation, June 1999

4 A toy locomotive of mass 0.50 kg is initially at rest on a horizontal track. The locomotive is powered by a twisted rubber band which, as it unwinds, exerts a force which varies with time as shown in the table.

Time/s	0.0	1.0	2.0	3.0	4.0	5.0	6.0	7.0	8.0
Force/N	0.20	0.18	0.15	0.12	0.10	0.08	0.05	0.02	0.00

a **i** Plot a graph of force against time for the rubber band power source.

  ii State what is given by the area between the graph and the time axis. (4)

b The rubber band is wound up and released to power the locomotive. Use your graph to show that the speed of the locomotive 8.0 s after the twisted rubber band is released is 1.6 m s^{-1}. Ignore the effects of air resistance and energy losses due to friction. (2)

c 8.0 s after release, the locomotive collides with and couples to a toy truck, initially at rest, which has a mass of 1.50 kg.

  i Calculate the speed of the coupled locomotive and truck after collision.

  ii Calculate the combined kinetic energy of the locomotive and truck immediately after collision.

  iii Show, with the aid of a calculation, whether or not the collision is elastic. (5)

AQA (NEAB), Mechanics and Electricity (PH01), Mar 1999

5 When engineers design a new car, they measure the resistive force which acts on it at different speeds. The table shows the results they obtained for a particular car.

Speed/m s^{-1}	Resistive force/N
10	140
20	260
30	460
40	740
50	1100

a Name two forces which cause this resistance to motion. (2)

b Explain why the resistive force increases as the car goes faster. (2)

c One version of the new car has a maximum speed of 40 m s^{-1} on a level road. Calculate the power used in overcoming the resistive forces at this speed. (2)

d A more expensive version of the car is designed to have a top speed of 50 m s^{-1}. Calculate the maximum useful power output of its engine. Explain your answer. (3)

OCR, Sciences, Physics of Transport, June 1999

6 Take the value of *g* to be 9.8 m s^{-2} or 9.8 N kg^{-1}.

a Explain, in general terms, why the front section of a car is designed to crumple in a frontal collision. Your answer should refer to the forces involved in the collision, the energy dissipated, the change in length of the crumple zone and the deceleration of the car while the crumple zone is being deformed. (4)

b A car of mass 2100 kg travelling at 15 m s^{-1} collides with the rear of a stationary car of mass 800 kg.

The two vehicles crumple for a time of 0.20 s during the collision and afterwards move as a single body. Neglect external forces.

Calculate the average acceleration experienced by the 800 kg car during the collision. (2)

c The combined wreckage of the two cars is brought to rest by friction with the road. Describe, without further calculation, the forces associated with the collision described in **b** and the subsequent deceleration experienced by the driver of the 800 kg car. Your answer should be restricted to the action of:

i the head restraint (headrest)

ii the seat belt. (2)

OCR, Physics, Transport, Mar 1999

IX
THERMAL
PROCESSES

33 Internal energy and thermal transfers of energy

THE BIG QUESTIONS

● How do energy transfers take place between bodies of material?
● What do these energy transfers tell us about **a** energy and **b** material?

KEY VOCABULARY

absolute zero   calibration   caloric theory   Celsius scale   centigrade scale
dynamic equilibrium   First Law of Thermodynamics   fixed point   greenhouse gas
heat capacity   ice point   ideal gas behaviour   kelvin   kinetic theory
latent heat of melting (or fusion)   latent heat of vaporisation   linear scale
negative feedback   specific heat capacity   specific latent heat of melting (or fusion)
specific latent heat of vaporisation   steam point   temperature scale
thermal contact   thermal equilibrium   thermal transfer
thermodynamic temperature scale   triple point of water   volatile
Zeroth Law of Thermodynamics

BACKGROUND

This chapter provides an introduction to the science of thermodynamics – the study of energy transfers that involve thermal as well as mechanical processes. Thermodynamics is about processes in devices like engines, turbines and refrigerators. It also provides us with some very fundamental ideas about what can and cannot happen in the Universe (Figure 33.1).

Figure 33.1
Thermal processes in the Universe.

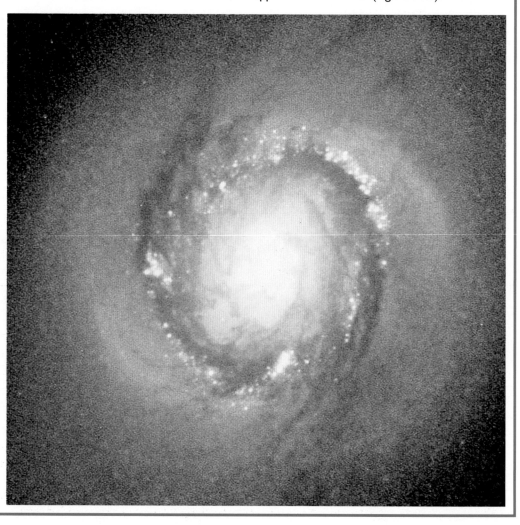

During gravitational collapse of a huge cloud of gas, particles smash together so hard that all atoms become ionised and their nuclei begin to experience fusion. If there is enough material this fusion sustains the high temperatures that can continue to throw particles together fast enough, so that fusion continues. A star is born, and thereafter it pours radiation, carrying energy, out into cold inter-stellar space. This thermal transfer of energy and the processes within the star itself, even those that involve mass–energy conversion, are subject to the laws of thermodynamics. In this galaxy new stars are being formed, their first radiations causing excitation of hydrogen atoms and hence the pink colour.

Active and passive systems

Figure 33.2
We can use an energy transfer diagram to show the relative amounts of energy transferred in a given time by heating and by doing work.

An engine can transfer energy to its surroundings by thermal processes and by mechanical processes, doing work on materials around it (Figure 33.2). An engine can be called an active system. A living thing can do work, so is also an active system.

A system of two or more bodies that exchange energy only by thermal processes is a passive system. A brick, a bucket of water or a statue, together with their immediate environments, can all be thought of as passive systems. They may do a little work on the world around them by expansion, pushing other material out of the way, but this is very small compared with thermal transfers of energy taking place at the same time. This chapter is concerned, mostly, with passive systems.

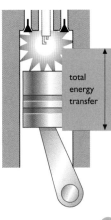

energy transfer by thermal processes (heating)

total energy transfer

energy transfer by mechanical processes (doing work)

Principal mechanisms of thermal transfer

Figure 33.3
Your coffee cools by thermal transfer processes.

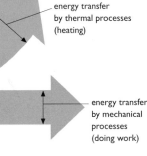

convection currents are a consequence of heating nearby air

evaporation is, like convection, a physical movement of energetic particles

uneven temperatures within the drink may produce internal convection currents

energy is carried towards and away from the drink and the mug by radiation, but not at equal rates

net conduction takes place away from the drink and the mug and into adjacent materials

If two bodies of material come into contact, their moving or vibrating particles can exert forces of attraction and repulsion on each other. As a result, **thermal transfer** of energy can take place between the bodies (Figure 33.3). Such particle interactions also take place within a single body, of course, and can transfer energy from one part of a body to another. The interactions result in the tendency for more energetic particles to lose energy, while others gain. This process of thermal transfer is called conduction.

Particles in the lattice of a solid are able to interact more effectively than those in liquids and gases. The particles may be the atoms themselves, the forces between them giving one atom influence over another and allowing energy to transfer between them. In metals, the free electrons also interact with each other and with the latticework of ions through which they move. Energetic electrons can migrate through a metal, taking their energy with them. It is the effect of these free electrons that makes metals such good conductors.

Thermal transfer of energy from place to place can be due to movement of material. The material carries its internal energy with it. Different densities of a fluid (liquid or gas), due to different temperatures, can result in such movements of material. Convection currents take place within the fluid. Evaporation also transfers energy, but since it involves physical movement of material it is often thought of as a special example of convection.

All bodies emit and absorb electromagnetic radiation. The rate of emission depends strongly on a body's temperature – the higher the temperature the more intense the radiation, and also the higher its average frequency. The radiation, as for other transfer processes, then carries energy away and into the surroundings. During absorption, energy is transferred to the body from its environment.

Any two bodies that are exchanging energy by one or more of these thermal processes are said to be in **thermal contact**. Energy transfer occurs in both directions between the bodies.

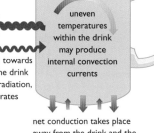

1 What are the dominant thermal transfer processes
 a in a chimney
 b when you toast bread
 c when you fry an egg?
2 Are you emitting electromagnetic radiation? Are there bodies around you that are not emitting electromagnetic radiation? Are you absorbing electromagnetic radiation?
3 Describe the differences and similarities between thermal electromagnetic radiation and ionising electromagnetic radiation.

Temperature difference and thermal transfer

You are in thermal contact with your surroundings – transfer of energy by thermal processes (conduction, convection and evaporation, radiation) is possible between you and the surroundings. You are an active system that can maintain a constant temperature, which is almost always higher than that of your surroundings. So the net transfer of energy is almost always from you to the surroundings.

The Angel of the North (Figure 33.4) is simpler than you are. It and its surroundings can be thought of as a passive system. There are no processes that significantly influence its temperature other than thermal transfers of energy between the metal and the environment. Unlike you, it is not always at a higher temperature than its surroundings.

Figure 33.4
The Angel of the North – a huge steel sculpture in Gateshead. Energy continually transfers in and out, depending on temperature differences between the metal and the surroundings.

Figure 33.5
If two passive bodies are at different temperatures, energy is transferred from the hotter body to the cooler body until thermal equilibrium is reached.

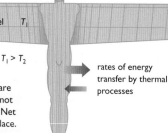

a

Angel T_1

air T_2 $T_1 > T_2$

rates of energy transfer by thermal processes

The statue and the air are in thermal contact but not in thermal equilibrium. Net energy transfer takes place. The temperatures of the bodies – statue and surroundings – changes until they are the same.

b

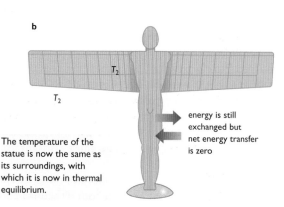

T_2

T_2

energy is still exchanged but net energy transfer is zero

The temperature of the statue is now the same as its surroundings, with which it is now in thermal equilibrium.

Day and night, there is energy transfer into and out of the Angel. Often, the Angel and its surroundings have the same temperature, and then energy is transferred outwards at the same rate that it is transferred inwards. There is a balance, even though processes such as absorption and emission of radiation are transferring energy inwards and outwards. Net energy transfer is zero. Such a balance, involving two-way activity, is an example of **dynamic equilibrium**. In the case of thermal transfers of energy, the balance is also called **thermal equilibrium** (Figure 33.5).

On an evening after a hot summer's day, however, the steelwork of the Angel is likely to be at a higher temperature than the surrounding air. The temperature difference between the metal and the air results in net energy transfer, from the Angel to the air. And, conversely, whenever the surrounding air is at a higher temperature than the metal then the net energy transfer is from the air to the Angel. Thus there is a relationship connecting the temperature difference between the Angel and the surrounding air and the net transfer of energy between them.

The Zeroth Law of Thermodynamics

Suppose that, at a time when the Angel is in thermal equilibrium with the air around it, a visitor to the Angel leaves an empty aluminium drinks can at its feet. Energy transfers can take place between the three bodies – between Angel and air, can and air, Angel and can.

It is possible for the Angel and the drinks can not to be in thermal equilibrium, but instead for there to be net transfer of energy from one to the other. This might happen if the can when first dropped was recently full of chilled liquid and was at a lower temperature than the Angel and the rest of the environment. The can will then accept more energy from its surroundings than it transfers outwards. The consequence is that the drinks can reaches the same temperature as its surroundings, and when it does so the net rate of energy transfer becomes zero. Thermal equilibrium is reached. Thermal transfers tend to take place so as to eliminate difference and replace it with uniformity.

More improbably, but still possible, the drinks can might have been dropped after having been heated. Again, energy transfers take place for as long as the can is at a different temperature to its surroundings, which include the air and the Angel.

But suppose that the drinks can's initial temperature was the same as that of the air. It will be in thermal equilibrium with the air. The Angel is also in thermal equilibrium with the air. It follows that the drinks can and the Angel are in thermal equilibrium with each other (Figure 33.6). The statement that:

> If two bodies are each in thermal equilibrium with a third body then they are in thermal equilibrium with each other

is called the **Zeroth Law of Thermodynamics**. It's called the Zeroth Law because when it was first seen as a fundamentally important statement there were already three laws of thermodynamics – the first, second and third.

The Zeroth Law helps us to understand what 'temperature' means. Temperature is a measurable quantity which determines whether or not passive systems are in thermal equilibrium. Differences in temperature result in energy transfer – from material at the higher temperature to that at the lower.

Figure 33.6
If we know that the Angel and the drinks can are both in thermal equilibrium with the air, then the Zeroth Law of Thermodynamics states that they are in thermal equilibrium with each other.

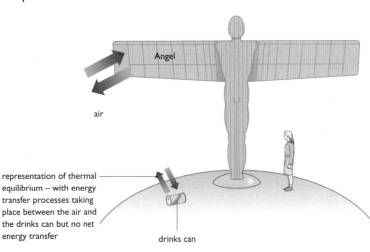

The Angel and the drinks can both emit and absorb radiation. At the same temperature, the rates of energy exchange between the two bodies are equal even though the sizes of the bodies are very different. It is temperature difference, not size, that determines net energy transfer.

air

representation of thermal equilibrium – with energy transfer processes taking place between the air and the drinks can but no *net* energy transfer

drinks can

energy emitted by the Angel and absorbed by the can

energy emitted by the can and absorbed by the Angel in the same time

4 Which way will thermal transfer of energy take place if you touch the Angel with your hand:
  a on a very hot day
  b on a very cold day?
5 What energy transfer takes place if you place a test tube of cold water in a beaker of hot water? When does the net energy transfer stop?
6 Two bodies with temperatures of 200 K and 300 K are placed in thermal contact. What can you predict about what will happen?

If two passive bodies are at the same temperature then they are in thermal equilibrium, regardless of whether they are both at a high temperature or both at a low temperature.

The absolute zero of temperature and the kelvin

Imagine a body that is at such a low temperature that all thermal transfers of energy take place *into* it, whatever other bodies it is in contact with. Its temperature is then as low as is possible for any body. It is at the **absolute zero** of temperature.

Scientists have reduced the temperature of samples of material to very close to absolute zero, but never to absolute zero itself. Any material at absolute zero must be colder than its surroundings, and energy transfer will inevitably take place into it, tending to increase its temperature. This rate of inward transfer will always be greater than the rate at which we can manage to make energy transfer outwards.

Figure 33.7
The three states of water. The triple point, at which the three states are in equilibrium, occurs at a unique temperature.

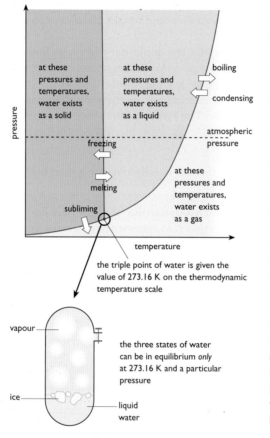

at these pressures and temperatures, water exists as a solid

at these pressures and temperatures, water exists as a liquid

boiling

condensing

atmospheric pressure

freezing

at these pressures and temperatures, water exists as a gas

melting

sublimities

temperature

the triple point of water is given the value of 273.16 K on the thermodynamic temperature scale

vapour

ice

liquid water

the three states of water can be in equilibrium *only* at 273.16 K and a particular pressure

For temperature measurement we need to create a complete **temperature scale** – a system of numbering with a standard unit. To establish this we look at the behaviour of matter to find some phenomenon that takes place at a specific temperature other than absolute zero. The phenomenon provides us with a **fixed point** with which we can establish a temperature scale.

Water is a convenient material to use. Under certain conditions, including a pressure much lower than atmospheric pressure, the three states of water can exist in contact with each other without any net change of the masses of each state. The three states are in one form of equilibrium. That is, the relative quantities of solid, liquid and gas stay the same. (It is a dynamic equilibrium, in that molecules are continuously entering and leaving each state, such as by evaporation and condensation, but these processes are in balance.) This equilibrium can only be achieved at one temperature and one pressure. This condition is called the **triple point of water** (Figure 33.7). The temperature of the triple point is defined to be 273.16 kelvin.

The number chosen, 273.16, is not a round number. This is because of the long and complicated history of establishing this way of defining a temperature scale. The name kelvin comes originally from the River Kelvin which flows through Glasgow. This name was borrowed by William Thomson, a leading 19th century physicist, when he was given a life peerage and became known as Lord Kelvin. The Kelvin is a river (and became the name of a person). The **kelvin** is a unit of temperature. (Note the difference in the initial letter.)

So the absolute zero of temperature is 0 kelvin, or 0 K, and the temperature of the triple point of water is 273.16 K. All other temperatures measured in kelvin are based on these fixed points (Figure 33.8), which define what is called the **thermodynamic temperature scale**.

7 Why is it impossible to reduce the temperature of any body to absolute zero?

8 What is the difference in the way we write the name of the Glasgow river and the name of the 19th century scientist, and the way we write the name for the SI unit of temperature? To what other SI units does the same convention apply?

Figure 33.8
The basis of the thermodynamic scale.

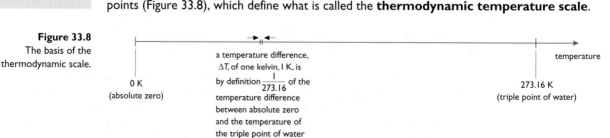

temperature

a temperature difference, ΔT, of one kelvin, I K, is by definition $\dfrac{1}{273.16}$ of the temperature difference between absolute zero and the temperature of the triple point of water

0 K
(absolute zero)

273.16 K
(triple point of water)

Measuring temperature

Measuring is a practical process. To measure temperature we look at the behaviour of samples of material, and of temperature-dependent variables – such as the length of a column of mercury inside a glass tube, the resistance of a wire or a thermistor, or the pressure exerted by a fixed volume of gas in a sealed flask (Figure 33.9).

Figure 33.9
Temperature-dependent variables used for temperature measurement.

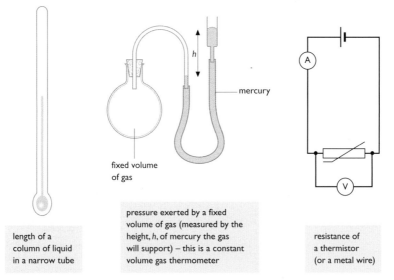

mercury

fixed volume of gas

length of a column of liquid in a narrow tube

pressure exerted by a fixed volume of gas (measured by the height, h, of mercury the gas will support) – this is a constant volume gas thermometer

resistance of a thermistor (or a metal wire)

For measuring temperature on the thermodynamic temperature scale it makes some sense to observe a variable that becomes zero at the absolute zero of temperature. The pressure exerted by a gas is such a variable. At 0 K the gas is incapable of transferring energy to its surroundings – it cannot get any colder. Its molecules no longer have energy and have ceased to move, so cannot exert pressure. This is a somewhat imaginary situation, since absolute zero cannot be achieved, and we refer to such behaviour as **ideal gas behaviour**.

Figure 33.10
Pressure versus temperature for an ideal gas.

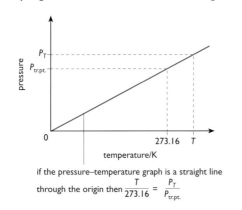

if the pressure–temperature graph is a straight line through the origin then $\dfrac{T}{273.16} = \dfrac{P_T}{P_{\text{tr.pt.}}}$

The pressure exerted by a gas increases with temperature, and for an ideal gas we can make the assumption that pressure is proportional to temperature provided that the volume of the gas does not change (see Figure 33.10). You can read more about ideal gases in the next chapter.

If the pressure exerted by an enclosed volume of an ideal gas at the triple point of water, 273.16 K, is $P_{\text{tr.pt.}}$ and the pressure exerted by the same gas at some other temperature, T, is P_T, then the simple proportionality allows us to say:

$$\frac{T}{273.16} = \frac{P_T}{P_{\text{tr.pt.}}}$$

So that we can give a value to the unknown temperature T, we rearrange this to:

$$T = \frac{P_T}{P_{\text{tr.pt.}}} \times 273.16$$

The behaviour of a real gas differs only a little from the simple proportionality of an ideal gas, and so this relationship gives a useful working way of measuring temperature – by means of measurement of pressure of a constant volume of a fixed mass of gas. It is not, however, a perfect way of measuring temperature, not least because of the practical difficulty of ensuring that the volume of the gas does not change, as it might if the container expands as temperature rises.

Centigrade and Celsius scales

The **centigrade scale** of temperature (now no longer used by scientists, although the principles used in establishing the scale are of interest) used the melting and boiling points – or the **ice point** and **steam point** – of water as its two fixed points. The values of *zero* degrees centigrade and *one hundred* degrees centigrade were given to the temperatures at which an ice–water mixture is in equilibrium (having unchanging mass of each of these two states) and at which a water–steam mixture is in equilibrium. In both cases the pressure must be fixed at atmospheric pressure. There was no reason for choosing the numbers 0 and 100 except that they are relatively easy to use.

A variable such as the length of a column of mercury in a glass tube was used as an indicator of temperature on the centigrade scale (Figure 33.11). The value of the chosen variable was observed when immersed in ice and water which were in a state of dynamic equilibrium, at standard atmospheric pressure. We could call this X_0. Then its value, X_{100}, when in contact with steam and water in equilibrium at standard atmospheric pressure, was observed. The total change in the variable is then $(X_{100} - X_0)$.

9 Why do we have to specify the pressure of the ice point of water but not the pressure of the triple point?

Figure 33.11
The temperature as read from the thermometer scale is the temperature of its surroundings with which it is in thermal equilibrium.

When a thermometer bulb is in thermal equilibrium with its surroundings, then it is at the same temperature, which we read on the scale marked on the tube following the calibration process.

The volume of mercury in the bulb is large compared with the volume in the narrow tube. The temperature measurement is made possible by the expansion of the mercury in the bulb.

For a linear scale, the tube must have a uniform diameter.

The value of the variable could also be observed at some intermediate temperature, θ. (It is common to denote temperature on the thermodynamic scale by using T, temperature on the centigrade scale by using θ, and temperature on the Celsius scale as t. Where scale and unit are not specified, then we most frequently use T.) We can call the value of the variable at this temperature X_θ. This is larger than X_0 by an amount $(X_\theta - X_0)$.

In the case of the length of a column of mercury, marks could be made on the outside of the glass tube corresponding to X_{100} and X_0, and then if required a scale running the length of the tube could be made. If this is made a **linear scale** – one with equal temperature differences equally spaced – then the assumption is being made that the change in the length of the column is proportional to the temperature. Care is needed here, because this may not be the case for all temperature-dependent variables. The marking of a scale in this way is called **calibration**.

Where the change in the value of the chosen variable *is* proportional to change in temperature above 0 degrees centigrade (Figure 33.12), then we can say

$$(X_\theta - X_0) = k\theta$$

and we can also say that

$$(X_{100} - X_0) = k \times 100$$

Figure 33.12
A linear temperature scale is based on the assumption of a proportional relationship between $(X_\theta - X_0)$ and temperature θ.

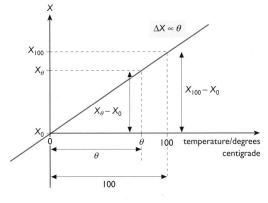

Dividing the first of the equations by the second,

$$\frac{(X_\theta - X_0)}{(X_{100} - X_0)} = \frac{k\theta}{k \times 100}$$

and so

$$\theta = \frac{(X_\theta - X_0)}{(X_{100} - X_0)} \times 100$$

This defines a temperature on the centigrade scale.

Instead of the centigrade scale we now use the **Celsius scale** for everyday temperatures, using °C as the abbreviation.

In developing the thermodynamic temperature scale, the size of the kelvin was chosen to be such that a temperature difference of 1 kelvin was the same as a temperature difference of 1 degree centigrade. The Celsius scale was developed from the thermodynamic temperature scale, so that this too provides the same value for the same *change* in temperature. Thus a temperature difference of 1 °C is the same as a temperature difference of 1 K and of 1 degree centigrade. For any temperature difference we can say

$$\Delta T = x \text{ K} = x \text{ °C} = x \text{ degrees centigrade}$$

A temperature in degrees Celsius is effectively the same as a temperature in degrees centigrade. The difference is a difference of definition (Figure 33.13). The centigrade scale was defined in terms of the ice point and steam point of water. The Celsius scale is defined in terms of the thermodynamic temperature scale, as follows:

temperature in Celsius (°C) = temperature in kelvin (K) − 273.15

which we can abbreviate to:

$$t = T - 273.15$$

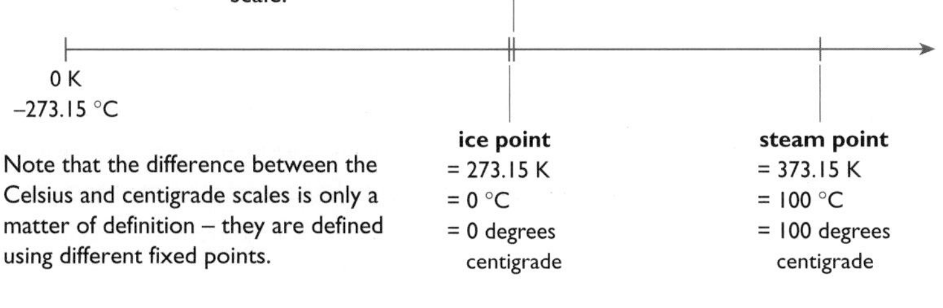

Figure 33.13
The defining temperatures on the thermodynamic (kelvin) scale, the Celsius scale and the centigrade scale.

triple point
= 273.16 K
= 0.01 °C

0 K
−273.15 °C

ice point
= 273.15 K
= 0 °C
= 0 degrees
  centigrade

steam point
= 373.15 K
= 100 °C
= 100 degrees
  centigrade

Note that the difference between the Celsius and centigrade scales is only a matter of definition – they are defined using different fixed points.

So the Celsius scale is not fundamentally based on the ice and steam points of water but, indirectly, on its triple point. Note that the triple point of water occurs at 273.16 K, but the ice point of water at atmospheric pressure is 273.15 K. The temperature of the ice point on the Celsius scale is 0.00 °C and the temperature of the triple point is 0.01 °C.

10 Human body temperature is 37 °C.
  a What is this temperature in kelvin?
  b What is the difference between this temperature and the boiling point of water at atmospheric pressure in
  i °C
  ii degrees centigrade
  iii K?
11 a What is the temperature of each of the following in degrees centigrade, °C and K?
  i the absolute zero of temperature
  ii the ice point of water at atmospheric pressure
  iii the triple point of water
  iv the steam point of water at atmospheric pressure.
  b Sketch a linear scale and mark these values on it.
12 The melting and boiling points of mercury are 234 K and 630 K. Devise a temperature scale, with units °X, such that these fixed points are 0 °X and 100 °X. On this scale, what is
  a the absolute zero of temperature
  b your body temperature
  c the boiling point of water?

Internal energy and thermal transfer in passive systems

The internal energy of a body of material is a combination of the total potential and kinetic energy of its particles. It is impossible to make measurements on all of the individual particles. But what we can do is measure the amount of energy that is transferred into or out of the material by thermal processes. We can denote such quantities of transferred energy as ΔQ:

ΔQ = energy transferred *into* a body by thermal processes

Note that if energy is transferred *out of* a body by thermal processes then ΔQ has a negative value.

For a passive body (that is, one that is not doing work and is not having work done on it) the thermal energy flowing in or out is equal to the change in internal energy (Figure 33.14). We can denote change in internal energy as ΔU:

ΔU = change in internal energy

For a passive body, such as the Angel of the North,

$$\Delta U = \Delta Q$$

Figure 33.14
Energy transfer into a body is equal to the increase in internal energy of the body, provided that no work is being done on or by the body.

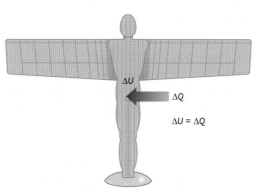

ΔU

ΔQ

$\Delta U = \Delta Q$

We need to take care with signs here. If energy transfers out of the Angel then ΔU and ΔQ both have negative values. (Note that some textbooks use an alternative sign convention, based on a positive value of ΔQ when the energy transfer is *out* of the body.)

The position is more complex for an active system (a body that is exchanging energy with its surroundings by doing work). We will return to this.

For a body whose internal energy, and temperature, are influenced by combustion or other processes (such as photosynthesis or nuclear fusion) taking place 'within' the body, we consider the energy transfer from the combustion as energy flow *into* the body. Thus energy transferred to a body by thermal processes resulting from combustion is denoted by ΔQ, just as thermal energy transfers from more remote locations. In the case of plant and animal respiration, an example of combustion, this seems to defy common sense, since most of us would think that once we've digested the food (fuel) we can count it as part of our own bodies. But, for the sake of simplicity we here consider respiration as a process that involves an external source of energy. The same applies to other 'internal' process, like nuclear fusion. It is for this reason that we've used the Angel of the North to illustrate thermal behaviour and not the much more difficult examples of the human body or the Sun.

13 For a test tube of cold water placed in a beaker of hot water, does ΔQ equal ΔU
   a for the cold water
   b for the hot water
   c for the test tube?

Working as well as heating – the First Law of Thermodynamics

Figure 33.15
A human body is more complex than a statue.

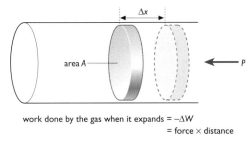

the body has a higher temperature than its surroundings, so energy transfers outwards by thermal processes

food (fuel) considered as external to the body

respiration (combustion) results in transfer of energy from food (fuel) to the body

human bodies also transfer energy to their surroundings by doing work

We have seen that a body can exchange energy with its surroundings by thermal processes, which we have described as passive energy transfer, and also by mechanical processes involving work being done, which we have described as active. By 'work' here we mean work as the product of force and distance, as defined in Chapter 15.

A body – a human body as a complex example (Figure 33.15), or a gas that's expanding as a simple example – can do work on its surroundings. It can exert a force over a distance. The expanding gas must do work to push air out of the way (assuming that the process is taking place on or near the Earth's surface). The gas must exert a force over a distance, and the work done is related to the atmospheric pressure.

For a cylinder of gas pushing a piston of area A through a distance Δx, against external pressure P, the external work that must be done (Figure 33.16) is:

$$\text{work} = -\Delta W = \text{force} \times \text{distance} = PA \, \Delta x = P \, \Delta V$$

ΔV is the increase in volume of the gas, and also the volume of external air that has been displaced. (Note that force $= PA$ because, by definition, pressure $=$ force/area, $P = F/A$.)

We use the negative sign because we are adopting the sign convention that work done *by* the gas has a negative value. It is also possible that the external air or some other external agency – such as a cyclist, in the case of a bicycle pump – could do work *on* the gas. We are saying that such work done *on* the body is positive.

ΔW = work done *on* the body, which tends to transfer energy *to* the body

When work is done *by* the body, tending to transfer energy *from* the body, then ΔW has a negative value. (Note again that some textbooks use the alternative convention.)

Figure 33.16
Work is done by an expanding gas.

area A

Δx

P

work done by the gas when it expands $= -\Delta W$
$= \text{force} \times \text{distance}$

force = pressure P the gas is applying (or pushing against)
$\times$ area A

distance $= \Delta x$

So there are two processes that can transfer energy to or from a body – passive heating processes, where the amount of energy involved is written as ΔQ, and active working processes, where the amount of energy involved is written as ΔW. These both affect the changes in internal energy of a body, ΔU. Since they are the *only* ways to change internal energy, we can write a simple sum:

change in internal energy of a body = energy transferred to the body by heating + energy transferred to the body by doing work on it

In symbols:

$$\Delta U = \Delta Q + \Delta W$$

As a sentence:

The change in internal energy of a body is the sum of the energy transferred to it by heating and the energy transferred to it by work done on it.

This statement is the **First Law of Thermodynamics**.

14 You go for a swim in the North Sea in February, and energy transfer takes place. Does ΔQ equal ΔU? Explain.
15 State whether ΔU, ΔQ and ΔW have positive or negative values or are zero in each of these situations:
  a the Angel of the North being heated by the Sun
  b air being compressed in a bicycle pump
  c a nail being hit several times by a hammer and becoming warmer
  d a balloon that expands (doing work both to stretch the rubber and to push air out of the way) when moved from a cooler to a warmer place
  e you, lying in bed
  f you, running up a hill.

Change in internal energy and temperature rise

Transfer of energy into a body and the resulting increase in internal energy have two possible consequences. The temperature can rise, or the material can change state. We need to treat these two effects separately.

For a body for which no change of state takes place, a change in internal energy, ΔU, results in a change in temperature, ΔT. The ratio of these, $\Delta U/\Delta T$, is called the **heat capacity** of the body (Figure 33.17):

$$\text{heat capacity of a body, } C = \frac{\Delta U}{\Delta T}$$

Since ΔU is measured in J and ΔT is measured in K, the unit of heat capacity, $C = \Delta U/\Delta T$, is $J\,K^{-1}$.

Figure 33.17
Heat capacity is a property of a *body*.

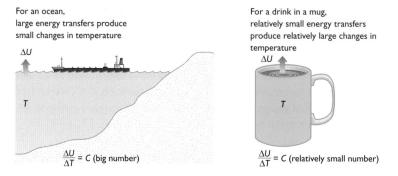

For an ocean, large energy transfers produce small changes in temperature

ΔU

T

$\dfrac{\Delta U}{\Delta T} = C$ (big number)

For a drink in a mug, relatively small energy transfers produce relatively large changes in temperature

ΔU

T

$\dfrac{\Delta U}{\Delta T} = C$ (relatively small number)

We can measure the heat capacity of a body. For this we need a body of material on which to work. Water in a lagged container is suitable. We can supply a measured quantity of energy, ΔQ, using an electric heater (for which $\Delta Q = VIt$, where V is voltage, I is current and t is time), and observe the consequent changes in temperature.

There is some expansion, and this involves pushing other material out of the way, and so doing work. So, strictly,

$$\text{energy supplied by heater} = \Delta Q = \Delta U - \Delta W$$

However, the work done by the expanding water and other materials is very small, so we can say

$$\Delta Q \approx \Delta U$$

The situation is made more complex because we cannot work with water in isolation – it must be in a container, and some of the energy supplied by the electrical heater will transfer to the container and to the surroundings. We must acknowledge that ΔU has three elements:

change in internal energy of the water, ΔU(water)
change in internal energy of container, ΔU(container)
change in internal energy of other surroundings, ΔU(surroundings)

$$\Delta Q \approx \Delta U\text{(water)} + \Delta U\text{(container)} + \Delta U\text{(surroundings)}$$

To minimise the influence of the energy taken by the container to raise its temperature, the simplest approach is to use a large mass of water and a container with relatively very low mass. The lagging or insulation minimises energy transfer from the water to the other surroundings. Then

$$\Delta Q \approx \Delta U\text{(water)}$$

Measurements reveal that for a fixed mass of material, heat capacity is very nearly constant. A graph of ΔU for the body against ΔT is close to a straight line through the origin (Figure 33.18).

Figure 33.18
The heat capacity of a body is very nearly constant, provided that mass is constant and there is no change of state.

In fact the gradient C is not *perfectly* constant, but increases slightly with temperature for most bodies.

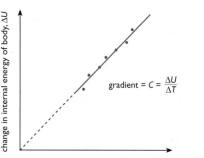

change in internal energy of body, ΔU

gradient $= C = \dfrac{\Delta U}{\Delta T}$

change in temperature of body, ΔT

Further investigation is possible. We can investigate the relationship between the heat capacity of a body of material and its mass. Again we find a simple relationship. Heat capacity is proportional to mass, and we have a graph with a constant gradient. If we repeat this investigation for different materials we find we still see proportionality, but that different materials produce graphs with different gradients (Figure 33.19).

Figure 33.19
The specific heat capacity of a material is approximately constant.

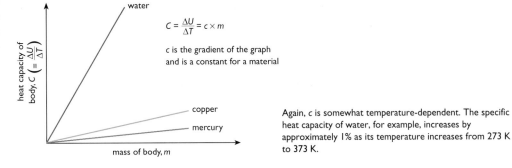

$$C = \frac{\Delta U}{\Delta T} = c \times m$$

c is the gradient of the graph and is a constant for a material

Again, c is somewhat temperature-dependent. The specific heat capacity of water, for example, increases by approximately 1% as its temperature increases from 273 K to 373 K.

The gradient of a graph of heat capacity against mass is a property of the *material* and not of the body. This gradient is a useful property of materials, and is called **specific heat capacity**, which is abbreviated as c.

$$\text{specific heat capacity of a material, } c = \frac{\Delta U}{m \, \Delta T}$$

It can be useful to write this equation as

$$\Delta U = mc \, \Delta T$$

Specific heat capacity is a material's change in internal energy per unit mass per unit temperature change. In the SI system of units, 'unit mass' means one kilogram and 'unit temperature change' means one kelvin. Specific heat capacity is measured in joules per kilogram kelvin, or $\text{J kg}^{-1} \text{ K}^{-1}$. Table 33.1 compares the specific heat capacities of some materials.

Table 33.1
Some specific heat capacities.

Material	Specific heat capacity /J kg^{-1} K^{-1}	Temperature rise of 1 kg of material due to a 1 J rise in internal energy/K
water	4170	2.4×10^{-4}
mercury	140	7.1×10^{-3}
copper	387	2.5×10^{-3}
nylon	1600	6.3×10^{-4}
iron	450	2.2×10^{-3}

Specific heat capacity is somewhat dependent on temperature. The values here apply at 293 K.

16 What is the change in internal energy when a kilogram of water is heated through
  a 10 K
  b 10 °C?
17 What is the difference between the unit of heat capacity and the unit of specific heat capacity?
18 If equal masses of water and mercury, both at room temperature, were poured into test tubes and placed in identical beakers of ice, why would the temperature of the water fall more slowly?
  (**Do not try this! Use of mercury should be avoided.** The consequences of even low levels of mercury poisoning are extremely unpleasant, and higher levels can be fatal.)
19 What will be the temperature rise of 1 kilogram of water when it is supplied with 1000 J of energy?

Note from Table 33.1 that water has a particularly high specific heat capacity. This means that the temperature of water changes less than that of the other materials in the table when its internal energy changes by a given amount. This has a number of consequences, described in the following paragraphs.

- When water and land lie adjacent, and receive energy from the Sun at the same rate, then (assuming that they absorb and reflect in approximately similar proportions) the temperature of the land will change much more quickly than the temperature of the sea. On a day-to-day basis this causes sea breezes, which are convection currents created by the temperature difference. On a month-by-month basis, it means that in, on and near the sea seasonal fluctuations in temperature are relatively small. Kiev, thousands of kilometres from the nearest large ocean, has very hot summers and very cold winters. Cardiff, at about the same latitude (as far north) faces the Atlantic and has a much gentler climate. Oceans play a major part in local and planetary temperature variations (Figure 33.20).

Figure 33.20
A typical storm in the South Pacific due to El Niño – changes in ocean currents.

In most years in the past a 'tongue' of cold water has spread upwards and outwards into the Pacific Ocean from the South American coast. When this doesn't happen we say an El Niño event is taking place. The change in ocean temperature affects weather not only in the Pacific region, but all around the world.

20 Our two bodies of question 6, initially at 200 K and 300 K, are again placed in thermal contact. The first body has a mass of 0.400 kg and specific heat capacity 387 J kg^{-1} K^{-1}, and the second has a mass of 0.600 kg and specific heat capacity 4170 J kg^{-1} K^{-1}.

a Which one loses energy?

b Which one gains energy?

c What can you say about these quantities of 'energy loss' and 'energy gain'? Write this down as concisely as possible.

d Suppose that the final temperature of both bodies, when they reach thermal equilibrium, is T. Write down expressions for

i the temperature changes of each body

ii the energy loss of one body

iii the energy gain of the other body.

e Calculate the value of T. (Make use of your answers to c and d.)

- Heating water through a certain temperature difference is more than 25 times more expensive than heating mercury through the same temperature difference. Our domestic energy bills would be much smaller if we could drink, wash and cook with mercury. Unfortunately, mercury is a powerful poison.
- Even a small mass of water can act as an effective energy store. A hot water bottle transfers an amount of energy into a bed that is four times bigger than would be made available by a lump of copper of the same mass and temperature.

Change in internal energy and change of state

There must be a net energy supply, ΔQ, to a solid to melt it. Assuming that the solid does not do work in melting, this transfer of energy is matched by a change in the material's internal energy, ΔU. That is, $\Delta Q = \Delta U$. Observation shows that such an energy transfer which produces a change in state does not produce a change in temperature. A change takes place in the internal energy of the material, but it cannot be detected by a thermometer. Energy transfer into a body that does not raise its temperature but makes it melt is called the **latent heat of melting (or fusion)** of the body:

$$\Delta Q = \Delta U = \text{latent heat of melting (or fusion)}$$

Since this is a quantity of energy it is measured in joules, J.

When a liquid body solidifies it loses internal energy and energy is transferred to the surroundings. In this case ΔQ and ΔU both have negative values, and are equal. Again, this change in internal energy is not observed as a temperature change.

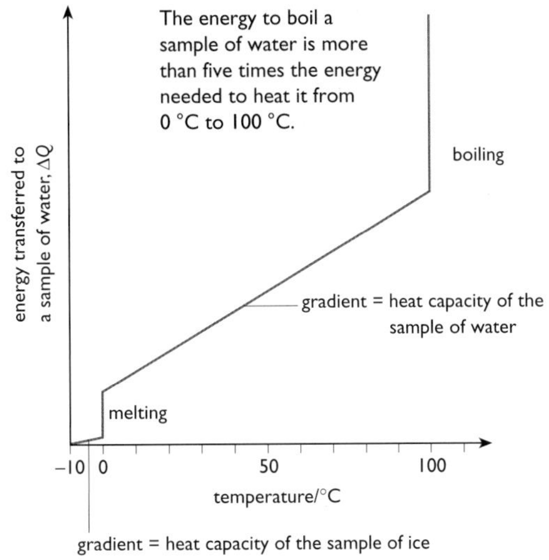

Figure 33.21
The changes in internal energy of a sample of water matched to changes in temperature. The relationship is quite complex.

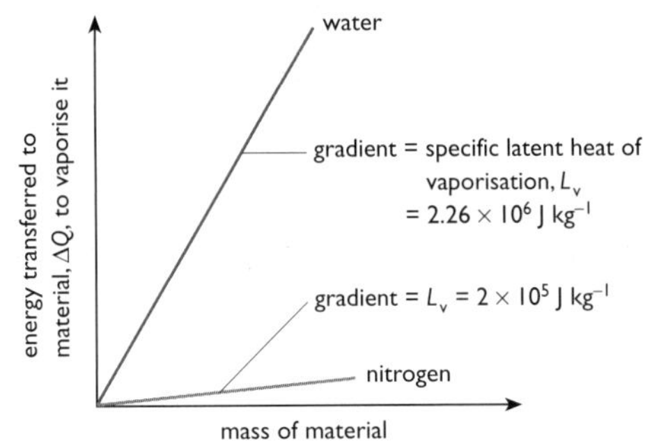

The energy to boil a sample of water is more than five times the energy needed to heat it from 0 °C to 100 °C.

boiling

gradient = heat capacity of the sample of water

melting

gradient = heat capacity of the sample of ice

Figure 33.22
The specific latent heat of vaporisation of a material is constant at constant pressure.

water

gradient = specific latent heat of vaporisation, L_v = 2.26×10^6 J kg^{-1}

gradient = L_v = 2×10^5 J kg^{-1}

nitrogen

21 Explain why steam at 100 °C causes more severe skin burns than the same mass of water at 100 °C.

22 On a winter's day, some warmer air moves over an area of bare cold ground and a similar area of cold snow. The bare ground and the snow both tend to cool the air. Explain why air temperature stays low for longer in the snowy area.

For a liquid to turn to gas an increase in internal energy, ΔU, is necessary. This can be seen by observing boiling water and measuring its temperature (Figure 33.21). Energy must continue to be transferred to the water (ΔQ must be positive) to keep it boiling. In the case of a liquid becoming a gas, the gas has a much greater volume than the liquid. Considerable expansion is involved, and in the Earth's atmosphere the gas must push air out of the way. To do this it must do work on the air. We say that the expanding gas does work against the external pressure. The energy change, ΔQ, is called **latent heat of vaporisation**. So, for boiling:

$$\Delta Q = \Delta U - \Delta W = \text{latent heat of vaporisation}$$

Since work done on the body, ΔW, has a negative value, work done here makes a positive contribution to the energy, ΔQ, that must be transferred to the material during boiling. Note that the amount of work that must be done depends on the pressure of the air into which the gas is expanding, so latent heat of vaporisation is very dependent on external pressure.

The total amount of energy that must be transferred, say, to water to boil it, depends on the mass of the water. We can investigate this experimentally using an electrical heater to transfer a measured amount of energy to a body of water in order to boil it. Again it is necessary to minimise the transfer of energy to the water's container and the surroundings beyond this. Careful experimentation at constant external pressure reveals that the energy required to turn a mass of water into steam is proportional to the mass. The gradient of the graph has the unit joule per kilogram, J kg^{-1}, and is constant for a specified material (Figure 33.22). It is called the **specific latent heat of vaporisation**, L_v, of the material. It is the amount of energy supplied per unit mass to turn a material from liquid to gas.

Similarly, the energy required to change a kilogram of a specified solid into liquid, without any change in temperature, is called its **specific latent heat of melting (or fusion)**. This quantity also has the unit J kg^{-1}.

Different materials have a wide range of different specific latent heats of vaporisation and melting.

Evaporation

Boiling and evaporation are different processes. Boiling is a change of state of a material at a particular temperature, involving all of the material that has this temperature. Evaporation is the escape of molecules one by one and can happen at any temperature due to the randomness of motion of molecules in the liquid.

Energy – latent heat of vaporisation – must be provided to a liquid for it to become a gas whether this is by boiling or by evaporation. The necessary energy can be provided by an external source of energy – by heating the liquid. But in rapid evaporation, where energy is not able to transfer into the liquid fast enough to provide energy for vaporisation, energy may be taken from the internal energy of the liquid itself (Figure 33.23). The temperature of the liquid then falls. During sweating where the sweat evaporates, for example, energy is then transferred from skin to liquid and there is a cooling effect on the skin. The cooling effect is strongest in a breeze, which carries away evaporated molecules and so increases the rate of evaporation.

23 Why does ethanol feel colder when rubbed on your skin than the same mass of water with the same initial temperature?

Cooling by evaporation can also be considered on a particle level. Particles evaporating from a liquid tend to be the more energetic particles, so evaporation tends to remove energy from the liquid and transfer it to the surrounding gases.

A **volatile** liquid is one with a low boiling point. Ethanol and most of the ingredients of petrol are volatile. Such a liquid tends to evaporate rapidly. The presence of petrol vapour makes it important not to smoke, for example, when filling a car's fuel tank.

Figure 33.23
Energy needs to be supplied for a liquid to evaporate.

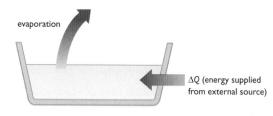

evaporation

ΔQ (energy supplied from external source)

Material that has evaporated in a given period of time has experienced a significant change in internal energy and has also expanded into the atmosphere, doing work. The energy needed for this can be supplied either by energy transferred into the liquid from outside, ΔQ, …

evaporation

ΔU (energy supplied by internal energy)

… or from the internal energy of the liquid itself, causing a fall in liquid internal energy, ΔU.

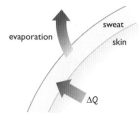

evaporation sweat

skin

ΔQ

During sweating, thermal energy transfer takes place from the skin to the sweat, supplying energy for evaporation.

● **Comprehension and application**

The Sun, the Earth and space

To extend our ideas of thermal equilibrium of a statue, a drinks can and the air around them, we could think about the Sun, the Earth, and the rest of space. The Sun is not in thermal equilibrium with the rest of space but transfers energy outwards. Energy transfer processes (nuclear fusion reactions) take place inside it so that its internal energy – the sum of the potential and kinetic energies of its particles – stays constant.

The Sun is not in thermal equilibrium with the Earth. They have different temperatures and there is a net energy transfer to the Earth. Yet day by day and even, usually, year by year, the average temperature of the Earth stays the same. To maintain a steady temperature the Earth must transfer energy away from it at just the same rate as it receives it (Figure 33.24). Since it is a net receiver of energy from the Sun it must be a net transmitter of energy to the rest of space. It does this by emitting thermal radiation.

Figure 33.24
The Earth radiates energy at the same rate as it absorbs energy.

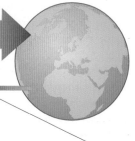

rate of energy transfer from Sun to Earth

rate of energy transfer from Earth to Sun (greatly exaggerated)

rate of energy transfer from Earth to space

rate of energy transfer from space to Earth (greatly exaggerated)

these transfer processes play little practical part and we will ignore them in what follows

Figure 33.25
Any imbalance is regulated
by negative feedback.

It may seem improbable that the Earth can maintain a steady temperature, but **negative feedback** is at work. The rate at which the Earth radiates energy into space depends on its temperature. If its temperature rises it tends to radiate energy more rapidly, losing energy and thus cooling. If its temperature falls it radiates more slowly, so that it receives energy from the Sun more rapidly than it transmits energy back to space. This has a heating effect. The effect of negative feedback is to keep the Earth's average temperature constant (Figure 33.25).

The Earth maintains a stable temperature when:
rate of energy transfer from the Sun = rate of energy transfer to space

Negative feedback is a tendency for a change in behaviour
(such as the rate of radiation) to oppose its own cause (such
as the temperature difference between the Earth and space).

If the Earth's temperature rises it radiates more strongly and
the imbalance has a cooling effect:

a hot Earth
has a natural
tendency to
cool

If the Earth's temperature falls it radiates more weakly and
the imbalance tends to increase its temperature:

a cool Earth
has a natural
tendency to
warm

24 The Sun and the planet Venus both have constant average surface temperatures (of 6000 °C and 400 °C respectively). Both emit and can absorb thermal radiation.
a Are they in thermal equilibrium with one another?
b If Venus receives a continuous net transfer of energy from the Sun, why does its temperature not rise continuously?
c The atmosphere of Venus contains a high level of greenhouse gases. With the aid of sketches, explain how the average temperature of Venus would change if these gases were removed from its atmosphere.

25 Carbon dioxide levels in the Earth's atmosphere are rising because of human activity and the Earth's climate is changing. Nevertheless, some scientists claim that the climate change is a natural fluctuation.
a Explain the 'greenhouse effect' that carbon dioxide in the atmosphere can cause.
b An increase in average global temperature can lead to
  i an increase in cloud cover (due to increased evaporation from the oceans), and
  ii a decrease in ice cover (due to melting).
Explain the influence of **i** and **ii** on the rate at which the Earth's atmosphere and surface absorb solar energy. Which is likely to be the stronger effect in the short term (over just a few years) and which in the longer term (over a decade or more)?
c Explain why the Earth's average temperature is very nearly constant throughout the year.
d Explain why the Earth's average temperature shows only small natural fluctuations (a few °C) over periods of millions of years.

26 DISCUSS
Will the Earth become a dead planet like Venus? Should we be worried? Should we be taking action to prevent this? What action? What effect would such action have on 'standards of living' in countries like the UK that have had developed industry (and have consequently polluted the atmosphere) for a very long time, and in countries which are trying to develop their industry to create new jobs and affluence?

Human activity is changing the Earth's atmosphere, and in particular is increasing the amount of carbon dioxide. This is a **greenhouse gas**. It allows the Sun's radiation to pass through it more easily than it allows the Earth's longer wavelength radiation to pass back to space. Increased absorption of this emitted radiation slows down the rate at which the Earth radiates energy through the atmosphere and into space. There is no rapid rise in the Earth's temperature, but the average temperature rises slowly until the rate of energy radiating into space is restored to the rate at which the Sun's energy arrives (see Figure 33.26, overleaf). Once this equality is achieved, the Earth can again have a stable, but higher, temperature.

If we were to stop generating carbon dioxide and other greenhouse gases now then a new stability might be reached that is not too different from the present. As it is, we are continuing to pour material into the atmosphere and the future of the planet cannot be predicted.

Figure 33.26
The greenhouse effect of the Earth's atmosphere. The effect is increased by the emission of carbon dioxide.

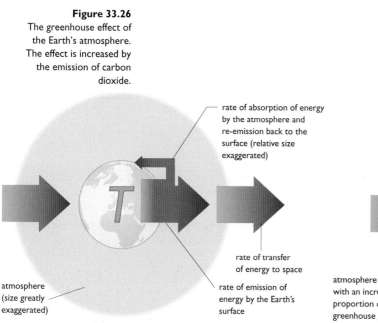

rate of absorption of energy by the atmosphere and re-emission back to the surface (relative size exaggerated)

rate of transfer of energy to space

atmosphere (size greatly exaggerated)

rate of emission of energy by the Earth's surface

higher rate of absorption and re-emission back to the surface (relative size exaggerated)

atmosphere with an increased proportion of greenhouse gases

higher average surface temperature to maintain a balance

For a stable temperature, energy must be transferred from Earth to space at the same average rate as it arrives from the Sun.

The atmosphere absorbs some of the energy that is emitted by the Earth's surface. Some of this energy is re-emitted into space but some is re-emitted back to the Earth's surface. To remain at a stable average temperature, therefore, the Earth's surface must radiate more rapidly than would be the case without an atmosphere. This requires that the Earth has a higher average temperature than it would have if there were no atmosphere. This is the greenhouse effect and without it the Earth would now be a very cold place.

Human activity is changing the Earth's atmosphere, making it more effective at absorbing the radiation that is emitted by the surface. So to ensure that a balance is maintained between the average rate of energy arriving from the Sun and the rate of energy transferred into space, the temperature of the Earth's surface must rise. The Earth can still have a stable average temperature, but this will be a higher temperature. This is the cause of global climate change.

The 'stable' average temperature will gradually drift upwards for as long as the amount of greenhouse gases in the atmosphere continues to increase. Other factors in the complex climate may cause sudden change. We do not know enough to make reliable predictions of this.

● ●

● **Comprehension and application**

Caloric theory versus kinetic theory

Is heat a substance that can be contained within other substances? Are there particles of heat that are different to particles of matter? Or is heat a result of motion within material? These were some of the big questions that scientists 200 years ago argued over, until ideas became established during the 19th century. While scientists nowadays might take sides in arguments over quantum physics, theories of gravity and details of the 'big bang', for many generations heat was an issue that generated hot debate!

During the 18th century many favoured the concept of heat as a substance, and they gave this substance the name of 'caloric'. They supposed that this caloric could flow in and out of visible material, though clearly it had no mass because hot bodies had the same mass as cold ones. The theory of heat as a flowing substance that is separate from the material it flows through was known as **caloric theory**.

The alternative theory is **kinetic theory**, which provides explanation of the thermal behaviour of the world not in terms of a separate flowing substance but as different states of the material itself, these different states being possible because the material is made of particles, which can have different states of motion. ('Kinetic' means 'of motion'.) In the 18th century there was not enough evidence for the existence of particles for this theory to win the argument.

In the middle of the 18th century mercury thermometers became available, allowing scientists to make measurements on the quantity of 'heat' that it was necessary to supply to masses of different materials in order to achieve a particular temperature change. The concept of heat capacity was developed as the ability of a material to 'hold heat', though in itself this development did little to solve the problem of the caloric/kinetic debate.

Figure 33.27
A contemporary cartoon of Count Rumford (1753–1814).

Rumford was born in America but joined the British side in the American War of Independence, and eventually became a minister in the Bavarian government. His work on the heating of cannon barrels and iron filings during boring undermined the caloric theory and increased support for the kinetic theory. The cartoon shows him (blue coat) at London's Royal Institution, where the annual Christmas lectures are still held.

It was known that friction causes heating, and the caloric theorists suggested that the friction induced the flow of caloric. A scientist called Count Rumford (Figure 33.27) investigated this in the 1790s, in particular the heating effect of boring the barrels of cannons. Scientists who favoured the caloric theory suggested that the caloric flowed from the smaller iron filings that were produced, causing the rise in temperature. This would mean that the filings would then have less heat or caloric than an equivalent mass of the solid iron. This would only be possible if the iron filings had lower specific heat capacity than the solid iron. Rumford made measurements and found that the specific heat capacity of the filings was the same as that of the solid iron. He also used a blunt borer which produced few iron filings but just as much heating effect. This made it difficult to keep to the idea of heat as a flowing substance.

In 1841 James Joule showed that the heating effect of friction is directly related to the mechanical work done (as calculated by force $\times$ distance) – a finding that is still called the 'mechanical equivalence of heat'. Again, this undermined the concept of heat as a flowing substance and supported the idea that doing work *or* heating *both* change the internal state of material.

Ludwig Boltzmann, in 1877, provided the statistical mathematics that supported the explanation of thermal behaviour in terms of the average energy of huge numbers of particles of a gas with a wide range of speeds, experiencing collisions with each other and with the walls of their containers. But even at the start of the 20th century there were physicists who did not accept that the internal energy of a body was related to the kinetic and potential energy of its individual particles. It was Einstein who showed that Brownian motion, first seen by Robert Brown in the early 1800s, was completely consistent with Boltzmann's statistics. This firmly established the kinetic theory as the better source of explanations of observed thermal behaviour of matter and indeed established the particle theory of matter itself.

Kinetic theory is developed further in the next chapter.

27 A red-hot iron filing rests on the cool barrel of a cannon.
   a Describe the energy transfers and the temperature changes that take place. The room temperature and the temperature of the cannon is 10 °C.
   b The initial temperature of the iron filing is 500 °C and its mass is 0.05 g. How much energy does it lose as it cools down? (Refer to Table 33.1.)
28 Show how the energy changes involved in the boring of cannon are consistent with the First Law of Thermodynamics.
29 Are you a caloric or a kinetic theorist? Defend your position. Why is it easier for you to do so than it was for scientists in the 19th century?
30 **a** Why did the developers of early thermometry develop the centigrade scale rather than the thermodynamic temperature scale?
   b Research the origins of the Fahrenheit scale of temperature.
31 **DISCUSS**
   The development of thermometry is a fine example of useful technology stemming from the desire of scientists to know the answer to fundamental questions about the Universe. Scientists who now work in, say, particle physics are also seeking fundamental answers, and again their work often has technological benefit that is not the primary motivation for their work. (The development of technology provides the chief reason why politicians are willing to provide money for expensive activity such as particle research.) What technologies exist in your home which have been made possible as a by-product of the desire of scientists to answer fundamental questions about the nature of the Universe?

32 **DISCUSS**
   Give an account of another example of competing theories which were settled by observation. (You could choose from the study of force and motion, the study of light, the study of aspects of space.) What was the fundamental behaviour of the Universe that was being considered by these theories? What different predictions did the theories make about how the Universe should behave? How were these predictions tested? What technologies have arisen as a result?

Examination questions

1 a State one similarity and one difference between conduction and convection of thermal energy. (2)

b By reference to thermal energy transfer, explain what is meant by
  i two bodies having the same temperature
  ii body H having a higher temperature than body C. (2)

c **i** Briefly describe how a physical property may be used to measure temperature on its empirical centigrade scale.
  ii Hence explain why two thermometers measuring temperature on their empirical centigrade scales do not agree at all temperatures. (5)

d The table shows data for ethanol.

density	0.79 g cm^{-3}
specific heat capacity of liquid ethanol	$2.4 \text{ J g}^{-1} \text{ K}^{-1}$
specific latent heat of fusion	110 J g^{-1}
specific latent heat of vaporisation	840 J g^{-1}
melting point	$-120 \,°\text{C}$
boiling point	$78 \,°\text{C}$

Use the data to calculate the thermal energy required to convert 1.0 cm^3 of ethanol at $20 \,°\text{C}$ into vapour at its normal boiling point. (6)

e **i** State the *first law of thermodynamics*.
  ii Suggest why there is a considerable difference in magnitude between the specific latent heats of fusion and vaporisation. (5)

OCR, Physics, Paper 3, June 1999

2 a Describe the process involved in
  i thermal conduction
  ii thermal convection. (3)

b

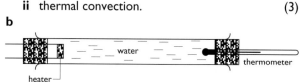

A long glass tube containing cold water is fitted with a heater at one end and a thermometer at the other, as shown in the diagram. During the period immediately after the heater has been switched on, describe and explain how the thermometer reading would change if the tube were
  i horizontal
  ii vertical, with the thermometer above the heater
  iii vertical, with the thermometer below the heater. (5)

AQA (NEAB), Physics (PH03), June 1999

3 a Define *specific heat capacity*. (2)

b An open-air swimming pool of surface area 60 m^2 has a uniform depth of 1.5 m and is heated by the Sun. The rate at which energy arrives from the Sun per unit area of water surface is 800 W m^{-2}. Of this energy, 20% is reflected and the rest is absorbed by the water. Water has specific heat capacity $4200 \text{ J kg}^{-1} \text{ K}^{-1}$ and density of 1000 kg m^{-3}. Calculate
  i the rate at which energy is absorbed by the pool (2)
  ii the mass of water to be heated (2)
  iii the mean rate of rise of temperature of the water (3)
  iv the time taken for the temperature of the water to rise by 3.0 K. (2)

c Suggest three reasons why, in cool climates, it is difficult to maintain a high water temperature in a swimming pool if it is only heated directly from the Sun. (3)

OCR, Sciences Basic 1, June 1999

4 An electric kettle, rated at 230 V 8.0 A, contains some water. It is placed on a balance.

The kettle is switched on and, when the water is boiling, the reading on the balance is found to decrease by 8.1 g in 10 s.

a Calculate
  i the power rating of the kettle
  ii the specific latent heat of vaporisation of water. (4)

b **i** State one source of error in this determination of the specific latent heat of vaporisation.
  ii Suggest briefly how this error may be reduced. (2)

OCR, Physics, Paper 2, Nov 1999

5 A thermometer has a thermal capacity of 1.3 J K^{-1}. The initial temperature of the thermometer is 20 °C. When used to measure the temperature of 40 g of water, it measures 37 °C.

  a **i** Determine the energy absorbed by the thermometer when it is placed in the water. (2)

    ii Calculate the temperature change of the water as a result of introducing the thermometer. (specific heat capacity of water = 4.2 × 10^3 J kg^{-1} K^{-1}) (2)

  b The thermometer consists of a bulb containing mercury of volume 9.8 × 10^{-8} m^3. When the temperature of 1.0 m^3 of mercury changes by 1 K, the mercury expands by 1.8 × 10^{-4} m^3.

    i Calculate the change in volume of the mercury when the thermometer is placed in the water. (2)

    ii The cross sectional area of the bore of the thermometer tube is 3.1 × 10^{-8} m^2. Calculate the distance the mercury moves up the thermometer for a temperature rise of 17 K. (2)

    iii The expansion of mercury for each kelvin of temperature rise is greater for higher temperatures. State and explain the effect this change will have on the accuracy of the mercury thermometer. (3)

AQA (AEB), Physics, Paper 1, June 1999

6 A small house uses a tank containing 1.2 m^3 of water as a thermal store. During the night its temperature rises to 98 °C. During the day, its temperature drops as the water is pumped round the house radiators to keep the house warm.

  a The density of water is 1000 kg m^{-3} and its specific heat capacity is 4200 J kg^{-1} K^{-1}. Calculate the energy given out by the water on a day when its temperature drops from 98 °C to 65 °C. (3)

  b The six radiators in the house give out an average power of 1.5 kW each. For how long can they all operate at this power before the water temperature drops to 65 °C? (3)

  c Explain why this heating system operates more effectively early in the morning than towards the evening. (2)

Edexcel, Physics PH3, June 1999

7 This question is about the caloric theory of heat.

  a Describe the 'caloric theory of heat', and mention **two** aspects of thermal behaviour that it seemed able to explain. (4)

  b Explain how the cannon-boring experiments of Count Rumford contributed to the demise of the caloric theory. (3)

International Baccalaureate, Physics, Higher Level, Nov 1998

34 Gases – kinetic theory and heat engines

THE BIG QUESTIONS

● What simple patterns are there in the behaviour of gases?
● What ideas can we use to consider gases in the simplest possible way?
● How do our simplifications match real behaviour?
● What makes an understanding of energy transfers to and from gases so extremely important to 21st century life?

KEY VOCABULARY

adiabatic expansion   amount of substance   atomic mass unit
Avogadro's constant   Boltzmann constant   Boltzmann distribution   Boyle's Law
Carnot cycle   Charles' Law   constant volume gas thermometer   empirical
equation of state of an ideal gas   gas laws   heat engine   ideal gas
ideal gas equation   isobaric expansion   isothermal expansion   mean square velocity
molar mass   mole   Pressure Law   relative atomic mass   relative molecular mass
rms velocity   Van der Waals equation   Van der Waals forces

BACKGROUND

One of the great scientific debates of the early 1600s was whether a vacuum could exist. There was at that time no way of knowing that space, above our atmosphere, was a near-vacuum.

People knew that it was impossible to pump water through a vertical pipe more than about 10 metres high. Italian experimenters built a lead pipe up the side of a tower, more than 10 metres high, with a glass flask at the top and a tub of water at the bottom. With the bottom of the pipe closed, they filled it from the top. Then they closed the top and opened the bottom. The level dropped and a space appeared in the glass flask.

In 1643, Vincenzo Viviani and Evangelista Torricelli hypothesised that the column of water was held in the tube by the pressure of the atmosphere acting on the water in the tube. Their suggestion was that the space in the glass flask was a vacuum. They predicted that it would be possible to make a similar column of mercury, but because of the much bigger density of the liquid the atmosphere would support a smaller column.

Torricelli performed the experiment that would test this hypothesis, inverting a glass tube of mercury over a bowl of the liquid. As predicted, a column of mercury of about 75 cm in height was produced, with a space above. Three years later Blaise Pascal used such tubes of mercury to compare the pressure at the top and bottom of a mountain, finding a real pressure difference. This suggested that pressure would continue to decrease with height, and that the atmosphere was a sea of air of limited depth.

● Discovery of a law

In 1660, an Irish scientist working in Oxford, the Honourable Robert Boyle, published a book called 'New Experiments Physico-mechanical Touching the Spring of the Air'. He described that he had found that 'electric and magnetic attractions go unimpeded in the void', and that liquids boiled at lower temperatures at reduced pressure. He also made measurements of the relationship between the volume of a gas and the pressure applied to it. (Note that the pressure applied to a gas – except during rapid compression and expansion – is the same as the pressure that the gas itself exerts. The term 'gas pressure' is sometimes used to mean the pressure acting on a gas, and sometimes the pressure exerted by it.)

Robert Hooke (who showed the proportionality of stress to strain of deformed material such as wires, beams and springs) worked as Boyle's assistant. It was Hooke who developed an air pump that Boyle used for experiments. Boyle's 'spring of the air' referred to its ability to expand into a vacuum and also the behaviour we can feel when we compress air, such as by squeezing a balloon or pressing the plunger on a syringe. Boyle, like Hooke, discovered a proportionality – in his case a proportionality of gas volume and gas pressure. As Boyle put it:

'This observation does both very well agree with and confirm our hypothesis that the greater the weight is, that leans upon the air, the more forcible is its endeavour of dilation and consequently its power of resistance (as other springs are stronger when bent by greater weights).'

The proportionality is known as **Boyle's Law**, stated formally and in present-day language as:

The volume of a fixed mass of gas at constant temperature is inversely proportional to its pressure.

Note the careful use of words – clearly the law does not apply when the mass of gas changes, such as by leakage of gas in or out of the sample being studied, or when the temperature changes.

The inverse proportionality means that we can say:

$PV = $ constant

But note that the value of this 'constant' will change if the mass or the temperature of the gas changes.

The laws of Hooke and Boyle share an important place in a general understanding of physics, in that they are early examples of the recognition of relatively simple patterns of behaviour (proportionality) in an apparently complex natural world. They worked in a tradition of the self-sufficient experimenter, creating their own equipment such as air pumps, glass flasks which Boyle described as 'of pretty bigness' to withstand pressure differences, and a 'slender glass pipe about the bigness of a swan's quill' which they graduated and in which they measured volumes of air. Their success has been repeated very many times in the 300 or so years since they lived and worked.

1 What do you think that Boyle meant, in modern terms, by
   a the weight that leans upon the air
   b its endeavour of dilation?
2 Sketch graphs to illustrate Hooke's Law and Boyle's Law.
3 **DISCUSS**
   Imagine that Boyle and Hooke could be transported through time from the 1660s to the 21st century. What would be the Big Questions in their minds? What answers would they find? What would amaze them most? Has all of our progress in physics, which they would see, done us more good or more harm? Is 'progress' a good word to use here?

Effects of temperature on gases

Study of the effects of temperature on gas pressure and volume did not reach straightforward conclusions for another hundred years after Boyle's work. Results were made possible then by the development of improved thermometers based on the expansion of liquids in glass tubes.

Alessandro Volta, Joseph Gay-Lussac and Jacques Charles independently found that the change in volume of a gas was proportional to the change in temperature, provided that pressure did not change. Now that we have an absolute or thermodynamic temperature scale – which has the benefit, unlike other temperature scales, of placing zero at the lowest possible temperature – we can state what is known as **Charles' Law** in its present-day form:

The volume of a fixed mass of gas at constant pressure is directly proportional to its temperature measured in kelvin.

Again, we can write the proportionality as an equation:

$V = $ constant $\times T$

This new 'constant' will change if either the quantity of gas or its pressure changes.

So if we can have a law describing the relationship between pressure and volume, and one describing the relationship between temperature and volume, then we can complete the set of **gas laws** by investigating the relationship between the pressure exerted by a fixed mass of gas and its temperature, when the volume is constant. Like the other gas laws, and like any law, we can only find the relationship by **empirical** processes – that is, by observation.

385

4 In examining Boyle's Law in a school or college laboratory, what limits the range of pressure over which measurements can be made?

5 Similarly, in looking at Charles' Law and the Pressure Law, what makes it impossible to make measurements over a wide temperature range?

6 In the investigation of the Pressure Law as shown in Figure 34.1, is the volume of the gas in the flask perfectly constant?

7 In the examination of Charles' Law (Figure 34.1), a bead of concentrated sulphuric acid seals the air into the glass tube. Evaporation of the acid into the space has less effect than evaporation of, say, water would have, but there is some effect and the total mass of the gas in the tube increases as temperature rises. What effect would you expect there to be on the shape of the graph?

We again find a proportionality, which is called the **Pressure Law**:

> The pressure exerted by a fixed mass of gas at constant volume is directly proportional to its temperature measured in kelvin.

As an equation,

$$P = \text{constant} \times T$$

again remembering that the 'constant' is dependent upon the volume and the mass of the gas. The three gas laws are summarised in Figure 34.1. The Pressure Law forms the basis of practical temperature measurements by means of the **constant volume gas thermometer** (Figure 34.2).

Figure 34.1

A summary of the three gas laws, in terms of the practical procedures for observing them, and the relationships shown in graphical form.

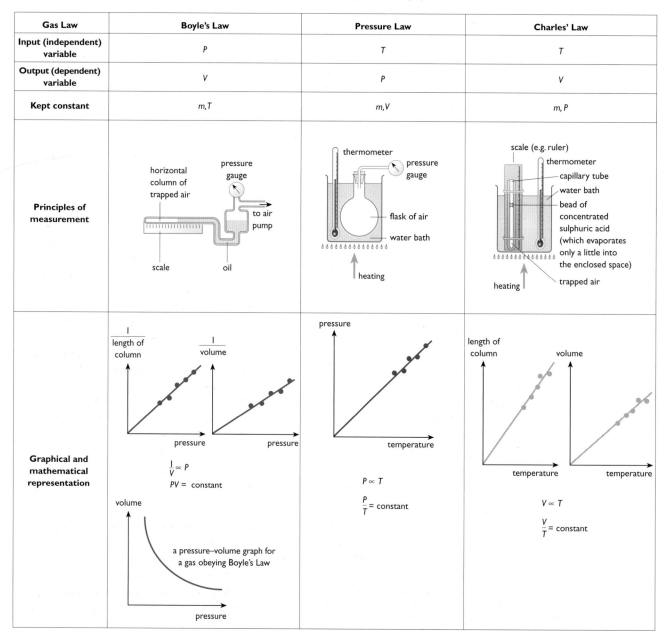

Gas Law	Boyle's Law	Pressure Law	Charles' Law
Input (independent) variable	P	T	T
Output (dependent) variable	V	P	V
Kept constant	m, T	m, V	m, P
Principles of measurement			
Graphical and mathematical representation	$\dfrac{1}{V} \propto P$ $PV = \text{constant}$ a pressure–volume graph for a gas obeying Boyle's Law	$P \propto T$ $\dfrac{P}{T} = \text{constant}$	$V \propto T$ $\dfrac{V}{T} = \text{constant}$

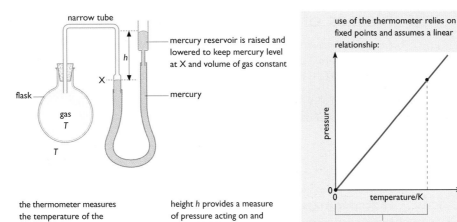

Figure 34.2
Principle of the constant volume gas thermometer. The thermometer can be calibrated using fixed points. For fixed points other than absolute zero, this involves immersing the flask in material of known temperature.

narrow tube

mercury reservoir is raised and lowered to keep mercury level at X and volume of gas constant

flask

mercury

gas *T*

T

h

X

use of the thermometer relies on fixed points and assumes a linear relationship:

pressure

temperature/K

fixed points

the thermometer measures the temperature of the flask's surroundings

height *h* provides a measure of pressure acting on and exerted by the gas

Amount of substance

Chemists have a useful way of measuring **amount of substance**. This is a measure of the number of particles in a sample. It is particularly useful to them because chemical reactions take place particle by particle, and so knowing the number of particles in a sample of material can be as important as knowing its mass. Chemists measure amount of reacting substance using the **mole**, and this unit is also useful when considering the behaviour of large numbers of molecules of gases (Figure 34.3). One mole is any collection of 6.02×10^{23} particles (e.g. molecules, atoms or ions). The formal definition of the mole is:

> The mole is the amount of substance that contains as many of its particles (normally molecules, atoms or ions) as there are atoms in 12 g of carbon-12.

The mole is a base unit of the SI system. It is often abbreviated to mol.

Figure 34.3
A mole of substance contains 6.02×10^{23} particles, in whatever form the particles of substance exist.

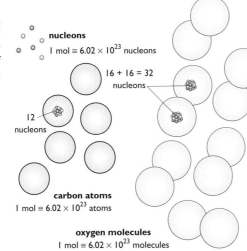

nucleons
1 mol ≡ 6.02×10^{23} nucleons

16 + 16 = 32 nucleons

12 nucleons

carbon atoms
1 mol ≡ 6.02×10^{23} atoms

oxygen molecules
1 mol ≡ 6.02×10^{23} molecules

The molar mass of carbon-12 is 12.000 g mol^{-1}. Each atom contains 12 nucleons. A mole of carbon-12 contains $12 \times 6.02 \times 10^{23}$ nucleons in its nuclei.

The molar mass of oxygen-16 can be written as 32 g mol^{-1} but not 32.000 g mol^{-1}. Although each molecule contains 32 nucleons, masses of nucleons vary according to the number of other nucleons with which they are gathered. This is a matter of nuclear physics, described in *Further Advanced Physics*.

The molar mass of naturally occurring oxygen is further different from 32.000 g, because of the presence of isotopes other than oxygen-16 (though this is by far the most abundant oxygen isotope in the environment).

For gases like oxygen and nitrogen the situation is made a little more complicated, in that, though they are chemical elements and not compounds, they normally exist as molecules, each with two atoms, and not as single atoms. In talking about gases we refer to the constituent particles as molecules. So though almost exactly 16 g of oxygen-16 contains 6.02×10^{23} atoms, it takes almost exactly 32 g to provide a total of 6.02×10^{23} molecules (Figure 34.3). The **molar mass** (the mass of one mole) of oxygen-16 is thus 32 g mol^{-1}.

Gases that are compounds, like methane and carbon dioxide, also exist as molecules, of course.

The number 6.02×10^{23} is called **Avogadro's constant**, which can be abbreviated to N_A. Since it is the number of particles in a mole, we can correctly write:

$$N_A = 6.02 \times 10^{23} \, \text{mol}^{-1}$$

Amedeo Avogadro was the first to suggest that equal volumes of all gases, at the same temperature and pressure, contain the same number of molecules. That was in 1811, though it took many years, and many observations of gases in chemical reactions, for the suggestion to win general support.

8 How many molecules are there in
  a 1 mol of nitrogen
  b 14 g of nitrogen?
9 Explain why the molar mass of carbon dioxide is the same for the solid state as for the gaseous state.

Masses of atoms and molecules

Atoms and molecules are very small, and their masses in kilograms are correspondingly very small numbers. For example, we know from the definition of the mole that 12.000 g of carbon-12 contains 6.02×10^{23} atoms and we can use this to find the mass in kilograms of one atom of carbon-12:

$$\text{mass of one atom of carbon-12} = \frac{12.000}{6.02 \times 10^{23}} \text{ g}$$

$$= 1.993 \times 10^{-23} \text{ g}$$

$$= 1.993 \times 10^{-26} \text{ kg}$$

To make life easier for ourselves, giving us easier numbers to deal with, we define this mass as 12 **atomic mass units**, u:

$$12 \text{ u} = 1.993 \times 10^{-26} \text{ kg}$$

$$1 \text{ u} = \frac{1.993 \times 10^{-26}}{12} \text{ kg}$$

$$= 1.66 \times 10^{-27} \text{ kg}$$

The atomic mass unit is by definition equal to one-twelfth of the mass of an atom of carbon-12. Now notice that an atom of carbon-12 contains twelve nucleons – that's six protons and six neutrons. So one atomic mass unit is *approximately* the same as the mass of nucleon. It is only approximately equal because:

- neutrons and protons do not have identical mass
- the total mass of a nucleus is very close to but not identical to the combined masses of its nucleons (mass is 'lost' when nucleons accumulate in nuclei – see *Further Advanced Physics*)
- 1 u is one-twelfth of the mass of an *atom* of carbon-12, and the atom contains electrons.

The **relative atomic mass** of an element is the average mass of its atoms relative to one-twelfth of the mass of an atom of carbon-12, measured in atomic mass units. A particular isotope of an element has identical atoms, and a fixed atomic mass. But the element might be found in nature as a mixture of several isotopes, and thus the relative atomic mass of a naturally occurring element takes account of the relative proportions of its isotopes.

Gases exist as molecules and not independent atoms, so for gases we refer to **relative molecular mass**. The relative molecular mass of a gas is the average mass of its molecules relative to one-twelfth of the mass of an atom of carbon-12.

Table 34.1 sums up the last two sections.

10 What is the mass, in kilograms and in atomic mass units, of
  a 1 mol of carbon-12
  b 1.4 mol of carbon-12
  c 2.8 mol of oxygen
  d 4.2 mol of carbon dioxide?

11 If the density of oxygen is 1.15×10^{-3} kg m^{-3} at 298 K and atmospheric pressure, what volume is occupied by
  a 1.15×10^{-3} kg
  b 1 kg
  c 0.032 kg
  d 1 mol
  e 1 molecule?

Table 34.1
Masses of some atoms and molecules.

Substance	Type of particle of which the substance is normally composed	Number of particles in one mole	Molar mass /g mol^{-1}	Molar mass to nearest whole number /g mol^{-1}	Relative atomic mass or relative molecular mass/u
carbon-12	atom	6.02×10^{23}	12.000	12	12.000
naturally occurring carbon	atom	6.02×10^{23}	12.011	12	12.011
hydrogen	molecule	6.02×10^{23}	2.058	2	2.058
nitrogen	molecule	6.02×10^{23}	28.014	28	28.014
oxygen	molecule	6.02×10^{23}	31.998	32	31.998
sodium	atom	6.02×10^{23}	22.989	23	22.989
carbon dioxide	molecule	6.02×10^{23}	44.009	44	44.009
uranium-238	atom	6.02×10^{23}	238.051	238	238.051
uranium extracted directly from ore	atom	6.02×10^{23}	238.029	238	238.029

The ideal gas

We have a problem. Gases show complications in their behaviour. One obvious complication is that they turn into liquids, and a gas in the presence of its own liquid behaves rather differently from a gas on its own, due to processes of change of state and particle collisions. And even in a gas there are electric forces, called **Van der Waals forces**, between particles that come close together. Another complication is that we should not really expect the volume of a gas to become zero at the absolute zero of temperature, 0 K, as Charles' Law predicts. The molecules themselves have volume and even if we could squash them all together we'd find that we couldn't make them merge to a single vanishing point.

The solution to the problem is to ignore these complications at first – to suppose that the gas *is* simple – and then to see how well this works at providing reliable predictions. If this works well for at least some of the time then it is a worthwhile exercise. That is, we idealise. We imagine an **ideal gas**.

An ideal gas is made of particles of negligible volume that have no effect on each other – the molecules exert no forces on each other. We can think of them as fast-moving point bodies that can collide with the walls of their containers but not with each other. When they collide with the walls of their containers they exert and experience force. The forces exerted by large numbers of molecules on surfaces are responsible for gas pressure.

The ideal gas and the equation of state

The gas laws are empirical laws – based on experiment. However, precise measurement shows that real gases do not obey them perfectly. The laws can only be thought of as being absolutely true for an ideal gas. They are descriptions of idealised behaviour.

For an ideal gas we can bring the three gas laws together, and note that the quantity PV/T is always the same for a fixed mass of a gas:

$$\frac{PV}{T} = \text{constant for fixed mass of a gas of specified relative molecular mass}$$

But if we want a universal law that applies to *any* ideal gas, whatever its relative molecular mass, then we need to take account of the amount of gas, for which we use the concept of the mole:

$$\frac{PV}{Tn} = \text{constant}$$

where n denotes the number of moles in the sample of gas. This constant is a 'true' constant – changes in the other variables in the equation might affect each other, but they do not affect the value of the constant. It is called the universal gas constant, R, and so we can write:

$$\frac{PV}{Tn} = R$$

The value of R is 8.31 J K^{-1} mol^{-1}.

The equation can be rearranged to:

$$PV = nRT$$

and in this form it is called the **ideal gas equation**, or the **equation of state of an ideal gas**.

Note that for one mole, $n = 1$, and

$$PV = RT$$

We will see later how this differs for a real gas.

12 State the units of each of the quantities on the left-hand side of

$$\frac{PV}{Tn} = R$$

What are the dimensions of each quantity?

13 Sketch a graph to show the relationship between the pressure exerted by an ideal gas (at fixed volume and temperature) and the amount of gas in moles. Explain why the presence of more gas molecules produces more pressure.

Pressure due to a single molecule of an ideal gas

We can go further with our ideal gas. We can consider how much pressure such a simple collection of point molecules can exert on its container. And since we are idealising we can, for the moment, simplify as much as we like, and start by thinking about just one molecule that is moving backwards and forwards between the opposite walls of a cuboid box (Figure 34.4a).

The molecule has mass, m, and velocity, c (to avoid any confusion of velocity with volume), and so it has momentum, mc. On each collision with a wall its velocity and its momentum change. The collision is elastic, so they simply reverse in direction:

$$\text{change in velocity on collision} = \text{final velocity} - \text{initial velocity}$$
$$= -c - c$$
$$= -2c$$

So

$$\text{change in momentum of a molecule on collision} = -2mc$$

In order for momentum to be conserved, there must be an equal and opposite change in momentum of the wall due to the collision. We will call this Δp, which is an impulse, defined in Chapter 14. (Note the lower case p is used for momentum here to avoid confusion with pressure, for which we are using upper case P.) So

$$\Delta p = 2mc \qquad (1)$$

Figure 34.4
The effect of a single gas molecule on one wall of its container.

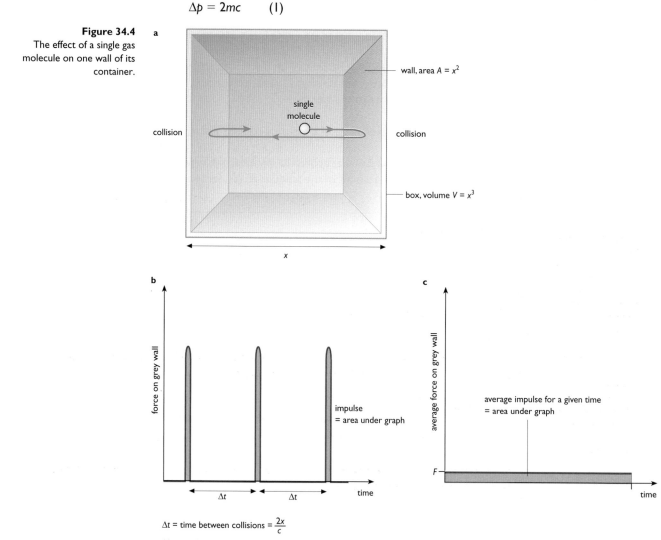

Δt = time between collisions = $\frac{2x}{c}$

(distance between two collisions with shaded wall = $2x$)

Such a change in momentum requires a force, F, to act on the wall. Note that since

$$F = ma = \frac{m\,\Delta c}{\Delta t} = \frac{\Delta(mc)}{\Delta t} = \frac{\Delta p}{\Delta t}$$

we can say that

impulse, $\Delta p = F\,\Delta t$

The duration of each collision is very short, but the time between one collision and the next is not so short. We can say:

Δp = average force acting during collision $\times$ duration of collision

but we can also say that

Δp = average force, F, over the time that passes between collisions with the wall $\times$ time, Δt, to cross the box and return

If collision time is negligible compared with journey time, then the time between one collision and the next collision with the same wall is equal simply to the journey time, which is related to speed:

$$\text{speed} = \frac{\text{distance}}{\text{time}}$$

$$c = \frac{2x}{\Delta t}$$

where x is the width of the box (see Figure 34.4a). So

$$\Delta t = \frac{2x}{c} \quad \text{(see Figure 34.4b)}$$

and

$$\Delta p = F\,\Delta t = \frac{2Fx}{c} \qquad (2)$$

where F, remember, is the force exerted by the molecule on the wall, averaged out over continuous time (Figure 34.4c).
   Combining equations (1) and (2),

$$\frac{2Fx}{c} = 2mc$$

and so

$$F = \frac{mc^2}{x}$$

The pressure on the wall is force/area. The total area over which our molecule is exerting force is the area of the wall, which, if the box is cubic, is x^2, so

$$\text{pressure} = \frac{mc^2}{x^3}$$

$$= \frac{mc^2}{V}$$

where V is the volume of the box. Remember that this is the pressure, averaged over time, that a single molecule exerts on one wall.

14 Explain
   a why the change in momentum of a molecule on collision is $-2mc$ and not $-mc$
   b why the change in momentum of the wall during the same collision is $2mc$
   c what $F = ma$ tells us about impulse
   d why the time between collisions with area A (Figure 34.4a) is $2x/c$ and not x/c
   e what, precisely, F stands for in $F = mc^2/x$.

Behaviour of *N* molecules of an ideal gas

Pressure exerted by *N* molecules moving in one dimension

Now we can think about a slightly more realistic situation in which our molecule is not alone, but is one of *N* molecules of an ideal gas, shuttling between collisions with the same two opposite walls. Suppose for the moment that they are all moving parallel to the Cartesian *x*-direction. The total pressure that the *N* molecules exert is a sum of the individual pressures. It is reasonable to suppose that the molecules are of the same kind, and have the same mass. But it is not reasonable to suppose that they all have the same velocity.

Our molecule number one can be said to exert a pressure, mc_1^2/V. Molecule number two exerts a pressure, mc_2^2/V, and molecule number *N* exerts a pressure, mc_N^2/V. So the total pressure on one wall is given by

$$P = \frac{mc_1^2}{V} + \frac{mc_2^2}{V} + \cdots + \frac{mc_N^2}{V}$$

which simplifies to

$$P = \frac{m(c_1^2 + c_2^2 + \cdots c_N^2)}{V}$$

We need an even simpler way of writing this, so we say

$$\frac{(c_1^2 + c_2^2 + \cdots c_N^2)}{N} = \overline{c_x^2}$$

$\overline{c_x^2}$ is the average, or mean, of the square of all the velocities, and it's called the **mean square velocity** in the *x*-direction. It is also often written as $\langle c_x^2 \rangle$.

Note that

$$(c_1^2 + c_2^2 + \cdots c_N^2) = N\overline{c_x^2}$$

so the total pressure on one wall due to the *N* molecules is

$$P = \frac{Nm\overline{c_x^2}}{V}$$

> **15** Three molecules have velocities of $c_1 = 200$ m s^{-1}, $c_2 = 220$ m s^{-1} and $c_3 = 320$ m s^{-1}.
> **a** What is the mean velocity, $\overline{(c_1 + c_2 + c_3)}$?
> **b** What is the square of the mean velocity, $\overline{(c_1 + c_2 + c_3)}^2$?
> **c** Calculate the square of each velocity, c_1^2, c_2^2 and c_3^2.
> **d** What is the mean of the squares of the velocities, $\overline{c_1^2 + c_2^2 + c_3^2}$?
> **e** Comment on the truth, or otherwise, of
> $$\overline{(c_1 + c_2 + c_3)}^2 = \overline{c_1^2 + c_2^2 + c_3^2}$$

Pressure exerted by *N* molecules moving in three dimensions

We are still simplifying, and so far have only been considering molecules that travel in the *x*-direction. The box will also contain molecules travelling in the *y*- and *z*-directions. And, of course, since the motion of the collection of particles is random, few will be travelling exactly along the *x*-, *y*- and *z*-directions. However, any molecule is likely to have an *x*-component, a *y*-component and a *z*-component to its motion. Of all the motions of all the particles, we can assume that one-third of the motion will be made up of *x*-components, one-third from *y*-components, and one-third from *z*-components (Figure 34.5).

For each molecule, the velocity *c* can be found from its three components using Pythagoras' Theorem:

$$c^2 = c_x^2 + c_y^2 + c_z^2$$

and the mean square velocity of a collection of molecules is given by

$$\overline{c^2} = \overline{c_x^2} + \overline{c_y^2} + \overline{c_z^2}$$

since the motions in the three perpendicular directions are independent.

There are a large number of molecules in random motion, so we can assume that

$$\overline{c_x^2} = \overline{c_y^2} = \overline{c_z^2}$$

Figure 34.5
For each molecule, velocity has three components. For many molecules moving in all directions, the maths is the same as for a set of particles of which one-third are moving in the *x*-direction, one-third in the *y*-direction, and one-third in the *z*-direction.

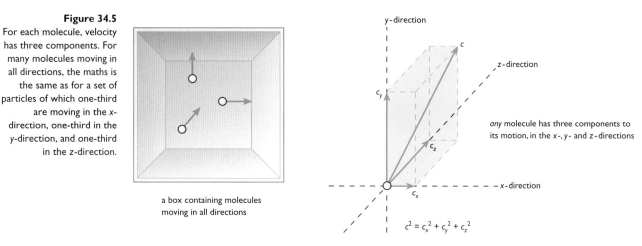

a box containing molecules moving in all directions

any molecule has three components to its motion, in the *x*-, *y*- and *z*-directions

$$c^2 = c_x^2 + c_y^2 + c_z^2$$

It follows that

$$\overline{c^2}/3 = \overline{c_x^2} = \overline{c_y^2} = \overline{c_z^2}$$

The pressure on one wall

$$P = \frac{Nm\overline{c_x^2}}{V} \quad \text{becomes} \quad P = \frac{Nm\overline{c^2}}{3V}$$

and is the same for all walls of the box. Note that *N* is the total number of particles and *m* is the mass of each one, so that *Nm* is the total mass of the gas in the box. *V* is volume, and so *Nm/V* is the gas density, ρ. The equation begins to look simpler:

$$P = \tfrac{1}{3}\rho\,\overline{c^2}$$

Figure 34.6
Gas pressure depends on the density of the gas and on the speed of its molecules.

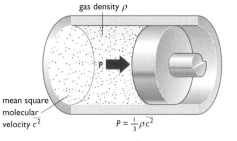

gas density ρ

mean square molecular velocity $\overline{c^2}$

$$P = \tfrac{1}{3}\rho\overline{c^2}$$

The pressure exerted by a gas depends on:

- density – the higher the density, the more molecules there are per unit volume and so the more collisions there are with the container walls; for different gases at a particular temperature, density increases with increasing molecular mass and so the greater the influence (impulse) of each collision
- mean square molecular velocity – the faster the molecules, the greater the impulse they exert on collision; mean square molecular velocity is related to temperature and to internal energy, as we shall see in detail in the following pages.

Figure 34.7
One method of measuring the speeds of gas molecules. Only molecules that pass through both spinning discs reach the target to be detected.

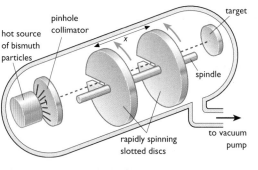

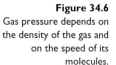

target

pinhole collimator

hot source of bismuth particles

x

spindle

rapidly spinning slotted discs

to vacuum pump

If the period of revolution of the discs is *T*, then

speed of particles hitting target $= \dfrac{n \times x}{T}$

where *n* is a whole number.

Working with an ideal gas we have derived a relationship relating the pressure exerted by a gas with its density and the mean square velocity of its molecules (Figure 34.6). The big question is whether the idealisation was justified and worthwhile – does the equation provide useful predictions of the behaviour of real gases? To answer that question we need to measure the speeds of gas molecules.

In the arrangement shown in Figure 34.7, only particles that travel the distance *x* in the time taken for the discs to spin a whole number of times reach the target. This enables a distribution of their speeds to be recorded, such as those in Figure 34.8.

$\bar{c^2}$ is the mean square velocity, and has units $m^2\,s^{-2}$. It is helpful to think about the root mean square velocity, or **rms velocity**. This has units $m\,s^{-1}$ and provides a much more satisfying indication of what is happening in a gas.

$$\text{rms velocity} = \sqrt{\text{mean square velocity}} = \sqrt{\bar{c^2}}$$

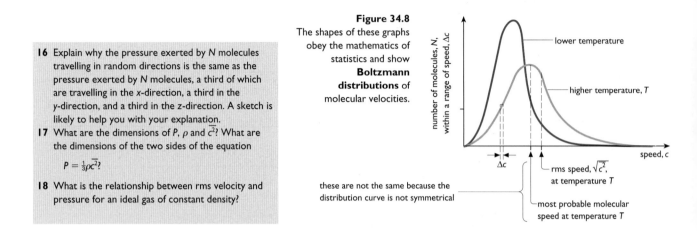

Figure 34.8
The shapes of these graphs obey the mathematics of statistics and show **Boltzmann distributions** of molecular velocities.

these are not the same because the distribution curve is not symmetrical

16 Explain why the pressure exerted by N molecules travelling in random directions is the same as the pressure exerted by N molecules, a third of which are travelling in the x-direction, a third in the y-direction, and a third in the z-direction. A sketch is likely to help you with your explanation.

17 What are the dimensions of P, ρ and $\bar{c^2}$? What are the dimensions of the two sides of the equation

$P = \frac{1}{3}\rho\bar{c^2}$?

18 What is the relationship between rms velocity and pressure for an ideal gas of constant density?

Molecular kinetic energy

We can combine the equation for the pressure of an ideal gas with the equation of state, and at the same time we can think about molecular kinetic energy. It is worth remembering that an ideal gas has no intermolecular forces, which means that the molecules have no potential energy, only their kinetic energy. The internal energy of an ideal gas, U, is simply the sum of the molecular kinetic energies. Normally we are cautious about quoting an actual value of internal energy, which is the sum of kinetic energy and complex potential energy, and we talk only about *changes* in internal energy, ΔU. But for the simple ideal gas we can throw caution to the wind and state that

$$U = \text{sum of molecular kinetic energies} = N \times \text{mean kinetic energy}$$

$$= N \times \frac{(\frac{1}{2}mc_1^2 + \frac{1}{2}mc_2^2 + \frac{1}{2}mc_3^2 + \ldots \frac{1}{2}mc_N^2)}{N} = N \times \frac{1}{2}m\bar{c^2}$$

For an ideal gas we can say that (page 389)

$$PV = nRT$$

and also that (page 393)

$$P = \frac{1}{3}\rho\bar{c^2}$$

We can combine these:

$$\frac{1}{3}\rho\bar{c^2}V = nRT$$

ρV is the mass of the gas, Nm, so

$$\frac{1}{3}Nm\bar{c^2} = nRT$$

We can multiply both sides of this by $\frac{3}{2}$ and divide by N, for reasons that will become clear in a moment:

$$\frac{3}{2} \times \frac{1}{3}\frac{Nm\bar{c^2}}{N} = \frac{3}{2}\frac{nRT}{N}$$

which simplifies to:

$$\frac{1}{2}m\bar{c^2} = \frac{3}{2}\frac{nRT}{N}$$

19 Why is the Boltzmann constant a very small number?
20 Sketch a graph of the mean molecular energy of an ideal gas against thermodynamic temperature. What is the significance of the gradient?
21 a What is the mean molecular kinetic energy for an ideal gas at 290 K?
    b Do oxygen and nitrogen have the same mean molecular kinetic energy at the same temperature? Explain.
22 Sketch a graph to show the relationship between molecular mass of a gas and mean molecular kinetic energy at a constant temperature.
23 Given that $N = nN_A$ and $k = R/N_A$, derive a relationship between U, R and T.
24 The mean kinetic energy of hydrogen molecules is 6×10^{-21} J at 300 K.
    a What is the mean kinetic energy at
    i 150 K
    ii 400 K?
    b If the relative molecular mass is 2.0 u, what is the rms speed of the hydrogen molecules at 300 K?
    c Sketch graphs to show the speed distributions of hydrogen molecules at 300 K and at a temperature in the region of 600 K.
    d Explain why hydrogen molecules are faster than any other molecules at any given temperature.
25 Escape velocity from the Earth is 11 km s^{-1}; molecules at the top of the atmosphere with velocities of more than this in an upwards direction may escape.
    a Use your answers to question 24 to help you to explain one reason why atmospheres of planets similar to Earth have little hydrogen but plenty of nitrogen. (Another reason is that hydrogen is much more chemically active than nitrogen.)
    b Escape velocity from the Moon is only 2.4 km s^{-1}. Use your knowledge of gases to explain why the Moon has no atmosphere.

The left-hand side of the last equation is mean molecular kinetic energy. Since n is the number of moles and N is the number of particles, N/n is the number of particles per mole. This is Avogadro's constant, N_A. So n/N is the inverse of Avogadro's constant, $1/N_A$. Therefore

$$\tfrac{1}{2}m\overline{c^2} = \tfrac{3}{2}\frac{RT}{N_A}$$

R and N_A are both constants, and their ratio is another constant. We will make life easier for ourselves if we give the ratio a name of its own. We call it the **Boltzmann constant**, k. This has the advantage of allowing us to write:

$$\tfrac{1}{2}m\overline{c^2} = \text{mean molecular kinetic energy}$$
$$= \tfrac{3}{2}kT$$

k has the value 1.38×10^{-23} J K^{-1}.

Note that the *total* kinetic energy in a sample of an ideal gas is given by

$$\tfrac{1}{2}Nm\overline{c^2} = \tfrac{3}{2}NkT = \text{total internal energy, } U$$

where N is again the total number of molecules.

The real gas and the ideal gas

We've seen that for an ideal gas
$$PV = nRT$$
If we have just one mole of gas then $n = 1$ and
$$PV = RT$$

V is the volume occupied by the gas, and according to the equation $V = 0$ when $T = 0$. This is only true for an ideal gas, which is made up of point molecules. But if the total volume of the molecules themselves is b, then at the absolute zero of temperature the total volume of the gas will be b. We can adjust the equation to:
$$P(V - b) = RT$$
so that at 0 K, $(V - b) = 0$, and the volume occupied is $V = b$.

We can also make an adjustment to allow for intermolecular forces. These forces result in a real gas exerting less pressure on the walls than an ideal gas would. The pressure exerted by the real gas is found to be reduced by an amount that is inversely proportional to the square of the volume. For our mole of gas we can write
$$\left(P + \frac{a}{V^2}\right)(V - b) = RT$$

This is an equation of state for a mole of real gas, and is called the **Van der Waals equation**. a and b are both constants, and both small.

26 What are the dimensions and hence the units of a in the Van der Waals equation?
27 If we suppose that $a = 0$, what assumption are we making about the behaviour of the gas?
28 A real gas at 273 K occupies a volume of 0.020 m^3 per mole at a particular pressure. If a can be assumed to be zero and $b = 3.2 \times 10^{-5}$ m^3 mol^{-1}, what is the ratio of the pressure exerted by a mole of the gas to the pressure that would be exerted by a mole of an ideal gas with the same volume and temperature?

When *V* is large compared with *b* ($V \gg b$) then ($V - b$) ≈ *V*. This is true for any gas at a temperature significantly above absolute zero (or significantly above its boiling point) unless the pressure is very high indeed. Also, when *V* is large then a/V^2 is very small. So to a reasonable *approximation*, we can say that for one mole of real gas, except at very low temperature or very high pressure, *PV* = *RT*.

Expansion of a gas at constant pressure

In the previous chapter we looked at the relationship between changes in internal energy of a gas and the energy supplied to it by thermal processes (that is by heating, written as Δ*Q*) and mechanical process (that is by doing work on the gas, written as Δ*W*). We also saw that the gas must do work in order to expand. An ideal gas has no intermolecular forces and so the only work that it needs to do is work to push other material – usually the air of the atmosphere – out of the way. In this case we can say that

work done by the gas on the atmosphere = −Δ*W* = *P* Δ*V*   (see page 373)

Figure 34.9
The graph of pressure against volume shows the work done by the gas.

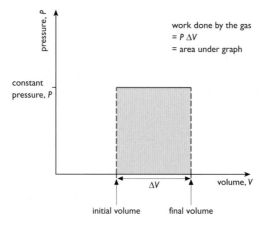

work done by the gas
= *P* Δ*V*
= area under graph

P is the external pressure and Δ*V* is the change in volume of the gas. The minus sign appears here, using the sign convention that we've adopted, because work is done *by* the gas and not *on* the gas.

If the expansion takes place slowly then it is possible that the pressure of the gas will be the same at all times as the pressure outside. The pressure outside is constant. So we can imagine expansion of a gas at a constant pressure, sometimes called **isobaric expansion**, for which a graph of pressure and volume is relatively simple (Figure 34.9). The area under the graph is a measure of the work done.

We know from the gas laws and from the ideal gas equation that during such an expansion temperature cannot stay the same:

PV = *nRT*

If *P*, as well as *n* and *R* stay the same, but volume *V* changes, then temperature *T* must change. Isobaric expansion of a gas is accompanied by an increase in temperature and an increase in internal energy, Δ*U*. We know that, in general, increase in internal energy is the sum of energy transferred thermally to the gas, Δ*Q*, and energy transferred mechanically to the gas, Δ*W*:

Δ*U* = Δ*Q* + Δ*W*

The work done *by* the gas is −Δ*W* and in an isobaric expansion Δ*U* is positive, so Δ*Q* > −Δ*W* (Figure 34.10).

Figure 34.10
A gas expanding at constant pressure.

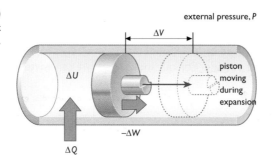

external pressure, *P*

Δ*U* = Δ*Q* + Δ*W*

During any expansion the gas does work on its surroundings.

During isobaric expansion the temperature must rise, so Δ*U* has a positive value.

Energy Δ*Q* must be supplied to the gas for this to be possible. This amount of energy must be bigger than the work done by the gas: Δ*Q* > −Δ*W*.

Isothermal changes

Figure 34.11
Isothermal expansion

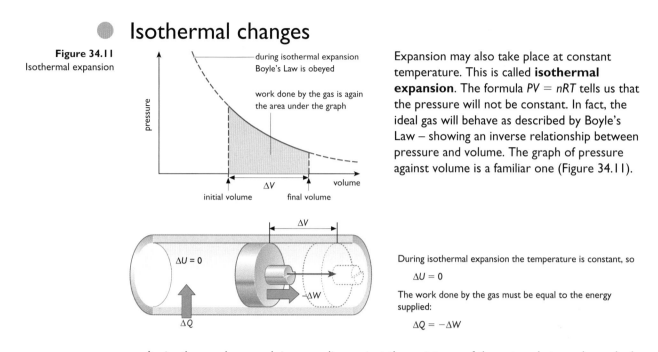

Figure 34.11
Isothermal expansion

during isothermal expansion Boyle's Law is obeyed

work done by the gas is again the area under the graph

pressure

ΔV

initial volume   final volume

volume

$\Delta U = 0$

ΔV

$-\Delta W$

ΔQ

Expansion may also take place at constant temperature. This is called **isothermal expansion**. The formula $PV = nRT$ tells us that the pressure will not be constant. In fact, the ideal gas will behave as described by Boyle's Law – showing an inverse relationship between pressure and volume. The graph of pressure against volume is a familiar one (Figure 34.11).

During isothermal expansion the temperature is constant, so

$$\Delta U = 0$$

The work done by the gas must be equal to the energy supplied:

$$\Delta Q = -\Delta W$$

Again, the gas does work in expanding against the resistance of the external atmosphere. And again, the area under the graph provides a measure of this. But if the temperature of the ideal gas stays the same then we are saying that its internal energy, U, stays the same. Change in internal energy, ΔU, is zero. Since

$$\Delta U = \Delta Q + \Delta W$$

we can say that, for an isothermal change,

$$\Delta Q = -\Delta W$$

That is, the work done by the gas ($-\Delta W$) is equal to the energy transferred to it by heating.

Adiabatic changes

Another scenario is possible, in which the energy needed to do work in expansion does not come from energy supplied from outside the gas, but all comes from the internal energy. In this case $\Delta Q = 0$ and

$$\Delta U = \Delta W$$

Since ΔW has a negative value, so does ΔU. That is, the work done to expand is made possible by a *fall* in the internal energy of the gas, and a fall in temperature. Such an expansion is called an **adiabatic expansion**.

Expansion of a gas is a perfect adiabatic expansion ($\Delta Q = 0$) if the gas is enclosed in a perfect insulator. An expansion *approximates* to a perfect adiabatic expansion if the change is rapid so that energy cannot transfer to it quickly enough to make a significant contribution ($\Delta Q \approx 0$). Processes that involve sudden release of pressure are, or approximate to, adiabatic expansions (Figure 34.12).

Figure 34.12
The release of vapour from an aerosol can approximates to an adiabatic expansion.

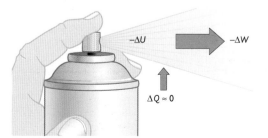

$-\Delta U$ $-\Delta W$

$\Delta Q \approx 0$

Work done by expanding gas = $-\Delta W$

This is matched by a fall in internal energy (and so a fall in temperature):

$$-\Delta U = -\Delta W$$

Figure 34.13
The graph of pressure against volume for an adiabatic change is not a simple Boyle's Law curve as it is for isothermal change.

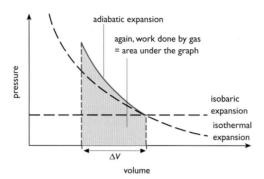

In an adiabatic expansion the pressure change, for the same volume change, is greater than in an isothermal expansion. This is because the pressure, given by $P = \frac{1}{3}\rho\overline{c^2}$, is affected by:

• change in volume and so change in density, ρ
• change in temperature and so change in mean square molecular velocity, $\overline{c^2}$.

In an adiabatic expansion a gas does more work on its surroundings.

In an adiabatic expansion of an ideal gas, the equation

$$PV = nRT$$

still applies, and

• V is changing – the gas is expanding
• T is changing – since internal energy must act as the energy source for the work done
• P is changing – the fall in temperature *and* the increase in volume both tend to reduce the pressure that the gas exerts.

This last point is very important. In an adiabatic change the gas does not obey Boyle's Law because the temperature is not constant. The fall in temperature on expansion means that the change in pressure is more than it would be at constant temperature (Figure 34.13).

29 a What is the relationship between volume and temperature during expansion of an ideal gas at constant pressure?
To which of the gas laws does this correspond?
30 Use Figure 34.13 to explain the different amounts of work done by a gas expanding adiabatically and one expanding isothermally, for the same volume change.

Cyclic processes and heat engines

Consider an imaginary four-stage process involving an ideal gas in which we . . .

1 expand it isothermally
2 expand it adiabatically
3 compress it isothermally
4 compress it adiabatically

. . . until we have the same pressure, volume and temperature as at the beginning. This is called an ideal **Carnot cycle**. The four stages of the process, plotted on a pressure–volume graph, enclose an area (Figure 34.14). This is a net amount of work done by the gas on its surroundings.

Figure 34.14
The Carnot cycle.

During the cycle the gas does a net amount of work, represented by the enclosed shaded area, on it surroundings.

Less work is done *on* the gas during compression than is done *by* the gas during expansion. The gas returns to its original state, ready to repeat the cycle, having done a net amount of work. This has been made possible by a net amount of energy being transferred to it thermally (e.g. by burning fuel).

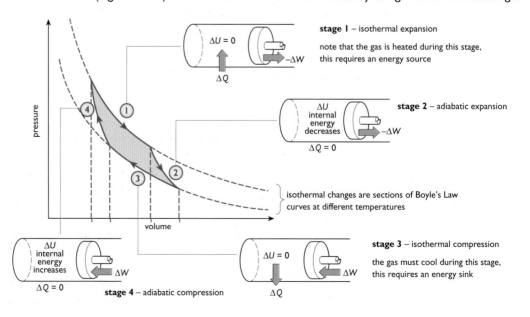

Such cycles provide a very useful way of studying processes of energy transfer involving gases – processes that take place in **heat engines**. All heat engines involve thermal transfer of energy between a hot 'source' and a cold 'sink', work being done in the process (Figure 34.15).

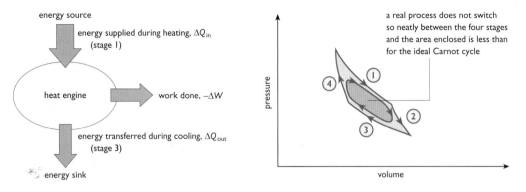

Figure 34.15
Energy transfers in a heat engine, related to stages of the Carnot cycle.

Real behaviour of a heat engine is not so simple as an ideal Carnot cycle, but the ideal concept is useful for the study of real heat engines. It shows, for example, that there is a maximum efficiency of any heat engine that is dependent on the temperature difference between the hot source and the cold sink.

$$\text{efficiency} = \frac{\text{work done by the engine}}{\text{energy supplied to the engine}} \times 100\%$$

$$= \frac{\Delta W}{\Delta Q_{in}} \times 100\%$$

For a heat engine, the maximum work it can do is the difference between ΔQ_{in} and ΔQ_{out} (because its internal energy is the same at the end of a cycle as at the beginning):

$$\Delta W_{max} = \Delta Q_{in} - \Delta Q_{out}$$

So

$$\text{maximum efficiency} = \left(\frac{\Delta Q_{in} - \Delta Q_{out}}{\Delta Q_{in}}\right) \times 100\%$$

$$= \left(1 - \frac{\Delta Q_{out}}{\Delta Q_{in}}\right) \times 100\%$$

The ratio of the energy transfers is equal to the ratio of the temperatures of the sink and source. (The mathematics that we would need to show this is complex.)

$$\frac{\Delta Q_{in}}{\Delta Q_{out}} = \frac{T_1}{T_2}$$

where T_1 is the (lower) temperature of the sink, and T_2 is the (higher) temperature of the source. So

$$\text{maximum efficiency} = \left(1 - \frac{T_1}{T_2}\right) \times 100\%$$

This is idealised of course, and the real efficiency will be less than this. The maximum efficiency, however, of a heat engine is dependent only on the ratio of temperatures of the energy source and the energy sink. So a power station, for example, uses steam at a very high temperature and then cools the steam, as it does work in turning the turbines, to as low a temperature as can be practically arranged. A power station (Figures 34.16 and 34.17) needs a hot energy source – the boiler – and a cool energy sink. Power station cooling towers are there to play their part in ensuring that the energy sink is cool.

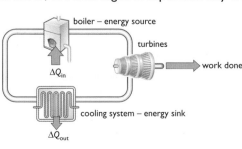

Figure 34.16
Schematic diagram of a power station system as a heat engine.

Power stations use fuel. They cost money and they pollute. The more efficient they are, the less they cost and the less they pollute. Engineers build power stations to be efficient, but they know that the ratio of the temperatures of the thermal energy source and the thermal energy sink sets a theoretical maximum that can never be beaten. Part of their job therefore, is to make the source hot and the sink cold.

Figure 34.17
The power station as a heat engine.

a Turbine-generator (heat engine)

b Fuel for the boiler (hot source) with cooling towers in the background (cold sink)

31 Explain why the temperature falls during stage 2 of the Carnot cycle.

32 What is the maximum efficiency of a turbine which is driven by steam that arrives at a temperature of 220 °C and leaves at a temperature of 50 °C? Why will actual efficiency be significantly less than this?

33 Comment on the feasibility of setting up a heat engine based on the cycle shown in Figure 34.18.

34 How, if at all, are the four strokes of the car engine (Figure 34.19) related to the four stages of the Carnot cycle?

Figure 34.18

Figure 34.19
The four strokes of a car engine.

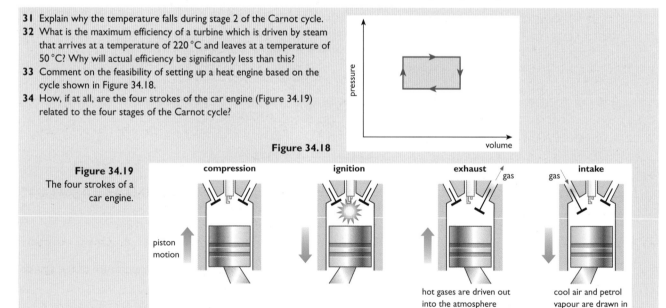

Extra skills task Application of Number

Atoms and molecules are very small. It is possible to estimate their size by a simple experiment. Take one small drop of oil, and estimate its volume. Drop the oil on to the surface of water, on which you have sprinkled some fine powder. The drop should spread out forming a layer, pushing the powder aside. Assuming the layer is one molecule thick, you should now be able to calculate the dimensions of an oil molecule. You will need to use the standard formulae for the volume of a sphere (drop) and the volume of a cylinder (layer). Compare your result with figures generally available for the size of molecules. Give reasons for any discrepancies. Estimate the number of molecules in the drop.

Examination questions

1 a State three basic assumptions of the kinetic theory of gases. (2)

b The product pV of pressure and volume of an ideal gas can be given by either

$$pV = \tfrac{1}{3}Nm\langle c^2\rangle$$

or

$$pV = NkT$$

where N is the number of molecules, m is the mass of each molecule, T is the thermodynamic temperature, and $\langle c^2\rangle$ is the mean square speed of the molecules.

Use these expressions to show that the average translational kinetic energy E_k of an ideal gas molecule is proportional to T. (4)

c Air consists principally of oxygen and nitrogen, both of which may be considered to behave as ideal gases at room temperature (290 K). The average translational kinetic energy of molecules of each of these gases is given by

$$E_k = \frac{3RT}{2N_A}$$

i Calculate E_k for a nitrogen molecule at a temperature of 290 K.

ii Oxygen molecules have greater mass than nitrogen molecules. Explain whether you would expect an oxygen molecule to have a greater value of E_k at 290 K than that calculated in **i**. (3)

OCR, Sciences, Basic 2, Nov 1999

2 a The equation of state of an ideal gas is

$$pV = nRT$$

For each of these symbols, state the physical quantity and the SI unit.

Symbol	Physical quantity	Unit
p		
V		
n		
R		
T		

b An ideal gas of volume $1.0 \times 10^{-4}\ \mathrm{m}^3$ is trapped by a piston in a cylinder as shown in the diagram. There is negligible friction between the piston and the cylinder. Initially, the temperature of the gas is 20 °C and the external atmospheric pressure acting on the piston is 100 kPa.

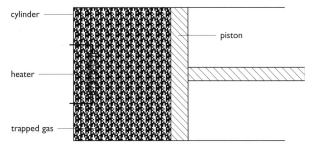

The gas expands slowly when heat is supplied by an electric heater inside the cylinder.

i Calculate the work done by the gas when its volume slowly increases by $5.0 \times 10^{-5}\ \mathrm{m}^3$, at a constant pressure, while being heated.

ii What is the temperature of the gas, in °C, following its expansion?

iii Describe *two* changes that occur in the motion of a typical molecule of the gas during the expansion. (8)

AQA (NEAB), Further Physics, (PH03), Mar 1999

3 a Define the term *density*. (1)

b Outline how molecular movement causes the pressure exerted by a gas. (2)

c One mole of oxygen has a mass of 32 g. Assuming oxygen behaves as an ideal gas, calculate

i the volume occupied by one mole of oxygen gas when at temperature 273 K and pressure $1.01 \times 10^5\ \mathrm{Pa}$.

ii the density of oxygen gas at this temperature and pressure. (5)

d i Explain what is meant by the root-mean-square speed $\sqrt{\langle c^2\rangle}$ of gas molecules.

ii Calculate the root-mean-square speed of four molecules travelling with speeds 300 m s^{-1}, 400 m s^{-1}, 500 m s^{-1} and 600 m s^{-1}. (4)

e Assuming ideal gas behaviour, calculate for oxygen at 273 K

i the root-mean-square speed of its molecules

ii the average kinetic energy of a molecule. (4)

f Oxygen has a boiling point of 90 K and a melting point of 55 K. Describe qualitatively how oxygen at 273 K and oxygen at 27 K differ in respect of

i density

ii spacing of the molecules

iii order in the pattern of molecules

iv motion of the molecules. (4)

OCR, Physics, Paper 3, Nov 1999

4 a i State what is meant by *thermal equilibrium*.
ii Explain thermal equilibrium by reference to the behaviour of the molecules when a sample of hot gas is mixed with a sample of cooler gas and thermal equilibrium is reached. (3)

b A sealed container holds a mixture of nitrogen molecules and helium molecules at a temperature of 290 K. The total pressure exerted by the gas on the container is 120 kPa.

molar mass of helium $= 4.00 \times 10^{-3}$ kg mol^{-1}
molar gas constant $R = 8.31$ J K^{-1} mol^{-1}
the Avogadro constant $N_A = 6.02 \times 10^{23}$ mol^{-1}

i Calculate the root mean square speed of the helium molecules.
ii Calculate the average kinetic energy of a nitrogen molecule.
iii If there are twice as many helium molecules as nitrogen molecules in the container, calculate the pressure exerted on the container by the helium molecules. (6)

AQA (NEAB), Further Physics, (PH03), June 1999

5 a Define:
i an *isothermal* expansion (1)
ii an *adiabatic* compression. (1)

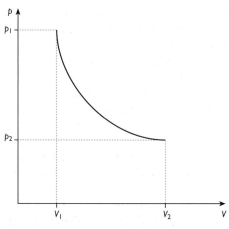

b The figure shows the variation of the pressure p with the volume V of a fixed mass of gas expanding isothermally from initial values p_1V_1 to final values p_2V_2.
i On the figure, sketch the graph for an adiabatic expansion from V_1 to V_2 from the same initial pressure p_1. Label the final pressure p_3. (1)
ii By referring to the first law of thermodynamics, explain why the final pressure p_3 for the adiabatic expansion is different from the final isothermal pressure p_2. (3)
iii Suggest what could be done to restore the gas to the pressure p_2 at volume V_2. (1)
c State and explain an everyday example of an adiabatic change. (2)

OCR, Thermal Effects and Fields, Mar 1999

6 Read the following passage.

Liquids, vapours and gases

When a small mass of water is introduced into a large evacuated vessel, the water evaporates completely to form what is called an unsaturated vapour. If the unsaturated vapour is then compressed slowly, the volume of the vapour decreases. During this change, the temperature is kept constant. This is known as an isothermal change and is shown by the line AB in the figure.

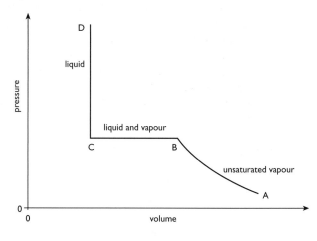

On reducing the volume further at constant temperature, the vapour condenses and the pressure remains constant. In the region BC, liquid and vapour are present and the vapour is said to be saturated. At C, all the vapour has condensed and only liquid is present. Any further reduction in volume is achieved only by applying very large pressures.

When values of the volume and the corresponding pressure are obtained at a number of different temperatures, a series of lines known as isotherms may be drawn as shown below. Each isotherm is drawn for the same mass of water.

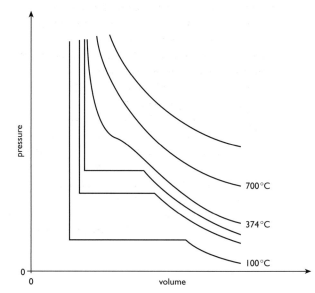

Answer the following questions.

a By reference to the passage:
  i State what is meant by
    1 'an isothermal change'
    2 'the vapour condenses' (3)
  ii Distinguish between an *unsaturated* vapour and a *saturated* vapour. (2)
  iii Distinguish between the behaviour of an unsaturated vapour and a saturated vapour when the volume is reduced at constant temperature. (2)
b State the feature of the line on the first figure which indicates that large changes in pressure are required to produce small changes in volume of a liquid. (1)
c In order to produce liquid by increasing the pressure alone, the vapour must be below a particular temperature which is different for different substances. This temperature is known as the critical temperature T_c of the substance. The table lists some substances and the corresponding critical temperatures, measured in kelvin.

Substance	T_c/K
hydrogen	33
nitrogen	126
oxygen	154
carbon dioxide	304
ammonia	406
sulphur dioxide	431
water	

Use the second figure to complete the table for water. (1)
d In early experiments to try to liquefy gases, increase in pressure alone was used. Gases which could not be liquefied at room temperature by pressure alone were known as *permanent* gases.
  i List the substances in the table which would have been known as permanent gases.
  ii Suggest, with a reason, which substance listed in the table proved to be most useful in early experimental work on the behaviour of a vapour near to its critical temperature. (4)
e Scientists investigating the behaviour of gases first used atmospheric air. As a result of their investigations, the gas laws were developed.
  i State the ideal gas equation.
  ii By reference to the figures and table, suggest why
    1 it was fortunate for the early investigators of gas laws that atmospheric air is composed mainly of nitrogen and oxygen
    2 when verifying the gas laws in a school laboratory, water vapour should be removed from the sample of air. (5)

OCR, Physics, Paper 2, Nov 1999

7 In this question we will make reference to a child's balloon filled with helium gas under various circumstances.

Assume such a balloon is spherical and is blown up from a helium cylinder such that the temperature inside the balloon finally reaches 25 °C and the diameter of the balloon reaches 38.0 cm. Take the temperature of the surrounding environment to be −5.0 °C and atmospheric pressure 102.5 kPa.

Remember that the volume of a sphere is given by $\frac{4}{3}\pi R^3$ and the surface area by $4\pi R^2$.

a The rate of heat conduction through the wall of the balloon is determined more by the thin air film attached to the rubber wall than by the rubber itself. Take the *total rate* at which heat is conducted through the balloon wall as 0.42 W m^{-2} K^{-1}.

Calculate the rate at which heat is initially conducted from the inside of the balloon out through the wall to the environment. (3)
b On axes of temperature against time *sketch* a graph to show how the temperature of the gas in the balloon would change with time as it cools down.

No calculations are necessary but indicate numerical values where possible. (2)
c Explain which heat transfer process is mainly responsible for getting heat from the gas *inside* the balloon to its *inner* surface. (2)
d As the balloon cools the pressure inside remains roughly constant at 5 kPa above atmospheric. When the balloon has cooled down to 0 °C, estimate the factor by which its volume has changed. (3)
e After a while, when the balloon has reached equilibrium at −5 °C, it is released and rises to a height where the atmospheric pressure is two thirds of that at ground level. Consider the following three possible, idealised, ways in which this could occur:
  A The temperature of the atmosphere remains constant at −5 °C and the balloon takes a long time to rise.
  B The temperature of the atmosphere remains constant at −5 °C and the balloon rises very quickly.
  C As the balloon rises and the pressure decreases, the temperature also decreases by just the right amount so that the volume of the balloon does not change.
  Answer the following questions:
  i Show these three processes on a *P–V* diagram and characterise them as being constant pressure, constant temperature, constant volume, adiabatic or isothermal. (5)
  ii For which of these would the work done by the gas in the balloon be greatest? Justify your answer. (2)
  iii For which of these would the work done by the gas in the balloon be least? Justify your answer. (1)

International Baccalaureate, Physics, Higher Level, May 1998 (part)

X
ADVANCED ELECTRICITY

35 Electrical conduction

THE BIG QUESTION

● What ideas does particle theory provide about electric current in different materials?

KEY VOCABULARY

action potential   conduction band   critical temperature   discharge lamps
drift velocity   energy gap   fluorescence   hole   ionic bond
ionic compound   ionic solution   magnetic resonance imaging (MRI)
negative gradient coefficient (NGC)   neuron   phosphor   plasma   resistivity
superconductivity   superfluidity   temperature coefficient of resistivity
valence band   valence electrons

BACKGROUND

In the 1830s, Michael Faraday experimented on electrical behaviour, and speculated on the relationship between these phenomena and the fundamental nature of matter:

'Although we know nothing of what an atom is, yet we cannot resist forming some idea of a small particle, which represents itself to the mind; and although we are . . . unable to say whether it is a particular matter or matters, or mere motion of ordinary matter, or some third kind of power or agent, yet there is an immensity of facts which justify us in believing that the atoms of matter are in some way endowed or associated with electrical powers, to which they owe their most striking qualities, and amongst them their mutual chemical affinity.'

Faraday was 'on the right lines' (that is, in agreement with present-day theory) by suggesting that electrical behaviour is related to the movement of particles of matter, and also that it is related to the forces that create chemical structures.

Faraday was an outstanding pioneer in work on both electrolysis and electromagnetism, and in the concept of the magnetic field in particular. Nowadays electrolysis of brine (salt water) is the starting point for many manufactured materials that we all use every day, and electromagnetic technologies provide us with electricity generators as well as electric motors.

Figure 35.1
Electromagnetic mobility – we all benefit from the magnetic effect of charged particles moving inside wires.

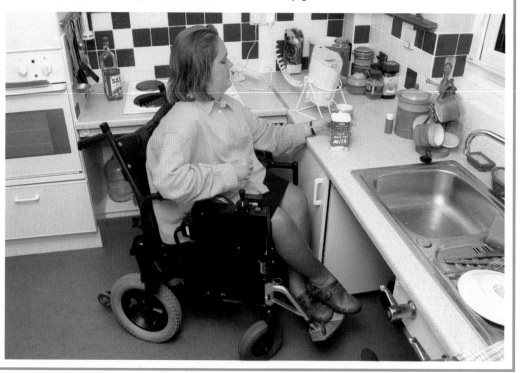

Particle beams and electric current in a vacuum

A beam of particles, whether a beam of electrons in a TV or a beam of protons in a huge particle accelerator, transfers charge and therefore can be thought of as carrying current. In the TV and the accelerator tunnel the current takes place in a vacuum. The presence of a gas would result in too many collisions between the beam of particles and the gas molecules, so that the beam would not be able to travel far. Any current is rate of flow of charge:

average I = rate of flow of charge = charge passing a fixed point per second

$$= \frac{Q}{t}$$

This formula gives average current when the current is unchanging. In most d.c. circuits we can assume current to be constant, but in more complex situations we may have to use the more general formula:

$$I = \frac{dQ}{dt}$$

In a TV, electrons travel from the electron gun to the screen. The charge on each electron is 1.6×10^{-19} C. The cathode emits about 5×10^{17} electrons per second. Thus the current is

$$I = \frac{Q}{t}$$

(the current is constant, so equal to the average current at all times)

with $Q = 5 \times 10^{17} \times 1.6 \times 10^{-19}$ C and $t = 1$ s, so

$$I = \frac{5 \times 10^{17} \times 1.6 \times 10^{19}}{1}$$

$$= 8 \times 10^{-2} \text{A}$$

$$= 80 \text{ mA}$$

1 In a TV, what is the current due to a single electron during its journey from electron gun to screen?

2 What mass of electrons hits a TV screen in each second? The electron mass (at rest) is 9.1×10^{-31} kg.

Conduction in gases

We think of air as a good insulator. But air can conduct if it contains free charge-carrying particles. Every physics teacher knows the frustrations of electrostatic experiments which work well on a day when the air is relatively dry, but don't work when the air is more humid. Droplets of water are often charged, and so if a plastic object is charged by friction then it will attract water droplets with the opposite charge, and the plastic and the water droplets tend to neutralise each other. Such a process generally involves very slow movement of small quantities of charge, and the currents involved are small.

Air can carry much larger currents if it contains free ions and electrons. This can be arranged by ionising atoms in the molecules of the air's gases, as in a spark. Some ionisation can also be achieved by heating the air, as in a flame. Thermal agitation then causes collisions of such intensity that electrons are set free from the atoms.

Temperatures in the Sun are so high that electrons and nuclei cannot form atoms – the material of the Sun is a **plasma**, a mix of positive ions and free electrons. The behaviour of a plasma is different from that of a non-ionised gas, and the plasma is sometimes called a fourth state of matter. A plasma, being composed of freely moving charged particles, is a good electrical conductor. A flame also has modest ability to conduct electric current due to ionisation and hence the presence of some free electrons and positive ions.

Conduction in liquids – ions and electrolysis

A proton is the same as a hydrogen ion, so a beam of protons is a beam of ions. Hydrogen ions are also on the move in the electrolysis of an acid such as sulphuric acid solution. Sulphuric acid is an **ionic compound**. In an ionic compound the forces that hold the solid crystal lattice together are forces between ions of opposite charge. The forces are called **ionic bonds**. In a solution of an ionic compound, an **ionic solution**, the ions become spread through the liquid. They are not bonded in molecules or in a crystal lattice but are free to move through the liquid – and so an ionic solution is a conductor of electricity (Figure 35.2).

Figure 35.2
As an ionic solution, sulphuric acid conducts electricity. It does so by movement of ions in the liquid, which results in movement of electrons within the metal electrodes. Combination of ions and electrons creates neutral atoms which are liberated from the solution.

At the anode, 'competition' between ions results in the sulphate ions, SO_4^{2-}, remaining in the solution. The hydroxyl ions, OH^-, give up their excess electrons to the anode:

$$OH^- \rightarrow OH + e^-$$

The result is the production of water and oxygen:

$$OH + OH \rightarrow H_2O + O$$

The oxygen atoms form pairs, or oxygen molecules:

$$O + O \rightarrow O_2$$

Bubbles of oxygen gas are released.

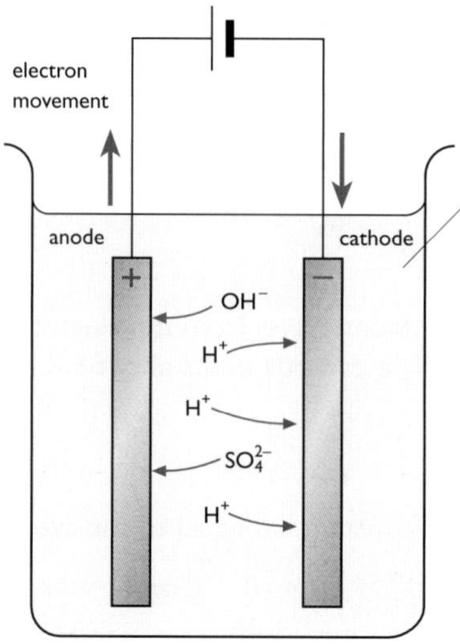

a solution of sulphuric acid contains positive and negative ions which are attracted to the opposite electrodes

At the cathode, hydrogen ions combine with electrons from the cathode:

$$H^+ + e^- \rightarrow H$$

And the hydrogen atoms form pairs, or hydrogen molecules:

$$H + H \rightarrow H_2$$

Bubbles of hydrogen gas are released.

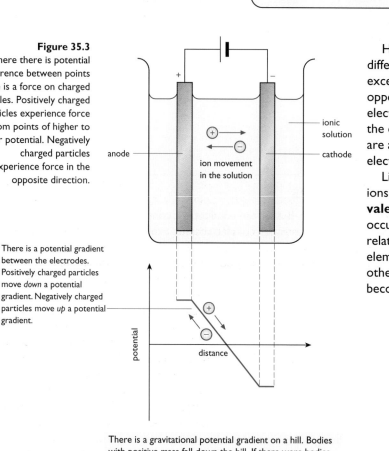

Figure 35.3
Where there is potential difference between points there is a force on charged particles. Positively charged particles experience force from points of higher to lower potential. Negatively charged particles experience force in the opposite direction.

There is a potential gradient between the electrodes. Positively charged particles move *down* a potential gradient. Negatively charged particles move *up* a potential gradient.

Hydrogen ions are influenced by potential differences in the same way as electrons, except that they experience force in the opposite direction. They are attracted to the electrode with the lower electrical potential – the cathode. Negative ions, just like electrons, are attracted to the electrode with the higher electric potential – the anode (Figure 35.3).

Like hydrogen, metals tend to form positive ions. Metal atoms have small numbers of **valence electrons** – electrons in their outer occupied energy levels – and these are relatively easily removed. Atoms of non-metal elements such as oxygen and chlorine, on the other hand, more easily gain electrons and become negative ions.

There is a gravitational potential gradient on a hill. Bodies with positive mass fall down the hill. If there were bodies with negative mass they would fall up the hill.

3 Are ions of the following elements more likely to be attracted to the anode or the cathode during electrolysis?
a chlorine
b copper
c hydrogen
d sodium

4 a During electrolysis of salt water (brine), Figure 35.4:
i where are electrons combined with positive ions to produce neutral atoms?
ii where are electrons liberated from atoms?
b Conduction in the power supply circuit is by flow of electrons. What are the charged particles that carry current through the brine? What is the direction of their flow, relative to that of the electrons?

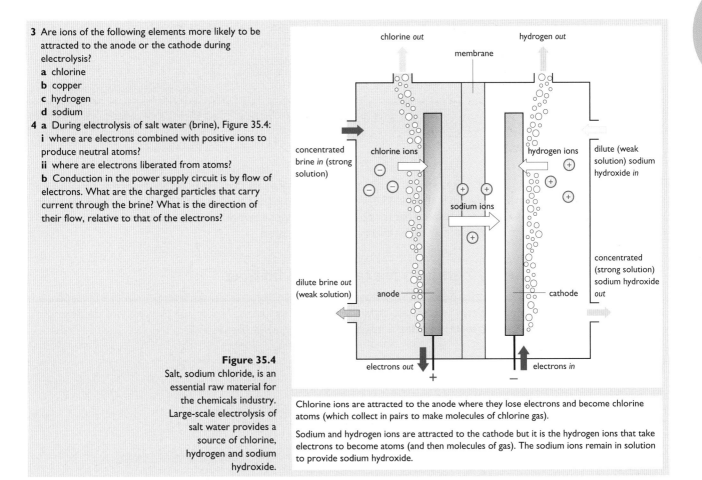

Figure 35.4
Salt, sodium chloride, is an essential raw material for the chemicals industry. Large-scale electrolysis of salt water provides a source of chlorine, hydrogen and sodium hydroxide.

Chlorine ions are attracted to the anode where they lose electrons and become chlorine atoms (which collect in pairs to make molecules of chlorine gas).

Sodium and hydrogen ions are attracted to the cathode but it is the hydrogen ions that take electrons to become atoms (and then molecules of gas). The sodium ions remain in solution to provide sodium hydroxide.

Conduction in metals

Electron energy levels and valence electrons

A solid metal can be thought of as a lattice of positive ions, each only able to vibrate about a fixed equilibrium position, surrounded by a 'gas' of electrons that are free to move within the lattice. The material as a whole is neutral.

If the metal is vaporised (as in a discharge lamp such as a yellow sodium street lamp) then the atoms are independent of each other. Each has an identical set of energy levels, between which electrons can move by the emission or absorption of electromagnetic radiation. Like balls in a bucket, electrons first fill the available space in which they have least potential energy. Electron numbers build up from the lower energy levels deeper in the atom.

When an atom is in its ground state the lower energy levels of each atom hold as many electrons as they are capable of holding, and there are higher energy levels that are completely empty. The outermost electrons are the ones that take part in ionisations and are involved with chemical changes. They are called the valence electrons. In lithium, the 'third' electron is its valence electron (Figure 35.5).

Figure 35.5
A *very simple* representation of a single lithium atom showing how its three electrons are distributed when it is in its ground state.

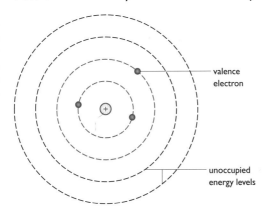

valence electron

unoccupied energy levels

Figure 35.6
Some of the energy levels available to the valence electron in a lithium atom.

In the ground state, the atom's other two electrons lie at much lower energy levels (not shown). These electrons play no part in the conduction process.

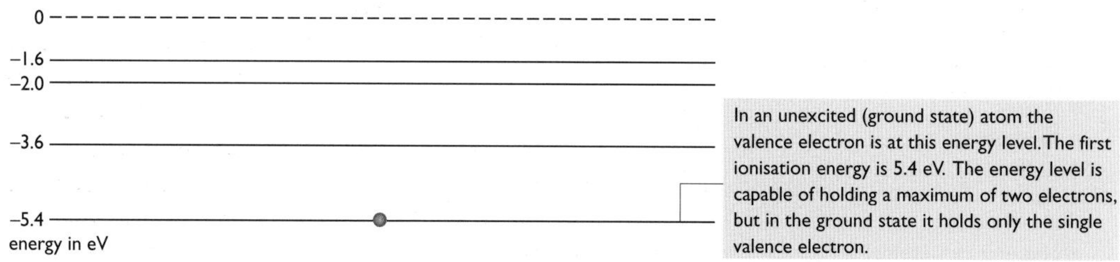

0 --

−1.6 --
−2.0 --

−3.6 --

−5.4 ------------------------------●------------------
energy in eV

In an unexcited (ground state) atom the valence electron is at this energy level. The first ionisation energy is 5.4 eV. The energy level is capable of holding a maximum of two electrons, but in the ground state it holds only the single valence electron.

In an excited atom it is usually the valence electrons that move further from the nucleus to a higher energy level (Figure 35.6). Valence electrons are the 'easiest' to remove from an atom to make it a positive ion. That is, it takes less energy to remove valence electrons than to remove electrons that are in lower energy levels of the atom.

Energy bands in metal crystal lattices

If two atoms are pushed very close together then, for every single energy level in an independent atom, the atom pair has *two slightly separated energy levels*. That is, the proximity of the atoms to each other seems to cause a 'splitting' of energy levels (Figure 35.7).

Figure 35.7
When atoms are close together the energy levels are split into multiple levels. A cluster of three lithium atoms, for example, has three closely spaced energy levels available for its three valence electrons (one valence electron from each atom).

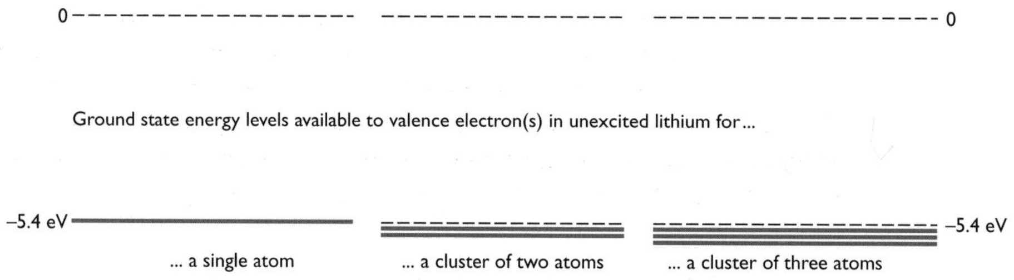

0 ------------------ ---------------- ------------------ 0

Ground state energy levels available to valence electron(s) in unexcited lithium for...

−5.4 eV ——————————   ================   ================ −5.4 eV

... a single atom   ... a cluster of two atoms   ... a cluster of three atoms

For very large numbers of atoms in a crystal this splitting produces bands of very many close energy levels (Figure 35.8). There is one energy level for each atom, so there are enormous numbers of levels in a band, and the energy difference between two levels is very small.

Figure 35.8
Splitting of an energy level as atoms come closer together creates an energy band in a crystal.

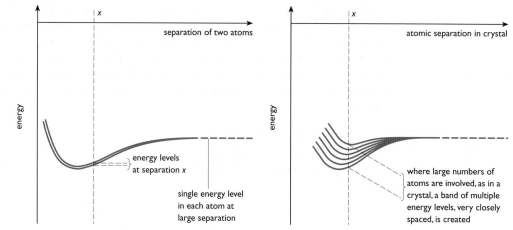

If there were no vacant energy levels then there would be nowhere for an electron to move to, except by changing places with another electron, a process which produces no net current. The valence electrons in a metal crystal exist in a band that is not completely full. There are vacant energy levels in the upper part of the band. An electron can move when there are vacant energy levels to move into (Figure 35.9).

Figure 35.9
In a metal there are plenty of vacant energy levels available and so valence electrons are free to move when an electric field is applied.

largely empty energy levels

largely full energy levels

Energy differences between one level and another within the same band are small, so an electron can move from the fully occupied lower energy levels of the band into the higher levels of the same band without a large 'quantum jump' of energy. There are plenty of unoccupied levels in the upper part of the band, so electrons in these higher levels are free to gain and lose energy from an applied electric field and to move through the metal without the need to change places with other electrons.

Electrons and thermal conduction

It is these 'free' electrons that are responsible for rapid transfer of energy through the metal when part of it is heated. The heating raises electrons to higher energy levels. They are free to move through the metal, taking their energy with them. Metals are good thermal conductors, as well as good electrical conductors, because of the presence and behaviour of free electrons.

Electron–ion collisions – heating effect and resistance

In a metal crystal, the mass movement of electrons that constitutes an electric current results in collisions between the electrons and the ions. That is, there is a transfer of energy from the electrons to the ions, resulting in increased thermal agitation of the ions. Electric current causes heating.

As the temperature rises so the level of vibration of ions in a metal crystal also increases. The probability of collisions between free electrons and ions is increased. This reduces the average distance travelled by each electron before transferring energy to ions and falling to a lower energy level where it is not free to move. It thus increases the electrical resistance of the sample of metal. (See also *Temperature coefficient of resistivity*, page 413.)

Insulators and semiconductors

Solid insulators, like metals, have energy bands that are made up of very large numbers of very similar but not identical energies. But, unlike metals, they do not have continuous bands that are partly full and partly empty. Instead they have bands that are completely full and *separate* bands of higher energy levels that are completely empty.

The lower, full band is called the **valence band** while the higher, empty band is the **conduction band**. There are no vacant energy levels in the valence band. Electrons in the valence band can move only by changing places with other electrons, which produces no net current. In the conduction band the energy levels are all vacant, but the valence band and the conduction band are separated by a large **energy gap** (Figure 35.10). The gap is too large for thermal processes or a moderate electrical field to provide enough energy to raise significant numbers of electrons to these levels, so there are never sufficient free electrons to carry current through the material.

Figure 35.10
The electrical behaviour of a material depends on the size of its energy gap.

In metals there is no energy gap at all, while in insulators the gap is too large for significant numbers of electrons to jump. Electrons in semiconductors can gain energy to cross the energy gap by heating the material or, sometimes, from incident light.

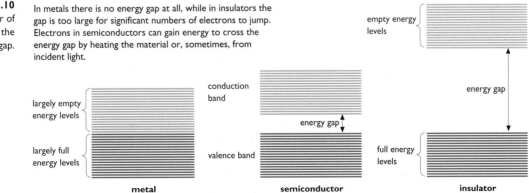

largely empty energy levels

largely full energy levels

metal

conduction band

valence band

energy gap

semiconductor

empty energy levels

energy gap

full energy levels

insulator

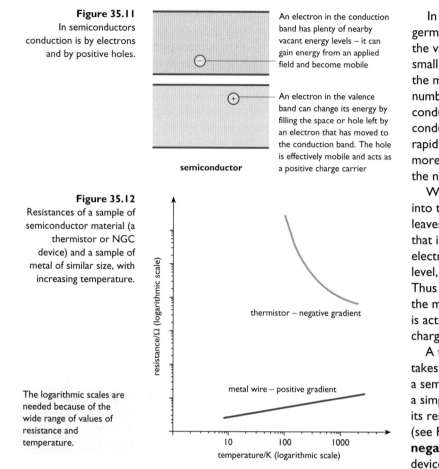

Figure 35.11
In semiconductors conduction is by electrons and by positive holes.

An electron in the conduction band has plenty of nearby vacant energy levels – it can gain energy from an applied field and become mobile

An electron in the valence band can change its energy by filling the space or hole left by an electron that has moved to the conduction band. The hole is effectively mobile and acts as a positive charge carrier

semiconductor

Figure 35.12
Resistances of a sample of semiconductor material (a thermistor or NGC device) and a sample of metal of similar size, with increasing temperature.

The logarithmic scales are needed because of the wide range of values of resistance and temperature.

thermistor – negative gradient

metal wire – positive gradient

resistance/Ω (logarithmic scale)

10   100   1000
temperature/K (logarithmic scale)

In semiconductors, like silicon and germanium, there is an energy gap between the valence and the conduction bands, but it is small (see Figure 35.10). Thermal excitation of the material is enough to allow significant numbers of electrons to cross the gap into the conduction band. For this reason, the conducting ability of a semiconductor increases rapidly as its temperature rises and more and more electrons become able to move without the need to change places.

When an electron jumps the energy gap into the conduction band of a semiconductor it leaves a vacancy behind; that is, an energy level that is no longer fully occupied. Another electron can move into this vacant energy level, with the effect that the vacancy moves. Thus the vacancy, or **hole**, can move around the material (Figure 35.11). When it does so it is acting as if it were a freely moving positive charge – it is a positive charge carrier.

A thermistor is a temperature sensor which takes advantage of the increased conduction in a semiconductor as the temperature rises. It is a simple piece of semiconductor material, and its resistance is highly temperature-dependent (see Figure 35.12). It is sometimes called a **negative gradient coefficient** (NGC) device.

5 **a** Describe the difference in patterns of electron energies in sodium vapour (as in a sodium lamp) and solid sodium.
   b When sodium vapour is subject to large potential difference the atoms become excited and then emit bright yellow light. What happens when a potential difference is applied to sodium metal?
6 Explain how electrons in the lower energy levels in a piece of semiconductor take part in carrying current.
7 Why is the ability of a semiconductor to carry current highly temperature-dependent?

Resistivity

Resistance is a property of a particular component in an electrical circuit – a length of wire, a piece of graphite, or a thermistor, for example. Resistance is dependent on the dimensions of the sample of material. **Resistivity** is not a property of a particular sample but a property of a material. It is the resistance of unit cross-sectional area per unit length. That is,

resistivity, ρ, of a material = resistance of unit cross-sectional area per unit length

$$\rho = \frac{RA}{l}$$

The unit of resistivity is the $\Omega\,m$.

The resistance of a sample depends on the resistivity of the material it is made of as well as on its dimensions. Note that the resistance of a sample of conductor, from rearrangement of the above, is given by:

$$R = \frac{\rho l}{A}$$

Temperature coefficient of resistivity

Over the range of temperatures normally found in a laboratory, the change in resistivity of a pure metal with change in temperature is approximately linear. That is,

$$\Delta\rho = \alpha\,\Delta T$$

The constant of proportionality, α, is the **temperature coefficient of resistivity** of the material:

$$\alpha = \frac{\Delta\rho}{\Delta T}$$

Some values of resistivity and temperature coefficients of resistivity are given in Table 35.1.

Table 35.1
The resistivity and temperature coefficient of resistivity of some conductors and semiconductors.

Material	Resistivity at 293 K / Ωm	Temperature coefficient of resistivity at 293 K / Ωm K^{-1}
aluminium	2.8×10^{-8}	3.9×10^{-3}
copper	1.7×10^{-8}	3.9×10^{-3}
iron	9.7×10^{-8}	5.0×10^{-3}
tungsten	5.6×10^{-8}	4.5×10^{-3}
graphite	4.0×10^{-5}	-0.5×10^{-3}
silicon	640	-75×10^{-3}

8 Calculate the expected resistance of an insulated copper wire 1 m long and with cross-sectional area 0.2 mm²
   a when immersed in melting ice
   b when immersed in boiling water
   (at atmospheric pressure in both cases).
9 A particular copper wire has a resistance of 1 Ω. What is the resistance of a tungsten wire of the same dimensions?
10 Sketch a graph of resistivity against temperature for copper, in the temperature range 250 K to 350 K, and give the value of the gradient.
11 What are the dimensions of the temperature coefficient of resistivity?
12 **a** Comment on the difference between the definition of temperature coefficient of resistivity and that of temperature coefficient of *resistance*, which can be expressed as

   temperature coefficient of resistance $= \dfrac{\Delta R}{\Delta T}$

   b If, for a pure metal, the temperature coefficient of resistivity is approximately constant over the temperature range 250 K to 350 K, can the same be said for the temperature coefficient of resistance of a wire made of this pure metal?

Note that the resistivity of a metal is significantly temperature-dependent, but less so than for a semiconductor. Also, the dependence is opposite. Metal resistivity increases with temperature while semiconductor resistivity decreases.

Superconductivity

Materials change their behaviours in some surprising ways when they are cooled to very low temperatures. At about 2 K, for example, helium (which is liquid at that temperature) can creep along surfaces. It can rise up the sides of a container and flow out of it, so that it looks as if the container is leaking. This behaviour is an example of **superfluidity**.

Figure 35.13
The resistivity of a metal (mercury in this example) drops to zero at very low temperature.

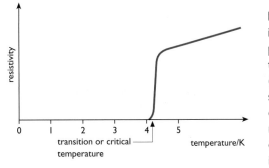

Metals also make sudden changes of behaviour at low temperatures. Most importantly, their resistivity drops to zero at a particular temperature known as the transition temperature or **critical temperature** (Figure 35.13). This phenomenon is called **superconductivity**. A superconducting wire can carry currents without resistance. There is no heating effect and no thermal transfer of energy to surroundings. Whether the current is large or small, the wire does not heat up.

Medical body scanners – **magnetic resonance imaging**, or MRI, devices – need very large currents to produce strong magnetic fields. Such currents would produce strong heating effects and wasteful dissipation of energy if they were to take place in wires at 'normal' temperatures. So their electromagnets are cooled, by immersing them in liquid helium, to a few kelvin so that they become superconductors.

There are potential applications for superconductivity in other devices, such as computers. Progress on this is limited by the need for very low temperatures. This requires cooling, such as by liquid helium, which must itself be cooled first of course, and it requires insulation to maintain the low temperatures. This is too expensive for practical application on a large scale. Technologists are now working on developing materials that become superconductors at higher temperatures – materials with high critical temperatures. The best that has been achieved so far is around 100 K, which is still a low temperature.

Drift velocity of electrons

The free electrons in a metal move randomly with speeds in the region of $10^5 \, \text{m s}^{-1}$. This is the thermal motion of the electrons – it contributes to the internal energy of the material and is dependent upon temperature. Note that, at 100 kilometres per second, this typical electron speed is quite high.

Figure 35.14
Electrons move up the potential gradient and gain kinetic energy.

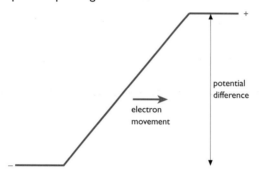

electron movement

potential difference

+

–

In a resistor, kinetic energy is transferred from the electrons by interactions with the ion crystal structure, and then transferred by thermal processes to the surroundings. Energy must be continually supplied (by the source of potential difference) to maintain conduction. The electrons reach an equilibrium, or steady, speed.

When a potential difference is applied to the metal it provides energy for mass movement of electrons along the potential gradient. This mass movement is in addition to the electron random movement, which continues but plays no part in the electric current. Remember that electrons naturally move *up* a potential gradient (Figure 35.14), while positively charged particles would fall down the gradient.

The electrons reach an average velocity of their mass movement at which the rate of supply of energy by the potential difference is equal to the rate at which work must be done to make progress. This equality creates an equilibrium – so that at a particular potential gradient a particular current flows in a particular conductor. This is analogous to the terminal velocity of a body such as a parachutist moving in a resistive medium. In that case the rate at which energy is provided by the gravitational field is equal to the rate at which work must be done to overcome resistance, so that the parachutist at terminal velocity does not gain or lose kinetic energy.

For electrons moving through a metal, their average velocity along the potential gradient is called the **drift velocity**. It is related to the steady current, I:

$$I = \frac{Q}{t}$$

where Q is the charge that passes a point in a circuit in time t. The charge is carried by electrons, each one carrying a charge e, which is $-1.6 \times 10^{-19} \, \text{C}$. The number of electrons per second passing a point in a circuit is the number of electrons per second that cross a particular cross-section, of area A, of the conductor (Figure 35.15).

Figure 35.15
The steady drift velocity of electrons is v, equal to x/t. It is useful to consider what happens when $t = 1$ s. A volume Ax of electrons passes through area A per second.

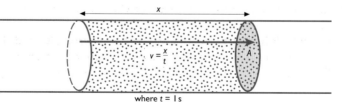

x

$v = \dfrac{x}{t}$

A

where $t = 1$ s

If we assume that all electrons have the same velocity – the drift velocity – then we can see that all of the electrons in a particular volume of conductor cross through area A in one second. The cross-sectional area of this volume is A and we can say that its length is x. Suppose that there are n free electrons in each unit of volume. (A 'unit' of volume is normally a cubic metre, even though that is a very large volume of conductor.)

$$n = \text{number of electrons per unit volume}$$

$$= \frac{\text{number of electrons in volume } V}{\text{volume } V}$$

If V is Ax, then

$$n = \frac{\text{number of electrons in volume } Ax}{Ax}$$

Therefore

$$\text{number of electrons in volume } Ax = nAx$$

The number of electrons in volume Ax is the same as the number of electrons passing through area A per second. But note that we should now add that drift velocity is related to x by $v = \frac{x}{t}$, where $t = 1$ s. So

$$\text{number of electrons crossing area } A \text{ per second} = nAx = nAv$$

The charge carried by the electrons across area A per second is the current, I, so

$$I = nAve$$

We can also write

$$v = \frac{I}{nAe}$$

Copper contains about 1.0×10^{29} free electrons per cubic metre. When a current of 1 A flows in a wire of cross-sectional area $7.5 \times 10^{-7} \text{m}^2$,

$$\text{drift velocity} = \frac{I}{nAe}$$

$$= \frac{1}{1.0 \times 10^{29} \times 7.5 \times 10^{-7} \times 1.6 \times 10^{-19}}$$

$$= \frac{1}{1.2 \times 10^4}$$

$$= 8.3 \times 10^{-5} \text{m s}^{-1}$$

Drift velocity is surprisingly small, but when a circuit switch is closed – as when you switch on an electric lamp – the effect is almost instantaneous throughout the circuit. This is because electric field spreads at the speed of light. So for example, when a simple torch is switched on, the negative terminal of the battery repels electrons, and the moving electrons repel those further along the conductor. Electrons throughout the circuit begin to move almost, but not quite, at the same instant.

In dealing with drift velocity we have assumed that current is steady in its size and its direction. If the applied potential difference varies sinusoidally, then the electrons will oscillate sinusoidally, and the current will also have a sinusoidal pattern (Figure 35.16).

Figure 35.16
Different patterns of applied potential difference result in different patterns of current and of electron drift velocity.

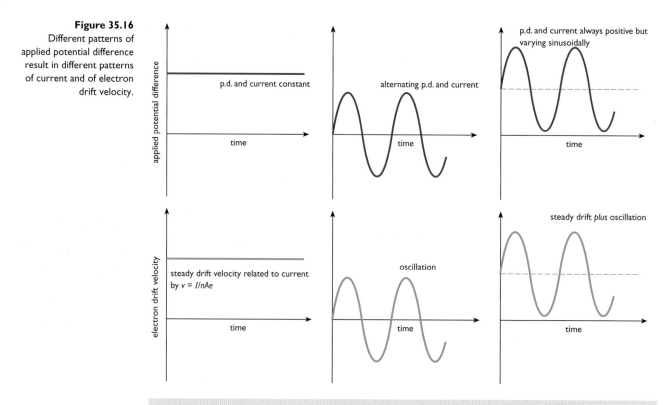

13 What is the ratio of typical thermal speed of electrons in a metal to typical thermal speed ('rms' speed) of molecules in the room you are in? (See Chapter 34.)

14 Assuming that the resistivity of iron varies uniformly with temperature, sketch a graph of its resistivity against temperature for the range 0 to 500 K.

15 If the current in a particular metal wire doubles, which of the following also doubles?
   a number of free electrons per unit volume
   b drift velocity

16 a How can the number of charge carriers per unit volume, n, in a sample of pure silicon be changed?
   b Why is it much easier to change n for silicon than for copper?

17 What is the drift velocity of electrons in a copper wire of radius 0.2 mm when the current is 0.5 A?

18 Compare the motion of electrons in a torch lamp filament and a mains kettle element.

19 What is a typical ratio of thermal speed of electrons in a wire to their drift velocity when carrying a current in the region of 1 A?

20 Sketch displacement–time graphs for electrons in
   a constant direct current
   b positive sinusoidal current
   c alternating current.

21 Why don't electrons all begin to move at *exactly* the same instant when a circuit switch is closed?

● **Comprehension and application**

Metal ions and the meaning of life

Metal ions carry positive charge, while non-metal ions are likely to have negative charge. Different metals have rather different electrical behaviour, due to the details of the electron structures of their atoms. Chemists know that they can list metals in order of their reactivity (their tendency to react). Metals like lithium and sodium are at one end of the list and non-reactive ones like tungsten and gold are at the other. Metallic sodium reacts violently with water, for example, while gold doesn't corrode and can be found in the ground as a metal, uncombined with other elements.

We have many useful applications of electrochemistry but nature, as always, got there first. Metal ions are vital in maintaining electrical equilibrium between living cells and the fluid around them. This controls the flow of material into and out of the cells. In the case of nerve cells (Figure 35.17), or **neurons**, it is the in-and-out movement of ions that is responsible for transmission of an impulse, called an **action potential**. As you read these words, such impulses are travelling into and within your brain. Neurons make contact with other neurons, by exchanges of chemicals that are stimulated by the electrical patterns. It is this hugely complex collection of impulses and neuron-to-neuron communications that provide what you call your vision and allow you to give meaning to these patterns of ink on a paper surface. Is your sense of self – your consciousness – just a matter of ion movements and chemical exchanges? That is an unresolved question, with many arguing that it is a question science can never answer.

Figure 35.17
Nerve action.

The transfer of charged particles through a 'membrane' is not just a matter of human industry. It is a process that exists in the natural world, for example through the membranes of the cells of your body, and is of special interest and importance in understanding how the nerves work. They are not like wires, carrying a continuous flow of charged particles, but transmit a travelling 'wave' of ion flow into and out of the nerve fibre through its surrounding membrane.

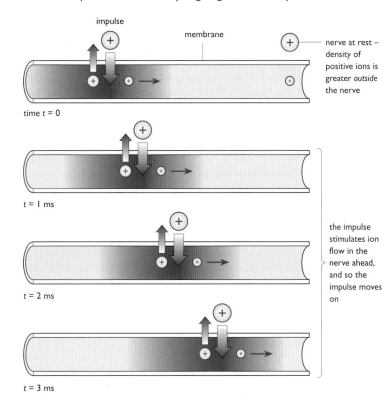

In the resting state, there are more positive ions in the fluid outside the nerve fibre than there are inside. The ability of the membrane at the end of a nerve fibre to allow different ions to pass though it is changed by a stimulus. There is then net inward movement of positive ions, followed by movement back out to restore the resting state. The impulse triggers ion flow in the next region of the nerve, so that the impulse moves along.

22 **a** Why do people and other animals take time to react to stimuli but when you switch a lamp on the effect seems to be instantaneous?
b Why is the concept of electron drift velocity of no relevance in considering nerve conduction?
23 The potential difference across the membrane of a resting neuron is about 70 mV and its thickness is about 10^{-7} m. Calculate the potential gradient and comment on the resistivity of the membrane.
24 By considering what happens to flow of charged particles as an impulse (action potential) passes along the nerve or neuron, describe what happens to the resistivity of the membrane at a point on a nerve fibre. In what ways is this behaviour very much more complex than flow of electrons in a wire?

25 Explain why a nerve cannot provide continuous stimulus to the brain but carries discrete signals.
26 MEG (magnetoencephalography) is a new method for detecting brain activity, based on detecting magnetic fields within the brain.
a Explain how magnetic fields arise from brain activity.
b Suggest why the detecting equipment must be extremely sensitive.
27 **DISCUSS**
Do you think that science
a can tell us nothing about human consciousness
b can provide us with relevant background to thinking about our sense of self
c will one day provide a complete physical explanation for self and consciousness?

● **Comprehension and application**

Discharge lamps

Excitation, ionisation and creation of a plasma can be achieved not only by thermal transfer of energy to a gas, but also by applying a large potential gradient (that is, a high potential difference per unit distance through the gas – this might also be described as a 'strong electric field'). This can do enough work on electrons to rip them from atoms and then to accelerate the pairs of ions and liberated electrons. The accelerated particles collide with atoms to produce more ionisation, and the result is an avalanche effect. Where this process is sudden and dramatic we see a spark.

In a fluorescent lamp, conduction by a gas is sustained and relatively stable. There are different kinds of fluorescent lamp but they all depend on ionisation of a gas and collisions between particles of the plasma that is produced. Some lamps use thermionic emission, and then collisions of the emitted electrons with particles of the gas in the tube create the plasma. Others use potential difference alone for this. The potential difference need not be very high – it can be mains 230 V, for example – if the pressure of the gas in the tube is suitable. Mercury can be used in such lamps as the gas at low pressure in the tube. Collisions excite and ionise the atoms. Electron transitions occur, and recombination of electrons and mercury ions also takes place. The result is the emission of photons of ultraviolet frequencies. These, of course, do nothing to help us to see, and exposure to them is unhealthy. So the inner surface of the lamp is coated with a **phosphor** – a material that absorbs the ultraviolet light and re-emits the energy at lower frequency. Different phosphors emit different frequencies, so for a fluorescent lamp we can use a mix of different phosphors to produce a mix of frequencies that our eyes and brains accept as white light. The action of a phosphor in absorbing light of higher frequency and emitting light of lower frequency is called **fluorescence**.

Neon lamps and sodium street lamps do not use a phosphor and do not rely on fluorescence. They are not fluorescent lamps, but they do rely on the same processes of excitation, ionisation and recombination. Together with fluorescent lamps they are called **discharge lamps** – all of which allow current to take place in a gas. Atoms at low pressure inside neon and sodium lamps are excited by an applied potential difference and emit their characteristic frequencies of light, producing red light in the case of neon and yellow light in the case of sodium.

Figure 35.18
Principle of a fluorescent discharge tube.

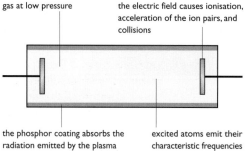

gas at low pressure

the electric field causes ionisation, acceleration of the ion pairs, and collisions

the phosphor coating absorbs the radiation emitted by the plasma and re-emits at lower frequencies

excited atoms emit their characteristic frequencies

28 a A lightning conductor on a tall building carries an average current, during a lightning strike, of 10 kA. If it transfers a charge of 10.0 C, how long did this take?

b A flash of lightning may be the result of a large potential difference between the ground and a cloud.
i What different mechanisms carry current through the air and through the metal lightning conductor?
ii What is the effect of the lightning strike on the potential difference?
c Compare a lightning strike to the processes taking place in a discharge lamp.

29 How would you expect the construction of a UV lamp to differ from that of an ordinary fluorescent lamp?

30 Explain the charge movements involved in the following processes:
a a comb is rubbed and gains a net negative charge
b the comb is held close to a small scrap of paper and attracts it
c the comb is held above a gentle flame and loses its ability to attract the paper
d operation of a mercury lamp.

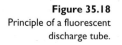

● **Extra skills task**   Information Technology and Application of Number

Create a database containing the resistivity and temperature coefficients of resistivity of different materials. Use a spreadsheet program to retrieve this information, and use it to calculate the resistance of different samples of the material. Where possible, extend your program to predict the changes in resistance of the samples as the temperature varies.

Examination questions

1 The figure illustrates a large conducting sphere on an insulating stand being charged by induction with a polythene strip.

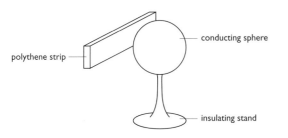

a Name a suitable material for
  i the conducting sphere
  ii the insulating stand. (2)

b The polythene strip is negatively charged and does not touch the sphere. The sphere becomes positively charged when it is earthed briefly.
  Explain why the sphere becomes positively charged. (3)

c The polythene strip is now removed. A small light conducting sphere, suspended from an insulating thread, is slowly brought close to the large sphere, as shown in the figure below.

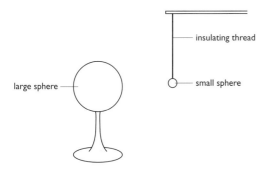

The small sphere eventually touches the large sphere.
  Describe and explain, in terms of charge, movement and electrical forces, what happens to the small sphere
  i as it approaches the large sphere (3)
  ii after it has touched the large sphere. (2)

OCR, Sciences, Further Physics, Mar 1999

2 a Explain, in terms of energy, what is meant by *a potential difference of 1 volt.* (1)

b The monitor of a computer makes use of an electron beam in a vacuum tube. The voltage accelerating the beam is 15 kV.

  charge of electron $= 1.6 \times 10^{-19}$ C
  mass of electron $= 9.1 \times 10^{-31}$ kg

  Calculate:
  i the energy given to each electron in the beam by the accelerating voltage (1)

ii the speed acquired by an electron. Assume that all the energy given to the electron appears as kinetic energy, and that the mass of the electron remains constant. (2)

c The current in the beam in **b** is 4.0 μA. Use your answer to **b ii** to calculate the number of electrons *per metre length* of the beam. (3)

d The drift velocity of electrons in a copper wire of the same cross-sectional area as the beam, carrying the same current as the beam, is 1.0×10^{-3} m s^{-1}. Without further calculation, state and explain how you would expect your answer to **c** to compare with the number of free electrons per metre length of the copper wire. (2)

OCR, Mechanics and Basic Electricity, June 1999

3 a A 100 W tungsten filament lamp operates from the 230 V mains. Calculate its resistance. (2)

b The drift speed of the electrons in the filament is much higher than the drift speed of electrons in the rest of the circuit. Suggest and explain a reason for this. (4)

Edexcel (London), Physics, PH1, June 1999

4 A heating coil is constructed from wire having a cross sectional area of 1.8×10^{-7} m^2 and 3.0×10^{29} free electrons per m^3. The current in the wire is 7.2 A. When the potential difference across the coil is 12 V, the temperature of the wire is such that it glows red.

a Calculate the mean drift velocity of electrons in the wire.

  charge on an electron, $e = -1.6 \times 10^{-19}$ C (3)

b The current in the wire is increased so that the temperature of the wire is raised. State and explain any changes you would expect to see in the appearance of the wire when the current is maintained at a significantly higher value. (4)

AQA (AEB), Physics, Paper 1, June 1999

5 a i Explain what is meant by an *electric current.*
  ii Explain why some solids are electrical conductors and some are insulators.
  iii Describe electrical conduction in a metal. (3)

b Explain why it is difficult to quote a value for the resistance of a filament lamp. (1)

c In a gas, conduction occurs as a result of negative particles flowing one way and positive particles flowing in the opposite direction, as illustrated below.

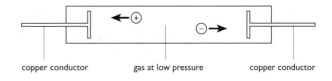

copper conductor   gas at low pressure   copper conductor

In this case, the copper conductors to the gas carry a current of 0.28 mA. The number of negative particles passing any point in the gas per unit time is $1.56 \times 10^{15}\,s^{-1}$ and the charge on each negative particle is $-1.60 \times 10^{-19}\,C$.

Calculate

i the negative charge flowing past any point in the gas per second

ii the positive charge flowing past any point in the gas per second

iii the number of positively charged particles passing any point in the gas per second, given that the charge on each positive particle is $+3.20 \times 10^{-19}\,C$. (6)

d By considering the significant figures available, explain why your answers to **c ii** and **iii** are unreliable. (2)

e In one practical example of electrical conduction of particles through a gas, a magnetic field is applied across the gas. Instead of travelling directly along the tube the positive and negative charges travel along different paths, as shown below.

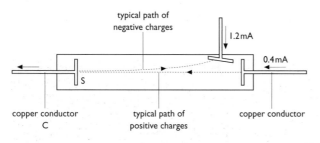

i State the direction of the applied magnetic field. Explain how you obtained your answer.

ii How is it possible for the positive particles still to travel in an apparently straight line?

iii State Kirchhoff's first law and use it to find the current in conductor C.

iv Suggest what is happening to the particles at surface S of the copper conductor C. (8)

OCR, Paper 3, Nov 1999

6 a State a formula for the resistance R of a sample of material in terms of the resistivity ρ. Explain any other symbols you use. (2)

b A student wishes to determine the thickness of a pencil line drawn on a sheet of paper. The line has length 9.0 cm and width 0.12 cm. The graphite of the pencil line has resistivity $8.0 \times 10^{-8}\,\Omega\,m$. Electrical connections are made to the ends of the pencil line.

i Complete the electrical circuit shown in the figure by including a voltmeter and an ammeter so that measurements may be made to determine the resistance of the pencil line.

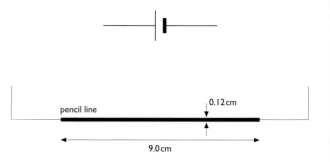

ii The voltmeter reads 1.4 V and the corresponding reading on the ammeter is $4.9 \times 10^{-3}\,A$. Calculate, for the pencil line,

1 the resistance

2 the thickness. (6)

OCR, Sciences, Physics, Foundation, June 1999

7 a Describe and explain, in terms of a simple energy band model, the essential differences between an insulator such as mica and a metallic conductor such as copper. (4)

b State and explain, in terms of a simple energy band model, how the conductivity of a semiconductor changes when its temperature is increased. (3)

c The cable shown in the diagram is used to transmit electricity and is made from strands of steel wire and strands of aluminium wire. The strands of wire are in electrical contact with each other along the length of the cable.

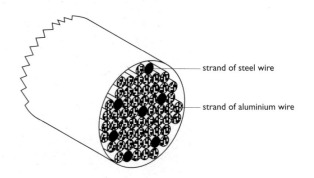

i Calculate the resistance of one strand of aluminium wire with a diameter of 3.2 mm and a length of 1.0 km.

ii The resistance of one strand of steel wire in a 1.0 km length of cable is 19.9 Ω. Calculate the resistance of 1.0 km of the cable made up of seven strands of steel wire and fifty four strands of aluminium wire. (5)

AQA (NEAB), Physics Advanced, Paper 1, Section B, June 1999

36 Circuit quantities

THE BIG QUESTIONS
● What universal patterns of behaviour can we see in the current and potential difference in all kinds of circuits?
● What do these patterns tell us about combinations of resistors in series and parallel, and what uses are there for some of these combinations?
● Can we be confident that a reading on a meter is a reliable indication of what is happening in a circuit?

KEY VOCABULARY
deflection   input transducer   Kirchhoff's First Law   Kirchhoff's Second Law
load circuit   load resistor   output voltage   potential divider
radial magnetic field   sink   strain gauge   time base   Wheatstone bridge
x-plates   y-plates

BACKGROUND
We can communicate through words and other signs, but with the aid of electrical devices one human mind has access to activity in another. The electrodes on the man's head (Figure 36.1) are used to detect voltage variations which are displayed as visual patterns on the screen. The process is called electroencephalography (or EEG) and is an example of sophisticated electrical technology.

Figure 36.1
Electroencephalography.

421

Units of current and charge

The amp, or ampere, is defined as

> the current flowing in each of two infinitely long straight parallel fine wires placed one metre apart in a vacuum when the magnetic force between each metre length of the wires is 2×10^{-7} N.

As definitions go it's fairly long. It might seem simpler to define the amp as the current that corresponds to a movement of one coulomb of charge in each second. (That's a movement of 6.28×10^{18} electrons per second, each electron carrying a charge of -1.6×10^{-19} C.) But the problem with using the coulomb as a starting point for defining the amp is that electrons are too small to count. The charge flowing in a circuit of wires is detectable only by its heating effect and by the magnetic field that surrounds it as it flows. That is, charge is observable through the effects of current. So we start with current and a phenomenon that we *can* detect. We measure the magnetic effect of a current, and we say that a certain level of effect in a certain arrangement corresponds to one unit of current, and then we give this unit the name of amp. We then define the coulomb from the amp. The coulomb is defined as the charge moved when a current of one amp flows for one second.

Patterns of current

In a simple d.c. circuit made of metal wires, in which electrons are the charge carriers, the rate at which electrons arrive at the positive terminal of the power supply must be the same as that at which they leave the negative terminal. Otherwise there would have to be a source or **sink** (the opposite of a source) of electrons somewhere in the external circuit (the circuit outside the battery) (Figure 36.2).

Figure 36.2
For the external circuit, a battery is a source and a sink of electrons.

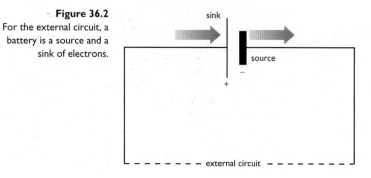

No point in an external circuit acts as a net source or sink of charged particles. So if we choose any point in an external circuit, the charge arriving in any period of time is the same as the charge leaving. Put differently, the current into any point in a circuit is the same as the current out. This is a statement of **Kirchhoff's First Law**. Applying this to every point in a simple series circuit, the current must be the same at *all* points in the circuit (Figure 36.3).

Figure 36.3
Kirchhoff's First Law applied to a series circuit.

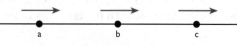

The current into any point in a circuit is the same as the current out:

$I_{in} = I_{out}$

This means that the current is the same at every point in a series circuit:

$I_a = I_b = I_c$ = circuit current, I

At a junction in a circuit, Kirchhoff's First Law still applies. The current may separate into the different available routes, or current from different routes may rejoin. But the current leaving the junction is always the same as the current arriving (Figure 36.4).

Figure 36.4
Kirchhoff's First Law applied to a junction between parallel branches.

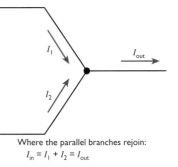

The current into this junction is the same as the total current in the parallel branches:
$$I_{in} = I_{out} = I_1 + I_2$$

Where the parallel branches rejoin:
$$I_{in} = I_1 + I_2 = I_{out}$$

---

1 Three parallel branches carrying currents of 0.05 A, 0.23 A and 0.28 A meet at a junction and the current then passes along a single wire. What is the value of the current flowing away from the junction in a single wire?
2 Explain, in terms of electrons, why the current leaving a circuit junction must be the same as the current arriving.
3 If the current in a torch lamp is 0.8 A, how many electrons
   a enter it
   b leave it
   every second?

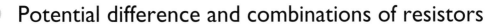

Potential difference and combinations of resistors

Note that it is meaningless to talk about potential difference at a single point. There have to be two points for there to be a difference. In following a complete circuit from any point within it and back to the same point we are bound to find that total potential difference is zero, however many different components exist in the circuit. We can say that the sum of the potential differences around any closed circuit must be zero. This statement is **Kirchhoff's Second Law**.

A given quantity of charge has more electrical potential energy when at one terminal of a battery than when it is at the other. As the charge flows through the circuit from one terminal to the other it transfers this energy, heating its surroundings or doing work through electromagnetic effects.

In this chapter we are considering circuits that have resistance only and do not transfer energy by doing work. If the resistance of the circuit is uniform all the way around the circuit then the energy is transferred uniformly to the surroundings. A uniform potential gradient exists around the whole circuit. If, however, much of the wire has negligibly low resistance so that no energy transfer takes place here, and only a given section has large resistance, the energy will be transferred in this part of the circuit – the resistor. The quantity of charge has much more potential energy when it is on one side of the resistor than on the other. There is a potential difference across the resistor. This matches the potential difference across the terminals of the battery (Figure 36.5a, overleaf).

Where there are just two resistors in series, each one transfers energy. The quantity of charge experiences a potential energy difference between the ends of both resistors. Each resistor has a potential difference between its ends. The *total* potential difference for the two resistors matches the potential difference between the terminals of the battery (Figure 36.5b, overleaf).

If the two resistors are in parallel, they are both connected directly to the battery so both have the same difference in potential as the battery does (Figure 36.5c, overleaf). This must be true, even if they have very different resistance.

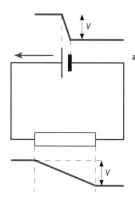

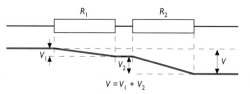

a An imaginary positive charge carrier travelling around the complete circuit, including through the battery, travels *down* a potential gradient as it passes through the resistor, but gains energy and travels *up* a potential gradient as it passes through the battery. The sum of the potential differences is zero. This is Kirchhoff's Second Law.

b For resistors in series, the total potential difference is the sum of the individual potential differences.

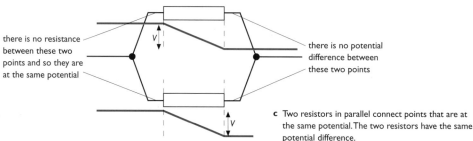

there is no resistance between these two points and so they are at the same potential

there is no potential difference between these two points

Figure 36.5
Kirchhoff's Second Law applied to resistors in a circuit.

c Two resistors in parallel connect points that are at the same potential. The two resistors have the same potential difference.

Patterns of potential difference for resistors in series

In a simple circuit with one battery and one resistor, the potential difference that is supplied by the battery is applied to the resistor. That is, the difference in potential across the battery terminals is the same size as the difference in potential across the resistor.

Where an external circuit consists of several resistors in series, the total potential difference across the terminals of the battery is applied across the ends of the group of resistors. Each resistor has its own potential difference, sometimes called potential 'drop', and its own potential gradient. The individual potential differences add up to equal the total, as they must do if they are to obey Kirchhoff's Second Law.

A key question now is – how big is the potential drop for each of the series resistors? Not surprisingly, it depends on its resistance. Consider the three resistors in Figure 36.6. We can call their resistances R_1, R_2 and R_3. The potential differences are V_1, V_2 and V_3, and the potential difference supplied by the battery can be just V.

Figure 36.6
Just as a single resistor in a simple circuit produces a potential difference which could be called a 'step' or 'drop', each of the series resistors has a potential drop. The sum of each individual drop is equal to the total potential difference across the battery.

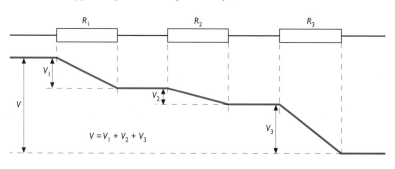

We know, from considering potential drops (or differences), that

$$V = V_1 + V_2 + V_3 \qquad (1)$$

We also know that all three resistors carry the same current, since current is always the same at all points in a series circuit. So we can say that

$$V_1 = IR_1 \qquad (2)$$
$$V_2 = IR_2 \qquad (3)$$
$$V_3 = IR_3 \qquad (4)$$

4 A 2 Ω and a 4 Ω resistor are connected in series, with a total potential difference of 3 V. Each resistor has a potential difference across it. Which one has the bigger potential difference? Can you say how big the two potential differences are?

Total or combined resistance of resistors in series

Suppose that we were to replace the three resistors of Figure 36.6 with a single resistor, R, which had exactly the same effect on the total current in the circuit as our three resistors. We could call R the 'combined' resistance, and we could say

$$R = \frac{V}{I} \qquad \text{or} \qquad V = IR \qquad\qquad (5)$$

R is the total circuit resistance, V the total applied potential difference, and I the current. Looking back at the equations at the foot of the previous page, we can substitute (2), (3), (4) and (5) into (1), to give:

$$IR = IR_1 + IR_2 + IR_3$$
$$= I(R_1 + R_2 + R_3)$$

Dividing both sides of this equation by I,

$$R = R_1 + R_2 + R_3$$

The total value of resistance of resistors in series is thus a simple sum of their individual resistances.

5 In this 'substitution' process, what was substituted for V_2?

Total or combined resistance of resistors in parallel

Three resistors in parallel have identical potential drops or differences, which we can call V.

If three resistors in parallel have different values, R_1, R_2 and R_3 (Figure 36.7) then they will carry different currents I_1, I_2 and I_3. And since current leaving a junction is always the same as the current arriving, we know that I_1, I_2 and I_3 add up to equal the total circuit current, I.

Figure 36.7
Even though they may have different resistances, R_1, R_2 and R_3, resistors in parallel must each have the same potential difference across them because each one is connected to the same two points.

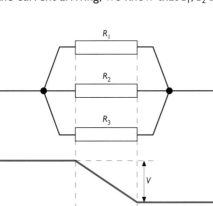

That is,

$$I = I_1 + I_2 + I_3$$

Since $I = V/R$, $I_1 = V/R_1$, $I_2 = V/R_2$ and $I_3 = V/R_3$, we can rewrite the above equation as:

$$\frac{V}{R} = \frac{V}{R_1} + \frac{V}{R_2} + \frac{V}{R_3}$$
$$= V\left(\frac{1}{R_1} + \frac{1}{R_2} + \frac{1}{R_3}\right)$$

which we can simplify to:

$$\frac{1}{R} = \frac{1}{R_1} + \frac{1}{R_2} + \frac{1}{R_3}$$

6 In rewriting $I = I_1 + I_2 + I_3$, what replaced I_3?

So for resistors in parallel the combined resistance, R, is related to the individual resistances by a more complex relationship than applies to resistors in series.

The moving coil current meter and its sensitivity

A moving coil meter responds to the magnetic effect of current. Current in the coil in the meter results in a magnetic field around it, and the coil also lies in the field of a small permanent magnet. The two fields interact, so that a force exists between the coil and the magnet. In most cases, the permanent magnet is fixed to the structure of the meter while the coil is pivoted and is able to turn.

The magnetic field provided by the permanent magnet is **radial**. That is, the field lines in the region of the coil lie along radii with their centre at the coil's axis. This means that as the coil turns the strength of the field it lies in, due to the permanent magnet, stays the same.

The moment experienced by the coil is then proportional to the current that flows. But if the coil is suspended freely by its axis then even a small current will cause it to turn, so that on its own the moving coil says very little of practical use about the size of the current. A spiral spring is attached to the coil to oppose its turning – that is, to provide a restoring moment (Figure 36.8).

If the spring obeys a rotational version of Hooke's Law – that the angle through which it is turned is proportional to the moment that is applied to it – then the total angle, θ, through which the coil turns is proportional to the current. A needle is fixed to the coil, and it experiences a **deflection** through an angle θ. The constant of proportionality is the sensitivity, S, of the meter:

$$\theta \propto I$$
$$\theta = SI$$
$$S = \frac{\theta}{I}$$

A high sensitivity corresponds to a large ratio of deflection, θ, to current, I (Figure 36.9). The unit of sensitivity is the radian per amp.

Figure 36.8 (left)
The coil rests at the position at which the moment due to the magnetic force is in balance with the restoring moment due to the spring.

Figure 36.9 (right)
Graph showing sensitivity of a moving coil meter.

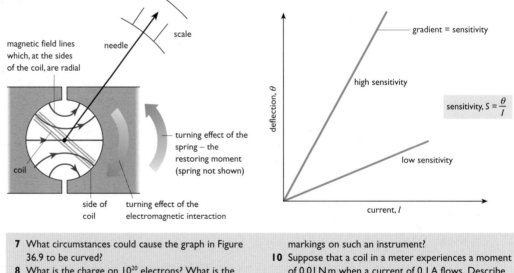

7 What circumstances could cause the graph in Figure 36.9 to be curved?

8 What is the charge on 10^{20} electrons? What is the current if 10^{20} electrons flow through the coil of a meter per second?

9 a Calculate the sensitivity of an instrument, in radian A^{-1}, that experiences a 0.18 radian deflection for a current of 20 mA.
   b What will be the angular separation of 2 mA scale markings on such an instrument?

10 Suppose that a coil in a meter experiences a moment of 0.01 N m when a current of 0.1 A flows. Describe what will happen:
   a if no spring is fitted to the coil
   b if a spring is fitted to the coil.

11 A circuit with resistance 10 Ω carries a current of 1.2 A. What current will it carry when an ammeter of resistance 2.8 Ω is added?

Effect of an ammeter on the current it measures

In order to respond to the magnetic effect of current, the coil must carry the current to be measured. Inevitably, the coil does not have zero resistance. The result is that the ammeter has an influence on the quantity it is measuring. This is rather like a ruler that has the magical property of changing the length of an object as it is brought up to it. The ruler, of course, is a passive measuring device and this does not happen. Unfortunately we have to accept that an ammeter is not a perfectly passive component of a circuit.

Changing ammeter sensitivity and range – calculation of shunt values

There is an inverse relationship between sensitivity of a meter and the range of values it can measure. A sensitive meter will experience a large deflection for a small current, and hence experiences full scale deflection of the needle for a small current. So a sensitive meter is unable to measure very large currents and has a relatively small range.

Figure 36.10
The connection of a shunt.

The meter coil and the shunt are resistors in parallel and they have the same potential difference, V.

$$V = I_c R_c \qquad V = I_s R_s$$

so

$$I_c R_c = I_s R_s$$

and

$$\frac{I_c}{I_s} = \frac{R_s}{R_c}$$

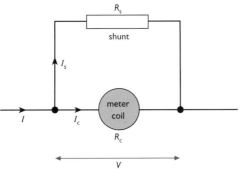

Figure 36.11
A microammeter with shunts to alter its range. A shunt is connected in parallel with the meter coil, so that coil and shunt carry fixed proportions of the total current.

Sensitivity and range can be changed by changing the properties of the spring that provides the restoring moment. A 'weaker' spring produces greater sensitivity and smaller range.

An alternative way to change the sensitivity and range of a meter is to arrange for only a fixed proportion of the circuit current to flow through the coil. The remaining current then 'by-passes' the coil. The ammeter is fitted with a by-pass or shunt resistor, in parallel with the coil, with a resistance that has a known ratio to that of the coil itself (Figures 36.10 and 36.11).

If the resistance of the coil and its connections is R_c and shunt resistance is R_s, then the ratios of the currents passing through the two routes is given by

$$\frac{I_c}{I_s} = \frac{R_s}{R_c} \quad \text{(total circuit current, } I = I_c + I_s)$$

If $R_c \gg R_s$ then $I_s \gg I_c$, and since only a small proportion of the total circuit current, I, passes through the coil, the sensitivity of the meter is much lower than it would be without the shunt, and the meter's range is correspondingly increased.

For example, suppose that a meter has a coil resistance of $2\,\Omega$ and reads currents up to $1\,\text{mA}$ with no shunt attached. It is then required to have an increased range to read currents up to $1\,\text{A}$.

$I_c = 0.001\,\text{A}, \ I_s = 1 - 0.001 = 0.999\,\text{A}$. So, using the formula above,

$$\frac{0.001}{0.999} = \frac{R_s}{2}$$

giving $\qquad R_s = 0.002\,\Omega$

R_s is a small resistance, and it is quite difficult to make resistors with a resistance precise to three decimal places. Note that the larger the coil resistance, the larger the shunt resistance and the easier it is to provide a suitable resistance. So large coil resistance, though it has a large effect on the current being measured when the meter is used without any shunt, has the benefit of allowing use of shunts without excessive loss of accuracy.

The ammeter as a whole now consists of the coil and its shunt. Total meter resistance, R, is given by

$$\frac{1}{R} = \frac{1}{R_c} + \frac{1}{R_s}$$

In the example above,

$$\frac{1}{R} = \frac{1}{2} + \frac{1}{0.002} = \frac{1}{2} + \frac{1000}{2} = \frac{1001}{2}$$

$$R = \frac{2}{1001} = 1.99 \times 10^{-3}\,\Omega$$

12 What is the ratio of the sensitivities of the meter in the sample figures given in the text, before and after the shunt is fitted?

13 a A $20\,\Omega$ microammeter has full scale deflection with a current of $100\,\mu\text{A}$. So that it can be used to give a full scale deflection in a circuit carrying $5\,\text{A}$, what value of shunt should be fitted?
b With the shunt fitted and the meter now showing a reading of $5\,\text{A}$, what current is actually flowing through the coil?

The moving coil meter as a voltmeter – calculation of multiplier values

There is a major difference between the principle of measuring current and measuring potential difference, or voltage. An ammeter must be placed into the circuit so that the current, or a known proportion of it, flows through the meter. A voltmeter, on the other hand, measures a difference between the potentials at two points in the circuit. It therefore must be connected to these two points – or 'across' the two points.

The same moving coil meter that is used for current measurements can be used to measure voltage. The deflection of the needle is an indication of magnetic force and hence of current. But the potential difference across two points and the current are proportional to each other as long as the resistance between the two points is constant. Use of the moving coil meter as a voltmeter takes advantage of this proportionality.

It is important that the voltmeter has as little impact as possible on what is happening in the circuit. Current must continue to flow through the circuit between the points to which the voltmeter is connected and not all be diverted through the voltmeter. (It is also important that excessive current through the coil of the meter doesn't cause a tendency for deflection beyond full scale, and cause the coil to overheat and melt.)

To act as a voltmeter, a moving coil meter has a large resistor, called a multiplier, connected in series with it (Figure 36.12). This limits the size of the current passing through the meter.

Figure 36.12
The connection of a multiplier to create a voltmeter.

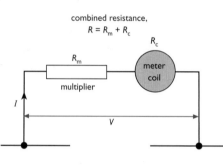

In use as a voltmeter, the moving coil meter provides a reading of voltage by showing the deflection of the needle that corresponds to a current in the coil. A high resistance multiplier ensures that only an appropriately small current passes through the meter.

The size of the resistance of the multiplier, relative to the resistance of the moving coil meter itself, determines the range of the voltmeter. Suppose that a moving coil meter when in use without a shunt or multiplier shows a full scale deflection with a current of 1 mA, and the meter (its coil and connections) has a resistance of $5\,\Omega$. To have the same full scale deflection when, with a multiplier fitted, it is connected to two points with a potential difference of 10 V, the *total* resistance of the meter and multiplier must be given by

$$R = \frac{V}{I} = \frac{10}{0.001} = 10\,000\,\Omega$$

14 What value of multiplier must be fitted to a meter of resistance $20\,\Omega$ and full scale deflection at $200\,\mu A$ in order for it to be used as a voltmeter reading up to 10 V?

15 A voltmeter consists of a meter, of 1 mA full scale deflection and coil resistance $12\,\Omega$, together with a $1000\,\Omega$ multiplier.
   a What voltage will produce full scale deflection?
   b Someone disconnects the voltmeter from the circuit, removes the multiplier and reconnects the voltmeter to the same points in the circuit. What happens next? Why?

16 A $270\,\Omega$ resistor is connected to a cell of e.m.f. 6.0 V and internal resistance $30\,\Omega$.
   a What current flows in the circuit? (Use $E = IR + Ir$.)
   b What is the potential difference across the cell terminals?
   c A $2430\,\Omega$ voltmeter is connected across the resistor. What is the total resistance of the external circuit?
   d What is the new current?
   e What does the voltmeter read?
   f Comment on the two potential differences you have calculated.

But $R = R_c + R_m$, where R_c is the resistance of the coil and its connections, and R_m is the multiplier resistance. So

$$R_m = R - R_c = 10\,000 - 5 = 9995\,\Omega$$

In a general algebraic form, multiplier resistance is given by:

$$R_m = \frac{V}{I} - R_c$$

where V is the voltage required to produce full scale deflection, and I is the current in the meter at full scale deflection.

The cathode ray oscilloscope for measuring potential difference

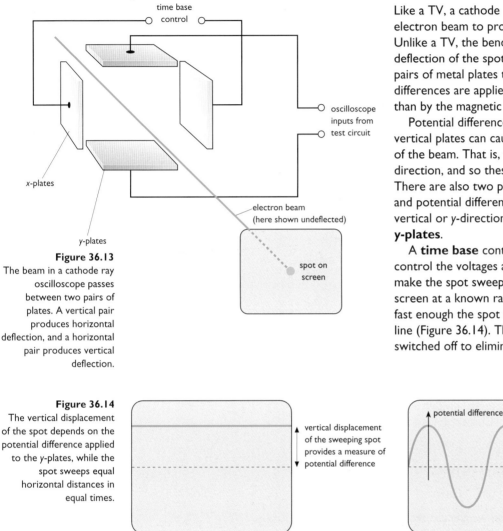

Figure 36.13
The beam in a cathode ray oscilloscope passes between two pairs of plates. A vertical pair produces horizontal deflection, and a horizontal pair produces vertical deflection.

Like a TV, a cathode ray oscilloscope uses an electron beam to produce a spot on a screen. Unlike a TV, the bending of the beam and the deflection of the spot are achieved by two pairs of metal plates to which potential differences are applied (Figure 36.13), rather than by the magnetic fields of coils.

Potential difference applied to a pair of vertical plates can cause horizontal deflection of the beam. That is, the deflection is in the *x*-direction, and so these are called the **x-plates**. There are also two plates fixed horizontally and potential difference applied to these causes vertical or *y*-direction deflection. These are the **y-plates**.

A **time base** control can be adjusted to control the voltages applied to the *x*-plates to make the spot sweep horizontally across the screen at a known rate, and when the sweep is fast enough the spot appears as a continuous line (Figure 36.14). The time base can also be switched off to eliminate horizontal motion.

Figure 36.14
The vertical displacement of the spot depends on the potential difference applied to the *y*-plates, while the spot sweeps equal horizontal distances in equal times.

17 a How does the method of deflection of the electron beam in a cathode ray oscilloscope differ from that in a TV?
b Why is a moving coil meter incapable of measuring battery e.m.f.?

18 With a sensitivity setting of 5 V cm⁻¹, what is the potential difference corresponding to a spot deflection of 2.45 cm?

The two input terminals of the oscilloscope are connected to the *y*-plates. The result is that the vertical deflection of the spot is dependent on the potential difference applied to these terminals. This makes the oscilloscope a device for measuring potential difference. With the time base switched on and no p.d. (or a steady p.d.) applied to the input terminals the screen shows a horizontally moving spot or a continuous straight line. If the input p.d. varies, the spot moves up and down accordingly, so that the screen can display a voltage–time graph.

To measure a steady potential difference it doesn't matter whether the time base is on or not – it depends on your preference for a spot or a line. The sensitivity of the oscilloscope, in terms of volts per centimetre of screen, can be adjusted. At 2 V cm⁻¹, for example, a 1 cm vertical deflection of the beam represents an input potential difference of 2 V.

No current need flow between the *y*-plates of an oscilloscope for deflection of the electron beam to take place. The resistance of the oscilloscope can be considered to be infinite. This means that, unlike a moving coil meter which does not have infinite resistance, it can be connected across the terminals of a cell and the reading it produces can be taken to be the cell's e.m.f. In general, an oscilloscope disturbs the circuit it is measuring less than a moving coil meter.

The potential divider

Two resistances, R_1 and R_2, in series in a circuit will each have a potential difference across them, and these two voltages will add together to give a combined or total potential difference.

$$V = V_1 + V_2$$

If the resistors both carry the same current, I, then

$$V_1 = IR_1 \quad \text{and} \quad V_2 = IR_2$$

so that

$$\frac{V_1}{V_2} = \frac{IR_1}{IR_2} = \frac{R_1}{R_2}$$

Figure 36.15
A potential divider system with a load resistor.

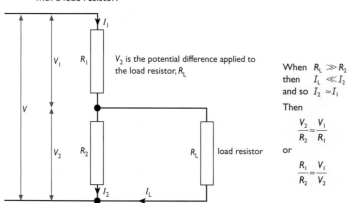

V_2 is the potential difference applied to the load resistor, R_L

load resistor

When $R_L \gg R_2$
then $I_L \ll I_2$
and so $I_2 \approx I_1$

Then

$$\frac{V_2}{R_2} \approx \frac{V_1}{R_1}$$

or

$$\frac{R_1}{R_2} \approx \frac{V_1}{V_2}$$

The ratio of the potential differences is the same as the ratio of the resistances. We can use two (or more, if we wish) resistors in series to share voltage in a circuit. The resistors act as a **potential divider** system (Figure 36.15).

Where there is no junction between the resistors we can always say that $V_1/V_2 = R_1/R_2$. However, when there is a junction then we have to be more careful. Suppose that a potential divider is used to provide a proportion of the power supply voltage to components connected in parallel with one of the resistors, as in Figure 36.15. For simplicity we can start by supposing that these components are made up of just a single resistor, which we can call the **load resistor**, R_L. There are now three different currents to consider: the current through the load resistor, I_L, and the currents, I_1 and I_2, through the two resistors of the potential divider.

Current splits at the junction such that

$$I_1 = I_2 + I_L$$

If $I_L \ll I_2$ then

$$I_1 \approx I_2$$

and we can use the simple potential divider ratio, $V_1/V_2 = R_1/R_2$.

If, however, I_L is of a more significant size then we must consider the combined resistance of R_2 and the load resistor. The resistance of this parallel arrangement is given by

$$\frac{1}{R} = \frac{1}{R_2} + \frac{1}{R_L}$$

That is

$$R = \frac{R_2 R_L}{R_2 + R_L}$$

and potential is shared according to:

$$\frac{V_1}{V_2} = \frac{R_1}{R} = \frac{R_1}{[R_2 R_L/(R_2 + R_L)]} = \frac{R_1(R_2 + R_L)}{R_2 R_L}$$

Note that when R_L is very large, in the numerator $R_1 R_L \gg R_1 R_2$, and so the formula can again be written as $V_1/V_2 \approx R_1/R_2$. It is important, however, to remember that this simple ratio can only be used when the load resistance is much bigger than the resistance of R_2.

The potential divider in electronic circuits

One or both of the resistors in a potential divider arrangement can be variable. A resistor can be manually variable, or it can be a sensor device that responds to its environment, such as a thermistor (as in Figure 36.16) whose resistance decreases as temperature increases, or an LDR whose resistance varies with light intensity.

Figure 36.16
Use of a potential divider system to control the voltage applied to a load circuit.

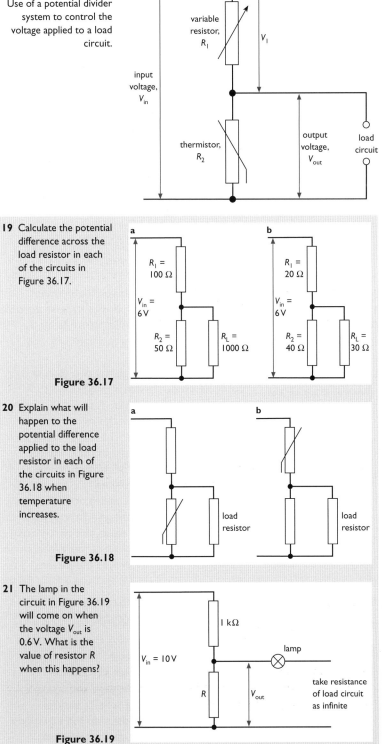

The voltage that is applied to the load circuit, the output voltage, is determined by the values of the resistors R_1 and R_2. If the current in the load circuit is small then we can say that

$$\frac{V_{out}}{V_1} = \frac{R_2}{R_1}$$

$$V_{out} = \frac{R_2 V_1}{R_1}$$

$$= \frac{R_2(V_{in} - V_{out})}{R_1}$$

$$R_1 V_{out} = R_2 V_{in} - R_2 V_{out}$$

$$V_{out}(R_1 + R_2) = R_2 V_{in}$$

$$V_{out} = \frac{R_2}{R_1 + R_2} V_{in}$$

If R_1 is set to a fixed value and R_2 decreases (due to a rise in temperature) then V_{out} decreases. The output voltage from this circuit decreases with increasing temperature.

The output potential difference or **output voltage** from such a potential divider is the voltage across one of the resistances. The voltage that is applied to a load resistor or other components in a **load circuit** can be made to be dependent on temperature or light intensity. The potential divider with its sensor resistor(s) is acting as an **input transducer** to the load circuit.

In electronic circuits the load circuit can consist of logic units, and in this way the circuits can provide digital outputs that are dependent on environmental conditions such as temperature or light intensity.

19 Calculate the potential difference across the load resistor in each of the circuits in Figure 36.17.

a

$R_1 = 100\ \Omega$

$V_{in} = 6\ V$

$R_2 = 50\ \Omega$ $R_L = 1000\ \Omega$

b

$R_1 = 20\ \Omega$

$V_{in} = 6\ V$

$R_2 = 40\ \Omega$ $R_L = 30\ \Omega$

Figure 36.17

20 Explain what will happen to the potential difference applied to the load resistor in each of the circuits in Figure 36.18 when temperature increases.

a

load resistor

b

load resistor

Figure 36.18

21 The lamp in the circuit in Figure 36.19 will come on when the voltage V_{out} is 0.6 V. What is the value of resistor R when this happens?

1 kΩ

$V_{in} = 10\ V$

lamp

R V_{out}

take resistance of load circuit as infinite

Figure 36.19

● Comprehension
and application

The Wheatstone bridge for medical measurement

Heart disease is the biggest killer in the developed world. Medical services have to devote a lot of their time, and money, to treating it and to preventing it. Where a patient has a damaged heart, either inside or in the blood vessels on its outer surface that supply the heart muscle itself with food and oxygen, then the pressure at different points in the system is key information. A way to compare the pressure at different points in the system is to insert a tube, or catheter, through the blood vessels all the way to the heart, usually from a small cut in the chest or leg.

Figure 36.20
Pressure measurement in the body using electrical strain gauges.

In many cases the catheter carries a pressure sensor directly to the site being investigated. Alternatively, the outer end of the catheter can be connected to a cylinder which contains a diaphragm that changes shape as the pressure varies (Figure 36.20). The change in shape is quantified by connecting **strain gauges** to the diaphragm.

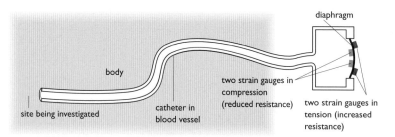

diaphragm

body

two strain gauges in compression (reduced resistance)

two strain gauges in tension (increased resistance)

site being investigated

catheter in blood vessel

Each strain gauge is a resistor, whose resistance depends on its length and cross-sectional area. The strain gauges on the diaphragm are subject to shape changes. Their change in resistance allows electrical measurement. In general, the sensitivity of a strain gauge is given by

$$S = \frac{\Delta R/R}{\Delta L/L}$$

where R is its unstrained resistance and ΔR its change in resistance due to strain; L is its unstrained length and ΔL its change in length.

What is required here is to increase the overall sensitivity of the pressure measurement process. That is, to achieve a change in measurable system output, ΔX, that is as large as possible relative to a change in pressure, ΔP, in the patient's body. We want to make $\Delta X/\Delta P$ as big as possible. To do this, four strain gauges are used together, all attached to the diaphragm, and connected into a **Wheatstone bridge** circuit.

Figure 36.21
A Wheatstone bridge circuit.

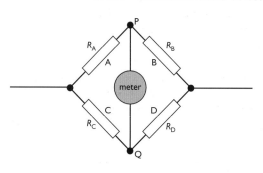

The meter has a zero reading when points P and Q are at the same potential. Taking $[R_A, R_B]$ and $[R_C, R_D]$ as two potential dividers in parallel, this will be true when

$$\frac{R_A}{R_B} = \frac{R_C}{R_D}$$

Changes in the values of the resistances result in a non-zero reading on the meter.

A Wheatstone bridge has two pairs of potential divider systems, each of which is made up of two resistors. A sensitive meter forms a bridge between the two potential dividers (Figure 36.21). The meter reads zero when the two mid-points of the potential dividers (P, Q) are at the same potential.

When not subject to pressure the four strain gauges A–D have approximately equal resistances. As the diaphragm curves, two of the resistances increase in value and two decrease. This changes the potential at each of the mid-points of the Wheatstone bridge, and so changes the potential difference across the meter.

Assume that in the unstrained state all resistors in Figure 36.21 have resistance 25 Ω and the meter reading is zero.

22 What is the potential difference across each resistor when the total applied potential difference is 3.0 V? Suppose that under strain, the resistances of A and D decrease by 0.2 Ω and the resistances of B and C increase by 0.2 Ω.

23 What is the potential difference across resistor B?
24 What is the potential difference across resistor D?
25 What is the potential difference across the meter?
26 If the meter is a moving coil meter with full scale deflection at 5 mA and resistance of 10 Ω, what fraction of full scale deflection does the meter show?

● Extra skills task

Information Technology and Application of Number

Create a spreadsheet that can calculate the total resistance of different combinations of resistors, including series and parallel arrangements. It should cope with complex arrangements including several parallel branches. You should develop the equations to be used yourself from first principles.

Examination questions

1 a Define the *volt*. (1)
  b i State one similarity between potential difference and electromotive force.
  ii Explain one difference between potential difference and electromotive force. (4)
  c The figure shows a battery of e.m.f. *E* and internal resistance *r* connected to an external resistor of resistance *R*.

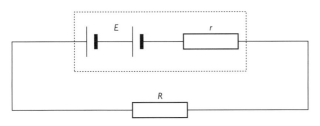

Explain, in terms of the energy transfer from the battery to the external resistor, how the internal resistance of the battery affects the potential difference across the external resistor. (4)

OCR, Sciences, Basic 1, Mar 1999

2 a State what is meant by an electric current. (1)
  b A car battery has a capacity of 40 ampere-hours. That is, it can deliver a current of 1.0 A for 40 hours. When delivering the current, the potential difference across the terminals of the battery is 12 V. Calculate the electrical energy available from the battery. (3)
  c An electric heater is rated at 1.0 kW, 240 V but the supply voltage is only 230 V.
  i Calculate the resistance of the heater if it were operating at 240 V.
  ii Assuming that the resistance of the heater is constant, calculate the output power of the heater when it is being used with the 230 V supply. (3)
  d You have available a number of resistors, each of resistance 100 Ω. Draw an arrangement of resistors which would give a total resistance of
  i 50 Ω **ii** 150 Ω **iii** 67 Ω. (5)

OCR, Sciences, Physics, Foundation, Mar 1999

3 A very high resistance voltmeter reads 20 V when connected across the terminals of a d.c. power supply. The high resistance meter is disconnected and a second voltmeter of resistance 1.0 kΩ is then connected across the supply. The second meter gives a reading of 16 V.
  a State the e.m.f. of the power supply.
  b Calculate the current which flows through the second meter.
  c Calculate the internal resistance of the power supply.
  d Show that the current is equal to 0.080 A when the supply is short circuited. (5)

AQA (NEAB), Mechanics and Electricity, (PH01), Mar 1999

4 a The figures show two circuits in which a supply of e.m.f. 6.0 V and internal resistance 5.0 Ω is delivering power to a pair of resistors.

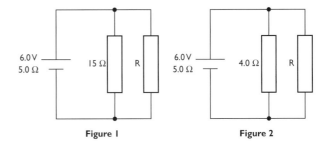

Figure 1   **Figure 2**

When maximum power is dissipated in an external circuit, the resistance of the external circuit is equal to the internal resistance of the supply.
  i For the circuit in Figure 1, determine the value of R which results in the maximum power being delivered to the external circuit. (3)
  ii Calculate the terminal potential difference when the supply is delivering maximum power to the circuit in Figure 1. (1)
  iii Calculate the power that will be dissipated by the 15 Ω resistor when the supply is delivering maximum power to the external circuit. (2)
  iv For the circuit in Figure 2, explain why the supply cannot deliver maximum power in this circuit for any value of the resistor R. (2)
  b i The 15 Ω resistor is made from wire of length 2.3 m. The wire has a diameter of 3.0×10^{-4} m. Calculate the resistivity of the material from which the wire is made. (3)
  ii Sketch a graph showing how the resistance of 2.3 m of wire made from this material varies with the diameter of the wire. The value for a wire of diameter 3.0×10^{-4} m has been plotted for you below. (2)

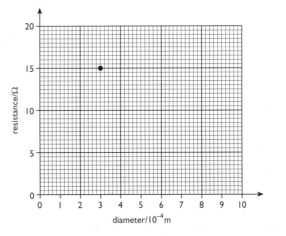

AQA (AEB), Physics, Paper 1, June 1999 (part)

5 A camping products manufacturer wants to market a 'two cup' mini kettle that could be powered by a 12 V car battery and engages a scientist to evaluate the idea.

The scientist's work is to be in two parts:

Part 1: To measure the electrical resistance of the wire which is to be used to make the heating elements.
Part 2: To determine the thermal requirements for the mini kettle.

Part 1

In order to measure the electrical resistance of the wire, the scientist connects a circuit to measure the current through a wire sample as the voltage across it is increased from 0 V to 14 V. The circuit she intends to use is shown below.

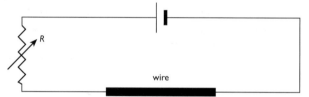

a Draw in an ammeter and voltmeter on the circuit above showing how they should be connected in order to measure the current through and the potential difference across the wire. (2)
b What is the purpose of the resistor R in the circuit? (1)
c In order for the measurements to be accurate, *explain briefly* whether the resistances of the ammeter and voltmeter should be much higher or much lower than the resistance of the sample. (3)

The data shown in the graph below are collected for one of the sample wires.

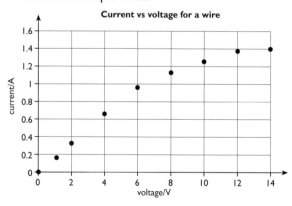

d Over what range of current and voltage does the wire sample obey Ohm's Law? (2)
e Explain why the slope of the graph changes at higher voltages. (2)
f From the graphed data calculate the resistance of the wire sample for *small* currents and voltages. (2)

Part 2

The figure below shows the main features of the proposed mini kettle. When filled to the 'maximum water level' the kettle holds 400 ml of water. (Take the specific heat capacity of water as 4190 J kg^{-1} °C^{-1}.)

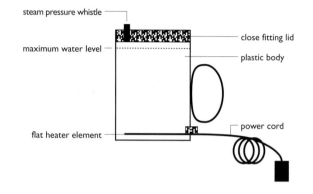

g Why is the heater element placed so close to the bottom? (1)
h Give *two* advantages of using plastic for the body of the kettle rather than metal. (2)
i Estimate the electrical energy required to bring a full mini kettle of water to the boil. State clearly *two assumptions* and *any estimates* you have made in your calculation. (4)
j Explain whether the electrical energy required in reality is likely to be more or less than that estimated above. (1)
k If, for safety reasons, the maximum current that may be drawn from the car battery is limited to 8.0 A, what is the maximum electrical power available for the mini kettle? (2)
l How long would it take to boil a full kettle of water? (2)
m If you were the scientist advising the camping goods manufacturer, explain briefly what would be your recommendation on the practicality of the 'two cup' mini kettle? (1)

International Baccalaureate, Subsidiary/Standard Level, May 1998

6 a State what is meant by
  i the *resistance* of a sample of metal
  ii the *resistivity* of the metal. (4)
b A strain gauge consists of a thin metal foil firmly bonded to a flexible plastic backing sheet, as illustrated below.

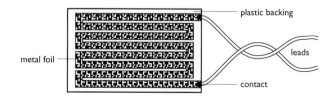

The foil has total length l and total resistance R between the contacts. When the foil is extended by an amount Δl, the resistance changes by an amount ΔR, such that

$$\Delta R/R = 0.485 \ (\Delta l/l)$$

i Suggest why the device illustrated is referred to as a *strain* gauge.

ii Explain, by reference to the definition of resistivity, why the resistance of the metal foil changes when it is extended. (4)

c In one particular application, the strain gauge S is connected in series with a fixed resistor F of resistance 400 Ω and a battery of e.m.f. 4.50 V and negligible internal resistance, as illustrated.

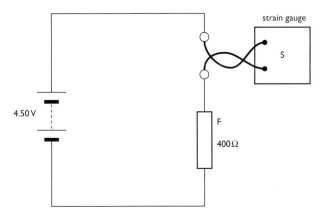

strain gauge

S

4.50 V

F

400 Ω

i Show that there is a potential difference of 2.40 V across the resistor F when the strain gauge has a resistance of 350 Ω. Explain your working carefully. (3)

ii The gauge is given a strain of 1.18%. Using the expression given in **b**, calculate the *change* in

1 the resistance of the strain gauge

2 the potential difference across the resistor F. (5)

d The strain gauge is firmly fixed to a structure so that the strain, as calculated from the change in potential difference, is used to determine the stress on the structure. State how the stress on the structure may be determined from the measured strain. (2)

e In practice, the method outlined in **b**, **c** and **d** may not give a reliable result for the stress. Suggest two factors, other than a change in strain, which might give rise to a change in the potential difference across the resistor F. (2)

OCR, Physics, Paper 3, June 1999

37 The nature of data

THE BIG QUESTIONS
- What are data?
- How and why do we gather data?
- How do we store and transmit data?

KEY VOCABULARY
bistable (flip-flop)   byte   carrier wave   corruption (of data)   data   false colour
frequency division multiplexing   hard disk   line of sight   logic circuit
mapping (of data)   modulation (of radio wave)   pulse code modulation   RAM
real time communication   ROM   sampling (of data)   sampling rate
time division multiplexing   transducer (input and output)   word (of digital data)
word length

BACKGROUND

Compared with other species, humans have got many advantages to aid survival and comfort. We have the use of our heads, our hands, our senses by which we know of the world, and most of all we've got each other. We communicate, passing ideas and skills from one to another. If each of us had to start alone with no help from others, not one of us would get very far. Communication matters. Language skills are built into people so strongly that it's impossible to stop children learning to understand words and to speak them.

And our language goes further. We can learn to extract information from visual symbols.

Figure 37.1
We are exposed to symbols that mean a lot to us, even though we are not always consciously aware of the meaning. (The study of symbols and how they influence us in general is called semiotics.)

Then we can scan across a page and decode the symbols as quickly as we can talk and listen. Of course, they have to be symbols that we've learnt. (Μοστ οφ υσ ονλψ τρουβλε το λεαρν ονε σετ οφ σψμβολσ.)

Language helps us to make sense of our observations. **Data** are streams of information that originate in observations or sensing of the world, by humans or by sensing machines. But ultimately the data must reach our brains; that is where data gains meaning and value.

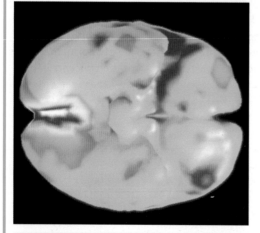

Neuroscientists are working to try to understand the physical process that happens in our brains. This is a PET scan (see page 312) constructed by computer from data emerging from a brain. Here the part of the brain that responds to visual stimulation is shown to be active.

Figure 37.2 (left)
When you hear or read a word it triggers a response in your mind. When you see a picture it does the same. Words and images change your mind, literally.

Figure 37.3 (right)
This is another icon – a familiar and meaningful image. It is one of the most famous faces of the last century. The owner of the face made many quotable statements, but this one fits well with this chapter: '... all our science, measured against reality, is primitive and childlike – and yet it is the most precious thing we have.'

A few words about numbers

Figure 37.4
The binary system works on the same principle as our more familiar counting system except that it uses just two symbols instead of ten.

= 5 = 101

| h | t | u | | 128 | 64 | 32 | 16 | 8 | 4 | 2 | 1 |

199 = 1 1 0 0 0 1 1 1

Binary has advantages – where there are just two symbols they can be represented by just two distinct values of voltage or light intensity. It also has disadvantages – writing out binary numbers requires the use of many digits.

Numbers can exist as symbols. For every whole number we can choose from a range of alternative symbols – from dots on a domino to binary code (Figure 37.4), to the familiar ten Arabic numerals (0, 1, 2, 3, 4, 5, 6, 7, 8, 9) and combinations of these.

The same numbers can be written in different ways, which means that numbers are not simply symbols. There is more to them than that. They exist separately from their symbols. They exist in our heads, as concepts. We can apply these concepts in useful ways. We can take the concept of 'three' and apply it to sheep, chairs, words, years, journeys, and so on. They are useful concepts.

The consistency of the behaviour of numbers when subjected to the operations we call division, multiplication, addition and subtraction is stunning. Five with another one added is always the same as two lots of three, for example. Nine divided by three is always the same as three divided by one. Amazing but true! And so familiar that it's easy to forget just how amazing it is.

> **1** Convert these into binary numbers:
> **a** 3   **d** 32
> **b** 7   **e** 100
> **c** 11
> **2** Express these binary numbers in Arabic numerals:
> **a** 101   **c** 1111
> **b** 1000   **d** 1000000
> **3 DISCUSS**
> What is learning? What happens to your brain during learning (in general terms)?
> What is the relative importance of the following in your learning processes?
> • friends and others of the same age   • books
> • people of other age groups   • magazines and newspapers
> • TV and radio   • practical experience
> Are any of your learning experiences independent of input by other people? When do you expect your learning processes to stop?

Quantitative representations of reality

Figure 37.5
Measurement creates representation of a reality by number.

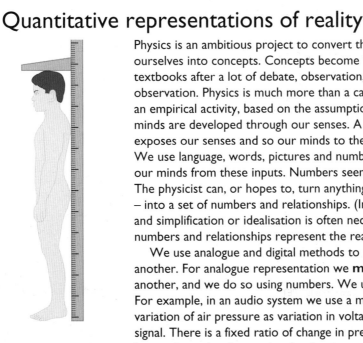

Physics is an ambitious project to convert the world outside (and inside) ourselves into concepts. Concepts become 'accepted' and written into textbooks after a lot of debate, observation, more debate and more observation. Physics is much more than a catalogue of information – it is an empirical activity, based on the assumption that the concepts in our minds are developed through our senses. A practical activity in physics exposes our senses and so our minds to the behaviour of the world. We use language, words, pictures and numbers to construct meaning in our minds from these inputs. Numbers seem to be especially useful. The physicist can, or hopes to, turn anything – any real object or event – into a set of numbers and relationships. (In fact the world is complex, and simplification or idealisation is often necessary.) We say that the numbers and relationships represent the reality (Figure 37.5).

We use analogue and digital methods to represent one reality by another. For analogue representation we **map** one reality on to another, and we do so using numbers. We use scales, which are ratios. For example, in an audio system we use a microphone to map a variation of air pressure as variation in voltage, creating an electrical signal. There is a fixed ratio of change in pressure to change in voltage, just as the scale of a map is a fixed ratio.

The device that does the mapping – converting variation in one variable to variation in another – is called a **transducer**. Sometimes the term is used only for devices that map the variation in or out of electrical signals. A microphone maps into electrical signals, and is called an input transducer. A loudspeaker maps from electrical to pressure variations, and is called an output transducer.

Other devices that convert one physical variation to another can also be called transducers, however. This includes a film-based camera, which maps light from a three-dimensional world onto a two-dimensional surface. Measuring instruments such as manometers and thermometers also represent one physical variable by another (Figure 37.6). Note that these instruments are based on an assumption that one variable – such as the length of a column of liquid in a tube – behaves like the one being measured. They are analogue instruments. They are not electrical transducers, but a simpler form of transducer.

Figure 37.6
A manometer converts one physical reality, pressure difference, into one that we can sense directly.

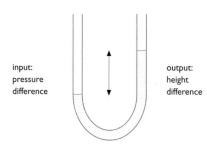

input:
pressure
difference

output:
height
difference

Circuits with thermistors, LDRs and strain gauges are all examples of electrical transducer circuits. They respond to the behaviour of some aspect of the world, and map this behaviour in the form of a voltage level. This new form can be transmitted as streams of information, combined, manipulated and stored. These electrical transducers convert behaviour into data.

In the audio system, a loudspeaker reverses the microphone's mapping process (Figure 37.7) to recreate patterns of pressure variation that match the original. The ratio of the amplitude of the new pressure variations, at a given distance from the loudspeaker, to the amplitude of the original pressure variations, at a given distance from the source of sound, may equal one. The sound we hear will then have the same loudness as the original sound. We can, of course, increase or decrease the amplitude ratio using a volume control on the amplifier. The volume control acts on the signal when it is in electrical form. Electrical signals are more easily controlled than sound signals.

Figure 37.7
A microphone is an electrical transducer – it maps pressure as voltage.

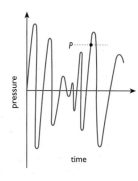

P

pressure

time

voltage

V

time

For the graphs, the ratio $\frac{P}{V}$ is constant.

A cathode ray oscilloscope can replace the amplifier and loudspeaker. It then re-maps the electrical signal not as sound but as deflection of an electron beam, and so as the displacement of a spot on a screen (Figure 37.8).

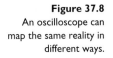

Figure 37.8
An oscilloscope can map the same reality in different ways.

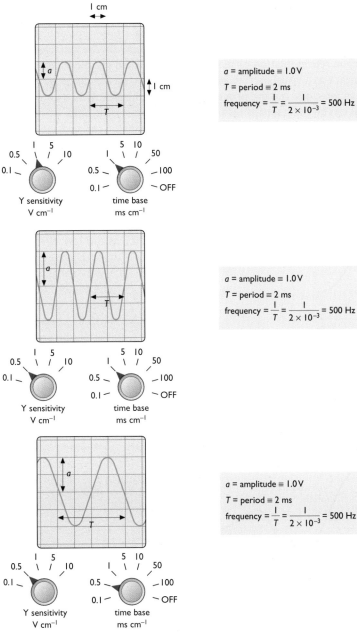

a = amplitude ≡ 1.0 V

T = period ≡ 2 ms

frequency = $\dfrac{1}{T} = \dfrac{1}{2 \times 10^{-3}} = 500$ Hz

4 What does the word 'analogy' mean? How is it related to the word 'analogue'? How is it related to the word 'model'?

5 a Taking the input to a microphone to be measured in terms of pressure changes and its output in terms of voltage changes, what is the unit of the ratio of output to input?

b The scale of a map, such as '1 in 25 000', is a ratio which can be written as

$$\frac{1}{25\,000}$$

Why does this have no unit?

6 Your
a eyes **b** ears
could be said to be transducers. What physical variables act as input and output in each case?

7 What is the effect of a time base adjustment of a cathode ray oscilloscope on the way it maps one variable as another?

8 The surface of an A4 sheet has dimensions 0.297 m by 0.210 m, which suggests an area of 0.0624 m². This suggestion is based on an idealisation of the nature of the surface and is only approximately true.
a Explain this last sentence.
b For what kind of surface is the answer a more reliable measurement?
c For what kind of surface is the answer less reliable? (Thinking about the difference between regular and diffuse reflection may be of some help to you.)
d In saying that the area is 0.0624 m², we are creating a numerical model of a sheet of paper. Do numerical models convey absolute truth?

Continuous and discontinuous monitoring

We may wish to measure a variable that changes over time and so to observe the pattern of changes. In some cases we can continuously monitor the variable and observe the patterns by analogue representation or mapping, such as when we show changing patterns of potential difference as a visual pattern on an oscilloscope screen. In other cases we make readings only at certain intervals, and the process of taking such readings is then called **sampling**. An electrical signal can be represented on a graph of voltage against time. At equal intervals of time the voltage can be detected or sampled. Each voltage can then be represented as a binary number (Figure 37.9).

Each reading provides a sample of the variable we are studying. The frequency of the readings is called the **sampling rate**. The more frequent the sampling the more we can conclude about the variable (see Figure 37.10). For any sampling process, if the sampling rate is too low, then we cannot reach reliable conclusions.

A hospital patient might have his or her temperature taken at certain intervals; the sampling rate might be twice each day. Or it might be continuously monitored and then the information stored in a way that can be quickly read, such as on a pen recorder system.

Figure 37.9
The principle of sampling.

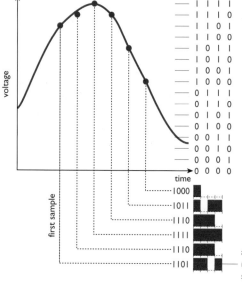

In the example shown here, the binary codes representing each voltage level have four 'digits' or 'bits' (see page 26). Each voltage level must be 'rounded' to the closest corresponding digital value.

a four bit digital signal, matching the first voltage sample

Figure 37.10
Measurement of a time-dependent variable at a low sampling rate provides only crude information; a higher sampling rate provides more detailed information.

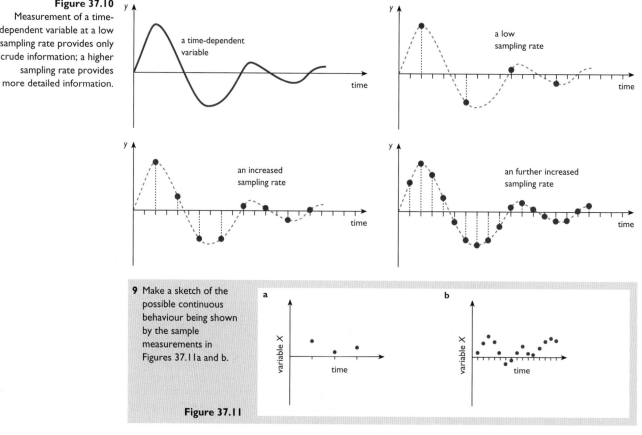

9 Make a sketch of the possible continuous behaviour being shown by the sample measurements in Figures 37.11a and b.

Figure 37.11

Data transfer

Streams of information, or data, can be transmitted from place to place in very many ways. This paper carries information that is encoded in symbols, and many realities are represented in this book by patterns of light reflections from the page. The human voice carries data encoded in patterns of pressure variation. Electrical signals, light signals and radio signals provide the basis for 'high technology' data transfer.

Person-to-person communication must use a system of inputting data that is human-friendly. It is a human being who inputs the information. The nature of the output must also be human-friendly. It is humans who will extract the information. In between the input and the output, the information may be converted out of human-friendly form and into machine-friendly form for storage and later retrieval, such as in computer circuitry or on a CD, or for transmission, perhaps as radio waves travelling long distances or as digital light signals along optic fibres. It is then converted back again to human-friendly form at the point of output.

Radio provides **real time communication**, while computer memory, books, tapes and CDs also communicate but not in real time. A radio set receives the data with a fixed delay after transmission. The fixed delay is very short, however, because the data are carried at the speed of light. Figure 37.12 shows a complete radio communication system in diagrammatic form. You can read more about radio communication and about signal bandwidth in Chapter 3.

Figure 37.12
Radio communication.

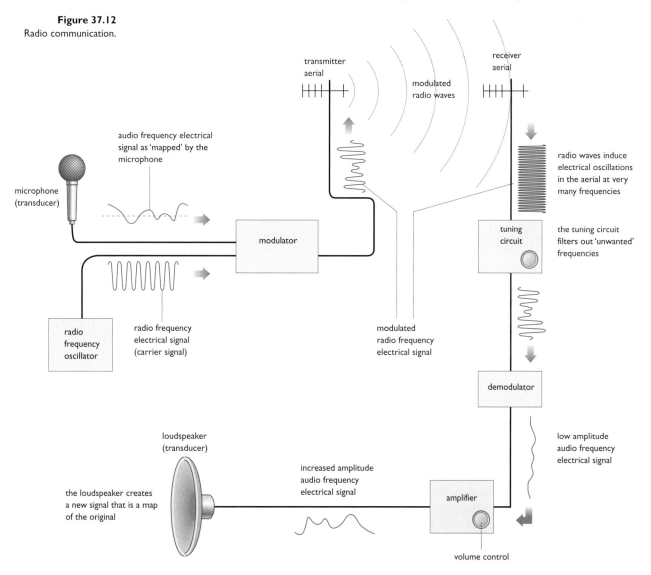

Analogue transmission

The electrical signal that carries the data in an audio system of microphone, amplifier and loudspeaker is a simple analogue signal. There is direct mapping from sound to signal to sound. In data transfer by AM and FM radio, the mapping is more complex. Patterns in the properties of the radio wave, called the **carrier wave**, carry the signal in encoded form. The changes to the carrier wave that represent the signal are called **modulation** of the wave.

In AM (amplitude-modulated) transmission the signal is not mapped simply on to the carrier wave but onto its amplitude. The variation of amplitude with time for the modulated wave matches the variation of air pressure with time that takes place in the radio studio (Figure 37.13).

Figure 37.13
Amplitude modulation.

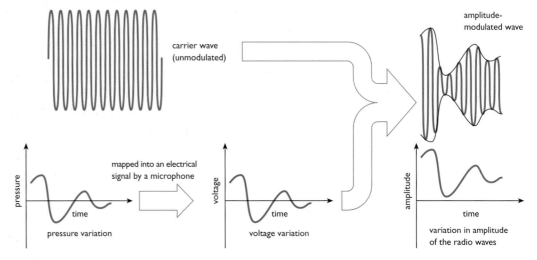

For FM (frequency-modulated) transmission the frequency of the carrier wave varies with time, matching and mapping the variations in air pressure that constitute the original sound wave (Figure 37.14).

Figure 37.14
Frequency modulation.

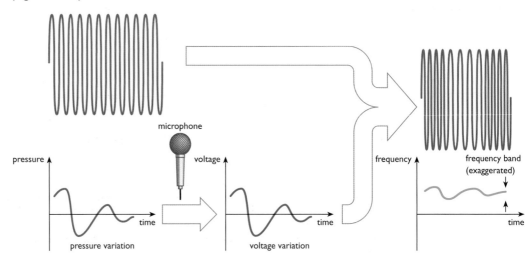

Digital transmission

Variables in the world, such as temperature or air pressure, change continuously, and transducers produce analogue representation. Digital information does not use the same simple constant ratio (or scale) mapping, but represents values of variables using binary numbers. Sampling is necessary to convert analogue data into digital data. The digital data can then be transmitted as binary 1s and 0s, by **pulse code modulation** of a radio wave. Modulation here consists of turning the carrier wave on and off, to create pulses (Figure 37.15). The 'code' is

Figure 37.15
Pulse code modulation.
Note that the size of the
radio wave period is
greatly exaggerated here
relative to the duration of
the pulse.

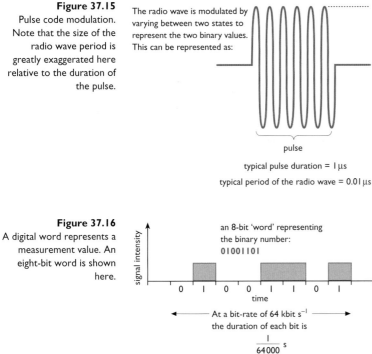

The radio wave is modulated by
varying between two states to
represent the two binary values.
This can be represented as:

typical pulse duration = 1 μs
typical period of the radio wave = 0.01 μs

usually simply the binary system of numbering,
though more complex codes exist.

A binary number can be transmitted as a
series of 'ons', or 1s, and 'offs', or 0s, of
voltage or of light level. Each 1 or 0 is a single
bit, and the series of bits that make up the
value to be transmitted is called a **word**. Each
measurement value, or word, therefore
requires several bits, so the rate at which bits
are transmitted, the bit-rate, is greater than
the sampling rate (Figure 37.16). For telecommunications a typical sampling rate would
be 8 kHz, or 8000 measurements per second.

To limit the number of bits that are
required for each voltage value, these values
are sorted into discrete voltage levels. Like
energy levels in an atom, in-between values are
not allowed – if two adjacent voltage levels are
1.5 and 1.6 V then a value of 1.54 will be read
as 1.5. Each sampled measurement then exists
as a voltage level which is described by a binary
code made up of a word of several bits. The
more bits it is made up of, the more closely
the overall pattern of transmitted voltage
values reflects the analogue pattern from
which it is generated (Figure 37.17).

Figure 37.16
A digital word represents a
measurement value. An
eight-bit word is shown
here.

an 8-bit 'word' representing
the binary number:
01001101

At a bit-rate of 64 kbit s⁻¹
the duration of each bit is
$$\frac{1}{64\,000}\ \text{s}$$
and the duration of each word is
$$\frac{8}{64\,000} = \frac{1}{8000}\ \text{s}$$
$$= 1.25 \times 10^{-4}\,\text{s}$$
$$= 125\,\mu\text{s}$$

Figure 37.17
The more bits per word,
the better the quality of
signal. These two signals
have the same sampling
rate, but different word
length.

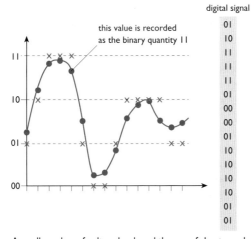

this value is recorded
as the binary quantity 11

A small number of voltage levels and the use of short words
– here they are two bits long – allows crude information to
be sent at high speed.

The digital signal that represents the analogue signal is
shown by the Xs.

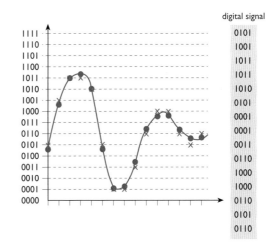

Longer word length allows the 'shape' of the digital signal,
shown by the Xs, to match more closely that of the analogue
signal.

Note that data need to be transmitted at a higher bit-rate.

Bit-rate and sampling rate are related by:

bit-rate = sampling rate × number of bits per sample

The number of bits per sample is the **word length**. If word length is 16 bits per word and the
sampling rate is 8 kHz, then

$$\text{bit-rate} = 8000 \times 16$$
$$= 128\,000\,\text{bit s}^{-1}$$
$$= 128\,\text{kbit s}^{-1}$$

Figure 37.18
The process of transmission can 'blur' a pulse but a digital signal can still be read from a stream of such pulses, provided they do not merge too much. A similar level of distortion of an analogue signal would greatly reduce the quality of the received information.

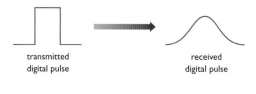

transmitted
digital pulse

received
digital pulse

Digital representation of reality is based on sampling, while analogue representation is based on direct mapping. The quality of digital data depends very much on the sampling rate, as we have seen (Figure 37.10). It might seem that analogue methods are superior. But analogue information is more likely to become **corrupted** – distorted beyond recognition – in the storage or transmission process (Figure 37.18). A scratch on a plastic record destroys the music. Analogue radio loses clarity; analogue telephone signals can fade into incoherence. A big advantage of a digital system is that a 1 is very different from a 0, and the entire message is just 1s and 0s. These can blur, as we've seen in Chapter 23, and eventually can become corrupted, but the information they carry is less affected by such changes than is the case for analogue signals. Also, a digital system has a greater capacity – it can transmit more data in the same time, using the same physical system (such as radio transmission at a given frequency range).

10 Why is no 'carrier wave' needed for a simple microphone–amplifier–loudspeaker audio system?

11 In Figure 37.12, what changes to the system would be needed for digital transmission?

12 When might bit-rate be equal to sampling rate?

13 Calculate the word length for a sampling rate of 16 kHz and a bit-rate of 256 kbit s^{-1}. Which provides more detailed transmission of information, this or the example given at the bottom of page 443?

14 What is the highest binary number that can be represented by a word length of 16 bits? What is this in Arabic numerals?

15 **DISCUSS**

a Is the binary system any more a 'code' than the decimal system based on Arabic numerals? Or is it just that we are more familiar with the decimal system? Why don't we use the binary system for everyday counting?

b In what way can the binary system be said to be based on the number 2, while the decimal system is based on the number 10? It would be possible to have a counting system based on the number 3. How many different symbols would it need? Why would you find such a system hard to use? Is it inherently harder than the decimal system?

c You learned to use the decimal system when you were quite young. Could you learn it now?

● Multiplexing

The air around you is filled with radio waves. All of them induce electron movement in an aerial, and yet a radio set can reproduce sounds from just one chosen radio studio. Each radio station has its own frequency or range of frequencies, and so information from one station can be separated from others by the radio set's 'filter' or tuning circuits. A similar process can be used with analogue signals that are transmitted along cables. Signals from different telephones, for example, can be transmitted at the same time along the same cable at different frequencies or frequency bands. This process is called **frequency division multiplexing**.

Time division multiplexing can only be used with digital signals. Each multiplexed signal is a series of bits which can be transmitted with significant time gaps between each one. Bits that are parts of other signals can then fit into the time gaps. Bits of several multiplexed signals each 'take their turn' to travel along the cable (Figure 37.19).

Figure 37.19
Time division multiplexing of digital signals.

16 Why can't time division multiplexing be used with analogue signals?

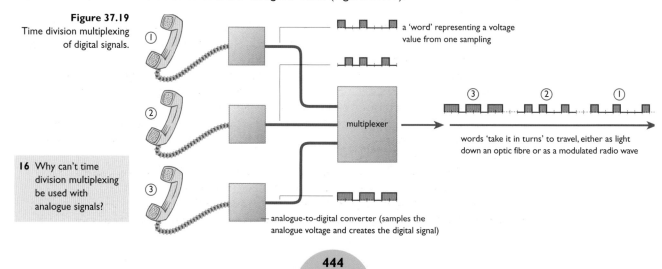

a 'word' representing a voltage value from one sampling

words 'take it in turns' to travel, either as light down an optic fibre or as a modulated radio wave

analogue-to-digital converter (samples the analogue voltage and creates the digital signal)

Transmission of radio waves

Radio waves travel through the air and out into space. Within the atmosphere there will be some absorption and some reflection, and the extent of these is dependent on the frequency (Figure 37.20).

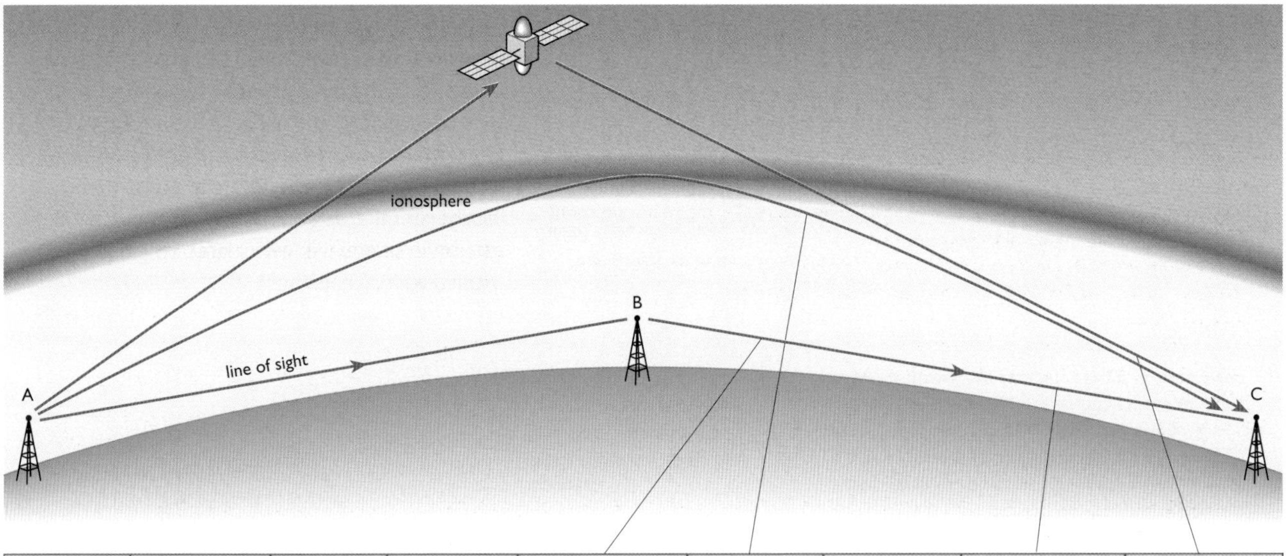

Frequency	30 Hz–3 kHz extra low frequency	3–30 kHz very low frequency	30–300 kHz low frequency (long wave)	300 kHz–3 MHz medium frequency (medium wave)	3–30 MHz high frequency (short wave)	30–300 MHz very high frequency	300–3000 MHz ultra high frequency	3000 MHz + microwave
Range	1000 km +	1000 km +	1000 km +	less than 1000 km (line of sight)	very long	line of sight	line of sight	line of sight but can be reflected from satellite
Uses	submarine	military		sound radio		sound/TV	TV	satellite-based communication

Figure 37.20
Radio wavebands and some of their modes of travel between stations.

Extra low and very low frequency waves (30 Hz to 30 kHz) experience little absorption, and can even travel significant distances in water. That makes them useful for submarine communication. However, because of the low frequency they can carry only simple information at speed. The bandwidth of several kHz that is required for rapid transmission of more complex information (see page 26) is not available.

High frequency waves (3 to 30 MHz), called short waves, can also travel long distances due to reflection from the ionosphere – a layer of the atmosphere. These, therefore, are used for long-distance global communication.

Medium frequency waves experience significant absorption and have ranges in the atmosphere of only a few hundred kilometres. They are not reflected by the ionosphere and so can be received only when there is an unobstructed straight line – or **line of sight** – between the transmitter and the receiver. (Some diffraction in the vicinity of the receiver may allow some signal to reach into what would otherwise be 'shadows'.) These waves are, however, adequate for radio broadcasting.

Very high frequency (VHF) waves have the advantage of potential for large bandwidth and so they can carry complex information at a high rate, though their range is relatively short.

Radio waves travelling through space experience no absorption. But unless they are travelling as a parallel beam their energy will become spread over a wider and wider area as they travel.

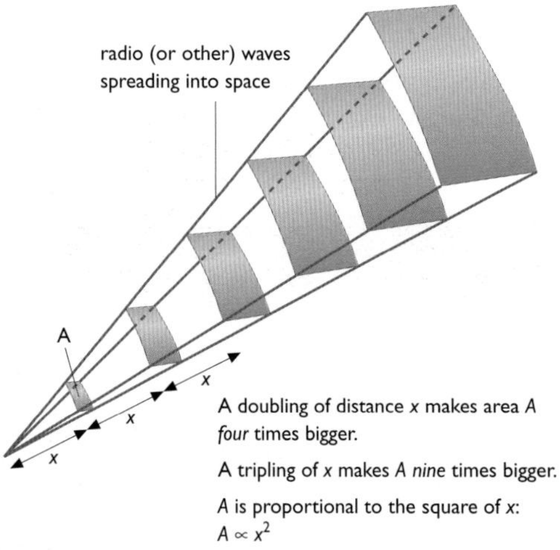

Figure 37.21
The intensity of radio (or other) waves spreading into space becomes weaker and weaker. It is inversely proportional to the area through which the waves pass.

radio (or other) waves spreading into space

A

x
x
x
x

A doubling of distance x makes area A *four* times bigger.

A tripling of x makes A *nine* times bigger.

A is proportional to the square of x:
$A \propto x^2$

For a point source transmitting in all directions, this area is in the shape of the surface of a sphere. The size of the area is proportional to the square of the radius (surface area of sphere $= 4\pi r^2$). The intensity of the wave is measured as the rate of transmission of energy per unit area, in W m^{-2}. As the wave spreads and the area grows, its intensity decreases. Intensity is inversely proportional to the area, and therefore inversely proportional to the square of the radius (Figure 37.21). We say that the intensity obeys an inverse square law (you can see how the inverse square law relates to gamma radiation in Chapter 8).

17 See Figure 37.22 and the text below it. If the Earth became a radio source in 1910, and aliens were only 19 light-years away, when would they have reached us if they'd come at the speed of light?

18 How is long-distance global radio communication possible without using satellites?

19 Why are medium frequency waves suitable for local radio but not for carrying telephone calls between continents?

20 Why does videophone need a larger bandwidth than ordinary voice telephone? (See Chapter 3.)

21 Why is the inverse square law of importance in receiving signals from missions to Mars but not for VHF communication on Earth?

Figure 37.22
Radio waves with many frequencies are spreading into space from our radio stations.

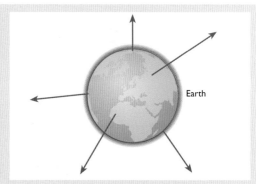

Earth

If there were aliens with sensitive radio receiving equipment, living 100 light-years from the Sun, this distant star will have seemed very ordinary until very recently. (We can't know how such 'aliens' measure time, but we'll stick with the Earth year as our unit of time here, and the light-year as our unit of distance.) Then, with increasing intensity, the star or something very close to it started to send out a fog of radio signals with a wide range of frequencies. The aliens may have worked out that since this new radio source was 100 light-years from them, the process must have begun 100 years ago. There is no evidence that the aliens are on their way to investigate – if they set out now and travel at near the speed of light then they'll arrive 100 years from now!

Data storage

A written letter such as 'h' carries little information, but in combination with strings of others it tells us a lot. This book contains something in the region of a million such symbols. Strings of letters, combined in words, sentences, paragraphs, chapters and books, carry meaningful information from person to person. You can close the book and open it later. The patterns will still be there. They are in storage.

Words on a page store information in a form that can be directly accessed by human vision. But direct human access to data is not always so important. Some data can be, and must be, stored first of all in machine-friendly form. Machine-friendly data storage can be simply physical, for example as the shape of a vinyl record surface (Figure 37.23a) or the tracks of pits on a CD (Figure 37.23b), or it can be magnetic, as in the magnetic patterns on a tape (Figure 37.23c).

Machine-friendly data storage can also be in the form of voltage levels. An unchanging voltage level within a circuit is, in effect, 'remembering' its own value. It 'remembers' for as long as that voltage is maintained. This is used as 'electronic' memory in computers, along with 'magnetic' memory provided by magnetic disks which store information in a similar way to magnetic tape.

Figure 37.23
Storage of information.

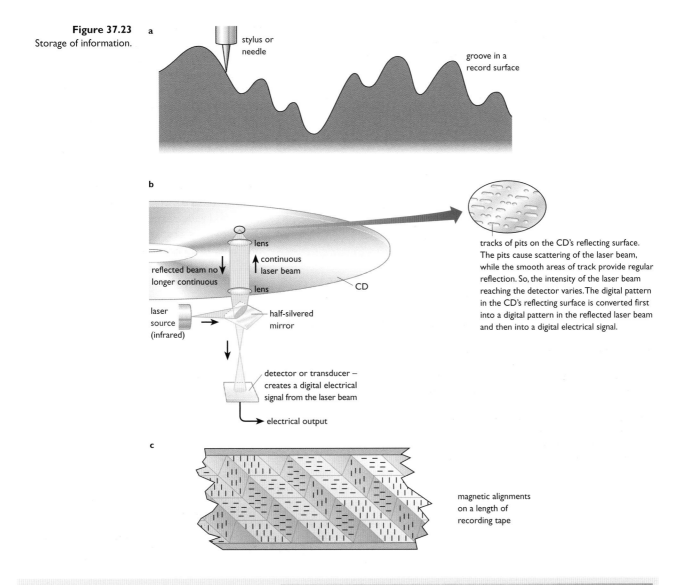

a stylus or needle

groove in a record surface

b lens

continuous laser beam

reflected beam no longer continuous

lens

CD

laser source (infrared)

half-silvered mirror

detector or transducer – creates a digital electrical signal from the laser beam

electrical output

tracks of pits on the CD's reflecting surface. The pits cause scattering of the laser beam, while the smooth areas of track provide regular reflection. So, the intensity of the laser beam reaching the detector varies. The digital pattern in the CD's reflecting surface is converted first into a digital pattern in the reflected laser beam and then into a digital electrical signal.

c

magnetic alignments on a length of recording tape

22 Which of these data storage systems
   a a vinyl record
   b magnetic disk
   c digital voltage levels
   involves direct analogue mapping of one variable by another, as described on pages 437–9?
23 If each of the systems in question 22 is used for storing sound data, what transducers are used to convert the data from storage and into an analogue electrical signal?
24 Explain why data stored on a vinyl record are easily corrupted.
25 **DISCUSS**
   'This is not a pipe'; but Figure 37.24 is a famous painting by René Magritte, painted in 1929. What important reminders does it provide to us about how we represent the world in physics?

Ceci n'est pas une pipe.

Figure 37.24

People-friendly data

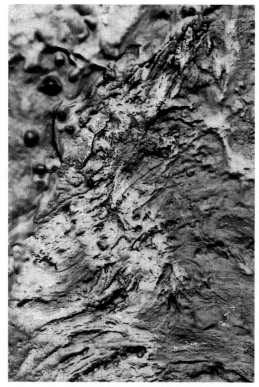

Images carry information. They are especially useful because people are very good at quickly extracting information from a scene or a visual image. Images are very human-friendly. So it is not surprising that we've built our computers so that much of the communication between them and us is in the form of imagery (and if you count letters of the alphabet as images, nearly all of the communication is in this form). The screen is a machine-to-human interface.

In the 1950s, some postgraduate geology students were sent to walk the hills of Scotland, carrying instruments which contained heavy weights. Small variations in the weight from place to place provided information about the structure of the rocks below. But there was so much data that they could not be usefully analysed – not until the 1980s when the data could be entered into computers, and the computers were programmed to show the data in visual form (Figure 37.25). The result was a new kind of geological map that is useful for activities such as mineral exploration and earthquake prediction. The computers even showed the variations as different colours, by adding **false colour** to the images. The colour does not exist in the real entity being imaged, but is added to make the image easier to analyse.

Computers can deal with large quantities of data. So can you, but the processes are very different. One consequence of this is that you make more mistakes in routine processes, but you are more flexible and much quicker to learn from complex situations.

Computers are not very imaginative, but they can help people to be imaginative. We can feed information into computers so that they produce models for us. The models may be visual, as in models of molecules that might be used by a medical researcher to investigate the behaviour of large and complex molecules or structures and to develop drugs accordingly.

Alternatively, computer models may be primarily mathematical – a collection of mathematical relationships that can be used to answer 'what if' questions, like 'what will happen to the world's climate if ...' or, on the money markets, 'what will happen to the price of X if ...'. Such models have to deal with large numbers of interdependent variables. The computers need large memories and fast data-handling.

26 For a personal computer, what are the human-to-machine interfaces? What transducers are involved?

Computer memories

Quantity of computer data is measured in **bytes**. A byte is a sequence of eight bits – a word eight bits long. That is, it is a sequence of eight binary 0s or 1s, corresponding to numbers up to 255. Computer memory capacity may be measured in megabytes or gigabytes. Here, a megabyte does not use the normal SI convention – megabyte is close to but not exactly a million bytes. Our normal counting system is based on powers of 10, but the binary system is based on powers of 2. So

$$1 \text{ million} = 10^6 = 1\,000\,000$$

but

$$1 \text{ megabyte} = 2^{20} \text{ bytes} = 1\,048\,576 \text{ bytes}$$

Information can be stored electronically as voltage across capacitors. A silicon chip can contain hundreds of thousands of capacitors, all of them able to store charge. A charged capacitor has a voltage across it, denoting a binary value of 1. A relatively uncharged capacitor denotes a binary value of 0. Such memory can be 'read' by allowing the charge to flow off the plates, though in the process the information is lost and must be rewritten onto the capacitor circuits. The system loses the information when the computer is switched off. So before switching off, any information to be stored for longer must be saved by being transferred to the **hard disk** within the computer or to a CD, floppy disk or zip disk which can be inserted by the user. Note that adding information to the hard disk or taking information from it involves motion – it is a mechanical process and is therefore slow compared with information exchange that is purely electronic.

The hard disk unit of a personal computer contains a stack of spinning platters (Figure 37.26). The faces of the platters are coated with material that can be locally magnetised. Information is placed on the disk by a head, which is an electromagnet whose polarity can be reversed. The platter rotates past the head, and in one polarity the head aligns the magnetism in one direction so that information is stored as a binary 'on' or 1, and in the opposite polarity the information acts as a binary 'off' or 0. To read the data the platter or disk spins past the head and a current is induced in the head.

Figure 37.26
The spinning platters in the hard disk unit of a personal computer.

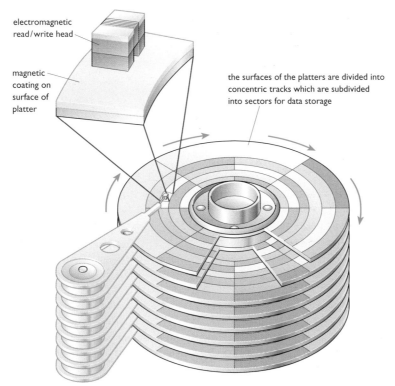

electromagnetic read/write head

magnetic coating on surface of platter

the surfaces of the platters are divided into concentric tracks which are subdivided into sectors for data storage

Electronic memory can also exist as potential differences in **bistable** or **flip-flop** circuits. These have two possible states, with two output voltages, and so are ideal for representing the two possible conditions of a binary information system. A bistable circuit can 'flip' or 'flop' between the two states. When it does so its state reverses, so it carries information about its previous state as well as its present one.

Both electronic and magnetic memories can be ROM or RAM types. **ROM** is 'read only' memory that cannot be added to. It stores information that the computer can access but which cannot be changed by new information. Much of this information is built into the computer at manufacture or it can be provided through software. An ordinary CD-ROM is a disk which carries data that can be accessed but not changed – it is 'read only'. **RAM** is 'random access' memory into which information can be added, by software or by inputs such as from a keyboard or mouse, as well as extracted. RAM may also be stored or saved on the hard disk, but for speed of access other methods are better.

27 Why is access to electronic information storage much faster than reading a disk?

28 Does a blank floppy disk provide RAM or ROM?

Automatic sensing – machine-to-machine data

Machines that can respond to their environment are clearly very valuable. We can place a baby in an incubator, for example, and not only leave the machine to monitor variables such as temperature and oxygen level, but also let its computer take action to maintain appropriate levels. In such a life-or-death situation we need to have other warning devices, such as independent monitoring equipment with audible alarms so that a human being can intervene if the computer fails for whatever reason. Such back-up mechanisms are an important part of any computer system, and without them there can be very serious problems.

Machines that respond to their environment need sensors. These are input transducers which convert information from the environment into voltage levels. Considering the transducer as a 'machine', it generates data specifically to be acted upon or manipulated by another machine.

Thermistors and light-dependent resistors (LDRs) form the basis of some sensor circuits. They are connected in series with another resistor to create a potential divider system. The output voltage is the voltage across one of the two resistors in the potential divider (Figure 37.27).

Figure 37.27
A light-sensing circuit whose output may be high or low.

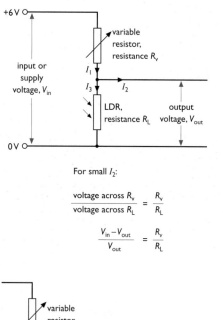

For small I_2:

$$\frac{\text{voltage across } R_v}{\text{voltage across } R_L} = \frac{R_v}{R_L}$$

$$\frac{V_{in} - V_{out}}{V_{out}} = \frac{R_v}{R_L}$$

Note that the required output from this circuit is a voltage level. The circuit is not required to provide power as its output and the current I_2 can be small.

If $I_2 \ll I_1$ then $I_1 \approx I_3$ and we can consider the two resistors as if they were simply in series. Two resistors in series share the voltage in the same ratio as their resistances, so

when $R_v = 0$, $V_{out} = 6\,V$

and when $R_v \gg R_L$, $V_{out} \approx 0$

The output voltage from such a sensor circuit can be fed to a **logic circuit** (Figure 37.28), which creates a digital output, 0 or 1, depending on whether its input voltage is more than or less than a certain value (usually approximately 1.2 V). Thus the real-world temperature or light level becomes a digital signal.

Figure 37.28
Creating a digital signal from a sensor circuit. The logic circuit acts as an analogue-to-digital converter.

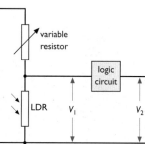

V_1 is continuously variable. The logic circuit, however, switches between two very different voltage levels. It does so at a particular value of V_1, as the graph shows

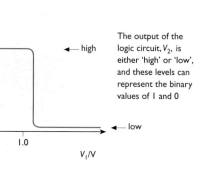

The output of the logic circuit, V_2, is either 'high' or 'low', and these levels can represent the binary values of 1 and 0

---

29 In Figure 37.27, explain whether the statement that when $R_v = 0$ then $V_{out} = 6\,V$ is consistent with the equation

$$\frac{V_{in} - V_{out}}{V_{out}} = \frac{R_v}{R_L}$$

30 What is the output voltage of the sensor circuit in Figure 37.27 when input voltage is 6 V, the value of the variable resistor is 1 kΩ and that of the LDR is 2 kΩ?

31 a For an input voltage of 6 V, what is the ratio of resistances of a potential divider for an output voltage of 1.2 V?
b Why is this value of output voltage particularly significant for an electronic sensor circuit?

Data gathering for development of scientific concepts

Figure 37.29
The Hubble Space Telescope is a passive receiver of light.

Note that to achieve high magnification the Hubble Space Telescope collects light from a tiny 'area' of space – a tiny proportion of the sphere of which it is the centre. It therefore has a small field of view. This picture is a true-colour image showing faint blue galaxies which are very far away. Due to the time taken for light to travel such long distances, they are seen as they were billions of years ago.

The extension of normal visual perception by the use of optical instruments had a big influence on people of the 1600s. It forced them to see things, literally and conceptually, in new ways. It encouraged them to change ideas – to change their minds.

Now, when we look at the moons of Jupiter or a galaxy beyond our own, we are still passive observers. Information flows to us, carried by the light. We take action to make ourselves as sensitive as possible to the light. To do this we use large telescopes including the Hubble Space Telescope which, carried by a satellite, has provided us with fascinating new details of objects in space (Figure 37.29), just as the earthbound optical telescopes did hundreds of years before. We can also diffract light so that it is splayed out according to its wavelength, and from this we read information about the elements that make up a distant star.

Figure 37.30
A radiotelescope passively receives data from the sky.

The Arecibo radiotelescope in Puerto Rico has a large surface area. This is not only so that it has maximum sensitivity to the faint radio waves from distant sources, but also to maximise resolution. Rayleigh's criterion (see Chapter 24) shows that for a circular aperture, which in this case is the dish diameter,

$$\theta = \frac{1.22\lambda}{d}$$

where θ is the angular separation of two points in space that the telescope can recognise as two separate points (resolve), λ is the radio wavelength, and d is the aperture (or dish) diameter. For good resolution (small value of θ), the formula tells us that d should be as large as possible.

We can step beyond the limits of our eyes, detecting radio waves (Figure 37.30), infrared, ultraviolet and X-rays from the stars. But we are still passive receivers of the information. We cannot go to the stars to tweak them to see what happens.

A similar passive gathering of data occurs when a seismic detector receives information from a distant earthquake (Figure 37.31). The information arrives as physical vibration, which is amplified to trace the changes taking place over time. The earthquake was not initiated by scientists in order to test their ideas. The scientists are again passive receivers. It is possible to examine such a stream of data that arrives naturally, in order to look for evidence to support or falsify a hypothesis. We can go further and become active rather than passive – we can deliberately manipulate a situation in order to see if nature provides a response to challenge or agree with the hypothesis. Such an activity is an experiment. A true experiment is a test of a hypothesis. Merely making a measurement or setting up practical arrangements to observe an established law in action is not testing a hypothesis. It is not a true experiment.

Figure 37.31
Active seismic
measurement – but it is
only a true experiment if it
is testing a hypothesis.

Data gathering for diagnosis

We need not receive seismic information only in passive ways. We can create our own seismic events, and control them to suit our purposes. This may not be to test a hypothesis about fundamental behaviour, but to infer specific information about structure that is hidden. It is diagnostic. During a seismic survey, the surveyors create explosions. They provide the input to the situation to see what nature does with this particular input.

Seismic surveying is an example of purposeful or diagnostic probing – an attempt to obtain information on the state of a hidden zone, working within the framework of scientific principles that are already accepted.

Diagnostic probing is also used a good deal in medicine (Figures 37.32 and 37.33). Input signal and output signal can take a variety of forms. To test the hearing of a baby, for example, the input signal is a sound signal and the output signal is the pattern of electrical voltages that is produced by brain activity. If there is brain activity that corresponds to the input sound then the hearing is satisfactory.

Figure 37.32
A beam of X-rays is sent
into a human body but not
all of them emerge. Much
of the energy is absorbed
or scattered, most strongly
by bone. So the beam of
emerging X-rays has
patterns from which we
can infer information about
internal structure.

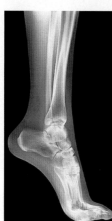

Figure 37.33
A beam of high frequency
sound (ultrasound) is sent
into a mother's body and
the pattern of reflections is
constructed into an image
of the unborn baby. The
frequency of the sound is not
changed, but, as with X-rays,
the distribution is affected.

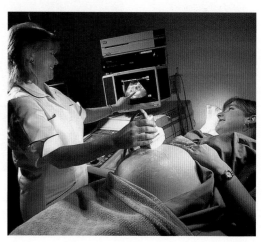

Figure 37.34
Reflected ultrasound signals can also be used to monitor velocity of the flow of blood. On reflection there is a change in frequency. The faster the flow of blood, the bigger the change in frequency.

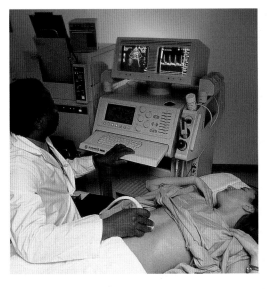

Ultrasound signals can be used to measure blood flow velocity (Figure 37.34). Reflection of the signals within the moving blood causes a frequency change in the ultrasound which depends on the velocity of the blood. This is an example of the Doppler effect, as explained in Figure 37.35 (see also Chapter 27).

Figure 37.35
The Doppler effect on reflection from a moving surface.

Consider three wavefronts, 1, 2 and 3, being reflected. For a stationary reflecting surface the situation, **A**, is quite simple. Wavelengths of incident and reflected waves are the same, λ_1. Their frequency is f_1.

If the surface is moving towards the incident waves, as in **B**, then reflection of wavefront 2 takes place at a shorter time interval after reflection of wavefront 1. The result is that the reflected wave has a reduced wavelength, λ_2, and an increased frequency, f_2:

$$\frac{(f_2 - f_1)}{f_1} = \frac{2v}{c}$$

where c is the speed of the waves and is constant.

$(f_2 - f_1)$ is the Doppler shift in frequency. If the surface were moving in the opposite direction, i.e. in the same direction as the oncoming waves, this would be negative: f_2 would be smaller than f_1.

Note that Doppler shift can also occur not by reflection but by relative movement of the source of waves and the observer/detector.

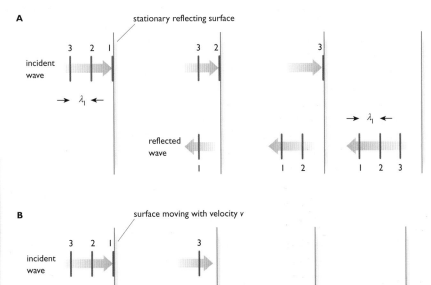

33 a In the Doppler measurement of blood flow, why does a reduced wavelength of reflected waves inevitably indicate an increase in their frequency?
b What would be the frequency shift for reflection of radio waves from the front of a car moving at $40\,\mathrm{m\,s^{-1}}$, using waves of initial frequency $300\,\mathrm{MHz}$? The speed of light is $3 \times 10^8\,\mathrm{m\,s^{-1}}$.

● **Comprehension and application**

Smart work

Figure 37.36
Jane Croucher – soon to be Dr Jane Croucher – in her lab at King's College.

Jane Croucher (Figure 37.36) is a researcher working at King's College in London. Optic fibres are her speciality, and she's researching into how to use them in the structures of airliners and buildings. Jane explains:

'Imagine a world where aeroplanes twist themselves into the best shapes for flying, buildings protect themselves against cracks, and water pipes find and report their own leaks. My research is all about uses of fibre optics to detect and respond to changes in pressure, temperature and movement in so-called smart structures – objects that can "feel".

'Fibre optics are mostly used for communications. These extremely fine threads of very pure glass, thinner than a fishing line and stronger than steel, can carry enormous amounts of data – over a million phone calls at the same time. In fact the bandwidth, or information-carrying capacity, of a fibre optic is limited by the electronics needed to decode data quickly enough.

'As well as having a much higher bandwidth than copper cables, fibre optics are smaller and lighter. That's important in aircraft design. By switching from copper cables to fibre optics an aeroplane can be up to 400 kg lighter. And because the fibres carry light and not electric current that generates magnetic field, there's no interference between one optic fibre system and another.

'In our research laboratory at King's College in London, we're working on a system of sensors based on optic fibre technology. We're developing a "nervous system" that can be built into the structure of an aircraft where it can detect tiny cracks that are invisible to the human eye. Corrosion cracking is a major problem that we're trying to solve, one that materials scientists don't fully understand, and a potential cause of air disasters.

'The sensors are based on a criss-cross network of optic fibres that are embedded into carbon composite materials. Composites are made of more than just a single material, combining the benefits of the different substances. An aircraft body can be made out of carbon composites instead of metal. Then, the embedded fibre optics can detect changes in pressure, temperature and movement – because these changes affect the wavelength of the light carried by the fibre.

'At the hearts of these sensing systems are Bragg gratings. These can be formed inside a single-mode optic fibre by exposing it to a repeating pattern of UV radiation, which permanently changes the refractive index of the glass. The fibre then has strips or layers of one refractive index and then another (Figure 37.37). At the boundaries of these layers, where one refractive index meets another, there is reflection of light travelling along the fibre. The reflected light interferes with the incident light. If white light travels into the fibre then at the Bragg grating some wavelengths experience destructive interference and some experience constructive interference. The result is that only a sharp wavelength range is transmitted to the end of the fibre.

Figure 37.37
The principle of a Bragg grating in an optic fibre.

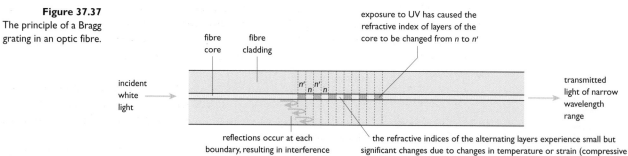

'The key point is that changes to the refractive indices of the layers of a Bragg grating change the value of the transmitted wavelength. The refractive indices of the Bragg grating are changed by temperature and by extension or compression of the glass. So the Bragg grating acts as a non-electrical remote temperature sensor and strain sensor. The wavelength of the transmitted light acts as the variable that maps the values of temperature or strain.

'One way to detect the change in wavelength of the light emerging from the optic fibre would be to diffract the light and detect the change in position of a diffraction maximum. This, however, would require a very small diffraction grating at the fibre end. It is easier to detect the change in intensity of a fixed wavelength of emerging light. The emerging wavelength does not have an absolute, discrete value, but has a narrow range of wavelengths with a sharp intensity peak (Figure 37.38). Changes in the Bragg grating cause this peak to "move" to the left or right along the intensity–wavelength spectrum. So the intensity of a single wavelength increases and decreases. It is this changing intensity that's detected.

Figure 37.38
The sensing system detects a change in intensity of the transmitted light at a particular wavelength.

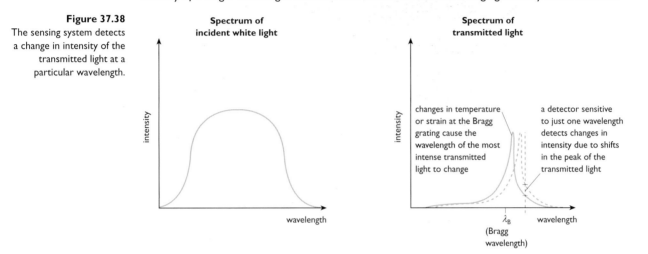

34 Explain whether or not a Bragg grating is a transducer.
35 What is a 'single-mode fibre'?
36 How might embedded optic-fibre sensing systems be of use in large concrete structures such as dams?
37 An optic fibre sensor itself is an analogue device.
   a Explain fully how the output from an optic fibre sensor could be converted into digital data.
   b In an application such as sensing cracks in aircraft structure, is there any advantage in either analogue or digital data?
38 Outline the differences between the use of optic fibres for sensing and their use for communications.
39 In what ways are fibre-optic sensors more useful than electrical sensors such as thermistors and strain gauges?
40 Explain why it is the capability of the electronics and not the optic fibres themselves that limits the bandwidth of a communications system.

'Structures with embedded optic fibre sensors are called "passive" smart structures. In future, the information provided by the sensors may be used to "feed back" to the structure and change its stiffness, shape or position. These are "adaptive" smart structures. Already, helicopter rotor blades use the Active Twist Rotor (ATR) concept to control blade twisting. Eventually, smart structures might be capable of learning. One day, aeroplanes could be built like human bodies – with a "nervous system", "muscles" and a "brain".'

· ·

● **Extra skills task**   Application of Number

An analysis of 'errors' or uncertainties is always important in physics. In general, you should never give your answer to more significant figures than the measurements that it's based on. Readings should be stated as $x \pm e$. It is also important to have a feel for the percentage error in your answer. This can be used to estimate the percentage error for each type of measurement.

Revisit one of your experiments. Estimate the percentage error in each of the quantities that you have measured. These percentage errors can be converted into absolute values for each point plotted on the graph and shown as error bars. Now draw your line of best fit. Your line of best fit should always lie within these error bars.

· ·

Examination questions

1 **a** State an example of the use in telecommunications of:
   i microwaves
   ii UHF signals. (2)
   b Give the approximate wavelength of each of the following:
   i a microwave signal
   ii a UHF signal. (2)
   OCR, Communications, Mar 1999

2 The diagram shows a trace on the screen of an oscilloscope. The Y-sensitivity of the oscilloscope is set at 5.0 V per division and the time base is set at 0.50 ms per division.

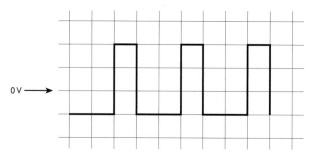

For the trace, determine
   a the maximum positive value of potential difference
   b the maximum negative value of potential difference
   c the frequency of the signal. (4)
   AQA (NEAB), Mechanics and Electricity, (PH01), Mar 1999 (part)

3 **a** Explain why high-quality sound reproduction using digital techniques requires a high sampling rate. (3)
   b i What is meant by *bandwidth*? (1)
   ii The figure shows the frequency band occupied by an AM signal.

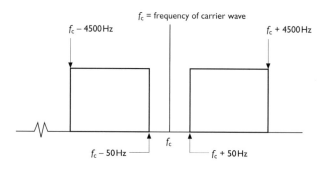

   Use the information in the figure to determine:
   1 the bandwidth of the AM signal
   2 the maximum frequency of the modulating signal. (2)
   c Suggest a reason why a restricted number of broadcasts are transmitted in each of the LW and MW bands. (2)
   OCR, Communications, Mar 1999

4 The graph shows an amplitude modulated signal received by a radio.

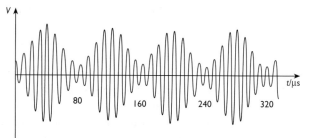

   a What frequencies are shown on this radio wave?

   The diagrams A and B show two radio signals.

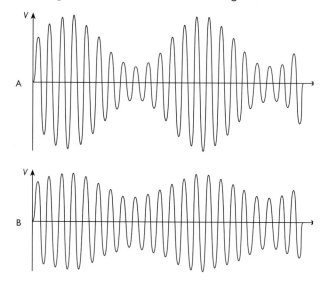

   b Explain one similarity between the audio waves they carry and one difference. (4)
   Edexcel (London), Physics, PH3, June 1999 (part)

5 A typical compact disc (CD) stores an hour of recorded and digitised music. The music is sampled at a frequency of 44.1 kHz. Each sample consists of two 16-bit words corresponding to the left and right stereo channels.
   a Calculate the total number of bits which are stored on this CD. (2)
   b It is not possible to hear a frequency of 44.1 kHz. Explain why such a high sampling frequency is used. (1)
   c Explain one advantage and one disadvantage of storing and transmitting musical information in a digital rather than an analogue format. (2)
   d Modern communication systems store and transmit written information using ASCII codes. In this system, each alphanumeric character (i.e. letter or number) is represented by an 8-bit word. Seven of these eight bits represent the character itself and the eighth bit is a parity or error checking bit.

i Calculate the maximum number of alphanumeric characters which can be represented by an ASCII code.

ii Estimate the total number of bits required to store a page of text in a novel. Explain how you arrive at your answer.

iii A CD-ROM is essentially a compact disc which stores text rather than musical data. Use your answers to **a** and **d ii** to estimate the total number of pages of text which could be stored by the bits available on the audio CD. (4)

e Early communication systems used Morse code to represent the letters of the alphabet. This code consisted of a series of short pulses (dots ·) or long pulses (dashes −). The Morse coded alphabet is shown.

A · −	G − − ·	M − −	S · · ·
B − · · ·	H · · · ·	N − ·	T −
C − · − ·	I · ·	O − − −	U · · −
D − · ·	J · − − −	P · − − ·	V · · · −
E ·	K − · −	Q − − · −	W · − −
F · · − ·	L · − · ·	R · − ·	X − · · −
			Y − · − −
			Z − − · ·

i State one way in which ASCII code is similar to Morse code.

ii Suggest two reasons why Morse code was abandoned for modern communication systems and ASCII code developed instead. (3)

OCR, Sciences, Telecommunications, Nov 1999

Answers to numerical questions

Chapter 1

Marginal questions
7 $20°$

Chapter 2

Marginal questions
6 $0.30\,\text{m s}^{-1}$
7 $2.0\,\text{Hz}$
8 a $\text{m}^2, [\text{L}]^2$ **b** $\text{m}^3, [\text{L}]^3$
9 $6 \times 10^{14}\,\text{Hz}$

Chapter 3

Marginal questions
1 b $7 \times 10^{14}\,\text{Hz}, 4 \times 10^{14}\,\text{Hz}$
2 blue light
3 violet light
4 a about 2000 **b** 2×10^{10} **c** 3×10^{13} **d** 3×10^{-14}
12 3000 to $0.3\,\text{m}$
13 1.7×10^{-4} or 17 parts in 100 000

Chapter 4

Marginal questions
4 501.5
5 $[\text{M}][\text{L}]^{-3}$
6 1×10^{44}
7 a 9 **b** 9×10^{30}
8 10^9
9 a $100\,\text{m}$ **b** $10\,\text{mm}$ **c** $100\,\text{mm}$ **d** $10\,\mu\text{m}$
10 $2.70\,\text{g cm}^{-3}$
11 $10^{-30}\,\text{g cm}^{-3}$
12 a $1\,\text{g}$ **b** $20\,\text{g}$ **c** $2\,\text{kg}$

Comprehension and application
16 a i $17.1\,\text{m}^3$ **ii** $14.2\,\text{m}^3$ **c** $13\,900\,\text{m}^3$

Extra skills task
1 $46.7, 57.3, 58.6, 60.2, 62.7, 66.1, 68.5, 70.4\,\%$
2 a 0.155% **b** 0.155%

Chapter 5

Marginal questions
4 b 2.4

5 $\Delta l = \dfrac{F}{k}$

6 $k = 50\,\text{N m}^{-1}$
7 $k = 5 \times 10^3\,\text{N m}^{-1}$
8 $[\text{L}]$
9 a $60\,\text{N mm}^{-1}$ **b** $6 \times 10^4\,\text{N m}^{-1}$

12 Young's modulus, $E = \dfrac{Fl}{A\Delta l}$ so $A = \dfrac{Fl}{E\Delta l}$

    a $1.5 \times 10^{-7}\,\text{m}^2$ **b** $0.15\,\text{mm}^2$

14 **a** 0.19 **b** copper wire   **c** 5.4

17 $\Delta l = \dfrac{Fl}{AE} = 1.8 \times 10^8 \times \dfrac{2}{1.3 \times 10^{11}} = 2.8 \times 10^{-3}\,\text{m}\ (2.8\,\text{mm})$

● Chapter 6

Marginal questions

9 **a** about $2 \times 10^5\,\text{N}$ (assuming skin area $= 2\,\text{m}^2$)

● Chapter 7

Marginal questions

1 green

Extra skills task

$K = 30.8$ (dimensions $[\text{T}]^{-1}$)

● Chapter 9

Marginal questions

3 $[\text{T}]^{-1}$

10 **a** 235, 92; 238, 92; 239, 92   **b** 143, 146, 147

13 **a** $^{234}_{92}\text{U} \rightarrow\ ^{230}_{90}\text{Th} + ^{4}_{2}\alpha$

   b $^{234}_{90}\text{Th}$: 234, 90, 144

     $^{230}_{90}\text{Th}$: 230, 90, 140

Comprehension and application

15 **a** $4.3 \times 10^{17}\,\text{Bq}$ **b** $4.3 \times 10^{14}\,\text{Bq}$

16 $7 \times 10^6\,\text{km}^2$

17 **a** $8 \times 10^{-4}\,\text{kg}$ **b** $3 \times 10^{11}\,\text{Bq}$ **c** $8 \times 10^{-10}\,\text{kg}$, $3 \times 10^5\,\text{Bq}$

● Chapter 10

Marginal questions

2 $120\,\text{km}\,\text{h}^{-1}$, $30\,\text{km}\,\text{h}^{-1}$

3 **a** $18.8\,\text{m}\,\text{s}^{-1}$

8 **b** 2 s, 5 m   **c** $2.5\,\text{m}\,\text{s}^{-1}$

● Chapter 11

Marginal questions

6 **a** $2.5\,\text{m}\,\text{s}^{-2}$ **b** $-2\,\text{m}\,\text{s}^{-2}$

7 *F* vector, *m* scalar, *a* vector

9 $[\text{M}][\text{L}][\text{T}]^{-2}$

10 $240\,\text{N}$

11 $a = \dfrac{F}{m} = \dfrac{27\,\text{N}}{4.5\,\text{kg}} = 6.0\,\text{m}\,\text{s}^{-2}$

Comprehension and application

20 **a** $1.7\,\text{m}\,\text{s}^{-2}$

● Chapter 12

Marginal questions

1 scalar, scalar, vector, vector

4 $1\,\text{m}\,\text{s}^{-1}$ up the escalator

5 **a** $18\,\text{N}$ **b** $2\,\text{N}$

6 **a** $1.6\,\mathrm{m\,s^{-1}}$ **b** $270\,\mathrm{N}$

7 resultant force $F = ma = 2 \times 10^6\,\mathrm{N} = 2000\,\mathrm{kN}$
add force to cancel weight ($1000\,\mathrm{kN}$)
total upward force $= 3000\,\mathrm{kN}$

9 $10.2\,\mathrm{m\,s^{-1}}$ at $11°$ to $10\,\mathrm{m\,s^{-1}}$ vector

10 $5\,\mathrm{N}$ at $37°$ to the normal

12 $69\,\mathrm{m\,s^{-1}}$, $12\,\mathrm{m\,s^{-1}}$

13 $188\,\mathrm{N}$, $68\,\mathrm{N}$

14 $7.88\,\mathrm{m\,s^{-1}}$, $1.39\,\mathrm{m\,s^{-1}}$

15 **a** $4.3\,\mathrm{N}$ **b** $24.6\,\mathrm{N}$ ($25\,\mathrm{N}$ to 2 sig. figs)

16 component of $3\,\mathrm{m\,s^{-1}}$ to right $= (3\cos 40°)\,\mathrm{m\,s^{-1}} = 2.30\,\mathrm{m\,s^{-1}}$
sum of velocities to the right $= 4.80\,\mathrm{m\,s^{-1}}$
component of $3\,\mathrm{m\,s^{-1}}$ down page $= (3\sin 40°)\,\mathrm{m\,s^{-1}}$
$$= 1.93\,\mathrm{m\,s^{-1}}$$

now use Pythagoras' theorem:

resultant $= \sqrt{4.80^2 + 1.93^2}\,\mathrm{m\,s^{-1}}$
$$= 5.2\,\mathrm{m\,s^{-1}}$$
angle (down page) to $2.5\,\mathrm{m\,s^{-1}}$ velocity

$$= \tan^{-1}\left(\frac{1.93}{4.80}\right)$$

$$= 22°$$

Comprehension and application

20 **a** $300\,\mathrm{rpm} = 5\,\mathrm{rps}$
so $t = 0.2\,\mathrm{s}$

b speed $= \dfrac{2\pi r}{t} = 140\,\mathrm{m\,s^{-1}}$ backwards

c $360\,\mathrm{km\,h^{-1}} = 100\,\mathrm{m\,s^{-1}}$ forwards
speed of retreating blade end $= 40\,\mathrm{m\,s^{-1}}$ backwards, relative to ground

d $240\,\mathrm{m\,s^{-1}}$ forwards, relative to ground

22 $\dfrac{\text{thrust}}{25\,\mathrm{kN}} = \tan 5°$

thrust $= 2.2\,\mathrm{kN}$

● Chapter 13   Marginal questions

8 vector

9 **a** $[M]$ **b** $[L]$ **c** $[T]$ **d** $[L][T]^{-1}$ **e** $[L][T]^{-2}$ **f** $[M][L][T]^{-2}$ **g** $[L][T]^{-2}$

10 $1.6\,\mathrm{N\,kg^{-1}}$

● Chapter 14   Marginal questions

5 momentum before $= (90 \times 8) + (50 \times 6) = 1020\,\mathrm{kg\,m\,s^{-1}}$
momentum after $(90 + 50)v = 1020\,\mathrm{kg\,m\,s^{-1}}$
so $v = 7.3\,\mathrm{m\,s^{-1}}$

9 **d** $3.0\,\mathrm{m\,s^{-1}}$

11 $\Delta(mv) = (-0.02 \times 2) - (0.02 \times 20)$
$$= -0.44\,\mathrm{kg\,m\,s^{-1}}$$

Comprehension and application

14 **a** $5 \times 10^{11}\,\mathrm{m}$, $3.2 \times 10^{11}\,\mathrm{m}$ **b** $3.2 \times 10^4\,\mathrm{m\,s^{-1}}$ **c** $1.9 \times 10^{29}\,\mathrm{kg\,m\,s^{-1}}$, $-1.9 \times 10^{29}\,\mathrm{kg\,m\,s^{-1}}$
d $5.2 \times 10^{11}\,\mathrm{m^3}$, $1.3 \times 10^{15}\,\mathrm{kg}$

16 **a** $-4.2 \times 10^{19}\,\mathrm{kg\,m\,s^{-1}}$ **b** $1.9 \times 10^{29}\,\mathrm{kg\,m\,s^{-1}}$ **c** $-2.2 \times 10^{-8}\%$

● Chapter 15 **Marginal questions**

2   $W_A = 2800\,kJ$
    $W_B = 1200\,kJ$

5   **a** W = area under graph
        = (average F) × x
        $\approx (1.2\,N) \times 10^{-3}\,m$
        $\approx 10^{-3}\,J$
    b $W \approx (0.15\,N) \times 10^{-3}\,m$
        $\approx 1.5 \times 10^{-4}\,J$
    c $W \approx (0.2\,N) \times 0.01 \times 10^{-3}\,m$
        $\approx 2 \times 10^{-6}\,J$

6   1 kJ

7   **a** $3 \times 10^{-18}\,J$ to $2.65 \times 10^{33}\,J$
    b 0.038 MJ
        0.000 15 MJ
        $2.65 \times 10^{27}\,MJ$
        $3 \times 10^{-24}\,MJ$
    c 38 MJ

    d $\dfrac{150}{3 \times 10^{-18}} = 5 \times 10^{19}$

    e $\dfrac{2.65 \times 10^{33}}{150} = 1.8 \times 10^{31}$

        factor = 1.8×10^{21} (i.e. seconds for 10^{10} to survive)
        length of time = $1.8 \times 10^{21}\,s$

        $\qquad = \dfrac{1.8 \times 10^{21}}{365 \times 24 \times 3600}$ years

        $\qquad \approx 6 \times 10^{13}$ years

8   0.03 J

10 **a** 17 MJ   **c** 23 km

14 $P = \frac{1}{2}F\dfrac{x}{t} = 0.73\,W$

15 **a** 1700 MW

16 **a** $\dfrac{P_2}{P_1} = 0.35,\ P_1 = 290\,MW$

    c $3.6 \times 10^5\,MJ$
    d $1.0 \times 10^6\,MJ$

17 **a** $9 \times 10^{16}\,J$
    b $1.5 \times 10^{-10}\,J$

Extra skills task
350 MJ, $1.2 \times 10^4\,MJ$

● Chapter 16 **Marginal questions**

1   **a** 800 N, for example
    b **i** 2.16 kJ   **ii** 17.28 kJ

    c $\dfrac{\Delta E_2}{\Delta E_1} = 8$

3   $\Delta E = 10^{11}\,J$, negative, as 'up' is positive

4   $\dfrac{g_s}{g_o} = \dfrac{r_o^2}{r_s^2} = 1.10$

5 a 2×10^5 J

  b i $+1.1 \times 10^5$ J **ii** -2×10^5 J

7 $E_T = \frac{1}{2} \times 3 \times 10^3 \times 10^2 = 1.5 \times 10^5$ J

  $E_C = \frac{1}{2} \times 10^3 \times 30^2 = 4.5 \times 10^5$ J

  the truck has more kinetic energy

8 a E_p loss $= mgh = 500 \times 10 \times 1.2 = 6000$ J $= E_k$ gain

  b $F_{av} = \dfrac{w}{x} = \dfrac{6000 \text{ J}}{0.07 \text{ m}} = 86$ kN

Comprehension and application

12 $r = 4.4 \times 10^9$ km

  $\Delta r = 2.9 \times 10^9$ km

13 $\Delta E_p = \dfrac{2.0 \times 10^{42} \times 2.9 \times 10^{12}}{4.4 \times 10^{12} \times 7.3 \times 10^{12}} = 1.8 \times 10^{29}$ J

  from furthest to closest, 1.8×10^{29} J loss

  from closest to furthest, 1.8×10^{29} J gain

14 from furthest to closest, $\Delta E_k = +1.8 \times 10^{29}$ J

  from closest to furthest, $\Delta E_k = -1.8 \times 10^{29}$ J

15 average $r = 5.85 \times 10^9$ km

  $c = 2\pi r$ where r is average

    $= 3.7 \times 10^{13}$ m

  average speed equals $\dfrac{c}{\text{orbit time}}$

  $= \dfrac{3.7 \times 10^{13}}{247.7 \times 365 \times 24 \times 3600}$

  $= 4700$ m s^{-1}

  average kinetic energy $= \frac{1}{2} \times 1.5 \times 10^{22} = 4700^2 \ = 1.7 \times 10^{29}$ J

  max. kinetic energy $= 1.7 \times 10^{29} + \frac{1}{2}(1.8 \times 10^{29}) \ = 2.6 \times 10^{29}$ J

  min. kinetic energy $= 1.7 \times 10^{29} - \frac{1}{2}(1.8 \times 10^{29}) \ = 0.8 \times 10^{29}$ J

  max. speed $= 5900$ m s^{-1}

  min. speed $= 3300$ m s^{-1}

● Chapter 17 Marginal questions

6 6.25×10^{18}

7 10^{14}

  charge $= -1.6 \times 10^{-5}$ C

8 -1.7×10^{11} C

10 a $F_g = mg = 8.9 \times 10^{-30}$ N

    $a_1 = 9.81$ m s^{-2}

    $F_E = qE = 1.6 \times 10^{-16}$ N

    $a_2 = \dfrac{F_E}{m} = 1.8 \times 10^{14}$ m s^{-2}

    $F_B = Bqv = 6.4 \times 10^{-15}$ N

    $a_3 = 7.0 \times 10^{15}$ m s^{-2}

15 a 16C

● Chapter 18 Marginal questions

4 $V = 5$ V

5 $q = \dfrac{W}{V} = 4$ C

7 a negative **b** $W = -2\,J$ **c** $V = 20\,V$
9 p.d. $= 25\,V$
10 $W = +0.2\,J$
11 a $0.0125\,J$ **b** $2.0 \times 10^{-19}\,J$
12 volt
13 $0.3\,V$

Chapter 19

Marginal questions

1 a $24\,C$ **b** 1.5×10^{20}
4 $1.7 \times 10^4\,J$
5 $4\,h\ 16\,min$

Comprehension and application

16 a i $W = IVt = 4.8 \times 10^4\,J$ **ii** $2.9 \times 10^6\,J$
b $0.23\,J$

Chapter 20

Marginal questions

3 $30\,\Omega$
4 $4.0\,V$
5 a $192\,J$ **b** $1.2 \times 10^4\,J$
6 a $30\,\Omega$

b $P = \dfrac{V^2}{R} = 19\,W$

7 a $W = I^2Rt = 2300\,J$ **b** $9200\,J$

8 a $t = \dfrac{W}{I^2R} = 5\,s$ **b** $28\,min$ **c** $8\,h\ 20\,min$

9 a $W = \dfrac{V^2 t}{R}$ **b** $160\,kJ$

11 $I = \dfrac{P}{V} = 10\,A$

21 $V = IR$ so $\varepsilon = IR + Ir$ and $r = \dfrac{\varepsilon - IR}{I} = 4.0\,\Omega$

23 a gradient $= -0.3\,\Omega$
b i zero **ii** $8\,A$

Comprehension and application

28 a approx. 80

Chapter 22

Marginal questions

11 $25.5 - x = 14.5 + x$
$25.5 - x + x = 14.5 + x + x$
$25.5 = 14.5 - 14.5 + 2x$
$11 = 2x$
$\dfrac{11}{2} = \dfrac{2x}{2}$
$x = 5.5$

12 a $a = \dfrac{b - c}{2}$ **b** $a = 2p - q$ **c** $a = \dfrac{5y - z}{x}$ **d** $a = 3vw$ **e** $a = \dfrac{3e}{4c + d}$

f $a = p - \dfrac{4q}{3}$ **g** $a = \dfrac{d}{6c} + b$ **h** $a = 15c - \dfrac{3b}{4}$ **i** $a = \dfrac{xy}{x - y}$ **j** $a = \dfrac{mnr}{np + mq}$

18 **a** $[M][L]^2[T]^{-3}[I]^{-2}$
19 **a** $1.1\,pm$ **b** 10^5
20 **a** 6×10^{10} **b** 3.2×10^{10} years
 c $1\,Gs$ (about 32 years)
22 (1.8 ± 0.3)
23 **a** **i** 3 **ii** 3
 b **i** $1300\,N$ **ii** $0.000\,57\,m$
24 **a** **i** $(39 \pm 2)\,mm$ **ii** $(0.050 \pm 0.003)\,mm$
 b **i** $(38.8 \pm 0.4)\,mm$ **ii** $(0.0500 \pm 0.0005)\,mm$

● Chapter 23

Marginal questions

1 **a** **i** 1.49 **ii** 0.663
 b **iii** 40.0° **iv** 60.0°
 c **v** 56.5° **vi** 39.3°
3 **a** 0.885 **b** 50.0°
4 1.48, 1.46
6 **a** **i** 30° **ii** 30.0° **iii** 29.99° **iv** 29.990°
7 **a** 47.8° **b** 24.0°
12 40.5°

16 **a** $v = \dfrac{3.00 \times 10^8}{1.38} = 2.17 \times 10^8\,m\,s^{-1}$

 b $\lambda = \dfrac{6.00 \times 10^{-7}}{1.38} = 4.35 \times 10^{-7}\,m$

 c $f = \dfrac{v}{\lambda} = 5.00 \times 10^{14}\,Hz$

17 $\lambda = \dfrac{3.7 \times 10^{-7}}{1.33} = 2.8 \times 10^{-7}\,m$

Comprehension and application

21 **a** $5.03 \times 10^{-5}\,s$, $5.10 \times 10^{-5}\,s$
22 **c** $1.41x$
 d $14.1\,km$
 e $7.23 \times 10^{-5}\,s$
 f $5.13 \times 10^{-5}\,s$

Examination questions

1 **c** **i** $2.26 \times 10^8\,m\,s^{-1}$
 ii $5.0 \times 10^{14}\,Hz$
 iii $\lambda_{air} = 6.0 \times 10^{-7}\,m$
 $\lambda_{water} = 4.5 \times 10^{-7}\,m$
 iv 48.8°

2 **b** **i** critical angle $= \sin^{-1}\dfrac{1}{1.60} = 39°$

 angle of incidence less than critical angle
3 **c** **i** 0.898 ($n =$ ratio of speeds) **ii** 63.9° (critical angle)
4 **b** **i** $D_m = (37.2 \pm 0.1)°$
 $\theta = (49 \pm 1)°$
 ii $n = 1.50$
 c **i** $D = (n - 1)A$ **ii** work in degrees: use false origin and large scale
5 **b** $10\,\mu s$
 c **i** $2.0 \times 10^8\,m\,s^{-1}$ **ii** 1.5 (ratio of speeds)

● Chapter 24

Marginal questions

7 **a** virtual, magnified, upright
　　b real, magnified, inverted
　　c real, diminished, inverted

8 **a** $-33.3\,\text{cm}$ 　**b** $175\,\text{cm}$ 　**c** $85.7\,\text{cm}$

9 $-8.6\,\text{cm}$

10 0.75
　　180, 360, 57
　　2π

11 **a** $-1.67, -8.35\,\text{cm}$
　　b $2.50, 12.5\,\text{cm}$
　　c $0.714, 3.57\,\text{cm}$

12 **a** $0.080\,\text{rad}$
　　b $0.24\,\text{cm}$
　　c $0.60\,\text{rad}, 1.8\,\text{cm}$
　　d 7.5

13 **a** 18
　　b 0.16, 0.00087, 18; equal
　　c $28\,\text{mm}$

15 **a** $5x\,\text{rad}$ 　**b** $120\,\text{cm}$

Comprehension and application

23 **a** **i** $3\,\text{cm}$
　　b **ii** $2.6\,\text{cm}$

24 **a** $13.4\,\text{cm}$ 　**b** at a distance of $47\,\text{cm}$

27 **a** $2 \times 10^{-4}\,\text{rad}$

29 $12.5\,\text{cm}, -12.5\,\text{cm}$

32 **c** $u = f$

34 1 calculate v for objective $(4.67\,\text{cm})$

　　linear magnification for objective $m_o = \dfrac{v}{u} = 8.34$

　　2 calculate u for eyepiece $(4.93\,\text{cm})$
　　3 calculate v for eyepiece $(-23.34\,\text{cm})$

　　4 linear magnification for eyepiece $m_e = \dfrac{v}{u} = -4.73$

　　5 $M = m_e m_o = -39$

Examination questions

2 **a** **ii** $6.25 \times 10^{-3}\,\text{rad}$

3 **b** $m = \dfrac{v}{u} = -5$

　　$v = -5u$ (virtual image)

　　$$\frac{1}{f} = \frac{1}{u} + \frac{1}{v}$$

　　$$\frac{1}{100} = \frac{1}{u} + \frac{1}{-5u}$$

　　$u = 80\,\text{mm}$

4 **a** **ii** $-60\,\text{mm}$
　　b converging lens image is real, inverted, magnified, to right of lens; diverging lens image is virtual, upright, diminished, to left of lens

5 **c** **i** converging 　**iii** $33\,\text{cm}$ $(v = -100\,\text{cm})$

● Chapter 25

Marginal questions

3 2

7 1, 0, −1, 0, 1, 0, 0

8 a 10π rad   **b** 1800°

9 2 rad

10 a 90°, 360°, 720°, 900°, 1080°, 57.3°, 114.6°

  b $\frac{1}{4}$, 1, 2, $2\frac{1}{2}$, 3, 0.159, 0.318

11 a $\pi/3$ or 1.05 rad, π or 3.14 rad

    $3\pi/2$ or 4.71 rad, 4π or 12.57 rad

  b $\frac{1}{6}$, $\frac{1}{2}$, $\frac{3}{4}$, 2 revolutions

16 a 40° or 0.7 rad   **b** 1 m (so no fifth max. as 1 > 0.8)   **c** 2 m

20 $\delta x = \dfrac{\lambda D}{d} = 0.3$ mm

22 a 0.04°

  b $\delta\theta = 7 \times 10^{-4}$ rad

    $\delta x = 0.2\delta\theta$ m = 0.1 mm

23 $\sin\theta_1 = \dfrac{\lambda}{d}$

  $\sin\theta_v = \dfrac{3.9 \times 10^{-7}}{1/500\,000}$

  $\theta_v = 11.2°$, $\theta_r = 20.5°$

  total angle = 20.5 − 11.2 = 9.3°

24 $n\lambda = d\sin\theta_n$

  $\theta_2 = 28.1°$

Examination questions

1 d B is $\pi/2$ rad ahead of A

2 b ii $n = \dfrac{\sin\theta}{\sin r}$

    $= \dfrac{\sin\theta}{\sin(90° - \theta)}$

    $= \dfrac{\sin\theta}{\cos\theta}$

    $= \tan\theta$

  $\theta = \tan^{-1} 1.33 = 53°$

6 b use Rayleigh criterion, $\theta = \dfrac{1.22\lambda}{d}$

  θ is smaller for shorter λ

7 a ii 78.6°

  c ii $f = \dfrac{\omega}{2\pi} = \dfrac{4100}{2\pi} = 653$ Hz

   iii $kx = 2\pi$ when $x = \lambda$ (phase change of 2π rad in one cycle)

    $\lambda = \dfrac{2\pi}{k}$

      $= 0.483$ m

    $v = f\lambda$

      $= 653 \times 0.483$

      $= 315$ m s^{-1}

    iv first wave $y_1 = A \sin(\omega t + kx)$

        second wave $y_2 = A \sin(\omega t - kx)$ (opposite travel)

   v $y = y_1 + y_2$

        $= A \sin(\omega t - kx) + A \sin(\omega t + kx)$

        $= 2A(\cos kx)(\sin \omega t)$

  vi distance between consecutive nodes $= \dfrac{\lambda}{2} = 0.242\,\text{m}$

Chapter 26 — Marginal questions

4 **a** $-1.8 \times 10^{11}\,\text{C}$ **b** $1.6 \times 10^{-19}\,\text{C}$

10 **a** $4.4 \times 10^{14}\,\text{Hz}$ **b** $1.8\,\text{eV}$

15 **a** $\dfrac{1}{2000}$

    b $\theta = \dfrac{1.22\lambda}{d}$

      $\dfrac{\theta_n}{\theta_e} = \dfrac{\lambda_n}{\lambda_e} = \dfrac{1}{2000}$

Comprehension and application

16 **a** **i** $1.3 \times 10^{-22}\,\text{kg m s}^{-1}$ **ii** $1.3 \times 10^{-27}\,\text{kg m s}^{-1}$

    b **i** $4.0 \times 10^{-19}\,\text{J}$ **ii** $2.5\,\text{eV}$

18 $1.3 \times 10^{-27}\,\text{kg m s}^{-1}$ (see **16 a ii**)

    $p = mv$

    $v \approx 10^4\,\text{m s}^{-1}$

Examination questions

1 **c** **i** $1.66 \times 10^{-27}\,\text{kg m s}^{-1}$ **ii** $4.97 \times 10^{-19}\,\text{J}$

2 **a** $4.86 \times 10^6\,\text{m s}^{-1}$ $\left(\lambda = \dfrac{h}{mv}, \text{ so } v = \dfrac{h}{m\lambda}\right)$

3 **b** $\lambda = \dfrac{d\,\delta x}{D}$

       $\delta x = \dfrac{\lambda D}{d}$

          $= 8.3 \times 10^{-4}\,\text{m}$

    c **i** $5.2 \times 10^{-10}\,\text{m}$ **ii** $\delta x = 7.8 \times 10^{-7}\,\text{m}$

Chapter 27 — Marginal questions

2 **a** $1.0 \times 10^6\,\text{Hz}$ **b** $6.6 \times 10^{-28}\,\text{J}$ **c** 9.7×10^8

5 **a** $6.6 \times 10^{-18}\,\text{J}$ **b** $1.5 \times 10^{13}\,\text{Hz}$ **c** $41\,\text{eV}, 0.063\,\text{eV}$

6 **a** $10.2\,\text{eV}$ **b** $2.5 \times 10^{15}\,\text{Hz}$

7 $1.9\,\text{eV}$: C to B

11 $2.9 \times 10^{15}\,\text{Hz}$

Comprehension and application

18 **a** $2.90 \times 10^{-7}\,\text{m}, 3.32 \times 10^{-7}\,\text{m}, 5.66 \times 10^{-7}\,\text{m}$

20 **a** $+2.5 \times 10^{12}\,\text{Hz}$ for B to A

    b $-2.5 \times 10^{14}\,\text{Hz}$

22 **a** $4.7 \times 10^{14}\,\text{Hz}$ **b** $640\,\text{nm}$: red

Examination questions

1 **a** $2.1 \times 10^6\,\mathrm{m\,s^{-1}}$

 b **i** $E_2 - E_0 = 1.94 \times 10^{-18}\,\mathrm{J}$
   so $2.0 \times 10^{-18}\,\mathrm{J}$ is more than enough

  ii $E_2 - E_1 = 3.06 \times 10^{-19}\,\mathrm{J}$

$$= h\nu = \frac{hc}{\lambda}$$

$$\lambda = \frac{hc}{E_2 - E_1} = 650\,\mathrm{nm}\ (\text{visible: red})$$

 c $13.6\,\mathrm{V}$

2 **b** $E = \dfrac{hc}{\lambda}$

   $= 4.1 \times 10^{-19}\,\mathrm{J}$

3 **a** **i** $2810\,\mathrm{eV}$

  ii $\lambda = \dfrac{hc}{E} = 4.4 \times 10^{-10}\,\mathrm{m}$

  iii borderline of UV and X-rays

 b $\dfrac{h}{mv} = 1.1 \times 10^{-13}\,\mathrm{m}$

5 **b** **i** $2.6 \times 10^{-18}\,\mathrm{J}$

  ii $1.0\,\mathrm{eV}$

  iii $E_3 - E_2 = h\nu = \dfrac{hc}{\lambda}$

    $= 5.6 \times 10^{-7}\,\mathrm{m}$

  iv visible: yellow

● Chapter 28

Marginal questions

3 $^{234}_{90}\mathrm{Th}$, thorium-234 (see periodic table)

4 zero

6 $^{14}_{6}\mathrm{C} \rightarrow\ ^{14}_{7}\mathrm{N} +\ ^{0}_{-1}\mathrm{e} + \bar{\nu} + Q$

9 $^{15}_{8}\mathrm{O}$, $^{120}_{42}\mathrm{Mo}$, $^{113}_{50}\mathrm{Sn}$

14 $^{216}_{84}\mathrm{Po} \rightarrow\ ^{212}_{82}\mathrm{Pb} +\ ^{4}_{2}\alpha$

  $^{13}_{8}\mathrm{O} \rightarrow\ ^{13}_{7}\mathrm{N} +\ ^{0}_{1}\mathrm{e} + \nu$

  $^{131}_{53}\mathrm{I} \rightarrow\ ^{131}_{54}\mathrm{Xe} +\ ^{0}_{-1}\mathrm{e} + \bar{\nu}$

16 **a** decrease **b** increase **c** increase **d** increase

20 $3.84 \times 10^{-12}\,\mathrm{s^{-1}}$
  $0.29\,\mathrm{s^{-1}}$
  $80\,\mathrm{s^{-1}}$
  $5.61 \times 10^{-3}\,\mathrm{s^{-1}}$
  $0.026\,\mathrm{s^{-1}}$
  $1.7 \times 10^{-17}\,\mathrm{s^{-1}}$
  $8.5 \times 10^{-9}\,\mathrm{s^{-1}}$
  $4.4 \times 10^{-3}\,\mathrm{s^{-1}}$
  $2.3 \times 10^{6}\,\mathrm{s^{-1}}$
  $46.2\,\mathrm{s^{-1}}$
  $1.2 \times 10^{-8}\,\mathrm{s^{-1}}$
  $1.4 \times 10^{-13}\,\mathrm{s^{-1}}$
  $4.9 \times 10^{-18}\,\mathrm{s^{-1}}$

22 **a** $\frac{1}{2}$

 b smaller

23 **b** 0.693

Comprehension and application

24 a 8.7×10^{19} u

  b 1.4×10^{-7} kg

25 a 38

  b 85

  c 50

  d Sr has one more proton

27 a $^{90}_{36}\text{Kr} \rightarrow ^{90}_{37}\text{Rb} + ^{0}_{-1}\text{e} + \bar{\nu} + Q$

  b i 9.5×10^{-7}

   ii 47

35 a i $1.28 \times 10^{-5}\,\text{s}^{-1}$

   ii $N = \dfrac{\text{activity } (A)}{\lambda} = \dfrac{10^{14}}{1.28 \times 10^{-5}}$

   $= 7.8 \times 10^{18}\,\text{atoms}$

   $\text{mass} = 7.8 \times 10^{18} \times 24.0 \times 1.66 \times 10^{-27}$

   $= 3.1 \times 10^{-7}\,\text{kg}$

  b $A = A_0 e^{-\lambda t}$

   $= 10^{14} \times e^{-1.28 \times 10^{-5} \times 10 \times 3600}$

   $= 6.3 \times 10^{13}\,\text{Bq}$

  c $\text{mass} = 3.1 \times 10^{-7}\,\text{kg} \times \dfrac{10^{8}}{10^{14}}$

   $= 3.1 \times 10^{-13}\,\text{kg}$

Examination questions

1 a i 82, 214

   ii lead $^{214}_{82}\text{Pb}$

  b i kinetic energy

   ii 9.6×10^{-30} kg (use $E = mc^2$)

2 c i $^{215}_{84}\text{Po} \rightarrow ^{211}_{82}\text{Pb} + ^{4}_{2}\alpha$

   $^{215}_{84}\text{Po} \rightarrow ^{215}_{85}\text{At} + ^{0}_{-1}\text{e} + \bar{\nu}$

   ii typical ionisation energy $\sim 1\,\text{eV} \sim 1.6 \times 10^{-19}\,\text{J}$

   for a million atoms, energy $\sim 2 \times 10^{13}\,\text{J}$

4 b i $^{14}_{6}\text{C}$

   ii $^{10}_{6}\text{C}$ ($^{11}_{6}\text{C}$ is the other positron emitter)

  c kinetic energy $\quad E_1 = 2.2\,\text{MeV} = 3.52 \times 10^{-13}\,\text{J}$

   mass energy $\quad E_2 = 2m_e c^2 = 1.64 \times 10^{-13}\,\text{J}$

   $E_{\text{total}} = 5.16 \times 10^{-13}\,\text{J}$

   $E_\gamma = \tfrac{1}{2}E_{\text{total}}$

   $= 2.6 \times 10^{-13}\,\text{J}$

6 b i $^{40}_{19}\text{K} \rightarrow ^{40}_{18}\text{Ar} + ^{0}_{1}\text{X}$

   ii positron (or beta$^+$)

  c i $\dfrac{N}{N_0} = e^{-\lambda t}, \lambda = \dfrac{0.693}{T_{\frac{1}{2}}}$

   $\dfrac{N}{N_0} = \tfrac{1}{8}, \lambda = \dfrac{0.693}{1.4 \times 10^{9}} = 4.95 \times 10^{-10}\,\text{year}^{-1}$

   $\ln \tfrac{1}{8} = -4.95 \times 10^{-10} t$

   $t = 4.2 \times 10^{9}\,\text{years}$

Chapter 29

Marginal questions

6 a 1.6×10^{-12} J   **b** 1.6×10^{-10} J

7 a i 5×10^{20} eV   **ii** 80 J

   b 80 W

12 a no   **b** no   **c** yes

13 a baryon number; strangeness

   b charge number

   c baryon number

14 a i no   **ii** yes

Comprehension and application

24 a 1.6×10^{-8} J

Examination questions

1 a d changes to u ($n \rightarrow p + \beta^+$)

   b not at all (annihilation of leptons which have no quark structure)

2 a 7, 14

   b antineutrino

   c neutron (to proton), electron, antineutrino

   d i up, down, down (udd)

       up, up, down (uud)

    ii a neutron d quark changes to a proton u quark

3 a conservation of charge number

   b conservation of lepton number

   c lepton numbers: $-1 = -1 + 1 - 1$

4 b π^+: longer lifetime (track length)

      ionising (has charge; will interact)

5 a $\pi^+ \rightarrow \Lambda^0 + K^0 + 2\pi^+$

Chapter 30

Marginal questions

7 1.82 m

13 20 N m

14 a Fx

   b Fx

Comprehension and application

16 a 900 N   **b** 900 N

   c 171 N   **d** 171 N

   e 178 N, 16° above horizontal

   f 178 N, 16° below horizontal

   g 890 N m

   h 890 N m

Examination questions

2 a i 6.9×10^9 Pa   **ii** 2.3×10^8 Pa   **iii** 0.044

3 b i 350 N   **ii** 550 N

4 b iii 1   10 N cm

        2   70 N cm

        3   300 N cm

    iv 38 N

5 b i 120 N   **ii** 280 N

   c 300 N m

● Chapter 31

Marginal questions

2 **a** none

   b 2v

3 **a** $\dfrac{ds}{dt} = 0$

   b no

   c average velocity

5 $1.67\,\mathrm{m\,s^{-2}}$

6 $1.33\,\mathrm{m\,s^{-2}}$

11 **a** $22\,\mathrm{m\,s^{-1}}$

12 11 s

13 **a** force is downwards

   b $-7\,\mathrm{m\,s^{-1}}$

15 4.4 m

16 **a** $3.9\,\mathrm{m\,s^{-2}}$

17 **a** negative

   b $12\,\mathrm{m\,s^{-1}}$

18 5 s

19 **a** 1.4 s

   b 2.2 s

20 **a** $v^2 = u^2 + 2as$, $s = 1.65\,\mathrm{m}$

   b $s = ut + \tfrac{1}{2}at^2$, $t = 1.13\,\mathrm{s}$

Comprehension and application

22 **a** stopped at Keaton for 50 s

24 **a** 0.77 s

   b $6.5\,\mathrm{m\,s^{-1}}$

25 train must stop in less than $(80 - 23 \times 0.7)\,\mathrm{m} = 64\,\mathrm{m}$

   $v^2 = u^2 + 2as$, $s = 59\,\mathrm{m}$

29 **a** draw graphs of v against t for each: 34 s

   b on Moon 920 m, on Earth 1650 m

30 on Earth

Examination questions

1 **b** **i** inelastic; $6.0 \times 10^7\,\mathrm{m\,s^{-1}}$ **ii** $4.0 \times 10^6\,\mathrm{m\,s^{-1}}$

2 **a** treat vertical and horizontal motion separately

   b $v = 29.4\,\mathrm{m\,s^{-1}}$

   c $7.00\,\mathrm{m\,s^{-1}}$

   d $29.4\,\mathrm{m\,s^{-1}}$

   e 13.4°

   f increase

3 **b** **i** 0.42 s   **ii** 0.84 s   **iii** 3.4 m

4 **c** **i** $7.00\,\mathrm{m\,s^{-1}}$

    ii 0.017 s

5 **a**

A	B
104	113
152	173
176	186 (all $\mathrm{km\,h^{-1}}$)

   b

13.8	12.7
9.47	8.32
8.18	7.74 (all s)

   c B

   e **i** B   **ii** A

Chapter 32

Marginal questions

1   **a** $-0.2\,\text{m s}^{-2}$
    b $+0.3\,\text{m s}^{-2}$
    d $34.5°$

3   **a** $400\,\text{N}$
    b $6.0\,\text{kN}$
    c $-6.0\,\text{kN}$

4   **a** $23\,\text{m s}^{-1}$
    b $1.4\times10^{4}\,\text{kg m s}^{-1}$
    c $2.7\times10^{5}\,\text{J}$

7   **a** $3.0\,\text{m s}^{-2}$
    b $7.3\,\text{s}$
    c $81\,\text{m}$

8   **a** $28\,\text{kJ}$
    b $111\,\text{kJ}$

9   **a** $980\,\text{kJ}$
    b $20\,\text{m s}^{-1}$
    d $8\,\text{m}$

12 **a** 4

Comprehension and application

15 $3.2\times10^{4}\,\text{kg m s}^{-1}$, $3.2\times10^{5}\,\text{J}$
16 **a** $1.6\times10^{4}\,\text{N}$
    b $1.4\times10^{4}\,\text{N}$
    c $1.8\times10^{4}\,\text{N}$
    d $700\,\text{N}$
    f $2.6\,\text{m}$

Examination questions

1   **b** **i** $8.4\,\text{N}$
      ii $600\,\text{W}$
   c **i** pressure drag force increases with speed, resistive friction force is constant
      ii 1 $2.8\,\text{N}$
          2 $200\,\text{W}$

2   **a** **i** $160\,\text{N}$
      ii 0.17
   b $3200\,\text{W}$
   c 4

3   **a** **i** 1 $0.020\,\text{s}$
          2 $14.4\,\text{m}$
   b **i** $u=39\,\text{m s}^{-1}$
      ii $2250\,\text{N}$

4   **a** **ii** momentum $(\text{N s}=\text{kg m s}^{-1})$
   b $mv=0.80\,\text{kg m s}^{-1}$
      $v=1.6\,\text{m s}^{-1}$
   c **i** $0.40\,\text{m s}^{-1}$
      ii $0.16\,\text{J}$
      iii inelastic (E_k before $=0.64\,\text{J}$)

5   **c** $30\,\text{kW}$
   d $55\,\text{kW}$

6   **b** use conservation of momentum to find v
      then use $v=u+at$
      $a=55\,\text{m s}^{-2}$

● Chapter 33   **Marginal questions**

10 **a** 310.15 K
   b i 63 °C
      ii 63 degrees centigrade
      iii 63 K
11 **a i** −273.15 deg. cent., −273.15 °C, 0 K
      ii 0 deg. cent., 0 °C, 273.15 K
      iii 0.01 deg. cent., 0.01 °C, 273.16 K
      iv 100 deg. cent., 100 °C, 373.15 K
12 **a** −59.1 °X   **b** 19.2 °X   **c** 35.1 °X
15 **a** positive, positive, zero
   b positive, zero, positive
   c positive, zero, positive
   d positive, positive, negative
   e zero, zero (heat from food = heat to environment), zero
   f positive, positive, negative
16 **a** 41 700 J   **b** 41 700 J
19 0.24 K
20 **a** second   **b** first
   d i first $(T − 200)$ gain
            second $(300 − T)$ loss
      ii $4170 × 0.6(300 − T)$ J
      iii $387 × 0.4(T − 200)$ J
   e 294 K

Comprehension and application
27 **b** 11 J

Examination questions
1 **d** 770 J
3 **b i** 38.4 kW
      ii 90 000 kg
      iii $1.0 × 10^{-4}$ K s^{-1}
      iv $3.0 × 10^4$ s
4 **a i** 1840 W
      ii $2.3 × 10^6$ J kg^{-1}
5 **a i** 22 J   **ii** 0.13 K
   b i $3.0 × 10^{-10}$ m^3 **ii** 9.7 mm
6 **a** $1.7 × 10^8$ J
   b $1.8 × 10^4$ s or 5.1 h

● Chapter 34   **Marginal questions**

8 **a** $6.02 × 10^{23}$
   b $3.01 × 10^{23}$
10 **a** 0.012 kg, $7.22 × 10^{24}$ u
   b 0.0168 kg, $1.01 × 10^{25}$ u
   c 0.0896 kg, $5.39 × 10^{25}$ u
   d 0.185 kg, $1.11 × 10^{26}$ u
11 **a** 1.00 m^3
   b 870 m^3
   c 27.8 m^3
   d 27.8 m^3
   e $4.62 × 10^{-23}$ m^3

12 P : Pa or Nm^{-2} : $[M][L]^{-1}[T]^{-2}$
V : m^3 : $[L]^3$
T : K : $[\theta]$
n : mol : $[mol]$

15 a $247\,ms^{-1}$

b $6.1 \times 10^4\,m^2s^{-2}$

c 4.00×10^4, 4.84×10^4, $10.24 \times 10^4\,m^2s^{-2}$

d $6.4 \times 10^4\,m^2s^{-2}$

17 $[P]\ = [M][L]^{-1}[T]^{-2}$
$\rho\ = [M][L]^{-3}$
$[c^2] = [L]^2[T]^{-2}$
$LHS = [M][L]^{-1}[T]^{-2}$
$RHS = [M][L]^{-1}[T]^{-2}$

18 rms velocity $\propto \sqrt{P}$

21 a $6.00 \times 10^{-21}\,J$

23 $U = \dfrac{3}{2}nRT$

24 a i $3 \times 10^{-21}\,J$

ii $8 \times 10^{-21}\,J$

b $\frac{1}{2}m\overline{c^2} = 6 \times 10^{-21}\,J$, where $m = \dfrac{0.002}{6.02 \times 10^{23}}\,kg$

rms speed $= 1900\,ms^{-1}$

26 $\dfrac{a}{V^2} = P$

$[a] = [M][L]^5[T]^{-2}$, $kg\,m^5s^{-2}$

28 $P_R(V - b) = RT = PV$

$\dfrac{P_R}{P} = \dfrac{V}{V - b} = 1.0016$

32 34%

Examination questions

1 b $\dfrac{3}{2}\dfrac{pV}{N} = \frac{1}{2}mc^2 = E_k$

$pV = NkT$

$E_k = \dfrac{3}{2}kT$

c i $6.00 \times 10^{-21}\,J$

2 a pressure   Pa
volume m^3
no. of moles   mol
molar gas constant $J\,mol^{-1}K^{-1}$
temperature   K

b i $5.0\,J$

ii $V \propto T$ at constant pressure; temperature $= 167\,°C$

3 c i $0.0225\,m^3$

ii $1.42\,kg\,m^{-3}$

d ii $464\,ms^{-1}$

e i $\langle c^2 \rangle = 3P/\rho$

$c_{rms} = 462\,ms^{-1}$

ii $5.67 \times 10^{-21}\,J\left(m = \dfrac{0.032}{N}\right)$

4 **b** **i** $1340\,\text{m\,s}^{-1}$
     ii equals E_k av. for He $= 6.00 \times 10^{-21}\,\text{J}$
     iii $80\,\text{kPa}$

6 **c** $647\,\text{K}\ (347\,°\text{C})$
     d **i** hydrogen, nitrogen, oxygen
       ii carbon dioxide (critical temp. close to room temp.)

7 **a** $5.7\,\text{W}$

     d factor $= \dfrac{T_1}{T_2} = \dfrac{273}{298} = 0.92$

Chapter 35   Marginal questions

1   $1.6 \times 10^{-19}\,\text{A}$
2   $4.6 \times 10^{-13}\,\text{kg}$
8 **a** $0.078\,\Omega$
   b $0.11\,\Omega$

9   ratio $\dfrac{R_w}{R_{Cu}} = \dfrac{5.6}{1.7}$, $R_w = 3.3\,\Omega$

11   $[\text{M}][\text{L}]^3[\text{T}]^{-3}[\text{I}]^{-2}[\theta]^{-1}$

13   ratio $\approx \dfrac{10^5}{500} \approx 200$

15   drift velocity
17   $2.5 \times 10^{-4}\,\text{m\,s}^{-1}$

19   ratio $\approx \dfrac{10^5}{5 \times 10^{-4}} \approx 2 \times 10^8$

Comprehension and application

23   about $7 \times 10^5\,\text{V\,m}^{-1}$
28 **a** $1\,\text{ms}$

Examination questions

2 **b** **i** $2.4 \times 10^{-15}\,\text{J}$
     ii $7.3 \times 10^7\,\text{m\,s}^{-1}$
   c 3.4×10^5
   d much greater in copper wire (by factor 7.3×10^{10})

3 **a** $530\,\Omega$
4 **a** $8.3 \times 10^{-4}\,\text{m\,s}^{-1}$
5 **c** **i** $2.5 \times 10^{-4}\,\text{C\,s}^{-1}$ $(0.25\,\text{mA})$
     ii $3.0 \times 10^{-5}\,\text{C\,s}^{-1}$ $(0.03\,\text{mA})$
     iii $9.4 \times 10^{13}\,\text{s}^{-1}$

6 **a** $R = \rho\,\dfrac{l}{A}$

   b **ii** 1 $290\,\text{V}$
         2   $2.1 \times 10^{-8}\,\text{m}$

7 **c** **i** $3.5\,\Omega$

     ii using $\dfrac{1}{R} = \dfrac{1}{R_1} + \dfrac{1}{R_2} + \dfrac{1}{R_3} \ldots$ for resistors in parallel

        $R = 0.063\,\Omega$

● Chapter 36

Marginal questions

1 0.56 A

3 **a** 5×10^{18}

   b 5×10^{18}

4 4 Ω resistor; 1 V, 2 V

8 16 C, 16 A

9 **a** $9.0 \, \text{rad A}^{-1}$

   b 0.018 rad

11 0.94 A

12 $\dfrac{S_1}{S_2} = 1000$

13 **a** $4.00 \times 10^{-4} \, \Omega$

   b 100 μA

14 49 980 Ω

15 **a** 1.0 V

16 **a** 0.020 A

   b 5.4 V

   c 240 Ω

   d 0.022 A

   e 5.3 V

18 12.3 V

19 **a** 2.0 V (use approx. formula)

   b 2.8 V (use complete formula)

21 64 Ω

Comprehension and application

22 1.50 V

23 1.51 V

24 1.49 V

25 0.02 V

26 $\frac{2}{5}$ of fsd

Examination questions

2 **b** $1.7 \times 10^6 \, \text{J}$

  c **i** 58 Ω

     ii 920 W

3 **a** 20 V

  b 16 mA

  c 250 Ω

  d short circuit current $= \dfrac{20}{250}$ A

4 **a** **i** 7.5 Ω

     ii 3.0 V

     iii 0.6 W

  b **i** $4.6 \times 10^{-7} \, \Omega \, \text{m}$

5 **i** water at 20 °C to start, no heat loss; approx. $1.3 \times 10^5 \, \text{J}$

  k 96 W

  l about 1400 s (23 min)

6 **c** **i** $\dfrac{V_F}{V_S + V_F} = \dfrac{R_F}{R_S + R_F}$

    ii **1** 2.00 Ω

       2 0.01 V

● Chapter 37 **Marginal questions**

I a 11
   b 111
   c 1011
   d 100000
   e 1100100

2 a 5
   b 8
   c 15
   d 64

5 a V Pa^{-1}

13 16

14 1111111111111111; 65535

17 1948

28 RAM

30 4 V

31 a 4

33 b 80 Hz

Examination questions

2 a 10.0 V
   b −5.0 V
   c 670 Hz

3 b ii 1 9000 Hz
          2 4500 Hz

4 a 100 kHz, 4500 Hz

5 a 5.1 × 10^9 bits
   d i 127
      ii assuming one page has 33 lines at 60 characters per line,
          total no. of characters per page ≈ 2000
          no. of bits ≈ 2000 × 8 ≈ 16 000

      iii no. of pages ≈ $\dfrac{5.1 \times 10^9}{16\,000}$ ≈ 320 000

Index